AF326845

Optical Effects in Solids

An overview of the optical effects in solids, this book addresses the physics of materials and their response to electromagnetic radiation. The discussion includes metals, superconductors, semiconductors and insulators. The book begins by introducing the dielectric function into Maxwell's macroscopic equations and finding their plane-wave solution. The physics governing the dielectric function of various materials is then covered, both classically and using basic quantum mechanics. Advanced topics covered include interacting electrons, the anomalous skin effect, anisotropy, magneto-optics, and inhomogeneous materials. Each subject begins with a connection to the basic physics of the particular solid, after which the measurable optical quantities are derived. It allows the reader to connect measurements (reflectance, optical conductivity, and dielectric function) with the underlying physics of solids. Methods of analyzing experimental data are addressed, making this an ideal resource for students and researchers interested in solid-state physics, optics, and materials science.

David B. Tanner is a distinguished professor of physics at the University of Florida. His research focuses on condensed matter physics and on astrophysics. He has previously served as department chair and chair of the Division of Condensed Matter Physics for the American Physical Society (APS). In 2016 he received the APS Frank Isakson Prize for Optical Effects in Solids and, in the same year, shared a Special Breakthrough Prize in Fundamental Physics for the discovery of gravitational waves.

Optical Effects in Solids

DAVID B. TANNER

University of Florida

CAMBRIDGE
UNIVERSITY PRESS

CAMBRIDGE
UNIVERSITY PRESS

University Printing House, Cambridge CB2 8BS, United Kingdom

One Liberty Plaza, 20th Floor, New York, NY 10006, USA

477 Williamstown Road, Port Melbourne, VIC 3207, Australia

314–321, 3rd Floor, Plot 3, Splendor Forum, Jasola District Centre, New Delhi – 110025, India

79 Anson Road, #06–04/06, Singapore 079906

Cambridge University Press is part of the University of Cambridge.

It furthers the University's mission by disseminating knowledge in the pursuit of education, learning, and research at the highest international levels of excellence.

www.cambridge.org
Information on this title: www.cambridge.org/9781107160149
DOI: 10.1017/9781316672778

© David B. Tanner 2019

This publication is in copyright. Subject to statutory exception and to the provisions of relevant collective licensing agreements, no reproduction of any part may take place without the written permission of Cambridge University Press.

First published 2019

Printed in the United Kingdom by TJ International Ltd, Padstow Cornwall

A catalogue record for this publication is available from the British Library.

Library of Congress Cataloging-in-Publication Data
Names: Tanner, David B., 1945– author.
Title: Optical effects in solids / David B. Tanner (University of Florida).
Description: Cambridge ; New York, NY : Cambridge University Press, [2019] | Includes bibliographical references and index.
Identifiers: LCCN 2018046542 | ISBN 9781107160149 (hardback)
Subjects: LCSH: Solids–Optical properties. | Light. | Materials–Optical properties. | Light–Scattering. | Reflection (Optics) | Electromagnetic waves.
Classification: LCC QC176.8.O6 T36 2019 | DDC 530.4/12–dc23
LC record available at https://lccn.loc.gov/2018046542

ISBN 978-1-107-16014-9 Hardback

Cambridge University Press has no responsibility for the persistence or accuracy of URLs for external or third-party internet websites referred to in this publication and does not guarantee that any content on such websites is, or will remain, accurate or appropriate.

for Marcia

Contents

This book is a discussion of optical effects in solids, addressing the physics of many types of solids (metals, superconductors, semiconductors, insulators, and others) and their response to electromagnetic radiation. I try to make a connection between what an experimenter can measure or extract from measurements (reflectance, transmittance, optical conductivity, and dielectric function) and the microscopic physics of the solid. Methods of analyzing experimental data are addressed: the optics of thin films and the Kramers–Kronig relations.

I begin with introducing the dielectric function into Maxwell's macroscopic equations and finding their plane-wave solution. Then I discuss (first classically and then using basic quantum mechanics) the dielectric function of various materials. Other topics include interacting electrons, the anomalous skin effect, anisotropy, magneto-optics, and inhomogeneous materials.

I've attempted to write as a relatively complete coverage of the subject, starting with Maxwell's equations for the electromagnetism and with the Schrödinger equation and Newton's laws for the solid-state physics. The level of the presentation is aimed at the first- or second-year experimental graduate student. The electromagnetism assumed is undergraduate (Griffiths or Marion); the level of quantum mechanics is about the same (Griffiths or Peebles). Finally, it is helpful to have gone through solid-state physics (Kittel or Burns).

When I was a postdoc, the book by Frederick Wooten, *Optical Properties of Solids* (1972), appeared in the University Bookstore. I bought a copy and went through it during the next few weeks. It was just what I needed to read at the time: It was clear, it covered topics of interest to me, and it was at a level I could follow. I refer to it still and recommend it to others; I have also used it several times as a text in a course on optical properties for first-year graduate students in physics, materials science, electrical engineering, and chemistry. Unfortunately, the book is long out of print. I've attempted to write at the same level as Wooten and cover many of the topics of his book. (The exception is photoemission, which is a huge subject of its own by now.) I have included a number of things that Wooten did not address, such as phonons, superconductivity, anisotropy, magneto-optics, and inhomogeneous materials.

This book started life as the 2013 lecture notes for a graduate class in the optical properties of solids. This class has been offered to University of Florida graduate students at two- to six-year intervals over more than 25 years. That these started as lectures may be responsible for a certain informality in presentation. I've written in the first person, trying to avoid the royal or inclusive "we" that in much exposition attempts to co-opt the reader or the listener. I think that whenever a person hears "and now we see" or " in the future we must," a quite proper response is "What do you mean by *we* Kemo Sabe." First person avoids this response. I have also tried to avoid second-person instructions: "If you look at the figure, you can see …" Well, a person looking at the figure may see what I want her or him to see, but may not. If not, it is my fault for not designing the right figure, not explaining it correctly, or not providing adequate background. In any event, it is better if I say what I see when I look at the figure and not presume more.

All solids may be divided into three classes based on conductivity: metals, semiconductors, and insulators. I use silver, silicon, and sodium chloride as examples of these classes. These materials are useful to persons other than solid-state physicists. The cover of the book shows one example. It shows the bowl of a toddy ladle made in England around 1760. Set into the bowl of the ladle is a silver sixpence that was minted in 1758. The coin was designed during the reign of George II by John Sigismund Tanner, Chief Engraver of the Royal Mint. The sixpence coin was called a "tanner" by the British right up until decimalization. (Six pennies may not seem a lot but the hourly wage at the time was 2–8 pence.)

I've had discussion with many persons over many years about the subject of optical effects in solids. I am grateful for all these discussions. A number of these discussions had direct influence on what I wrote here. For such discussions I thank Larry Carr, Jim Garland, Alan Heeger, Claus Jacobsen, Kati Kamarás, Ricardo Lobo, Frank Marsiglio, Dmitri Maslov, Jan Musfeldt, Michael Rice, Danilo Romero, Al Sievers, David Stroud, Lila Tache, Tom Timusk, Axel Zibold, the students who took PHY7097 and made comments and found typos, and the editors at Cambridge University Press for patience and accommodation.

1 Introduction

The way in which light interacts with material objects is determined by the optical properties of the materials. Why might you want to think about these optical properties? There are at least two reasons. First, you can make use of known optical materials to design and build devices to manipulate light: mirrors, lenses, filters, polarizers, and a host of other gadgets. Second, you can measure the optical properties of some new material and obtain a wealth of information about the low energy excitations that govern the material's physics. Figure 1.1 is a chart that identifies some of these excitations and indicates the part of the spectrum where they might be expected to appear.

In common parlance, "optical" can be a synonym of "visual," and hence related to human eyesight. This interpretation would restrict discussion to the visible part of the electromagnetic spectrum, light with wavelengths of 390–780 nm, indicated by the little rainbow in Fig. 1.1. (I will need to introduce the variety of measures used for the wavelength or the frequency or the photon energy of electromagnetic waves. Four are shown in Fig. 1.1: λ in μm, f in THz, E in meV, and another frequency unit, wavenumbers or cm^{-1}. The latter may be unfamiliar to persons who have not worked in the field of optical effects in solids; it is the inverse of the wavelength in cm. The visible spectrum spans 385–770 THz, 1.59–3.18 eV, and 12,800–25,600 cm^{-1}. Units used in papers about the optical properties of solids are discussed in Appendix A.)

Of course I will *not* in this book make the interpretation that optical means visible; instead, materials properties will be considered over a wide range of frequencies or wavelengths.* Plausible ranges are discussed toward the end of this chapter.

Even if you were constrained to the use of your own eyes, you would see that solids have a wide range of optical properties. Silver is a lustrous metal used for centuries in coins and fine tableware, with a high reflectance over the whole visible range. Silicon is a crystalline semiconductor and the foundation of modern electronics. With its surface oxide freshly etched off, silicon is also rather reflective, although not as good a mirror as silver.[†] Salt (sodium chloride) is a transparent ionic insulator, is necessary for life, and makes up about 3.5% (by weight) of seawater. A crystal of salt is transparent over the entire visible spectrum; because the refractive index is about 1.5, the reflectance is everywhere about 4%.

If you had ultraviolet eyes, you would see these materials differently. Silver would be a poor reflector, with at most 20% reflectance and trailing off to zero at the shortest wavelengths. In contrast, the reflectance of silicon would be better than in the visible,

---

* I like the notion of "DC to daylight," used widely in the amateur radio community and also as the name of a symposium honoring Professor A.J. Sievers, at Cornell University, June 14, 2003.

[†] Silver reflects about 98% of red light and about 80% of violet light; silicon reflects about 33% of red and 50% of violet.

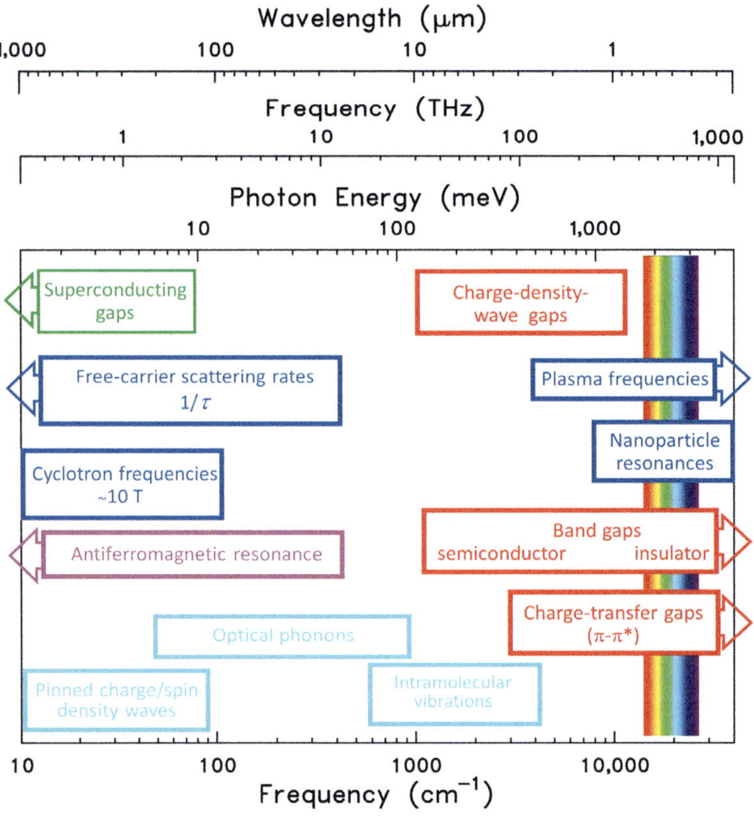

Fig. 1.1 Chart showing optical processes in solids, with an indication of the wavelengths or frequencies where these processes typically may be studied. The uppermost scale shows the vacuum wavelength in μm, 10^{-6} m. Frequencies (or energies) are given on three scales. Top to bottom they are: THz, 10^{12} cycles/s; photon energies, meV; and wavenumbers, $\tilde{\nu}$ (in cm^{-1}), defined by $\tilde{\nu} = 1/\lambda$ with λ the wavelength measured in cm.

reaching up to 75%. Sodium chloride would be opaque over much of the spectrum, with a reflectance a bit higher than in the visible. Those with infrared eyes would also see things differently from visible or UV-sensitive individuals. Silver would have a reflectance above 99%. Silicon would appear opaque at the shortest infrared wavelengths but would then become transparent, so that you could see through even meter-thick crystals.* Sodium chloride remains transparent over much of the infrared, but an opaque and highly reflecting "reststrahlen" (German for residual ray) region occurs at long wavelengths. In the reststrahlen band, NaCl has a reflectivity not much below that of a metal.

If you put on your solid-state-physics hat, you can understand the optical properties of these materials, at least qualitatively. Silver is a nearly free-electron metal, with one electron per atom in the metallic Fermi surface. These mobile electrons give the high electrical conductivity; they form a plasma that makes silver opaque and highly reflective

---

* Here ultra-high-purity is assumed. Moreover, in the middle infrared region there is a band caused by lattice vibrational effects – multiphonons in this case – where silicon is opaque unless rather thin.

below the plasma frequency.* Silicon is a semiconductor with a gap between the filled valence band and the empty conduction band. Photons with energies below the gap can propagate without loss in silicon. Photons with energies above the gap are absorbed, generating electron–hole pairs. This absorption renders silicon opaque and, as mentioned, increases the reflectance. Sodium chloride is an insulating crystal, with a band gap in the ultraviolet. Similar to silicon, photons with energy larger than the gap are absorbed. Sodium chloride has two atoms per unit cell; these occur as ions, Na^+ and Cl^-; an electric field displaces these ions, producing induced dipoles in the solid. With a two-atom basis, the lattice vibrations have an optical branch, and the reststrahlen band is a result of the light exciting this optical branch.

Now let me return to the question of the range of wavelengths (or the range of light frequencies or of photon energies) over which I can discuss the optical properties of solids. The electromagnetic spectrum extends over a huge range; a representative cartoon illustrating the "electromagnetic spectrum" is shown in Fig. 1.2.

This chart shows wavelengths from km to pm along with corresponding frequencies and photon energies. So the question is: What part of this spectrum might be used to study the optics of solids?

To start, I'll want to use continuum electrodynamics, so the short wavelength limit is set by a requirement that the wavelength be larger than the spacing between atoms. When the wavelength is less than the interatomic distances, diffraction effects dominate. X-ray diffraction is essential for determining crystal structure but beyond my scope. At somewhat longer wavelengths, continuum electrodynamics is fine, but the materials properties are essentially a superposition of atomic transitions. Solid-state effects contribute of course but minimally for wavelengths shorter than something on the order of 50 nm.[†]

As wavelengths get longer and longer, there is of course no problem with continuum electrodynamics. However, practically speaking, the physics that govern the electromagnetic response at dc and audio frequencies is the same as the physics at ultra-high radio frequencies and even microwaves. So the lowest frequencies that I will consider are around a few GHz.

There is a second reason for setting a long-wavelength limit. One GHz (4 μeV photon energy) corresponds to 30 cm wavelength, and this large scale raises an experimental issue: to shine light on a solid, one sends a beam that one would like to consider to be composed of plane electromagnetic waves. Cartoons of typical experimental setups are shown in Fig. 1.3. The left panel shows a reflectance ($\mathscr{R}$) experiment and the right a transmittance ($\mathscr{T}$) experiment. Light comes from a source (of known properties) that can emit a range of wavelengths, encounters the sample, and goes to a detector where it is converted to

---

* The connection between high absorption, high conductivity, and high reflectance is not intuitive. For the moment, I'll just assert that all three go together. So at wavelengths where a material is opaque, it also has increased reflectance. The more intense is the absorption, the higher the conductivity and also the higher the reflectance. A hand-waving argument says that high conductivity means large currents in response to applied electric fields; the power loss or absorption goes as $\mathbf{j} \cdot \mathbf{E} = \sigma E^2$. See Section 4.5 for further discussion.

[†] Using $\lambda f = c$, $\tilde{\nu} = 1/\lambda$, and $E = hf$ with λ the wavelength, f the frequency in Hz, $\tilde{\nu}$ the frequency in cm^{-1}, or wavenumber, E the photon energy, c the speed of light, and h Planck's constant, 50 nm corresponds to a frequency of 6 PHz (petaHertz), a wavenumber of 200,000 cm^{-1}, and photon energies of 25 eV.

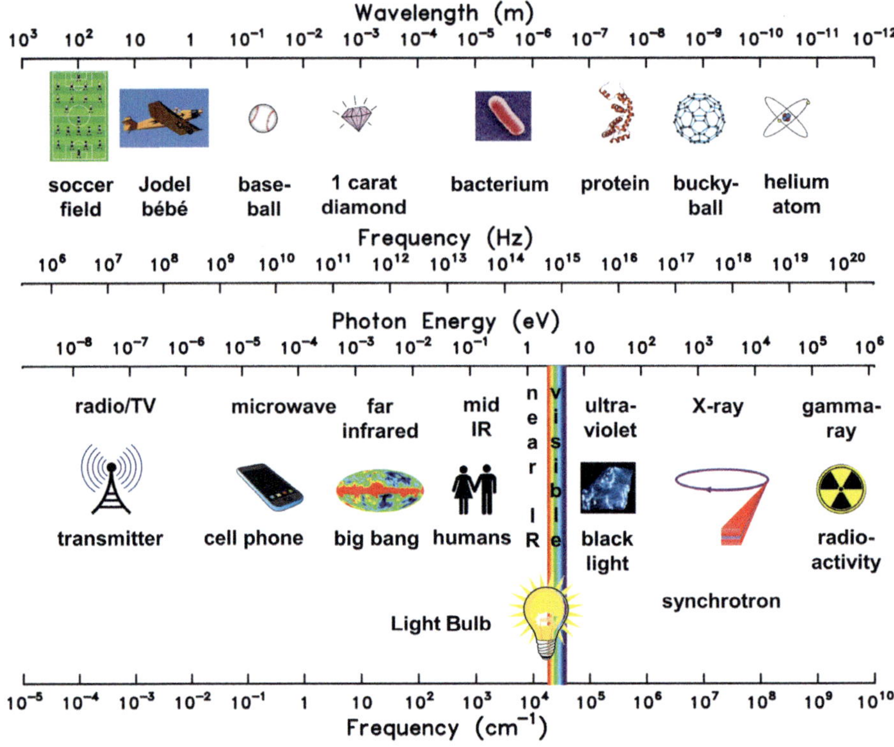

Fig. 1.2 The electromagnetic spectrum over a broad range of wavelengths. Things of various sizes are placed where their dimension equals the wavelength. Spectral ranges are identified and example sources are shown. ("IR" is short for infrared.)

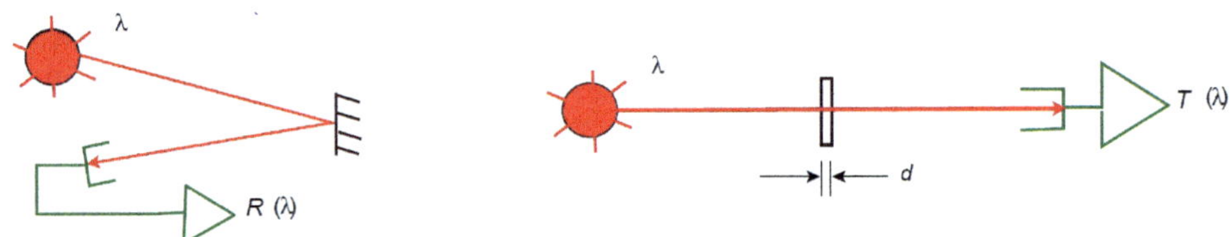

Fig. 1.3 Cartoon of an experiment where one measures reflectance (left) or transmittance (right).

an electrical signal, is amplified, and recorded, yielding a spectrum of $\mathscr{R}$ or $\mathscr{T}$ *vs.* the wavelength or the frequency. These ideas are only a good approximation to experiment when the wavelength is small compared to the size of the sample or of the experimental apparatus. When this condition is no longer the case, diffraction effects (by the sample not by the atomic lattice) as well as waveguide effects in the surrounding apparatus become important.*

---

* One may of course measure materials properties all the way to zero frequency (infinite wavelength) by electrical means: apply contacts and measure resistance, capacitance, etc. I will discuss the connection of optical measurements to dc electrical properties a number of times.

It makes little sense to be very precise in specifying the wavelength or frequency limits over which optical concepts are important for the physics of solids. So I'll say that I will consider the wavelength range to be several cm to several tens of nm, the frequency range to be a few GHz to a few PHz, and the photon energy range to be tens of μeV to tens of eV. This range covers the bands labeled microwaves, far infrared, midinfrared, near infrared, visible, and ultraviolet in Fig. 1.2. There is a factor of a million between one end and the other; that should be enough for everybody.*

In the following chapters, I'll remind you a little bit about electromagnetism, including Maxwell's equations and their plane-wave solutions. I'll then restrict myself for some time to local, nonmagnetic, isotropic, homogeneous, and linear solids.[†] The motivations for these restrictions are that the materials should be local, so that the current at point $\mathbf{r}$ is a function only of the fields at $\mathbf{r}$; nonmagnetic, because most solids are nonmagnetic and because even magnetic materials only show themselves to be magnetic at rather low frequencies; isotropic, so that the properties do not depend on the direction or polarization of the light; homogeneous, so that the response functions do not depend on spatial position; and linear, so that I may make a Fourier decomposition of the fields and treat each component independently. With these approximations, I'll introduce the idea of a complex dielectric function, discuss classical theories of free carrier response in metals, interband absorption in semiconductors and insulators, and lattice vibrations (phonons). Next, I will show data from the literature to illustrate some of the concepts of these theories.

I return to electromagnetism to calculate the reflection and transmission by a thin film or slab as a way to link experimental measurements to the optical properties of the material making up the film. The next step is to introduce simple quantum mechanics, leading to a discussion of free-electron metals, followed by a presentation of the quantum-mechanical perturbation-theory of optical absorption, culminating in an important sum rule for the conductivity. The sum rule result motivates an interlude about causality where I obtain the Kramers–Kronig relations between the absorptive and dispersive parts of the response function, discuss the analysis of reflectivity by Kramers–Kronig methods, and derive other sum rules. My focus shifts back to materials, with a simple treatment of the optics of superconductivity, a distinctly quantum-mechanical phenomenon. Next comes the band structure of simple solids and the interband absorption edge in semiconductors, followed by materials with strong correlations and interactions.

After this, it will be time to relax the initial conditions and discuss (one at a time!) nonlocal properties, mostly the anomalous skin effect in a pure metal, wave propagation in anisotropic materials, magneto-optics, and randomly inhomogeneous materials.[‡] Several appendices discuss units, some mathematics, and other things "optical."

---

* In fact there are few materials that have been studied over the entire range. A much more typical range is in wavelength from, say, 0.3 mm to 300 nm, far infrared to ultraviolet, a range of 10^3.

[†] Much of this discussion can apply to liquids as well, and even to a dilute gas, but the physics discussion will rely on solid-state physics ideas: Fermi surfaces, band structure, etc.

[‡] I'll leave the huge subjects of quantum optics and nonlinear optics to others; I think it is better to say nothing than to make short and probably superficial treatments.

2.1 Optical Constants

The response of materials to light is described by a number of quantities, often called "optical constants." Among these are:

- ϵ, the dielectric constant.
- σ, the electrical conductivity.
- χ, the susceptibility.*
- n, the refractive index.
- κ, the extinction coefficient.
- δ, the electromagnetic skin depth.
- Z, the surface impedance.

and many others. See Appendix F for a longer but still incomplete list.

These quantities are neither constant nor independent. They are functions of the frequency, temperature, pressure, external magnetic field, and many other things. By knowing two of these, one that describes the absorption in the solid (such as the electrical conductivity or the extinction coefficient) and one that describes dispersion (such as the dielectric constant or the refractive index), all of the others can be calculated.

2.2 Maxwell's Equations

In my initial Electricity and Magnetism course, the professor said that the subject is governed by equations written in the nineteenth century by Maxwell [1] and that some day a teacher would come in, write these equations on the blackboard on the first day of the class and then proceed to develop a theory based on these equations. Of course in a junior-level class, he did not do this, but I can do it here.

There are two versions: Maxwell's microscopic equations and Maxwell's macroscopic equations. The first are more fundamental, because they describe the microscopic fields arising from every charge in the Universe and from the motion of these charges as well.

---

* Electric *and* magnetic.

The second are more fun, because they average over the charges in macroscopic media*
and allow great simplification of the subject.

After restricting myself to macroscopic charges and currents, I have ρ_{ext}, the external
charge density, and $\mathbf{j}_{\text{free}}$, the free current density as sources in Maxwell's equations. The
external charge density is essentially the charge imbalance in the medium; it is zero for
electrically neutral objects.[†] The free current density is the result of the motion of free
charges in the metal. The standard example of these mobile charges is the free electrons in
a metal, but could also be doped or thermally excited free carriers in a semiconductor or
the diffusion of ions in an electrolyte. Note that the free carriers are compensated by bound
ions in the electrically neutral material so these charges do not generate any external charge
density.

So let me begin by writing Maxwell's equations for macroscopic media [2–8]. I'll use
cgs-Gaussian units; the translation to SI units appears in Appendix B.

$$\nabla \cdot \mathbf{D} = 4\pi \rho_{\text{ext}} \tag{2.1a}$$

$$\nabla \cdot \mathbf{B} = 0 \tag{2.1b}$$

$$\nabla \times \mathbf{E} = -\frac{1}{c}\frac{\partial \mathbf{B}}{\partial t} \tag{2.1c}$$

$$\nabla \times \mathbf{H} = \frac{4\pi}{c}\mathbf{j}_{\text{free}} + \frac{1}{c}\frac{\partial \mathbf{D}}{\partial t}, \tag{2.1d}$$

I must add a connection to classical mechanics to these equations. The force $\mathbf{F}$ on a
particle with electric charge q satisfies the Lorentz force law,

$$\mathbf{F} = q\left(\mathbf{E} + \frac{\mathbf{v}}{c} \times \mathbf{B}\right), \tag{2.2}$$

where $\mathbf{v}$ is the particle's velocity vector.

The definitions of the auxiliary fields are

$$\mathbf{D} = \mathbf{E} + 4\pi\mathbf{P}$$

$$\mathbf{H} = \mathbf{B} - 4\pi\mathbf{M}. \tag{2.3}$$

The quantities in Eqs. 2.1, 2.2, and 2.3 are all functions of space and time. The vector $\mathbf{E}$
is called the "electric field" and the vector $\mathbf{D}$ is called the "electric displacement field."
In vacuum, $\mathbf{D}$ and $\mathbf{E}$ are proportional to each other, with the multiplicative constant
ϵ_0 depending on the physical units.[‡] Inside a material they are different on account of
the polarization of the material. The vector $\mathbf{P}$ is the "polarization field," also known as
the "electric polarization," "electric polarization density," or "electric dipole moment/unit
volume." The vector $\mathbf{B}$ is called the "magnetic field," and it and $\mathbf{E}$ are defined as the vector

---

* In the absence of external fields, the positive charges (the nuclei) in a uniform medium are perfectly screened
by the negative charges (the electrons) so that both may be omitted completely from the charges and currents
in Maxwell's equations. External fields may polarize these charges (pushing $+$ in one direction and $-$ in the
opposite direction); the effects of such currents and dipole moments are the subject of our optical properties
studies.

[†] And it does *not* include any charge imbalance caused by external fields. This polarization gets included in the
dipole moment/unit volume $\mathbf{P}$.

[‡] Unity in cgs-Gaussian units.

fields necessary to make the Lorentz law correctly describe the forces on a moving charged particle. **B** is also called the "magnetic flux density," or the "magnetic induction." The vector **H** is also sometimes called the "magnetic field." Other names include the "magnetic field intensity," the "magnetic field strength," and the "magnetizing field." In vacuum, **B** and **H** are proportional to each other, with the multiplicative constant μ_0 depending on the physical units.* Inside a material they are different on account of the magnetization of the material. The scalar ρ_{ext} is the "free volume charge density" and the vector $\mathbf{j}_{\text{free}}$ is the "free electric current density." The vector **M** is the "magnetization field" or the "magnetic dipole moment/unit volume."

Three of the four Maxwell equations have names: Equation 2.1a is Gauss' Law; Eq. 2.1c is Faraday's law of induction; and Eq. 2.1d is Ampère's law with Maxwell's correction. I'll just call it Ampère's law. Eq. 2.1b is sometimes called Gauss' law for magnetism but is sometimes called the no-magnetic-monopole law. Others say that it has no name. I like "the no-monopole law."

2.3 Total, Free, and Bound Charges and Currents

If I view charges as being small point-like particles moving here and there in space, then a classical-physics-style definition of the microscopic charge density in a solid might be

$$\rho_{\text{micro}}(\mathbf{r}, t) = \sum_i q_i \delta(\mathbf{r} - \mathbf{r}_i(t)),$$

whereas the current would be written

$$\mathbf{j}_{\text{micro}}(\mathbf{r}, t) = \sum_i q_i \mathbf{v}_i(t) \delta(\mathbf{r} - \mathbf{r}_i(t)).$$

Here, the sum i runs over all charged particles, q_i is the charge of the ith particle, located at (time-dependent) coordinate $\mathbf{r}_i$ and moving with (time-dependent) velocity $\mathbf{v}_i$.

There are quantum mechanical versions of the charge density and current density also. To use them requires a solution of Schrödinger's equation in the solid. With the solution in hand, I could then find the contribution of the electrons[†] (with charge $-e$) to the total charge density to be $\rho_e = -e\Psi_e^*\Psi_e$ where Ψ_e is some total wave function of the electrons, and, similarly $\mathbf{j}_e = -(\hbar e/2mi)\left[\Psi_e^*\nabla\Psi_e - (\nabla\Psi_e^*)\Psi_e\right]$.

Accounting for all the particles in a material with 10^{22} particles per cubic centimeter is of course too hard, so I'll average over some volume ΔV. The scale of V is taken to be large with respect to the interatomic spacing a and small with respect to the electromagnetic wavelength λ, or

$$a^3 \ll \Delta V \ll \lambda^3.$$

---

* Unity in cgs-Gaussian units.

[†] Because of a choice made by Franklin, the electron is negative. I'll take e to be a positive number and put the sign in explicitly as needed.

The inequalities are easy to satisfy because $a \sim 0.1$ nm and, according to the discussion in Chapter 1, λ is bigger than 10 nm, so there is a factor of 10^6 between their cubes. The charge and current then are written as averaged quantities, $\rho(\mathbf{r},t) = qn(\mathbf{r},t)$ and $\mathbf{j}(\mathbf{r},t) = \rho(\mathbf{r},t)\mathbf{v}_d(\mathbf{r},t)$, where q is the average charge of the particles, n is their number density, and $\mathbf{v}_d$ the average velocity in the volume ΔV, known as the "drift velocity."

The electrons and nuclei do not know that I would like to average them out, so they still respond to applied fields, producing dipole moments in the material, $\mathbf{P}$ and $\mathbf{M}$. These bound or polarization charges, electric polarization currents, and magnetization currents are determined by the physics of the solid. Then, the macroscopic polarization charge density ρ_{pol}, the polarization current density $\mathbf{j}_{\text{pol}}$, and the magnetization current density $\mathbf{j}_{\text{mag}}$ are defined in terms of polarization $\mathbf{P}$ and magnetization $\mathbf{M}$ as $\rho_{\text{pol}} = -\nabla \cdot \mathbf{P}$, $\mathbf{j}_{\text{pol}} = \partial \mathbf{P}/\partial t$, and $\mathbf{j}_{\text{mag}} = c\nabla \times \mathbf{M}$.

Maxwell's macroscopic equations reduce to the microscopic equations if one recognizes that the microscopic, free, and bound charge and current density are related by $\rho_{\text{micro}} = \rho_{\text{ext}} + \rho_{\text{pol}}$, and $\mathbf{j}_{\text{micro}} = \mathbf{j}_{\text{free}} + \mathbf{j}_{\text{pol}} + \mathbf{j}_{\text{mag}}$, and then uses Eq. 2.3 to eliminate $\mathbf{D}$ and $\mathbf{H}$.

2.4 Maxwell's Equations for Solids

I am almost to a point where I can begin to address optical effects in solids. But there are two things still to do. First, the world is in general electrically neutral,* so I will take $\rho_{\text{ext}} = 0$. For notational convenience, I will drop the "free" subscript on the current, with the understanding that it is the free current that is meant.† Thus, $\mathbf{j}_{\text{free}} \rightarrow \mathbf{j}$.

Henceforth, I'll use the form of Maxwell's macroscopic equations written just below. I'll claim that these cover the most general cases that occur in the studies of optical effects in solids. I'll make further simplifications in the next several pages, but can always return to these as a starting point.

$$\nabla \cdot \mathbf{D} = 0 \tag{2.4a}$$

$$\nabla \cdot \mathbf{B} = 0 \tag{2.4b}$$

$$\nabla \times \mathbf{E} = -\frac{1}{c}\frac{\partial \mathbf{B}}{\partial t} \tag{2.4c}$$

$$\nabla \times \mathbf{H} = \frac{4\pi}{c}\mathbf{j} + \frac{1}{c}\frac{\partial \mathbf{D}}{\partial t} \tag{2.4d}$$

The four parts of Eq. 2.4 have a certain pleasing symmetry. After I've defined and used the complex dielectric function, this symmetry will become more perfect.

---

* Anyone who has been shocked by a static discharge after crossing a rug on a dry day or who has seen a lightning strike from a thunderstorm knows (1) that this statement is not always true but (2) that the charge imbalance does not last.

† And see the discussion in Section 3.1 where the distinction between free and bound currents will be blurred as well.

2.5 Plane-Wave Solutions

I write the electric field as a complex exponential

$$\mathbf{E}(\mathbf{r},t) = \mathbf{E}_0 e^{i(\mathbf{q}\cdot\mathbf{r}-\omega t)} \tag{2.5}$$

and also the magnetic field

$$\mathbf{H}(\mathbf{r},t) = \mathbf{H}_0 e^{i(\mathbf{q}\cdot\mathbf{r}-\omega t)}, \tag{2.6}$$

where $\mathbf{E}_0$ and $\mathbf{H}_0$ are constant vectors giving the amplitudes of the fields (complex quantities in the most general case) and their directions,* $\mathbf{q}$ is the wave vector of the field (measured in cm^{-1} in cgs), and ω is the angular frequency (in radians/s or s^{-1}). I choose to use $\mathbf{H}$ rather than $\mathbf{B}$, as does Jackson [2].

The quantity $i = \sqrt{-1}$ does not appear in Eqs. 2.1–2.4. Hence the only mechanism to have equations that contain real and imaginary quantities is through writing the fields as complex quantities, as in Eqs. 2.5 and 2.6. It would be perfectly valid to write $\mathbf{E} = \mathbf{E}_0 \cos(\mathbf{q} \cdot \mathbf{r} - \omega t + \phi_e)$ and $\mathbf{H} = \mathbf{H}_0 \cos(\mathbf{q} \cdot \mathbf{r} - \omega t + \phi_m)$, where the phases ϕ_e and ϕ_m allow for differing phases in electric and magnetic fields.† In this case, all quantities in electromagnetism would be purely real quantities. I could do this; at a minimum it would give fine training in the use of trigonometric identities, because as soon as I take a derivative, there will be both sines and cosines in the math.

It is conventional to say that one writes the fields as complex quantities but when one wants to evaluate the observable fields, the real part should be taken. This statement is basically true, though one has to be careful in cases where two complex fields are multiplied together.

I know from freshman physics that the crests and valleys of the wave repeat in space (at fixed time) by a translation of the wavelength λ. From this assertion I get $|\mathbf{q}|\lambda = 2\pi$ or $|\mathbf{q}| = 2\pi/\lambda$. At a point in space, the wave repeats every time the time advances by the period T; hence $\omega = 2\pi/T = 2\pi f$ with f the frequency. Moreover, $\lambda f = v$ and $\omega = qv$ with v the wave speed ($= c$ in vacuum). See Appendix A for a discussion of units used for length, time, frequency, energy, fields, and many other quantities used in optical studies.

The use of plane-wave fields may seem arbitrary or restrictive, but Fourier tells us that

$$\mathbf{E}(\mathbf{r},t) = \int d^3 q \int d\omega\, \mathbf{E}_0(\mathbf{q},\omega) e^{i(\mathbf{q}\cdot\mathbf{r}-\omega t)},$$

with

$$\mathbf{E}_0(\mathbf{q},\omega) = \int d^3 r \int dt\, \mathbf{E}(\mathbf{r},t) e^{-i(\mathbf{q}\cdot\mathbf{r}-\omega t)}.$$

I can write an arbitrary field in terms of the Fourier integral, do the usual trick of exchanging order of integration and differentiation, and find (for linear, local materials) that Maxwell's equations apply to each Fourier component.

---

* Hence, to be explicit I could write $\mathbf{E}_0 = \hat{e}\mathcal{E}_0 e^{i\phi_0}$, where $\hat{e}$ is a unit vector pointing in the field direction, $\mathcal{E}_0$ is the field magnitude, and ϕ_0 is a constant phase which, when combined with $(\mathbf{q} \cdot \mathbf{r} - \omega t)$, specifies where the zeros, crests, and valleys of the wave occur.

† Alternatively, write $\mathbf{E} = \mathbf{E}_c \cos(\mathbf{q} \cdot \mathbf{r} - \omega t) + \mathbf{E}_s \sin(\mathbf{q} \cdot \mathbf{r} - \omega t)$ and a similar equation for $\mathbf{H}$.

2.6 Converting Differential Equations to Algebraic Ones

Now, I insert plane-wave fields into Eq. 2.4. See Appendix C for the rules of how vector and partial differential operators act on plane wave functions. For example, Eq. 2.4a becomes

$$\nabla \cdot \mathbf{D} = i\mathbf{q} \cdot \mathbf{D} = 0$$

After doing this type of thing for all four of Maxwell's equations, cleaning up the signs, and canceling as many of the appearances of i as I can, I get

$$\mathbf{q} \cdot \mathbf{D} = 0 \tag{2.7a}$$

$$\mathbf{q} \cdot \mathbf{B} = 0 \tag{2.7b}$$

$$\mathbf{q} \times \mathbf{E} = \frac{\omega}{c}\mathbf{B} \tag{2.7c}$$

$$\mathbf{q} \times \mathbf{H} = -\frac{\omega}{c}\mathbf{D} - \frac{4\pi i}{c}\mathbf{j} \tag{2.7d}$$

with

$$\mathbf{D} = \mathbf{E} + 4\pi\mathbf{P} \qquad \mathbf{H} = \mathbf{B} - 4\pi\mathbf{M} \tag{2.8}$$

(I exchanged the order on the right-hand side of Eq. 2.7d for convenience in the steps coming up.)

2.7 Vector Directions

Vectors whose inner product is zero are orthogonal. Hence, $\mathbf{q}$ is perpendicular to both $\mathbf{D}$ and $\mathbf{B}$. The cross product produces a vector normal to the plane containing the other two vectors. Hence, Eq. 2.7c tells me that $\mathbf{B}$ is perpendicular to $\mathbf{E}$ (and to $\mathbf{q}$, which I already knew). Equation 2.7d says that $\mathbf{H}$ is perpendicular to a particular linear combination of $\mathbf{D}$ and $\mathbf{j}$ and, moreover, that $\mathbf{j}$ is perpendicular to $\mathbf{q}$ because $\mathbf{D}$ is.

2.8 Electromagnetic Waves in Vacuum

If there is no medium, there are neither electric nor magnetic dipoles nor free currents. Hence Eq. 2.8 gives $\mathbf{D} = \mathbf{E}$ and $\mathbf{H} = \mathbf{B}$ and Eq. 2.7 becomes

$$\mathbf{q} \cdot \mathbf{E} = 0 \tag{2.9a}$$

$$\mathbf{q} \cdot \mathbf{H} = 0 \tag{2.9b}$$

$$\mathbf{q} \times \mathbf{E} = \frac{\omega}{c}\mathbf{H} \tag{2.9c}$$

$$\mathbf{q} \times \mathbf{H} = -\frac{\omega}{c}\mathbf{E}. \tag{2.9d}$$

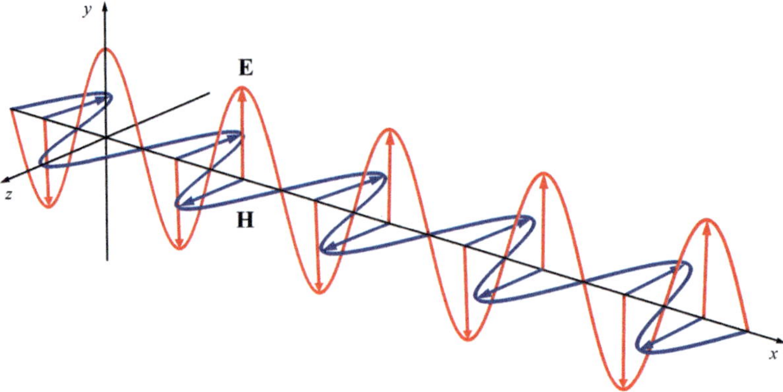

Fig. 2.1   Electromagnetic wave in vacuum.

Here, $\mathbf{q}$, $\mathbf{E}$, and $\mathbf{H}$ form a right handed orthogonal set, which I may orient respectively along the x, y, and z Cartesian axes. Then* $\hat{\mathbf{q}} \times \hat{\mathbf{e}} = \hat{\mathbf{h}}$ and $\hat{\mathbf{q}} \times \hat{\mathbf{h}} = -\hat{\mathbf{e}}$, making the magnitudes of Eqs. 2.9c and 2.9d be $qE = \omega H/c$ and $qH = \omega E/c$. Solving the second of these for H and substituting in the first, I find $q = \omega/c$; either of these then give $H = E$. So in vacuum, electromagnetic waves travel at the speed of light and have equal (cgs-Gaussian) amplitudes for the electric and magnetic components. The energy density [2] $u = (\mathbf{E} \cdot \mathbf{D} + \mathbf{B} \cdot \mathbf{H})/8\pi$ becomes in vacuum $u_{\text{vac}} = (E^2 + H^2)/8\pi$. Half the energy is carried by the electric component and half by the magnetic component.

Figure 2.1 shows a cartoon of the wave in vacuum.[†] The wave is traveling to the right, with the $\mathbf{E}$ field (red) in the vertical plane and the $\mathbf{H}$ field (blue) in the horizontal plane. These waves are plane waves, so the fields have spatial variation in the propagation direction while being constant in amplitude for perpendicular directions. Figure 2.2 is an attempt to illustrate this condition. It shows the $\mathbf{E}$ field at its crests (red arrow and red plane) and troughs (blue arrow and blue plane). The propagation vector (black) is normal to the planes of constant phase.

2.9 Five Easy Simplifications

Now, I return to Eqs. 2.7 and 2.8 and discuss how they change as simplifying assumptions about the material's properties are made.

2.9.1 Local Response

If I write Ohm's law as $\mathbf{j} = \sigma \mathbf{E}$, I am making an assumption that the response is local, that the current $\mathbf{j}$ at point $\mathbf{r}$ depends only on the electric field $\mathbf{E}$ at that point and the

---

* I'll define unit vectors parallel to any vector $\mathbf{A}$ as $\hat{\mathbf{a}}$ and the magnitude (length) of the vector as A.

† If you look on the web for such cartoons you may note that about half of them show the fields incorrectly, failing to satisfy $\hat{\mathbf{q}} \times \hat{\mathbf{e}} = \hat{\mathbf{h}}$.

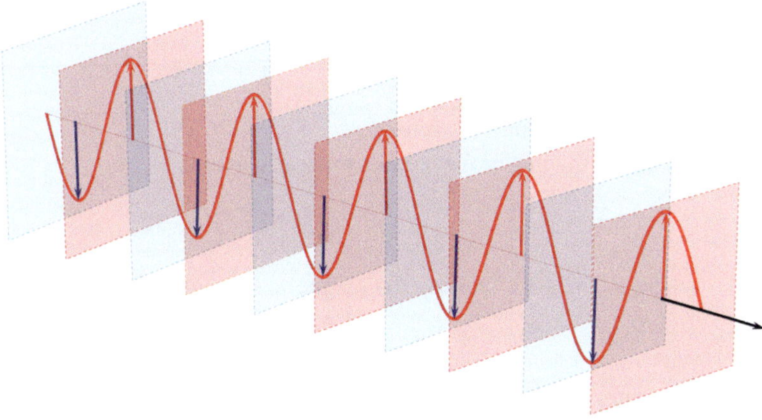

Plane waves.

conductivity σ. This statement is never completely true; the current possesses a certain momentum and, once set in motion, takes a certain time (e.g., the relaxation time τ) or distance (e.g., the mean free path ℓ) to relax back to zero.* Even so, local electrodynamics is the usual case, because the mean free path is typically short compared to the lengths over which the field itself varies and so the current is indeed related to the field by a local relation. However, in pure metals at low temperatures, the mean free path can become longer than the electromagnetic penetration depth (i.e., the skin depth). In that case, currents generated within the skin depth can travel deep into the metal, where the electric field is essentially zero. Nonlocal effects appear also in superconductors. I'll discuss nonlocal electrodynamics in Chapter 14. Meantime, local behavior will be assumed.

2.9.2 Nonmagnetic Materials

If the material is nonmagnetic, $\mathbf{M} = 0$ everywhere and $\mathbf{B} = \mathbf{H}$. Then Eq. 2.7 shows that $\mathbf{q}$, $\mathbf{D}$, and $\mathbf{H}$ (which is parallel to $\mathbf{B}$) form a right-handed set. $\mathbf{E}$ is perpendicular to $\mathbf{H}$, lying in the plane containing $\mathbf{D}$ and $\mathbf{q}$. $\mathbf{E}$ need not be perpendicular to $\mathbf{q}$, although it cannot be exactly parallel to it either. Analogous to the vacuum case, I'll set $\hat{\boldsymbol{q}}\|\hat{\boldsymbol{x}}$, $\hat{\boldsymbol{j}}\|\hat{\boldsymbol{y}}$, and $\hat{\boldsymbol{h}}\|\hat{\boldsymbol{z}}$.

2.9.3 Linear Materials

For many materials, the electric dipole moment/unit volume obeys

$$\mathbf{P} = \overset{\leftrightarrow}{\chi}_e \cdot \mathbf{E} \tag{2.10}$$

where $\overset{\leftrightarrow}{\chi}_e$ is a 3×3 tensor, the electric susceptibility. Then

$$\mathbf{D} = \mathbf{E} + 4\pi \overset{\leftrightarrow}{\chi}_e \cdot \mathbf{E} = (1 + 4\pi \overset{\leftrightarrow}{\chi}_e) \cdot \mathbf{E} \equiv \overset{\leftrightarrow}{\epsilon} \cdot \mathbf{E} \tag{2.11}$$

where $\overset{\leftrightarrow}{\epsilon} = 1 + 4\pi \overset{\leftrightarrow}{\chi}_e$ is the dielectric tensor.

---

* Actually, in time τ it relaxes back to $1/e$ of what it was at $t = 0$ and it requires a number of relaxation times to become unnoticeable.

Similarly

$$\mathbf{j} = \overset{\leftrightarrow}{\sigma} \cdot \mathbf{E}. \tag{2.12}$$

with $\overset{\leftrightarrow}{\sigma}$ the conductivity tensor.

And (if the material becomes magnetic again)

$$\mathbf{M} = \overset{\leftrightarrow}{\chi}_m \cdot \mathbf{H} \tag{2.13}$$

where $\overset{\leftrightarrow}{\chi}_m$ is a 3×3 tensor, the magnetic susceptibility. Then

$$\mathbf{H} = \mathbf{B} - 4\pi \overset{\leftrightarrow}{\chi}_m \cdot \mathbf{H}$$

and (after a bit of algebra)

$$\mathbf{B} = \overset{\leftrightarrow}{\mu} \cdot \mathbf{H} \tag{2.14}$$

where $\overset{\leftrightarrow}{\mu} = 1 + 4\pi \overset{\leftrightarrow}{\chi}_m$ is the permeability tensor.

Equations 2.10–2.14 indicate linear relations between applied fields and materials response. The relation is linear because if the $\mathbf{E}$ field is doubled, the polarization $\mathbf{P}$ and current $\mathbf{j}$ are doubled. Linearity is not a big restriction. Everything is linear if the fields are small enough. Everything is *non*linear if the fields are big enough. Most optical fields are actually rather weak; the exception being high power and short pulsed lasers.

Substituting Eqs. 2.11 and 2.12 into Eq. 2.7d, I get

$$\begin{aligned}
\mathbf{q} \times \mathbf{H} &= -\frac{\omega}{c} \overset{\leftrightarrow}{\epsilon} \cdot \mathbf{E} - \frac{4\pi i}{c} \overset{\leftrightarrow}{\sigma} \cdot \mathbf{E} \\
&= -\frac{\omega}{c}\left[\overset{\leftrightarrow}{\epsilon} + \frac{4\pi i}{\omega} \overset{\leftrightarrow}{\sigma}\right] \cdot \mathbf{E}
\end{aligned} \tag{2.15}$$

I will spend quite a bit of time on the consequences of this equation in Chapter 15, which is about anisotropic media (crystals). For now let me just note that $\mathbf{q} \times \mathbf{H}$ produces a vector that is perpendicular to $\mathbf{q}$ and $\mathbf{H}$ but that $\mathbf{E}$ is not parallel to that vector, because the multiplication of a tensor with a vector produces a new vector that is in general oriented different from the initial vector.*

2.9.4 Isotropic Materials

Amorphous solids, polycrystalline solids, and liquids are isotropic: their properties are uniform for all directions of fields and propagation. In the case of crystals, there is no requirement for isotropy. Indeed, the optical properties must reflect the symmetry of the underlying crystal structure. On the one hand, the majority of all possible crystals are anisotropic; of the seven crystal families only the cubic is by symmetry required to have isotropic optical properties. On the other hand, a large fraction of commonly encountered materials (silver, silicon, and sodium chloride, for example) are cubic and isotropic. On

---

* Only if the tensor is in diagonal form *and* the field is oriented along one of the principal axes of the tensor is the result of the multiplication not rotated.

the remaining hand, many of the most interesting solid state systems, such as the high-temperature superconductors, graphene, organic solids, etc., are highly anisotropic. Many persons think their remarkable properties are a consequence of this anisotropy.

The optics of anisotropic crystals are worked out in Chapter 15. For the time being, I will assume isotropy. In this case, the tensors of Eqs. 2.10–2.13 become scalars. Formally, one writes

$$\overset{\leftrightarrow}{\sigma} = \sigma \overset{\leftrightarrow}{\mathbf{1}}$$

with $\overset{\leftrightarrow}{\mathbf{1}}$ the identity tensor (diagonal, with all diagonal elements equal to 1). In isotropic media, $\mathbf{D}$, $\mathbf{E}$, and $\mathbf{P}$ are parallel, as are $\mathbf{H}$, $\mathbf{B}$, and $\mathbf{M}$.

2.9.5 Homogeneous Media

There are many materials whose properties vary with position. One class of such materials is exemplified by paint or ink, a collection of particles, small with respect to the wavelength, suspended in a host material. The optical properties of each type of grain and of the host are different but, because of the small grain size, the material has an apparent (or effective) uniform response. I'll go over some aspects of inhomogeneous materials in Chapter 17. A second class consists of a plane surface where the optical properties change from, say that of a metal to vacuum or a thin film of one or more layers. Simple layered materials are covered in Chapter 7.

At present, I will consider the material to be spatially homogeneous, so that the response functions, such as the dielectric function, satisfy

$$\epsilon \neq \epsilon(\mathbf{r}),$$

so that $\nabla \cdot (\epsilon \mathbf{E}) = \epsilon \nabla \cdot \mathbf{E}$.

2.10 Maxwell's Equations for These Easy Cases

With the assumptions of the previous section, I can write for local, nonmagnetic, linear, isotropic, and homogeneous materials:

$$\mathbf{j} = \sigma_1 \mathbf{E} \tag{2.16}$$

and

$$\mathbf{D} = \epsilon_1 \mathbf{E}. \tag{2.17}$$

The meaning of the subscripts will become clear in a moment.

Equation 2.17 replaces $\mathbf{D} = \mathbf{E} + 4\pi\mathbf{P}$, Eq. 2.3. With it and Eq. 2.16, I can rewrite Eq. 2.7:

$$\mathbf{q} \cdot \mathbf{E} = 0 \tag{2.18a}$$

$$\mathbf{q} \cdot \mathbf{H} = 0 \tag{2.18b}$$

$$\mathbf{q} \times \mathbf{E} = \frac{\omega}{c}\mathbf{H} \tag{2.18c}$$

$$\mathbf{q} \times \mathbf{H} = -\frac{\omega}{c}\left[\epsilon_1 + \frac{4\pi i}{\omega}\sigma_1\right]\mathbf{E} \tag{2.18d}$$

Only the fourth one has changed.

I am almost to the starting point where I can discuss the optical properties of simple* materials. But first I have to say what the subscripts on ϵ_1 and σ_1 mean, given that I am using electric fields like

$$\mathbf{E} = \mathbf{E}_0 e^{i(\mathbf{q}\cdot\mathbf{r}-\omega t)}$$

so that the field amplitudes – and the materials related coefficient $\epsilon_1 + 4\pi i\sigma_1/\omega$ – contain real and imaginary parts. I will use the first part of the next chapter to explain these subscripts.

---

* Simple = local, nonmagnetic, linear, isotropic, and homogeneous solids.

The Complex Dielectric Function and Refractive Index

3.1 Conductivity and Dielectric Constant

In this this chapter, I'll define the complex dielectric function as I'll use it for the remainder of the book. The first topic is the relation between conductivity (σ) and dielectric function (ϵ). For a moment, let me return to Eq. 2.4d, substituting into it Eqs. 2.16 and 2.17:

$$\nabla \times \mathbf{H} = \frac{4\pi}{c}\sigma_1 \mathbf{E} + \frac{\epsilon_1}{c}\frac{\partial \mathbf{E}}{\partial t}. \tag{3.1}$$

As discussed in Section 2.5, the Maxwell equations are only complex if complex fields are used with them. But how am I to think about σ_1 and ϵ_1? Are they real numbers or complex numbers? Well, if I use sines and cosines for the fields, both must be real numbers. If I use complex fields (as I plan to do), I can take one of two philosophical positions.

The first position omits the subscripts and identifies $\sigma\mathbf{E}$ as the free current density. Similarly $\epsilon\partial\mathbf{E}/\partial t$ contains the bound current density and Maxwell's displacement current. One can defend this distinction. For example, the conduction electrons in a metal are free carriers and their motion in response to an electric field is the free current. Because the carriers have inertia, their response will lag the field as frequency increases. This phase delay can be represented by having σ be complex, with the imaginary part being an inductive response that results from this inertia. Similarly, the remaining electrons in the solid are bound to their atoms, and their response is the polarization current. Inertial and dissipative effects will ensure that ϵ also is complex. I have four independent quantities (the real and the imaginary parts of these functions) but can identify them with the free and with the bound carriers.

The second position asserts that the conductivity and dielectric function in Eq. 3.1 are both real quantities, just as they would be if real-valued fields were used. It goes on to note that the distinction between free carrier and bound carrier response is clear in the case of well understood and well characterized systems such as simple metals. However, for some new material, perhaps with no good theoretical understanding, it would not be possible to make such a clear distinction. It is better to reduce the number of materials "constants" to two.

I'll take the second position. There are two independent quantities, the real conductivity σ_1 and the real dielectric function ϵ_1. The first leads to a current that is in phase with $\mathbf{E}$ (and hence is dissipative) and the second to a current that is $90°$ out of phase (and hence dispersive). I will combine them into a single complex dielectric function.

3.2 The Complex Dielectric Function

Using this approach, I can substitute our plane-wave fields into Eq. 3.1; indeed, I did that to derive Eq. 2.18d (the plane-wave-generated algebraic equation) which I copy here:

$$\mathbf{q} \times \mathbf{H} = -\frac{\omega}{c}\left[\epsilon_1 + \frac{4\pi i}{\omega}\sigma_1\right]\mathbf{E}.$$

and then I *define* a complex dielectric function ϵ as

$$\epsilon \equiv \epsilon_1 + \frac{4\pi i}{\omega}\sigma_1. \tag{3.2}$$

With this definition, I arrive at the form of Maxwell's equations which is appropriate for local, nonmagnetic, linear, isotropic, and homogeneous media. I'll put them in a box to make them stand out.

$$\mathbf{q} \cdot \mathbf{E} = 0 \tag{3.3a}$$

$$\mathbf{q} \cdot \mathbf{H} = 0 \tag{3.3b}$$

$$\mathbf{q} \times \mathbf{E} = \frac{\omega}{c}\mathbf{H} \tag{3.3c}$$

$$\mathbf{q} \times \mathbf{H} = -\frac{\omega}{c}\epsilon\mathbf{E} \tag{3.3d}$$

These equations have a very agreeable symmetry. The materials properties enter only through the presence of the complex dielectric function ϵ in Eq. 3.3d.

Equations 3.3a and 3.3b tell me that $\mathbf{q}$ is perpendicular to $\mathbf{E}$ and $\mathbf{H}$, just as it was in vacuum. Equation 3.3c tells me that $\mathbf{H}$ is perpendicular to $\mathbf{q}$ and $\mathbf{E}$; the three vectors form a right-handed set, with unit vectors $\{\hat{\mathbf{q}}, \hat{\mathbf{e}}, \hat{\mathbf{h}}\}$. Then, with $\hat{\mathbf{q}} \times \hat{\mathbf{e}} = \hat{\mathbf{h}}$, the vector part of Eq. 3.3d gives $\hat{\mathbf{q}} \times \hat{\mathbf{h}} = -\hat{\mathbf{e}}$ as it should, telling nothing new about vector directions.

I'll generally write complex quantities in the form $Z = Z_1 + iZ_2$, with Z_1 the real part and Z_2 the imaginary part.* The dielectric function is one such complex quantity, $\epsilon = \epsilon_1 + i\epsilon_2$. From the definition, Eq. 3.2, I see that the imaginary part of the dielectric function is related to the real part of the conductivity,

$$\epsilon_2 = \frac{4\pi}{\omega}\sigma_1. \tag{3.4}$$

3.3 The Optical Conductivity

There are times when it is convenient to work with a complex conductivity $\sigma = \sigma_1 + i\sigma_2$ rather than a complex dielectric function. There is no difference in the physics when I use

---

* Some authors use primes instead, $Z = Z' + iZ''$.

σ rather that ϵ, just a change of perspective. When using the conductivity, I have to get the case where $\sigma = 0$ correct, keeping the vacuum displacement current of Eq. 2.9d. This I do by starting with Eq. 2.7d, writing $\mathbf{D} = \mathbf{E} + 4\pi\mathbf{P}$, using $\mathbf{j}_{pol} = \partial\mathbf{P}/\partial t = -i\omega\mathbf{P}$. Then, as before, I write Ohm's law, $\mathbf{j} = \sigma_1\mathbf{E}$ and then define $\mathbf{j}_{pol} = \sigma_2\mathbf{E}$. With all this, I get

$$\mathbf{q} \times \mathbf{H} = -\frac{\omega}{c}\mathbf{E} - \frac{4\pi i}{c}\sigma\mathbf{E} = -\frac{\omega}{c}\left[1 + \frac{4\pi i}{\omega}\sigma\right]\mathbf{E}, \tag{3.5}$$

where $\sigma \equiv \sigma_1 + i\sigma_2$ is complex. Equation 3.5 preserves Maxwell's displacement current in the vacuum.* By equating the right sides of Eqs. 3.3d and 3.5, I see that the relation between ϵ and σ is

$$\epsilon \equiv 1 + \frac{4\pi i}{\omega}\sigma, \tag{3.6}$$

and then the complex conductivity is

$$\sigma = \frac{i\omega}{4\pi}(1 - \epsilon). \tag{3.7}$$

Finally, I put $\epsilon = \epsilon_1 + i\epsilon_2$ and $\sigma = \sigma_1 + i\sigma_2$ into Eqs. 3.6 and 3.7 and equate real and imaginary parts to obtain Eq. 3.4 and

$$\sigma_2 = \frac{\omega}{4\pi}(1 - \epsilon_1). \tag{3.8}$$

Note that σ_2 is positive when ϵ_1 is negative, or at least less than than +1. I will show in Chapter 4 that metals and insulators both may have ϵ_1 negative over certain frequency ranges and positive over others. A negative ϵ_1 or σ_2 should not worry you. Recall that the reactance of inductors ($i\omega L$, positive imaginary) and capacitors ($1/i\omega C$, negative imaginary) have opposite signs, allowing resonant circuits to be constructed.[†]

When considering a specific type of material (metal, insulator, ionic solid, superconductor) it is sometimes better to work out the conductivity and other times better to derive the dielectric function. Equations 3.6 and 3.7 allow one to translate back and forth between them.

3.4 The Complex Refractive Index

With the directions removed, Eqs. 3.3c and 3.3d become $qE = \frac{\omega}{c}H$ and $qH = \frac{\omega}{c}\epsilon E$ with both E and H varying as $e^{i(\mathbf{q}\cdot\mathbf{r}-\omega t)}$. I'll solve the first for H and substitute into the second to find

$$q = \frac{\omega}{c}\sqrt{\epsilon}, \tag{3.9}$$

thus motivating the definition of the complex refractive index,

$$N = n + i\kappa \equiv \sqrt{\epsilon}, \tag{3.10}$$

---

* The vacuum has $\sigma = 0$! In fact, the vacuum has $\epsilon = 1 + i0$ and $\sigma = 0 + i0$.
† A negative σ_1 *should* worry you.

so that $q = N\omega/c$. The real part of N, n, is the refractive index and the imaginary part, κ, is the extinction coefficient.*

The electric field for plane waves becomes

$$\mathbf{E} = \mathbf{E}_0 e^{i(\mathbf{q}\cdot\mathbf{r}-\omega t)} = \mathbf{E}_0 e^{i(N\frac{\omega}{c}\hat{\mathbf{q}}\cdot\mathbf{r}-\omega t)}. \tag{3.11}$$

For definiteness, let $\hat{\boldsymbol{q}} = \hat{\boldsymbol{x}}$, so $\hat{\boldsymbol{q}} \cdot \mathbf{r} = x$, and

$$\mathbf{E} = \mathbf{E}_0 e^{i(n\frac{\omega}{c}x-\omega t)} e^{-\kappa\frac{\omega}{c}x}. \tag{3.12}$$

Recall that in vacuum, $q = \omega/c$, so the wave vector (wavelength) in the material is larger (smaller) by a factor n. The wave decays with a decay length $\delta = c/\omega\kappa$. (In terms of the vacuum† wavelength $\lambda_0 = 2\pi c/\omega = c/f$, the wavelength inside is $\lambda = \lambda_0/n$ and the decay length is $\delta = \lambda_0/2\pi\kappa$.) The wave crests move at c in vacuum and at $v = c/n$ in the medium. Equation 3.12 can then be rewritten as

$$\mathbf{E} = \mathbf{E}_0 e^{2\pi i(\frac{x}{\lambda}-ft)} e^{-\frac{x}{\delta}}. \tag{3.13}$$

Figure 3.1 the shows observable (real part) of the electric field,‡ $\mathbf{E} = \mathbf{E}_0 \cos[2\pi(x/\lambda - ft)]e^{-x/\delta}$. Three cases are shown. The curve marked "strong absorber" has the inside wavelength (λ) and the damping length (δ) equal to each other and equal to about 90% of the vacuum wavelength. The amplitude dies away in about 5–7 vacuum wavelengths. The curve marked "weak absorber" has the inside wavelength (λ) also equal to about 90% of the vacuum wavelength while damping length (δ) is 10 times the vacuum wavelength. The amplitude decays slowly, requiring 140 vacuum wavelengths to be reduced to 1 ppm of the initial amplitude. The curve marked "metal" has a wavelength inside that is 100 times *longer* than the vacuum wavelength and a damping length around 10% of λ_0. The field amplitude decays exponentially to zero with no oscillations at all.§

With directions removed, Eq. 3.3c is $H = (cq/\omega)E$, and, then, with $q = \omega N/c$, I get

$$H = NE = \sqrt{\epsilon}E. \tag{3.14}$$

N and ϵ are both generally complex, so the phase of $\mathbf{H}$ is different from the phase of $\mathbf{E}$.

Squaring Eq. 3.10 gives

$$N^2 = n^2 - \kappa^2 + 2in\kappa = \epsilon = \epsilon_1 + i\epsilon_2$$

so that it is easy to see that

$$\epsilon_1 = n^2 - \kappa^2 \tag{3.15}$$

---

* The refractive index is a counterexample to the $Z = Z_1 + iZ_2$ convention. Note also that some (older) texts write $N = n(1 + i\kappa)$. Furthermore, it is more common to use k than κ as the symbol for the extinction coefficient. I choose the latter because I want to use k as a wave vector.

† The word "wavelength" sometimes means the vacuum wavelength, c/f, and sometimes means the wavelength inside the medium, c/nf. If not specified or obvious from context, take it to mean the vacuum wavelength.

‡ Here, I take $\mathbf{E}_0$ to be real.

§ To get the long wavelength inside, I use $n = 0.01$. A refractive index less than unity may seem impossible, but I shall show in Chapter 4 that both free carriers and weakly damped bound carriers may have such values of refractive index. It happens when $\epsilon_1 < 0$ and $\epsilon_2 \ll 1$.

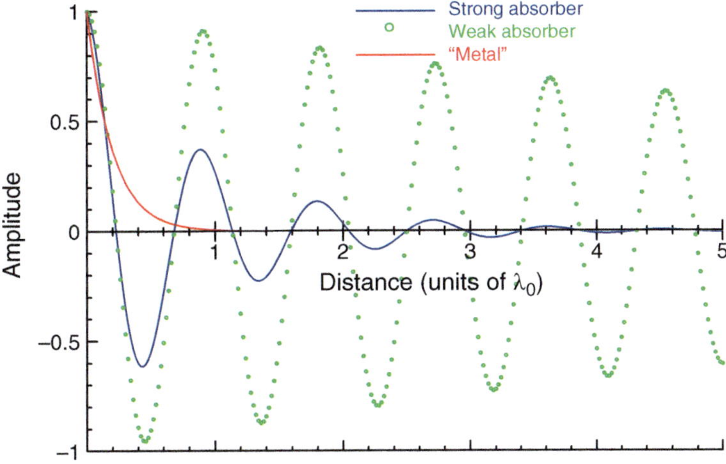

Fig. 3.1 Three damped cosinusoidal waves.

and

$$\epsilon_2 = 2n\kappa. \tag{3.16}$$

It takes a bit more labor to calculate things the other way, but after this labor I get

$$n = \sqrt{\frac{|\epsilon| + \epsilon_1}{2}} \tag{3.17a}$$

$$\kappa = \sqrt{\frac{|\epsilon| - \epsilon_1}{2}} \tag{3.17b}$$

where $|\epsilon| = \sqrt{\epsilon_1^2 + \epsilon_2^2}$.

3.5 Energy Density, Poynting Vector, and Intensity

Electromagnetic waves carry energy and can deposit that energy as heat in absorbing materials.* So here I will work out energy aspects of electromagnetic waves in local, linear, isotropic, and homogeneous materials. I'll start with a quick derivation of Poynting's theorem. I note that this theorem will be derived from Maxwell's equations, so that it adds no real new physics, only some additional insight.

I start by taking the dot (or inner or scalar) product of $\mathbf{H}$ with Eq. 2.4c and $\mathbf{E}$ with Eq. 2.4d. After doing so, I subtract the results, obtaining

$$\mathbf{H} \cdot \nabla \times \mathbf{E} - \mathbf{E} \cdot \nabla \times \mathbf{H} = -\frac{1}{c}\mathbf{H} \cdot \frac{\partial \mathbf{B}}{\partial t} - \frac{4\pi}{c}\mathbf{E} \cdot \mathbf{j} - \frac{1}{c}\mathbf{E} \cdot \frac{\partial \mathbf{D}}{\partial t}. \tag{3.18}$$

---

* I know this because if I stand in the sun or near a fire, I become warm through radiative heat transfer. Indeed, humans have known of radiative heat transfer at least as early as the discovery of fire, and physicists investigated this process for centuries, culminating in the blackbody theory of Planck and the birth of the quantum theory.

Now, I use the vector identity $\nabla \cdot (\mathbf{E} \times \mathbf{H}) = \mathbf{H} \cdot \nabla \times \mathbf{E} - \mathbf{E} \cdot \nabla \times \mathbf{H}$. Moreover, $\mathbf{B} = \mu \mathbf{H}$ and* $\mathbf{D} = \epsilon_1 \mathbf{E}$ so that quantities on the right-hand side can be manipulated to give

$$\mathbf{H} \cdot \frac{\partial \mathbf{B}}{\partial t} = \mu \mathbf{H} \cdot \frac{\partial \mathbf{H}}{\partial t} = \frac{\mu}{2} \frac{\partial H^2}{\partial t}$$

and

$$\mathbf{E} \cdot \frac{\partial \mathbf{D}}{\partial t} = \epsilon_1 \mathbf{E} \cdot \frac{\partial \mathbf{E}}{\partial t} = \frac{\epsilon_1}{2} \frac{\partial E^2}{\partial t}.$$

Therefore, with a modest rearrangement of the right-hand side, I get

$$\nabla \cdot (\mathbf{E} \times \mathbf{H}) = -\frac{1}{2c} \left(\epsilon_1 \frac{\partial E^2}{\partial t} + \mu \frac{\partial H^2}{\partial t} \right) - \frac{4\pi}{c} \mathbf{j} \cdot \mathbf{E}.$$

Almost done! I define the Poynting vector

$$\mathbf{S} \equiv \frac{c}{4\pi} \mathbf{E} \times \mathbf{H}, \tag{3.19}$$

and the electromagnetic energy density

$$u \equiv \frac{1}{8\pi}(\epsilon_1 E^2 + \mu H^2) = \frac{1}{8\pi}(\mathbf{D} \cdot \mathbf{E} + \mathbf{B} \cdot \mathbf{H}), \tag{3.20}$$

and then

$$\nabla \cdot \mathbf{S} = -\frac{\partial u}{\partial t} - \mathbf{j} \cdot \mathbf{E} \tag{3.21}$$

is Poynting's theorem.

The meaning of Poynting's theorem is most easily seen by integrating Eq. 3.21 in some volume V and using the divergence theorem to convert the volume integral on the left-hand side to an integral of $\mathbf{S} \cdot \hat{\boldsymbol{n}}$ over the surface of V. Then $\mathbf{S} \cdot \hat{\boldsymbol{n}}$ represents the energy flowing into or out of V, $\partial u/\partial t$ is the change of energy density within V, and $\mathbf{j} \cdot \mathbf{E}$ is the Joule heating ($I^2 R$ heating) due to currents in the lossy medium within V. Energy is conserved, which I find reassuring.

My next worry is how to handle the complex fields that I have been using within Maxwell's equations. Whenever I am thinking about energy, I better make sure when I insert $e^{i(qx - \omega t)}$ that I end up with pure real quantities when done. I of course will get pure real quantities if I write my fields as sines and cosines but there are good reasons – already invoked – to eschew sines and cosines and instead to use complex exponentials. Moreover, it is generally the case that I am interested in time-averaged energy flows and not the instantaneous energy flow. This consideration is true even when I have the ability to measure the time-instantaneous energy flux.[†]

---

* When I backed up to Eqs. 2.4c and 2.4d, I reverted to the state before I defined the complex dielectric function, so I must use $\mathbf{D} = \epsilon_1 \mathbf{E}$ and $\mathbf{j} = \sigma_1 \mathbf{E}$. Moreover, I'll take μ as real. Were it not, I'd need a term like the $\mathbf{j} \cdot \mathbf{E}$ term to account for the dissipation done by the magnetic field.

[†] Many authors argue at this point in the Poynting discussion that the reason to be interested in the time-averaged flux rather than the instantaneous flux is that it is impossible to follow the instantaneous flux. This statement is of course a canard. If I have a laser oscillating at 300 THz, I can mix it on a photodiode with a local-oscillator laser oscillating at 300 THz + 10 MHz and look at the 10 MHz beat note to determine the amplitude and phase of the primary laser. I can shine light at GHz frequencies on a fast photodiode and see the oscillaing output on an

Let me write the manifestly real fields $(\mathbf{E}+\mathbf{E}^*)/2$ and $(\mathbf{H}+\mathbf{H}^*)/2$, with $\mathbf{E}$ and $\mathbf{H}$ complex functions of space and time, specifically the electric and magnetic fields of Eqs. 2.5 and 2.6. Moreover, as in the footnote on that page, I'll write the constant describing the direction, amplitude, and phase (at $\mathbf{r} = 0$ and $t = 0$) of the electric field as $\mathbf{E}_0 = \hat{e}E_0 e^{i\phi_0}$. The electric and magnetic fields (and their complex conjugates) are solutions to Maxwell's equations. I look back at my solutions to see that $\mathbf{E}$ and $\mathbf{H}$ are orthogonal and that $\hat{e} \times \hat{h} = \hat{q}$. Thus, $\mathbf{S}$ is in the direction of $\hat{q}$. Moreover $\mathbf{H} = N\mathbf{E}$ where N is the complex refractive index. I use these fields in Eq. 3.19 and

$$\mathbf{S} = \hat{q}\frac{c}{4\pi}\left(\frac{E + E^*}{2}\right)\left(\frac{H + H^*}{2}\right),$$
$$= \hat{q}\frac{c}{16\pi}\left(EH^* + E^*H + EH + E^*H^*\right),$$
$$= \hat{q}\frac{c}{16\pi}\left[(N^* + N)EE^* + NE^2 + N^*(E^*)^2\right].$$

As always $N = n + i\kappa$, so that $N + N^* = 2n$. The complex electric field has an amplitude $|E|$ and a phase ϕ, so that $E = |E|e^{i\phi}$. If the wave is traveling along the x axis, the time and space dependent phase ϕ is of the form

$$\phi = n\omega x/c - \omega t + \phi_0,$$

while the amplitude is $|E| = |E_0|e^{-\omega\kappa x/c}$. This amplitude is a function of x in absorbing media but is not a function of time. With these, I get

$$\mathbf{S} = \hat{q}\frac{c}{8\pi}\left[n + n\cos 2\phi - \kappa\sin 2\phi\right]|E|^2. \tag{3.22}$$

$\mathbf{S}$ has a time-independent and a time-dependent component. The time-dependence is embedded in ϕ; $\mathbf{S}$ oscillates at twice the frequency of the electric field.

Being mostly interested in the time-independent part, I calculate the average

$$\langle\mathbf{S}\rangle = \frac{1}{T}\int_0^T \mathbf{S}\,dt$$

where T is either the period of the wave or a time large compared to many periods. The oscillating terms average to zero, leaving only the constant-in-time part,

$$\langle\mathbf{S}\rangle = \hat{q}\frac{c}{8\pi}n|E|^2 = \hat{q}I, \tag{3.23}$$

where I is the wave intensity, the energy flux/unit area. Equation 3.23 gives the energy flow in local isotropic, homogeneous, and linear media. One upshot of this discussion is that the usual definition of the time-averaged Poynting vector for harmonically varying fields [2],

$$\langle\mathbf{S}\rangle = \frac{c}{8\pi}\mathbf{E}\times\mathbf{H}^* \tag{3.24}$$

is only correct for nonabsorbing media. It does give the direction and take care of the factor $e^{i(qx-\omega t)}$ in both $\mathbf{E}$ and $\mathbf{H}$ because the effect of the complex conjugation is to eliminate this

oscilloscope. It is interesting to set up a photodiode looking up at 50 Hz or 60 Hz fluorescent lights illuminating an office or lab and observe that the intensity follows $I \sim \cos^2(2\pi ft) = (1 + \cos 4\pi ft)/2$. Of course my eyes cannot quite follow this 100 or 120 Hz flicker but electronics can.

term. But $H = NE$ and N is a complex function. If however I decide that the real part is the observable, Eq. 3.24 agrees with Eq. 3.23.

Returning to Eq. 3.12 for the field, I have at last the simple result for the Poynting vector as a function of distance traveled in the solid:

$$\langle \mathbf{S}(x) \rangle = \hat{q}\, \frac{c}{8\pi} n |\mathbf{E}(0)|^2 e^{-2\frac{\omega}{c}\kappa x},$$

where $\mathbf{E}(0)$ is the value of the electric field at $x = 0$. The only part of the field that I need to care about is its amplitude. I'll simplify by writing the intensity as

$$I = I_0 e^{-2\frac{\omega}{c}\kappa x} \equiv I_0 e^{-\alpha x}, \tag{3.25}$$

where I_0 is the intensity in the y–z plane at $x = 0$ and I define

$$\alpha = 2\frac{\omega}{c}\kappa, \tag{3.26}$$

the absorption coefficient (with units cm^{-1}) of the material. It is understood that I follow the ray traced by the light wave from $x = 0$ in calculating the intensity. Equation 3.25 is known as Lambert's law, Beer's law, or the Beer–Lambert law.* Note that because $\epsilon_2 = 2n\kappa$, one can also write

$$\alpha = \frac{\omega}{c}\frac{\epsilon_2}{n} = \frac{4\pi\sigma_1}{nc}. \tag{3.27}$$

Eq. 3.27 is exact but is often most useful when $\epsilon_2 \ll \epsilon_1$ so that $n \approx \sqrt{\epsilon_1}$. Another useful result is that $\alpha = 2/\delta$ with δ the penetration depth of the field (the "skin depth").

If $\epsilon_2 = 0$ and $\epsilon_1 > 0$, then $n = \sqrt{\epsilon_1}$ and $\kappa = 0$. Because κ governs the damping of the wave, a material with positive ϵ_1 and $\epsilon_2 = 0$ is nonabsorbing; the electromagnetic wave propagates to infinity without loss of amplitude. If $\epsilon_1 < 0$ and $\epsilon_2 = 0$, then $\kappa = \sqrt{-\epsilon_1}$ and $n = 0$. The electric field is damped as $e^{-\kappa\frac{\omega}{c}x}$ with no oscillatory part at all. (The wavelength is infinite.)[†]

3.6 Normal-Incidence Reflectance

I'll discuss reflection and transmission of a sample (with a front and a back surface) in Chapter 7. For the moment I want to consider the simpler case of Fig. 3.2. Light is incident normally on an interface between two semi-infinite media, one with complex dielectric function ϵ_a (refractive index N_a), occupied by the incoming and reflected fields, and a second with complex dielectric function ϵ_b (refractive index N_b), occupied by the transmitted field.

---

* Beer's law is conventionally written in terms of the concentration c of a molecule in solution, and breaks the absorption coefficient into two parts: $\alpha = c\varepsilon$ where ε is a molecular constant called the molar extinction coefficient.

[†] It is not immediately obvious whether this damping in the negative ϵ_1 plus $\epsilon_2 = 0$ case is accompanied by energy absorption. I need to consider energy flow across an interface to determine that, in fact, there is no absorption. I can argue here that the current $\mathbf{j}$ is 90° out of phase with the electric field $\mathbf{E}$, so the time average dissipation, $\langle \mathbf{j} \cdot \mathbf{E} \rangle$, is zero.

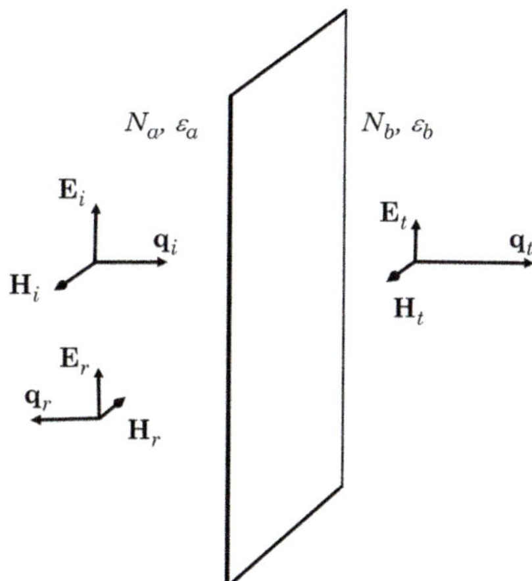

Fig. 3.2 Field vectors for incoming, reflected, and transmitted waves at an interface between two media.

Let me orient the $\hat{\boldsymbol{q}}$ vector of the incoming field along the $\hat{\boldsymbol{x}}$ axis and its $\mathbf{E}$ field along the $\hat{\boldsymbol{y}}$ axis, so that the $\mathbf{H}$ field is along $\hat{\boldsymbol{z}}$. The interface is in the y–z plane at $x = 0$. The incident fields are

$$\mathbf{E}_i = E_0 \hat{\boldsymbol{y}} e^{i(qx-\omega t)}$$

and

$$\mathbf{H}_i = N_a E_0 \hat{\boldsymbol{z}} e^{i(qx-\omega t)},$$

where I've used Eq. 3.14 to write the magnetic field amplitude.

The reflected field is traveling along the $-x$ axis, so $\mathbf{q} \cdot \mathbf{r} = -qx$ or $\hat{\boldsymbol{q}} = -\hat{\boldsymbol{x}}$. Its amplitude is proportional to E_0 and to the amplitude reflection coefficient r.

$$\mathbf{E}_r = r E_0 \hat{\boldsymbol{y}} e^{i(-qx-\omega t)}$$

and

$$\mathbf{H}_r = -N_a r E_0 \hat{\boldsymbol{z}} e^{i(-qx-\omega t)}.$$

Reversing $\hat{\boldsymbol{q}}$ means that one of $\hat{\boldsymbol{e}}$ and $\hat{\boldsymbol{h}}$ (but not both!) must reverse also. I chose $\hat{\boldsymbol{h}}$. Both fields have $q = \omega N_a/c$.

The transmitted field is in the second medium (b). Hence, it has $q = \omega N_b/c$. It also is linear in E_0 and in the amplitude transmission coefficient t.

$$\mathbf{E}_t = t E_0 \hat{\boldsymbol{y}} e^{i(qx-\omega t)}$$

and

$$\mathbf{H}_t = t N_b E_0 \hat{\boldsymbol{z}} e^{i(qx-\omega t)}.$$

All fields are parallel to the interface, so the boundary conditions are that the tangential components of the total fields are continuous across the interface.* So, let's dot with the appropriate unit vector, cancel $\mathbf{E}_0$, which appears in every field, and evaluate the boundary conditions at the interface, $x = 0$. I can evaluate the boundary conditions at any convenient time, zero being the most convenient. The boundary conditions produce two equations:

$\mathbf{E}$ is continuous ($\mathbf{E}_i + \mathbf{E}_r = \mathbf{E}_t$):

$$1 + r = t. \tag{3.28}$$

$\mathbf{H}$ is continuous ($\mathbf{H}_i + \mathbf{H}_r = \mathbf{H}_t$):

$$N_a(1 - r) = N_b t. \tag{3.29}$$

Substitute t from Eq. 3.28 into Eq. 3.29 and solve for r to find:

$$r = \frac{N_a - N_b}{N_a + N_b}. \tag{3.30}$$

This clearly gives the intuitive answer when $N_a = N_b$.

Now that I know r, it is easy to find t using Eq. 3.28:

$$t = \frac{2N_a}{N_a + N_b}. \tag{3.31}$$

Now, I'll specialize to the case where the incident and reflected waves are traveling in vacuum, so $N_a = 1$. I can let $N_b \equiv N$ and write Eq. 3.30 as

$$r = \frac{1 - N}{1 + N}. \tag{3.32}$$

It seems that for $N > 1$ and mostly real (a common case) the amplitude reflection coefficient is negative. This fact is usually expressed by saying the reflection from less-dense to more-dense media has a $180°$ phase shift.

In most experiments, one measures the intensity of incident and reflected light.† The normal-incidence reflectance is proportional to $|\mathbf{E}_r|^2$, i.e., $\mathscr{R} = rr^*$. With $N = n + i\kappa$, the reflectance is

$$\mathscr{R} = \left| \frac{1 - N}{1 + N} \right|^2$$
$$= \frac{(n - 1)^2 + \kappa^2}{(n + 1)^2 + \kappa^2}. \tag{3.33}$$

The range of $\mathscr{R}$ is $0 \leftrightarrow 1$.

To see how $\mathscr{R}$ depends on the refractive index, let me first look at the case where the material is weakly absorbing, so that κ can be neglected and $\mathscr{R} \approx (n - 1)^2/(n + 1)^2$. Then, if n is not large (say in the range of 1.1–1.9, common values for glass and other materials

---

* The condition on $\mathbf{H}$ is actually that the fields are discontinuous by the surface current density $4\pi \mathbf{K}/c$ but right now let me assume no surface charges or currents.

† Detectors typically can respond to amplitudes at low frequencies (up to GHz) but not at optical frequencies (THz to PHz and more). Instead they respond to the intensity or power in the optical beam.

in the visible), $\mathscr{R}$ is in the range 0.002–0.1. It requires $n = 6$ to raise $\mathscr{R}$ to 0.5 or so. $\mathscr{R} = 1$ only when n is very, very large

If κ is large, n can be close to unity and the reflectance still will be large. In the case[*] when $n \ll 1$, $\mathscr{R}$ approaches 1 independent of κ.

3.7 What If My Solid *Is* Magnetic?

If I keep the requirements of linear, isotropic, and homogeneous materials, but allow there to be an induced magnetization $\mathbf{M}$, I will have a nonzero magnetic susceptibility χ_m. The magnetization in a linear material is proportional to the magnetizing field $\mathbf{H}$: $\mathbf{M} = \chi_m \mathbf{H}$. Then, using $\mathbf{B} = \mathbf{H} + 4\pi \mathbf{M}$, I see that $\mathbf{B} = \mu \mathbf{H}$ (rather than $\mathbf{B} = \mathbf{H}$).

In a microscopic picture, the magnetic dipoles within the material are rearranged by the external field. Materials with $\mu > 1$ are "paramagnetic" (aluminum, platinum, MnO_2); those with with $\mu < 1$ are "diamagnetic" (bismuth, silver, water, benzene).[†] Typical deviations from unity are small.

I have to modify Eq. 3.3 when $\mathbf{M} \neq 0$. Using $\mathbf{B} = \mu \mathbf{H}$, Eq. 3.3c becomes

$$\mathbf{q} \times \mathbf{E} = \frac{\omega}{c}\mu\mathbf{H}. \tag{3.34}$$

Having the permeability in Eq. 3.3c increases the symmetry; now two equations have materials properties in them. Note that μ is complex, with $\mu = \mu_1 + i\mu_2$.

Continuing with the algebra which gave Eqs. 3.9 and 3.10 I find

$$q = \frac{\omega}{c}\sqrt{\epsilon\mu} \tag{3.35}$$

and

$$N = \sqrt{\epsilon\mu}. \tag{3.36}$$

The algebra which gave Eq. 3.14 leads to

$$H = \sqrt{\frac{\epsilon}{\mu}}E.$$

Finally, the normal incidence amplitude reflectivity (Eq. 3.30) becomes

$$r = \frac{\sqrt{\frac{\epsilon_a}{\mu_a}} - \sqrt{\frac{\epsilon_b}{\mu_b}}}{\sqrt{\frac{\epsilon_a}{\mu_a}} + \sqrt{\frac{\epsilon_b}{\mu_b}}} \tag{3.37}$$

and the reflectance (intensity) is $\mathscr{R} = rr*$.

---

[*] That case occurs in metals. In the extreme limit of materials with $n = 0$ and κ finite, $\mathscr{R} = 1$. Any value for κ suffices here. This result is the proof for the assertion in the footnote at the end of Section 3.5 that there is no absorption when $\epsilon_2 = 0$.

[†] The conduction electrons of a metal exhibit Pauli paramagnetism, a definite quantum mechanical effect. The core electrons or filled shells exhibit core diamagnetism, another quantum effect. The sum of these terms determines whether the material is overall paramagnetic or diamagnetic.

3.8　Negative Index Materials

The refractive index, Eq. 3.36, is the square root of the product of the dielectric function and the permeability. It is interesting to ask if there is ever a case where I should use the negative square root? The answer to the question is yes [9–13].

To simplify the discussion of negative refractive indices, let me restrict myself to materials where ϵ and μ are pure real numbers and allow them to be either positive or negative. In this case, the material is nonabsorbing. Start by writing $N^2 = \epsilon\mu$. In the usual case of transparent materials, $\epsilon > 0$ and $\mu > 0$. Then I have a pure real and positive refractive index, $N = n = \sqrt{\epsilon\mu}$. If one or the other of ϵ and μ is negative and the other one positive,* then I have a pure imaginary refractive index, $N = i\kappa$.

Now let me consider the case where both ϵ and μ are negative. Their product is positive so their square root is real. I'll start my discussion with Eqs. 2.7a–2.7d. I'll rewrite them using constitutive relations $\mathbf{D} = \epsilon\mathbf{E}$ and $\mathbf{B} = \mu\mathbf{H}$. I take $\mathbf{j} = 0$ because $\mathbf{j} = \sigma_1\mathbf{E} = (\epsilon_2/4\pi\omega)\mathbf{E}$ and ϵ_2 is zero for real ϵ. Then,

$$\epsilon\mathbf{q}\cdot\mathbf{E} = 0 \tag{3.38a}$$

$$\mu\mathbf{q}\cdot\mathbf{H} = 0 \tag{3.38b}$$

$$\mathbf{q}\times\mathbf{E} = \mu\frac{\omega}{c}\mathbf{H} \tag{3.38c}$$

$$\mathbf{q}\times\mathbf{H} = -\epsilon\frac{\omega}{c}\mathbf{E} \tag{3.38d}$$

The first two show as always that $\mathbf{q}$ is orthogonal to $\mathbf{E}$ and $\mathbf{H}$. The third shows that $\mathbf{E}$ and $\mathbf{H}$ are orthogonal. (The fourth is redundant as to direction; it also shows $\mathbf{H}$ and $\mathbf{E}$ are orthogonal.)

To be specific, I can without loss of generality orient once again $\mathbf{E}$ along $\hat{\mathbf{y}}$ and $\mathbf{H}$ along $\hat{\mathbf{z}}$. Then, according to Eq. 3.23, the Poynting vector $\mathbf{S}$ is along $\hat{\mathbf{x}}$. Indeed, $\hat{\mathbf{e}}\times\hat{\mathbf{h}} = \hat{\mathbf{s}}$, so that $\mathbf{S}$, $\mathbf{E}$, and $\mathbf{H}$ form a right-handed set. Energy flows along the positive x axis.

Now, both μ and ϵ are negative, so that the directional parts of Eqs. 3.38c and 3.38d read $\hat{\mathbf{q}}\times\hat{\mathbf{e}} = -\hat{\mathbf{h}}$ and $\hat{\mathbf{q}}\times\hat{\mathbf{h}} = +\hat{\mathbf{e}}$. These equations *both* show that $\mathbf{q}$ is oriented along $-\hat{\mathbf{x}}$. Thus $\mathbf{q}$, $\mathbf{E}$, and $\mathbf{H}$ form a left-handed set. For this reason the term "left-handed materials" is sometimes used. Wave motion is along the negative x axis. I can achieve this by writing Eq. 3.12 (with $\kappa = 0$)

$$\mathbf{E} = \mathbf{E}_0 e^{i\left(n\frac{\omega}{c}x - \omega t\right)}$$

and defining $n = -\sqrt{\epsilon\mu}$ to be a negative number. Thus I've shown that I must choose the negative square root when both ϵ and μ are negative.

Finding a material with both ϵ and μ negative at the same frequency is not easy. Moreover, many applications want the magnitudes not to be too large. Therefore, studies in this area have relied on synthetic materials, usually called "metamaterials." A common

---

* I have not yet shown evidence that negative dielectric functions or permeability exist. In Chapter 4 I will discuss classical models of the Drude metal and the Lorentz oscillator. In both, there are ranges of frequencies where the dielectric function is negative. Oscillator models for the permeability also give frequencies where it is negative.

example is a periodic array of circuit boards with unit cells made up of conducting straight wires for the negative ϵ elements and metallic split-ring resonators (looking like a c inside a reversed C) for the negative μ elements [14]. The materials are, however, far from nonabsorbing, requiring both ϵ and μ—to be complex. The theory can handle this, but at some cost in complexity [9, 10, 12].

The normal incidence amplitude reflection is still given by Eq. 3.37. More interesting things happen for oblique incidence. Snell's law (Appendix E) for refraction across an interface separating an ordinary material with refractive index $n_o > 0$ and a negative index material with refractive index $n_n < 0$ reads $n_o \sin \theta_o = n_n \sin \theta_n$, appearing identical to the ordinary Snell's law. However, the negative n_n then requires a negative θ_n, and the refracted ray is on the same side of the interface normal as is the incident ray, rather than crossing over it as happens in the usual case.

4.1 A Polarizable Medium

In my junior E&M class the lecturer worked out the case of a capacitor with a dielectric inside and found that the capacitance is increased by a factor of ϵ_1, the dielectric constant. (Dissipation was not considered at this initial stage.) The enhancement was attributed to the polarization of the atoms or molecules in the dielectric medium, typically viewed as little spheres or ellipsoids, each with a net dipole moment $\mathbf{p}$. This dielectric constant was not taken to be a function of frequency.*

I can quantify this a bit more by considering a single atom between the capacitor plates, with the plates charged up to produce a field $\mathbf{E}$ at the site of the atom. The atom thereby acquires a dipole moment, whose magnitude is determined by the atomic polarizability α_e:

$$\mathbf{p} = \alpha_e \mathbf{E}. \tag{4.1}$$

As long as the atom has at least one excited state (of the right symmetry) the polarizability can be calculated using second-order perturbation theory [15]. In cgs-Gaussian units, the polarizability has dimensions of volume, cm^3. Hence, the polarizability is often considered as a measure of the size of the charge-distribution. (Recall [2] that for a conducting sphere, $\alpha_e = r^3$.)

Now consider placing N atoms, all the same, in the volume V between the capacitor plates. At *low densities* of atoms, the dipole moment per unit volume so produced is

$$\mathbf{P} = \frac{1}{V} \sum_{i=1}^{N} \alpha_{ei} \mathbf{E} = n\alpha_e \mathbf{E}, \tag{4.2}$$

where $n = N/V$ is here the number density of atoms. Equation 4.2 ignores the effects that the polarization of one atom has on its neighbors. I'll work this out in some detail in Section 4.4.2 of this chapter. For now, because $\mathbf{P} = \chi_e \mathbf{E}$, I will write the dielectric susceptibility as $\chi_e = n\alpha_e$ and

$$\epsilon_c = 1 + 4\pi n\alpha_e. \tag{4.3}$$

The subscript (c) appearing in Eq. 4.3 stands for "core" and is meant to represent the contribution of tightly-bound core electrons to the optical properties. As long as $\hbar\omega \ll E_0 - E_1$ where E_0 is the ground-state binding energy of these core electrons and

---

* Although it is. But if I neglect the frequency dependence, I get to call it the dielectric "constant."

E_1 is their first available excited state* ϵ_c can be taken as a real constant. It is often called ϵ_∞, the limiting high frequency value of the dielectric function because, as I shall show, other contributions to ϵ are observable at specific frequencies; these contributions are not important as $\omega \to \infty$. As I work out simple models for the dielectric function of solids, I'll include ϵ_c where appropriate.

4.2 Drude Absorption by Free Carriers

The first serious model for optical properties that I will introduce is the one due to Drude [16, 17]. Drude's model of metallic conduction predates the quantum theory of solids and so omitted several important considerations. The most notable of these are (1) the fact that electrons in a metal obey Fermi–Dirac statistics and are at a temperature low compared to the Fermi temperature and (2) Bloch's theorem, which says that electrons can effectively propagate without dissipation in a perfect periodic potential. I'll introduce the theory keeping these concepts in mind, but at the beginning–before worrying about values of things like the mean free path or the effective mass–they make little difference.

4.2.1 Components of the Drude Model

The Drude model is based on the following ideas:

- The metal has a density n mobile carriers, where n is a number/cm^3 (number density, not mass density).
- These carriers are typically electrons,[†] with charge $q = -e$.
- The carriers are free. This statement means that the internal restoring force is zero or the potential is constant.
- The carriers do not interact; the electron–electron potential is neglected.
- If set in motion, the carriers relax to equilibrium due to damping from collisions. This damping is taken as a frictional or viscous force, proportional to the drift velocity of the carriers.[‡]
- Newton's second law for an electron whose coordinate is at $\mathbf{r}$ is

$$\sum \mathbf{F} = m\mathbf{a} = m\ddot{\mathbf{r}}.$$

- The forces are the external electric field $-e\mathbf{E}$ and the damping $-\Gamma\mathbf{v} = -\Gamma\dot{\mathbf{r}}$.
- Linear response is assumed, so coordinates, velocities, accelerations have the same time dependence as the field $\mathbf{E}$.

---

* The first excited state is at the Fermi energy (the chemical potential) or higher.

† But not always. Holes are easy to accommodate: change the sign of the charge carrier to positive. The equation for the conductivity will be the same whether one has holes or electrons.

‡ This is where the Fermi–Dirac statistics first rears its head. The electrons at the Fermi energy have very high speeds; the drift velocity is the deviation from equilibrium. Of course Drude took the equilibrium speed as zero or as a thermal speed for classical particles.

- The electric field is the applied field $\mathbf{E}_{ext}$ and not some local field. It is the applied field because the electrons sample all parts of the crystal and average over the dipole fields of the atoms. See Section 4.4.2 of this chapter for more on this point.

4.2.2 Equation of Motion and Its Solution

With the above, the equation of motion of one electron is

$$m\ddot{\mathbf{r}} = -e\mathbf{E}_{ext} - \Gamma\dot{\mathbf{r}}. \tag{4.4}$$

The electric field is the plane wave field of Eqs. 2.5 or 3.11. Note that there are two sets of coordinates here. The electron has a location ($\mathbf{r}$) which varies in time as it moves. The electric field has a value everywhere in space according to Eq. 2.5. It is the value of $\mathbf{E}$ at the location of the particle that matters. (I am considering local response.) For simplicity I'll orient coordinates so that $\hat{q} = \hat{x}$, making $\mathbf{E}$ be only a function of the x coordinate. It is constant in value for all y and z:

$$\mathbf{E}_{ext} = \mathbf{E}_0 e^{i(qx-\omega t)}.$$

To maintain a nice right-handed set of fields, I'll take $\hat{e} = \hat{y}$, so the force on the electron is in the $\hat{y}$ direction.*

Because the equation of motion contains only the velocity and acceleration of the electron, it makes sense to calculate the current, which, as was discussed in Section 2.3, is

$$\mathbf{j} = -ne\mathbf{v}, \tag{4.5}$$

where n is the number density of electrons. So let me write

$$\mathbf{v} = \mathbf{v}_0 e^{i(qx-\omega t)} = \dot{\mathbf{r}}, \tag{4.6}$$

with $\mathbf{v}_0$ a complex constant vector. Taking a time derivative of $\mathbf{v}$ gives the acceleration as $\ddot{\mathbf{r}} = -i\omega\mathbf{v}_0 e^{i(qx-\omega t)}$. Note that all three terms in Eq. 4.4 have the same complex exponential factor $e^{i(qx-\omega t)}$, allowing that factor to be hidden. The equation of motion is now an algebraic one,

$$-i\omega m\mathbf{v} = -e\mathbf{E}_{ext} - \Gamma\mathbf{v}.$$

It is convenient (and conventional) to make the replacement $\Gamma = m/\tau$, where τ is the relaxation time of the charge carriers.† With this definition I can solve for the velocity. At the same time I drop the subscript "ext"; that the electric field is the external applied field is understood when I use Ohm's law.

$$\mathbf{v} = \frac{-e}{m(1/\tau - i\omega)}\mathbf{E}. \tag{4.7}$$

The velocity is in phase with the applied field at low frequencies ($\omega \ll 1/\tau$); it also is independent of frequency. When $\omega \gg 1/\tau$ it is $90°$ out of phase with the electric field and the amplitude of the motion decreases as $1/\omega$.

---

* $-\hat{y}$ actually, since the electron has a negative charge.
† Suppose that when $\mathbf{v} = \mathbf{v}_0$ the electric field is turned off. Equation 4.4 with Γ replaced becomes $m\dot{\mathbf{v}} = -m\mathbf{v}/\tau$. Its solution is $\mathbf{v} = \mathbf{v}_0 e^{-t/\tau}$. τ is the time required to relax the velocity to $1/e$ of the initial value.

4.2.3 Ohm's Law and the Drude Conductivity

Now, substitute this velocity into Eq. 4.5 to find

$$\mathbf{j} = \frac{ne^2}{m(1/\tau - i\omega)}\mathbf{E}.$$

This equation is in the form of Ohm's law, $\mathbf{j} = \sigma\mathbf{E}$. The Drude conductivity is

$$\sigma = \frac{ne^2\tau/m}{1 - i\omega\tau}. \tag{4.8}$$

At dc ($\omega = 0$) the conductivity is

$$\sigma_{dc} = ne^2\tau/m. \tag{4.9}$$

σ_{dc} is proportional to the carrier density n, the relaxation time τ and inversely proportional to the mass m of the carriers.

4.2.4 Some Numbers

Silver or copper have a dc conductivity of $\sigma_{dc} = 6.2 \times 10^5\ \Omega^{-1}\,\text{cm}^{-1}$ (i.e., a resistivity of $1.6\ \mu\Omega$-cm). After reading Appendix A, I know to multiply σ_{dc} by $30c$ to get the conductivity into esu units; it is $5.6 \times 10^{17}\ \text{s}^{-1}$. This is a fairly high (angular) frequency; it corresponds to about 370 eV photon energy. The approximation $\sigma_{dc} \gg \omega$ is certainly valid at low frequencies ($\omega\tau < 1$).

I can estimate the mean free time from the dc conductivity. Solving Eq. 4.9 for τ, I obtain $\tau = m\sigma_{dc}/ne^2$ (with σ_{dc} in s^{-1}!). Using $n = 5.9 \times 10^{22}\ \text{cm}^{-3}$ (silver or gold), $e = 4.8 \times 10^{-10}$ esu, and $m = 9.1 \times 10^{-28}$ g, I get $\tau = 3.8 \times 10^{-14}$ s. If I convert this to a frequency (by inversion) $1/\tau = 2.6 \times 10^{13}\ \text{s}^{-1}$. I need to be clear about the units here, because there has been occasional confusion about what this time and rate mean [18]. τ is the mean free time between collisions and $1/\tau$ is the rate at which collisions occur; it also represents the angular frequency at which $\omega\tau = 1$. If the static field driving the current were suddenly to be extinguished, the currents would decay to zero as $j \propto e^{-t/\tau}$. Hence, I divide by 2π to convert to ordinary frequency and then by c to convert to cm^{-1}. I multiply by $\hbar$ to convert to an energy equivalent. Hence, to two digit precision,* I have

$$1/\tau = 2.6{\times}10^{13}\ \text{s}^{-1} = 4.2\ \text{THz} = 140\ \text{cm}^{-1} = 17\ \text{meV}/\hbar.$$

4.2.5 The Real and Imaginary Parts of σ

The complex conductivity ($\sigma = \sigma_1 + i\sigma_2$) may be split into its real (σ_1) and imaginary (σ_2) parts by multiplying by unity, written as $(1 + i\omega\tau)/(1 + i\omega\tau)$:

$$\sigma_1 = \frac{ne^2\tau/m}{1 + \omega^2\tau^2} = \frac{\sigma_{dc}}{1 + \omega^2\tau^2} \tag{4.10a}$$

$$\sigma_2 = \omega\tau\frac{ne^2\tau/m}{1 + \omega^2\tau^2} = \frac{\omega\tau\sigma_{dc}}{1 + \omega^2\tau^2}. \tag{4.10b}$$

---

* Probably more precision than the calculation deserves.

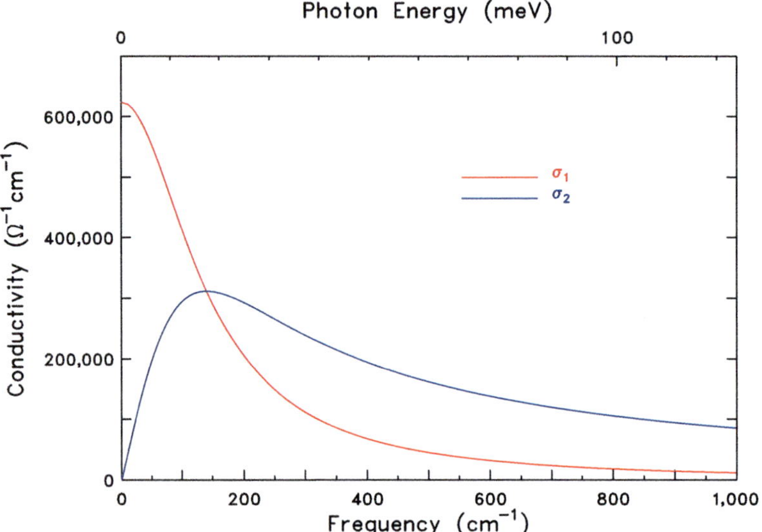

Fig. 4.1   Real and imaginary parts of the conductivity of a Drude metal like silver.

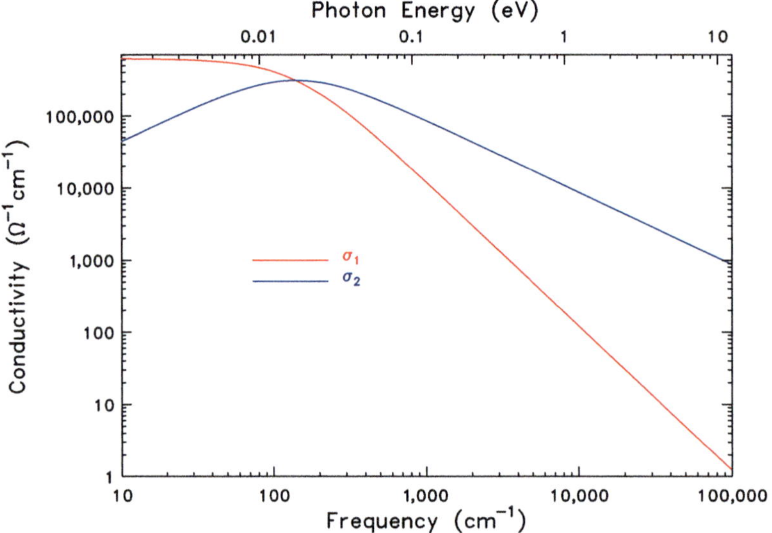

Fig. 4.2   Real and imaginary parts of the conductivity of a Drude metal like silver. The curves are plotted on a log-log scale.

Figures 4.1 and 4.2 show the real and imaginary parts of the conductivity of a Drude metal. The first show these functions on linear scales and the second on logarithmic scales. The parameters are $\sigma_{\mathrm{dc}} = 6.2 \times 10^5\ \Omega^{-1}\ \mathrm{cm}^{-1}$ and $1/\tau = 140\ \mathrm{cm}^{-1}$ (19 meV). These are meant to mimic the optical conductivity of silver.

The real part, σ_1, is constant at low frequencies but, as frequency increases, it rolls off, falling as $1/\omega^2$ at frequencies above $1/\tau$. The imaginary part starts off linear in ω and has a maximum at $\omega = 1/\tau$ (where $\sigma_1 = \sigma_2$) and falls as $1/\omega$ at high frequencies.

The shape of the Drude conductivity curve is always the same: a peak in $\sigma_1(\omega)$ at zero frequency and a hump-like feature in $\sigma_2(\omega)$. Two quantities control the details. One is the area under the $\sigma_1(\omega)$ curve, which (as I shall show) is proportional to ne^2/m and hence is often called the "spectral weight." The second is $1/\tau$, the full width at half maximum.

4.2.6 Complex Dielectric Function, $\epsilon(\omega)$

I've already written Eq. 3.6 relating the complex dielectric function to the complex conductivity: $\epsilon = 1 + 4\pi i\sigma/\omega$. But the Drude conductivity, Eq. 4.8, considers only the conduction electrons in the metal. As discussed in Section 4.1, there is also the polarization of the ion cores.* So, I'll include ϵ_c, modifying Eq. 3.6 to read

$$\epsilon = \epsilon_c + 4\pi i\sigma/\omega.$$

And then, substituting Eq. 4.8 and doing a little algebra, I get

$$\epsilon(\omega) = \epsilon_c - \frac{4\pi ne^2/m}{\omega^2 + i\omega/\tau}. \tag{4.11}$$

The Drude contribution comes in with a minus sign in the real part (but a positive imaginary part).

Look at the numerator in the second term: $4\pi ne^2/m$. This quantity has dimension of [esu^2/cm^3-g]. But Coulomb's law in cgs-Gaussian units, $F = q_1 q_2/r^2$, tells me that [esu^2 = dyne-cm^2]. And, the cgs unit of force is [dyne = g-cm/s^2], making the units of $4\pi ne^2/m$ [1/s^2], a frequency squared.† So I'll define the frequency, which I will call the "plasma frequency," ω_p, to be

$$\omega_p \equiv \sqrt{\frac{4\pi ne^2}{m}}, \tag{4.12}$$

and write

$$\epsilon(\omega) = \epsilon_c - \frac{\omega_p^2}{\omega^2 + i\omega/\tau}. \tag{4.13}$$

Moreover, Eq. 4.9 becomes $\sigma_{dc} = \omega_p^2 \tau/4\pi$.

4.2.7 The Real and Imaginary Parts of ϵ

The real and imaginary parts of ϵ are obtained by multiplying the top and bottom of the second term by $\omega - i/\tau$ and doing a small amount of algebra:

$$\epsilon_1 = \epsilon_c - \frac{\omega_p^2}{\omega^2 + 1/\tau^2} \tag{4.14a}$$

$$\epsilon_2 = \frac{\omega_p^2}{\omega\tau(\omega^2 + 1/\tau^2)}. \tag{4.14b}$$

---

* There is, of course, a lot more than core polarizability. There can be a number of interband transitions, motion of ion cores (lattice dynamics) and other interesting solid-state physics. These processes, which must be included in a complete dielectric function, are the subject of many later chapters.

† I knew this, of course, because ϵ is just a number and the denominator of Eq. 4.11 has units of frequency squared.

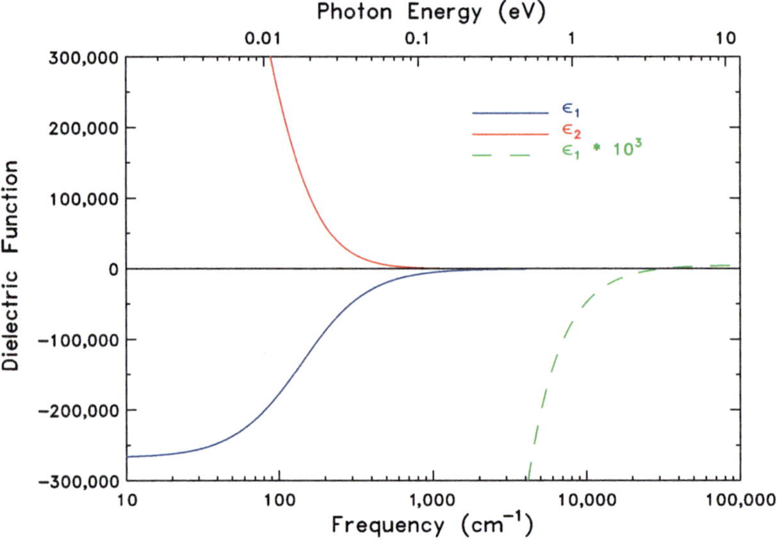

Fig. 4.3   Real and imaginary parts of the dielectric function of a Drude metal. The dashed line shows ϵ_1 multiplied by 1,000, allowing the zero crossing at 32,000 cm^{-1} (4 eV) to be seen.

These functions are plotted in Fig. 4.3. The parameters used, $\omega_p = 73{,}000$ cm^{-1} (9.0 eV), $1/\tau = 140$ cm^{-1} (17 meV), and $\epsilon_c = 5.1$, are meant to mimic the dielectric function of silver. The real part is negative and large in magnitude at low frequencies. Indeed, the low frequency limit of Eq. 4.14a is $\epsilon_1 \approx -\omega_p^2\tau^2$ (neglecting ϵ_c, which is almost always appropriate in good metals). As the frequency becomes large with respect to $1/\tau$, the dielectric function becomes that of pure free carriers, $\epsilon_1 \approx \epsilon_c - \omega_p^2/\omega^2$.

The zero of ϵ_1 occurs near $\omega_p/\sqrt{\epsilon_c}$. I'll call this frequency the "screened plasma frequency." The imaginary part goes as $1/\omega$ at low frequencies, diverging as $\omega \rightarrow 0$.* At high frequencies, $\epsilon_2 \sim 1/\omega^3$.

4.2.8 Some More Numbers

Plasma frequencies are typically in the visible to ultraviolet for metals but can be in the infrared or far infrared for low-density metals and for doped semiconductors. Using the same parameters as in Section 4.2.4 of this chapter ($n = 5.9 \times 10^{22}$ cm^{-3}, $e = 4.8 \times 10^{-10}$ esu, and $m = 9.1 \times 10^{-28}$ g), I calculate $\omega_p = 1.4 \times 10^{16}$ s^{-1}. If I convert angular frequency (s^{-1}) to regular frequency (Hz), wavenumber, and electron volts, I get

$$\omega_p = 2.2 \times 10^{15}\text{ Hz} == 72{,}000\text{ cm}^{-1} == 9.0\text{ eV}/\hbar.$$

Using the value for the mean free time estimated in Section 4.2.4 of this chapter, $\tau = 2 \times 10^{-14}$, the low frequency value for ϵ_1 is $-\omega_p^2\tau^2 = -250{,}000$.

---

* This behavior makes ϵ_2 not a very useful function for conducting materials.

If the mass is the free-electron mass, $m = m_e$, and the number density n is measured in cm^{-3}, then the plasma frequency is

$$\omega_p = 8980\sqrt{n} \text{ Hz} = 2.993 \times 10^{-7}\sqrt{n} \text{ cm}^{-1} = 3.711 \times 10^{-11}\sqrt{n} \text{ eV}/\hbar. \qquad (4.15)$$

If $m \neq m_e$, then one replaces $\sqrt{n}$ with $\sqrt{nm_e/m}$.

4.2.9 Factors that Affect the Optical Conductivity

Three materials-dependent quantities come into the conductivity of a metal: the carrier density n, the dynamical or effective mass m, and the mean free time τ. The first two combine to give the plasma frequency; the result is that only n/m can be extracted from optics alone.*

The mean free time in the Drude model represents the time between electron collisions (or more correctly [19], the time the electron has before it suffers its next collision). The electron moves under the influence of the field between the collisions and contributes to current flow. I'll say a lot more about this in Chapter 8. For now, I'll assert that the mean free time is affected by the purity and crystal quality of the metal and by the temperature. The mean free time is longer when purity is high and temperature is low. Empirically, these scattering rates are independent and add, following Matthiessen's rule:

$$\frac{1}{\tau} = \frac{1}{\tau(T)} + \frac{1}{\tau_i}$$

where $\tau(T)$ is the temperature-dependent scattering from thermal excitations in the metal and τ_i is the temperature-independent scattering from impurities or other defects. $\tau(T)$ has a complex temperature dependence, but at high temperatures many metals have $1/\tau(T) \sim T$; at low temperatures the power law is stronger: T^2, T^3, or T^5. At the lowest temperatures, the scattering rate (and the resistivity) saturates at the impurity-limited value. The ratio between the room-temperature resistivity and the low-temperature resistivity is called the residual resistivity ratio (RRR) and is a measure of purity. It is also a measure of relative scattering times, because the carrier density and mass in simple metals do not change much with temperature.†

As an example, Fig. 4.4 shows a simulation at three temperatures of the real parts of the conductivity (left panel) and dielectric function (right panel) for a metal of modest purity, RRR = 10. (I choose this value rather than 100 or 1,000, values which are possible in high-purity metals, to avoid entering the anomalous skin effect regime. See Chapter 14.) Relative to room temperature, the mean free time is assumed $3\times$ longer for the 100 K calculation and $10\times$ longer for the 10 K calculation. The dc conductivity goes up and the width of the peak becomes narrower with decreasing temperature; the areas under the three conductivity curves stay the same. The dielectric function follows $-\omega_p^2/\omega^2$ at high frequencies; with

---

* If I measure the Hall effect or other magneto-optical parameters such as cyclotron frequency, I can determine n or m individually and then the optical parameters will provide the other.

† There are other systems with other behaviors. Semiconducting materials have a much stronger temperature dependence to the carrier density than to the mean free time. Heavy-Fermion metals have a huge increase in carrier effective mass at low temperatures.

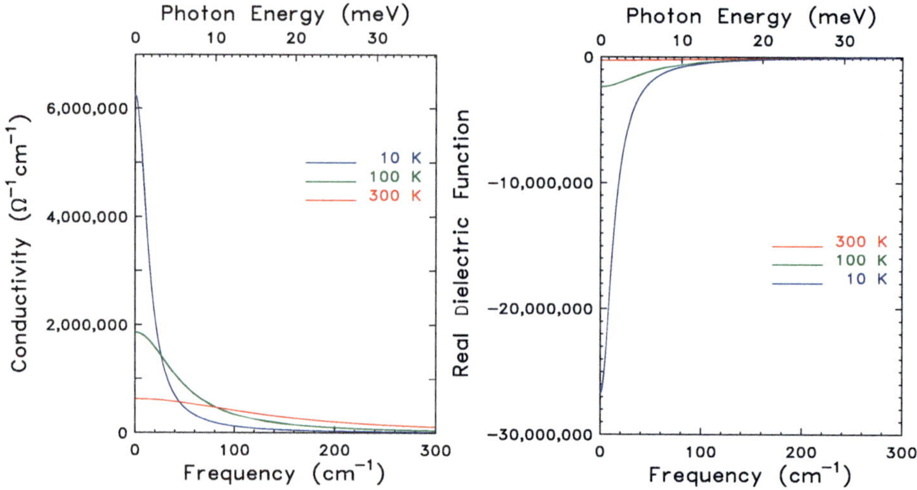

Fig. 4.4 Left: Optical conductivity of a hypothetical metal with modest purity. Curves are shown for room temperature, 100 K, and low temperatures. Right: Dielectric function of the same metal.

decreasing temperature the $1/\omega^2$ behavior continues to lower frequencies. (The curve marked 300 K occupies only a bit of the panel here; it is identical to the ϵ_1 shown in Fig. 4.3.) The limiting low frequency value of $-\omega_p^2\tau^2$ increases $100\times$ as the mean free time becomes 10 times longer.

4.2.10 Fields and Currents

This is a good time to take a look at the fields and currents in the metal. Recall that the physically observable field is the real part of the complex field. Hence, if I could put probes in and measure the field, I would see

$$\mathbf{E} = \mathrm{Re}[E_0\hat{\mathbf{y}}e^{-i\omega t}] = E_0\hat{\mathbf{y}}\cos(\omega t)$$

and if I could put a current meter in, I would see

$$\mathbf{j} = \mathrm{Re}[\sigma\,\hat{\mathbf{y}}\mathbf{E}e^{-i\omega t}] = \sigma_1 E_0\hat{\mathbf{y}}\cos(\omega t) + \sigma_2 E_0\hat{\mathbf{y}}\sin(\omega t)$$

so if $\sigma_1 \gg \sigma_2$ (low frequencies), the current is in-phase with the field and their time average $\langle\mathbf{j}\cdot\mathbf{E}\rangle$ equals half of the product of peak values. This is the Ohmic loss in the metal. If, however, $\sigma_1 \ll \sigma_2$ (high frequencies), the field follows the cosine and the current mostly follows the sine; the time average of their product is nearly zero. The electrical analogy is a nearly ideal inductor.

4.2.11 The Refractive Index

It is the complex refractive index ($N \equiv \sqrt{\epsilon}$) which most directly governs wave propagation in the medium. Figures 4.5 and 4.6 show the real and imaginary parts of the refractive index

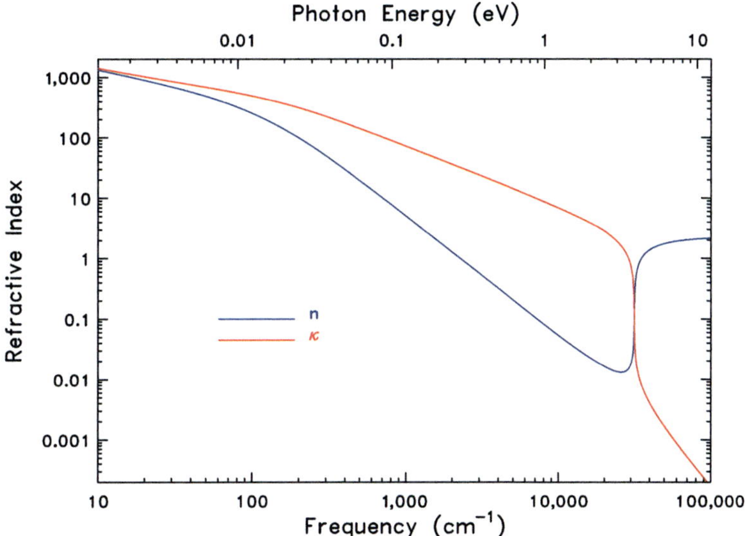

Fig. 4.5 Real and imaginary parts of the refractive index of a Drude metal. Data are shown on a log-log scale. The real and imaginary curves cross when $\omega \approx \omega_p / \sqrt{\epsilon_c}$.

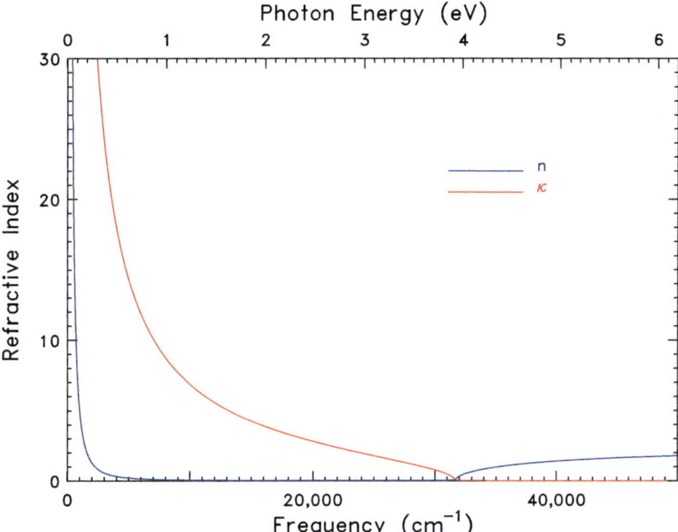

Fig. 4.6 Real and imaginary parts of the refractive index of a Drude metal. Data are shown on linear scales.

of a Drude metal. These are calculated using the same parameters as for the dielectric function.

There are several regimes to discuss. At low frequencies, the refractive index n and the extinction coefficient κ are both large and relatively close in magnitude. (See Fig. 4.5 for the low-frequency region.) At very low frequencies, where $\epsilon_1 \ll \epsilon_2$ (i.e., where $\omega \ll 1/\tau$), Eq. 3.17 gives $n \approx \kappa \approx \sqrt{\epsilon_2/2}$. See Figures 4.3–4.6 for the behavior of ϵ and N. For the Drude metal, then,

$$n \approx \kappa \approx \sqrt{\frac{\omega_p^2 \tau}{2\omega}} = \sqrt{\frac{2\pi \sigma_{dc}}{\omega}} \quad \text{when } \omega \ll \frac{1}{\tau}. \tag{4.16}$$

Both n and κ fall with increasing frequency, with n falling faster than κ.

At intermediate frequencies, the refractive index of a metal can be less than one, sometimes much less than one. This behavior is seen in the calculation shown in Figures 4.5 and 4.6 once $\omega \gg 1/\tau$ up until the screened plasma frequency. In this regime, the phase velocity, $v = c/n$ is bigger than c. I leave it to the reader to show that the group velocity is smaller than c. The waves in the material are of course strongly damped; with the electric field penetration length, $\delta = \lambda_0/2\pi\kappa$, smaller than the vacuum wavelength.

At a certain frequency (32,000 cm^{-1} in this case) κ falls rapidly and n jumps up to a value bigger than 1. This jump occurs exactly where ϵ crosses zero, the screened plasma frequency. Above this frequency, $n \approx \sqrt{\epsilon_c}$ and κ is small.

The behavior of N is easy to understand in the extreme case when I can neglect the scattering rate $1/\tau$. (Even if $1/\tau$ is finite, this is a good first approximation when $\omega \gg 1/\tau$.) Then

$$\epsilon = \epsilon_c - \frac{\omega_p^2}{\omega^2}$$

and

$$N = \begin{cases} 0 + i\sqrt{\dfrac{\omega_p^2}{\omega^2} - \epsilon_c}, & \text{for } \omega < \omega_p/\sqrt{\epsilon_c} \\[2em] \sqrt{\epsilon_c - \dfrac{\omega_p^2}{\omega^2}} + i0, & \text{for } \omega > \omega_p/\sqrt{\epsilon_c}. \end{cases}$$

Below the screened plasma frequency, this approximation gives a pure imaginary value $i\kappa$ for N. The real part of the complex refractive index is zero. Above the screened plasma frequency, $N = n$ is purely real and has the high-frequency limit of $n = \sqrt{\epsilon_c}$. Metals should be transparent above $\omega_p/\sqrt{\epsilon_c}$, and some are.

4.2.12 Reflectance

Once I have N, I can calculate the reflectance $\mathcal{R}$ from Eq. 3.33. At low frequencies, both parts of N are very large, so the difference between n and 1 is large. In this case, it is sometimes convenient to write Eq. 3.33 as

$$\mathcal{R} = 1 - \frac{4n}{(n+1)^2 + \kappa^2}. \tag{4.17}$$

Note that although Eq. 4.17 might look like an expansion, it is exact. The calculated reflectance, using parameters identical to those for Figures 4.1–4.3, 4.5, and 4.6, is shown in Fig. 4.7 as the red curve, marked "300 K."

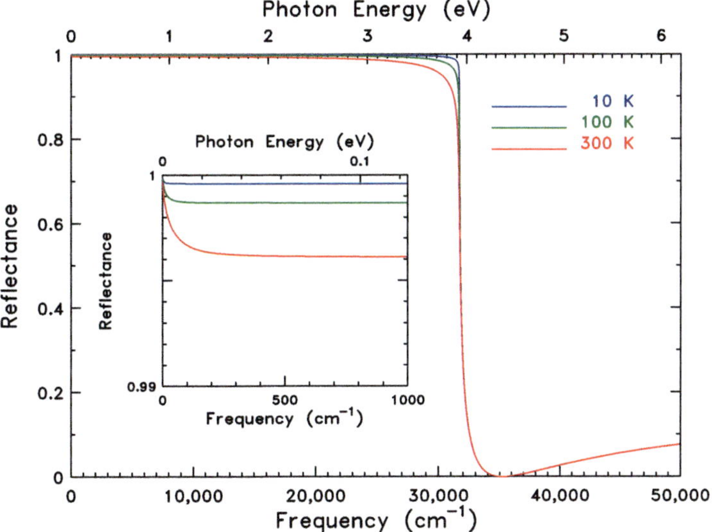

Fig. 4.7 Normal incidence reflectance of a Drude metal.

A good approximation in the far infrared, where $n \approx \kappa \gg 1$, is to write $n + 1 \approx n$ and then, using Eqs. 4.16 and 4.17, obtain the so-called Hagen–Rubens [20] equation

$$\mathscr{R} \approx 1 - \sqrt{\frac{2\omega}{\pi \sigma_{\mathrm{dc}}}}. \tag{4.18}$$

The reflectance is 100% at $\omega = 0$ and falls in accord with Eq. 4.18 with increasing frequency. Once ω is bigger than $1/\tau$ the reflectance flattens out. This behavior is shown in the inset of Fig. 4.7, where one can see the Hagen–Rubens regime at the left and the flattening out at higher frequencies. The reflectance is above 99.5% in each case.

At high frequencies, the reflectance falls rapidly from a value close to unity to a value close to zero. This drop occurs around 30,000 cm^{-1} (3.5 eV) for the parameters used for Fig. 4.7. This "plasma edge" occurs at the same frequencies where the rapid fall of κ and rise of n occur in Fig. 4.6. There is a minimum value of the reflectance, corresponding to the conditions* $n \approx 1$. (The approximation becomes an equality when $\kappa \ll 1$.)

The simulations for 100 K and 10 K (actually τ increased by $3\times$ and $10\times$) are shown also in Fig. 4.7. As the mean free time becomes longer, the turnover between the flat and high reflectance and the steep plasma edge becomes sharper and sharper.

---

* The exact location of this "plasma minimum" is a balance of the frequency where $n - 1 \sim 0$ and the value of κ near this frequency. To have a deep minimum, κ must have a relatively small value. It is *not* the same as the frequency where $\epsilon_1 = 0$ although these two frequencies are close to each other. In the case where $\omega_p \gg 1/\tau$, then I can write $\epsilon(\omega) = \epsilon_c - \omega_p^2/\omega^2$. With ϵ_c a real constant, the zero of $\epsilon(\omega)$ is at $\omega_p/\sqrt{\epsilon_c}$ and the reflectance minimum is when $n = \sqrt{\epsilon} = 1 = 1^2 = \epsilon$, making the minimum occur at $\omega_p/\sqrt{\epsilon_c - 1}$. If $\epsilon_c = 1$, the minimum is at infinity. If ϵ_c has dispersion (and an imaginary part), the minimum is still near where $\mathrm{Re}(\epsilon_c(\omega)) = 1$ but the exact location can be pushed up or down depending on the functional form of $\mathrm{Im}(\epsilon_c)$.

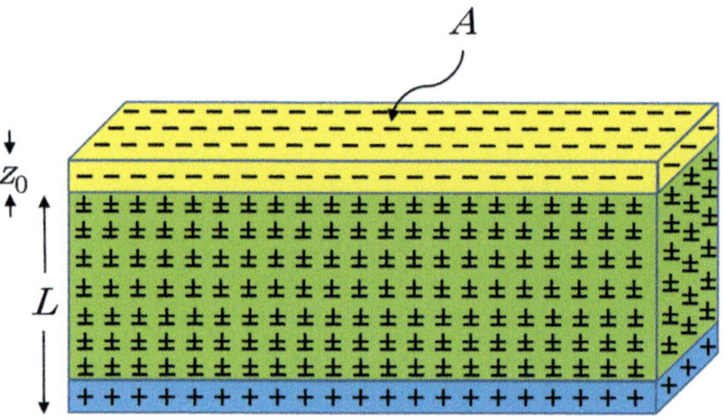

Fig. 4.8   A slab of free electrons and fixed positive ions. I pretend that the electrons are yellow and the ions blue. The region where both exist is therefore green. Each electron has been displaced upwards by a small amount z_0, creating a yellow region of negative charge on the top of the slab and a blue region of positive charge on the bottom. The central region remains neutral.

4.3 The Plasma Frequency

In Eq. 4.12 I defined the plasma frequency as $\omega_p = \sqrt{4\pi n e^2/m}$. What is the meaning of this frequency? There are no peaks in σ_1 (or ϵ_2) at ω_p. The dielectric function ϵ_1 changes sign near $\omega = \omega_p/\sqrt{\epsilon_c}$ but does so smoothly.

To see why the name is what it is, consider a slab of metal of thickness L. The metal contains positive ions bound to their lattice positions by the interatomic forces in the metal. Permeating the entire slab is the gas of conduction electrons. The metal is electrically neutral, with a number density of electrons of $n_e = N_{\text{electrons}}/V$ and a number density of ions of $n_i = N_{\text{ions}}/V$, where V is the volume of the slab. For conceptual simplicity, let me assume that the ions are singly charged,* so that the number of charges is $N_{\text{electrons}} = N_{\text{ions}} \equiv N$ and the number density is $n_e = n_i \equiv n$

Now, displace the electrons an amount z_0 in the upward ($\hat{z}$) direction. Take $z_0 \ll L$. Fig. 4.8 shows the slab with the displaced electrons. Because every electron is displaced by the same amount, there is a net dipole moment in the slab. The dipoles all point downward, from the negative electrons to the positive ions. The dipoles are $\mathbf{p} = -QL\hat{z}$ where Q is the magnitude of the charge on each surface. This quantity is the product of the number density of charges n, their individual charge e and the volume of uncompensated charge Az_0 with A the area of the slab surface: $Q = neAz_0$. Hence the dipole moment/unit volume, $\mathbf{P} = \mathbf{p}/AL$, is

$$\mathbf{P} = -\frac{neAz_0 L}{AL}\hat{z} = -nez_0\hat{z}.$$

At the same time, the uncompensated surfaces generate an electric field $\mathbf{E}$, just as in the case of a charged capacitor. $\mathbf{E}$ points upwards, from the positive charge on the bottom to the negative charge on top. Gauss' law tells me that $\mathbf{E} = 4\pi Q/A\hat{z} = 4\pi nez\hat{z} = -4\pi\mathbf{P}$.

---

* This assumption is by no means essential; a solid of trivalent ions plus electrons would have $N_{\text{electrons}} = 3N_{\text{ions}}$.

Table 4.1 Experimental plasma frequencies.			
Material	ω_p s^{-1}	ν_p cm^{-1}	E_p eV
K	6×10^{15}	31,000	3.9
Na	8×10^{15}	48,000	6.0
Al	2.4×10^{16}	140,000	18.0
"Si"	2.6×10^{16}	160,000	20.0

The field exerts a force on the electrons, $\mathbf{F} = -e\mathbf{E} = m\ddot{\mathbf{z}}$. The equation of motion of the electrons is thus

$$m\ddot{\mathbf{z}} = -4\pi ne^2\mathbf{z}.$$

The solution to this equation is that of a harmonic oscillator, $\mathbf{z} = \mathbf{z}_0 e^{-i\omega_0 t}$ with ω_0 the resonant frequency of the oscillator. Substitution gives

$$-m\omega_0^2\mathbf{z}_0 = -4\pi ne^2\mathbf{z}_0.$$

So, I get the frequency of the free oscillations to be

$$\omega_0 = \sqrt{\frac{4\pi ne^2}{m}} \equiv \omega_p.$$

This equation is identical to Eq. 4.12, the plasma frequency I defined when calculating the Drude dielectric function of the metal. This plasma oscillation is a normal mode of the electrons in the metal slab. The frequency of oscillation is the plasma frequency.

If the atomic core polarizability is included in the problem, it becomes analogous to a capacitor with a dielectric material between the plates. The Gauss' law problem finds $\mathbf{D}$; the electric field $\mathbf{E}$, dipole moment/unit volume $\mathbf{P}$, and the restoring force on the electrons are all reduced by a factor of ϵ_c, and the resonant frequency becomes

$$\omega_0 = \frac{\omega_p}{\sqrt{\epsilon_c}}.$$

A measurement of reflectance over the appropriate frequency range can provide me with the plasma frequency. There are four methods in common use. I can fit the reflectance to a model dielectric function that includes a Drude term; ω_p is a parameter from the fit. I can carry out Kramers–Kronig analysis (Chapter 10) and analyze the $1/\omega^2$ term in ϵ_1. Alternatively, a sum-rule analysis of $\sigma_1(\omega)$ (also Chapter 10) provides an estimate of ω_p. Finally, I can measure the reflection and transmission of a thin film (Chapter 7) and extract the optical constants and plasma frequency from those data. Each method then gives the ratio of carrier density to effective mass, n/m.

Metal plasma frequencies are typically rather high, in the ultraviolet for ordinary metals. Table 4.1 lists a few experimental values while Table 4.2 gives a calculated plasma frequency for a number of elemental metals. The assumptions are: (1) the carrier density is the atomic density times the valence, and (2) the carriers have the free-electron mass.*

* The charge is e, in esu!

Table 4.2 Calculated plasma frequencies for a variety of metals.

Valence	Metal	n 10^{22} cm^{-3}	ω_p 10^{15} s^{-1}	v_p cm^{-1}	E_p eV	c/ω_p nm
1	Li	4.7	12.23	64,800	8.03	24.5
	Na	2.65	9.18	48,700	6.04	32.7
	K	1.4	6.67	35,400	4.39	44.9
	Rb	1.15	6.05	32,000	3.97	49.6
	Cs	0.91	5.38	28,500	3.53	55.7
	Cu	8.45	16.40	86,900	10.77	18.3
	Ag	5.85	13.64	72,300	8.96	22.0
	Au	5.9	13.70	72,600	9.00	21.9
2	Be	24.2	27.75	147,200	18.25	10.8
	Mg	8.6	16.54	87,700	10.87	18.1
	Ca	4.6	12.10	64,100	7.95	24.8
	Sr	3.56	10.64	56,400	6.99	28.2
	Ba	3.2	10.09	53,500	6.63	29.7
	Zn	13.1	20.42	108,300	13.43	14.7
	Cd	9.28	17.19	91,100	11.29	17.5
3	Al	18.06	23.97	127,100	15.76	12.5
	Ga	15.3	22.07	117,000	14.51	13.6
	In	11.49	19.12	101,400	12.57	15.7
4	Pb	13.2	20.50	108,700	13.48	14.6
	Sn(w)	14.48	21.47	113,800	14.11	14.0
	1×10^{21}	0.1	1.78	9,460	1.17	168
	1×10^{22}	1	5.64	29,900	3.71	53.2

These assumptions are excellent for the monovalent alkali and noble metals and poor for multivalent metals. The last two rows show the numbers for two carrier densities a factor of 10 apart, taking the mass as the free electron mass. The frequency for any other density can be calculated by multiplying by the square root of the carrier density ratio.

4.4 Lorentz Model: Absorption by Bound Electrons

The Lorentz model is designed for insulators or at least for electrons which are bound by some force to an atom or ion in the solid. I'll also use it for the contribution to the dielectric function from optically active vibrational modes (phonons). This latter role for the model plays out in Chapter 5.

I will view the potential for these bound electrons as harmonic, i.e., the electrons are attached to their equilibrium positions by a spring. The spring constant will be the same for all the electrons in a particular orbital. On the one hand, the harmonic potential (a spring) is quite different from the Coulomb potential that–as is well known–governs the electronic properties of an atom, molecule, or solid. On the other hand, whatever the potential is, I know that it will look harmonic for small displacements from equilibrium. My assumption of linear response suggests that the displacements are indeed small.

4.4.1 Dilute Limit

To begin, I'll consider the dilute limit, just as I did in Section 4.1. The reason I want to start with the dilute limit is that an electric field applied* to the molecule polarizes the molecule. Near the location of each polarized molecule, there is also a dipolar field [2]

$$\mathbf{E}_d(\mathbf{r}) = \frac{3(\mathbf{p} \cdot \hat{\mathbf{r}})\hat{\mathbf{r}} - \mathbf{p}}{r^3} \qquad (4.19)$$

where $\mathbf{r}$ is a vector from the origin to the field point, $\hat{\mathbf{r}} = \mathbf{r}/r$, $\mathbf{p} = \alpha_e \mathbf{E}_{\text{loc}}$ and $\mathbf{E}_{\text{loc}}$ is the local field at the position of the molecule. I want the strength of these dipolar fields to be negligible at the other nearby molecules so as to make the local field (the field responsible for the polarization) equal to the field at the molecule from the external electromagnetic wave. The dipolar field varies in space, looping from the $+$ end of the dipole around to the $-$ end. Its magnitude is of order $E_d \approx (\alpha_e/r^3)\mathbf{E}_{\text{applied}}$. I'll take for the polarizability α of the molecule the polarizability of a conducting sphere, $\alpha_e \sim a_0^3$ with a_0 the sphere's radius. To keep the field from the polarized molecule smaller than, say, 0.1% of the external field, the nearby molecules have to be separated by $\sim 10\times$ the molecular size, whereas in close packing the separation is of order the molecular size. The density is 1,000 times smaller than the close-packed density; hence, the system is a gas.

The molecule has zero dipole moment without an external field. The electron is considered to be bound to the position where $\mathbf{p} = 0$ by a harmonic force.

$$\mathbf{F}_h = -K\mathbf{x} = -m\omega_0^2\mathbf{x}$$

where K is the spring constant, m is the electronic mass,† and $m\omega_0^2$ is a complicated way to write the spring constant.‡

A second force acting on the electron is a damping force proportional to the velocity, just as in the case of the Drude model, $\mathbf{F}_v = -\Gamma\mathbf{v} \equiv -m\gamma\dot{\mathbf{x}}$, where $\gamma = 1/\tau$ is the damping rate. This damping is formally identical with the Drude model for a metal but the

---

* I think it is easy to be confused about the meaning of the "applied field." In continuum electrodynamics it of course means the electric field applied to the medium from external sources (lamps, lasers, generators). From the point of view of a microscopic picture of the medium as a body consisting of polarizable entities, it is of course the field seen by the polarizable entities. These fields are not necessarily the same. The second one is the superposition of the external field and the fields emitted by all the other polarizable entities. So I need to be careful with the use of "applied field;" perhaps "external field" and "local field" are better terms. I also need to be aware that when I send light into the material from an external source, some of it will be reflected at the surfaces of the material and some will be absorbed on its way to the point where I want to evaluate the field. I can measure the total field at any point by inserting a (small) test charge at that point and seeing what force the test charge feels. Dilute media are those in which the external field *is* the applied field.

† It is the reduced mass in fact, but the ion mass is big enough that it can be ignored. I'll take m to be the effective mass – which can be quite different from the free-electron mass – in the solid.

‡ I write $-m\omega_0^2$ rather than K because I know how it will work out, anticipating that the resonant frequency $\omega_0 = \sqrt{K/m}$. I have solved the harmonic oscillator problem in the past.

underlying physics is a bit different. In a metal the damping is associated with collisions of the Fermi surface electrons with impurities, phonons, etc. In semiconductors and insulators there is no Fermi surface; the chemical potential lies in a forbidden band gap between valence and conduction band. The damping may be considered as due to collisions of photoexcited carriers with impurities, phonons, or other electrons, leading to relaxation to the ground state. Damping also can come from the emission of excitations like phonons, with a corresponding scattering of the electron. The remaining force is the force due to the external electric field, which I'll take as $\mathbf{F}_e = -e\mathbf{E} = -e\mathbf{E}_0 e^{-i\omega t}$. Newton's law is then written as

$$m\ddot{\mathbf{x}} = -m\omega_0^2\mathbf{x} - m\gamma\dot{\mathbf{x}} - e\mathbf{E}_0 e^{-i\omega t}. \tag{4.20}$$

In linear response, the location of the electron is $\mathbf{x} = \mathbf{x}_0 e^{-i\omega t}$. Taking the first derivative (bringing down $-i\omega$) and the second derivative (bringing down $-\omega^2$), Eq. 4.20 for the position of the electron becomes

$$m\omega_0^2\mathbf{x} - i\omega m\gamma\mathbf{x} - m\omega^2\mathbf{x} = -e\mathbf{E}$$

or

$$\mathbf{x} = \frac{-e/m}{\omega_0^2 - \omega^2 - i\omega\gamma}\mathbf{E}. \tag{4.21}$$

Clearly, $\mathbf{x}$ is resonant at ω_0; it diverges at this frequency if $\gamma = 0$.

With this displacement, the dipole moment is $\mathbf{p} = -e\mathbf{x}$, and the polarizability is

$$\alpha_e = \frac{e^2/m}{\omega_0^2 - \omega^2 - i\omega\gamma}. \tag{4.22}$$

The dipole moment/unit volume $\mathbf{P} = n\mathbf{p}$, where n is the number of molecules per unit volume. Thus

$$\mathbf{P} = \frac{ne^2/m}{\omega_0^2 - \omega^2 - i\omega\gamma}\mathbf{E} = \chi_e\mathbf{E},$$

with χ_e the dielectric susceptibility of the semiconductor or insulator. I then write $\epsilon = 1 + 4\pi\chi_c + 4\pi\chi_e$ where χ_c is the core polarizability from Section 4.1 ($\epsilon_c \equiv 1 + 4\pi\chi_c$) and find

$$\epsilon(\omega) = \epsilon_c + \frac{4\pi ne^2/m}{\omega_0^2 - \omega^2 - i\omega\gamma}. \tag{4.23}$$

Eq. 4.23 contains a "plasma frequency" which I'll call $\omega_{pe} \equiv \sqrt{4\pi ne^2/m}$. The e subscript stands for "electron, bound." With this definition I get the final form of the Lorentz dielectric function for dilute gases.

$$\epsilon(\omega) = \epsilon_c + \frac{\omega_{pe}^2}{\omega_0^2 - \omega^2 - i\omega\gamma}. \tag{4.24}$$

As $\omega \to 0$, the dielectric function becomes the static dielectric constant,

$$\epsilon(0) = \epsilon_c + \frac{\omega_{pe}^2}{\omega_0^2}.$$

Unlike a metal, the static dielectric constant of this nonconducting gas is positive.

Table 4.3 "Optical" dielectric constant for some gases	
Molecule	ϵ_1
He	1.000068
O_2	1.00054
HCl	1.0009
CS_2	1.003

Values of this dielectric constant for common gases at STP, and for visible wavelengths (which are below the first electronic transitions in the molecule and above any molecular vibration or rotational modes) are given in Table 4.3. The values are not much different from unity.

4.4.2 The Depolarizing Field

As the density increases, the difference between the local field and the external field grows rapidly. I'll write

$$\mathbf{p} = \alpha_e \mathbf{E}_{\text{loc}},$$

where α_e is the polarizability and $\mathbf{E}_{\text{loc}}$ is the value of the field at the molecule, including the external field and the field from the neighboring dipoles, in principle at distances all the way to infinity.

This problem was studied in the nineteenth century [21]. The solution is known as the Lorentz–Lorenz model or the Clausius–Mossotti model. I'll start by writing the electric field at some point as a sum of three terms

$$\mathbf{E}_{\text{loc}} = \mathbf{E}_{\text{app}} + \mathbf{E}_{\text{neighbor}} + \mathbf{E}_{\text{medium}}.$$

In this equation, $\mathbf{E}_{\text{app}} \equiv \mathbf{E}$ is the external (light) field. $\mathbf{E}_{\text{neighbor}}$ is the sum of the fields from the nearby polarized molecules, and $\mathbf{E}_{\text{medium}}$ is the contribution from the uniformly polarized medium far from the molecule.

To calculate the field from the polarization of the medium, one carves out a hollow spherical cavity (shown in Fig. 4.9) in the polarized medium centered on the polarizable molecule being considered and computes the field inside the cavity. The uniform polarization $\mathbf{P}$ outside the cavity leads to a surface charge density $\Sigma = \mathbf{P} \cdot \hat{n}$. This charge density varies as $\cos(\theta)$, and leads to a uniform field in the cavity. This field, found in electricity and magnetism textbooks [2], is

$$\mathbf{E}_{\text{medium}} = \frac{4\pi}{3}\mathbf{P}. \tag{4.25}$$

For a crystalline material, the fields of the individual dipoles in the cavity, given by Eq. 4.19, all contribute to the field at the center. The individual fields point in a set of

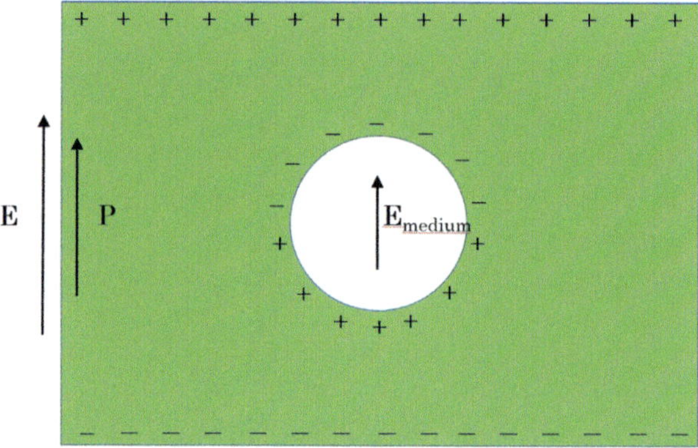

Fig. 4.9 A sphere is carved out from a uniformly polarized material. There is a surface charge density on the surface of the spherical cavity that is the source of a uniform depolarizing field in the cavity.

specific directions which depend on the coordinates of the dipoles. Their sum depends on the crystal symmetry. For a cubic system, the fields from the neighbors add to zero. Essentially the fields from the two dipoles behind and in front of the origin are canceled by the oppositely directed and half-strength fields from the four on either side, above, and below. This cancellation continues as one moves out from the center.*

Consequently, the local field is

$$\mathbf{E}_{\text{loc}} = \mathbf{E}_{\text{app}} + \frac{4\pi}{3}\mathbf{P}, \tag{4.26}$$

where

$$\mathbf{P} = \frac{1}{V}\sum_i \mathbf{p}_i = \frac{1}{V}\sum_i \alpha_{ei}\mathbf{E}_{\text{loc}} = n\alpha_e \mathbf{E}_{\text{loc}},$$

and (now writing $\mathbf{E}_{\text{app}}$ as just $\mathbf{E}$),

$$\mathbf{P} = n\alpha_e \left(\mathbf{E} + \frac{4\pi}{3}\mathbf{P}\right),$$

or

$$\mathbf{P} = \frac{n\alpha_e}{1 - \frac{4\pi}{3}n\alpha_e}\mathbf{E}. \tag{4.27}$$

As an aside, there is a possibility of a catastrophe here. Consider the denominator in the equation above. It contains n, the number of atoms per unit volume. Clearly $n = 1/\Omega_0$, with Ω_0 the volume of one molecule. Now the polarizability of a metal sphere is $\alpha_e = r^3$, with r the radius of the metal sphere. Then

$$\frac{4\pi}{3}\alpha_e \sim \frac{4\pi}{3}r^3 = \Omega_0,$$

---

* Clausius and Mossotti actually used a "tight" cavity that closely jacketed the molecule. The crystal structure was brought in by making the cavity an ellipsoid representing the anisotropy of the crystal.

and

$$\mathbf{P} = \frac{n\alpha_e}{1 - 1}\mathbf{E}$$

is finite when $\mathbf{E} \rightarrow 0$. Variations on such a model have been used in the theory of ferroelectrics, solids which have a spontaneous polarization.*

So now, making sure to keep α_e below r^3, I can write the complex dielectric function with local field corrections, and with a core polarizability, $\epsilon = \epsilon_c + 4\pi \chi_e$ as

$$\epsilon = \epsilon_c + \frac{4\pi n\alpha_e}{1 - \frac{4\pi}{3}n\alpha_e}. \tag{4.28}$$

This equation can be turned around. I write $\epsilon_c = 1$ for simplicity, and then solve for α_e in terms of ϵ.

$$\frac{4\pi}{3}n\alpha_e = \frac{\epsilon - 1}{\epsilon + 2} \tag{4.29}$$

in which form it is known as the Clausius–Mossotti equation.† It is approximately obeyed by simple molecular solids.

4.4.3 The Lorentz Dielectric Function in the Dense Limit

My model for the polarizability, Eq. 4.22, is unchanged, so I can substitute that expression into Eq. 4.28:

$$\epsilon = \epsilon_c + \frac{\frac{4\pi ne^2/m}{\omega_0^2 - \omega^2 - i\omega\gamma}}{1 - \frac{4\pi ne^2/3m}{\omega_0^2 - \omega^2 - i\omega\gamma}}.$$

Using the definition of the "plasma frequency" ω_{pe} and doing a little algebra, I get

$$\epsilon = \epsilon_c + \frac{\omega_{pe}^2}{\omega_0^2 - \omega^2 - i\omega\gamma - \frac{1}{3}\omega_{pe}^2}. \tag{4.30}$$

Let me combine the two constants in the denominator into a single frequency, ω_e. Local field corrections make this frequency smaller than the bare resonant frequency.

$$\omega_e \equiv \sqrt{\omega_0^2 - \frac{1}{3}\omega_{pe}^2}.$$

With this definition I finally arrive at the Lorentzian dielectric function for solids:

$$\epsilon = \epsilon_c + \frac{\omega_{pe}^2}{\omega_e^2 - \omega^2 - i\omega\gamma}. \tag{4.31}$$

---

* It is not enough to have a polarizability catastrophe; the crystal symmetry must allow the polarization to exist.
† If $\epsilon_c \neq 1$ then this becomes $(4\pi/3)n\alpha_e = (\epsilon - \epsilon_c)/(\epsilon + 3 - \epsilon_c)$.

4.4.4 The Real and Imaginary Parts of ϵ

The real and imaginary parts of ϵ are obtained by multiplying the top and bottom of the second term by $\omega_e^2 - \omega^2 + i\omega\gamma$ and doing a small amount of algebra:

$$\epsilon_1 = \epsilon_c + \frac{\omega_{pe}^2(\omega_e^2 - \omega^2)}{(\omega_e^2 - \omega^2)^2 + \omega^2\gamma^2} \tag{4.32a}$$

$$\epsilon_2 = \frac{\omega_{pe}^2\omega\gamma}{(\omega_e^2 - \omega^2)^2 + \omega^2\gamma^2}. \tag{4.32b}$$

These functions are plotted in Fig. 4.10. The parameters are $\omega_{pe} = 50{,}000$ cm^{-1} (6.2 eV), $\omega_e = 5{,}000$ cm^{-1} (0.62 eV), $\gamma = 1{,}000$ cm^{-1} (0.12 eV) and $\epsilon_c = 3.0$. These are meant to mimic* the optical properties of Ge or perhaps AlGaAs.

The real part is positive at low frequencies. The imaginary part is very small at low frequencies. As the frequency approaches ω_e, the real part of the dielectric function has a derivative-like shape while the imaginary part shows a resonance. The full width at half maximum is $\Delta\omega = \gamma$. Just above the resonance, $\epsilon_1 < 0$. It crosses zero above the resonance. (See the green dashed line, which shows ϵ_1 scaled by 100.)

I'll now calculate the actual "screened" plasma frequency $\tilde{\omega}_p$ where $\epsilon_1(\omega)$ equals zero. Setting the left side of Eq. 4.32a to zero and doing some considerable algebra, I get

$$\tilde{\omega}_p^2 = \left(\frac{\omega_{pe}^2}{\epsilon_c} - \gamma^2\right)\left(\frac{1}{2} + \sqrt{\frac{1}{4} - \frac{\omega_e^2\gamma^2}{(\omega_{pe}^2/\epsilon_c - \gamma^2)^2}}\right) + \omega_e^2. \tag{4.33}$$

This equation is not very illuminating, so let me evaluate it in the case $\omega_{pe} \gg \omega_e \gtrsim \gamma$. This is the usual case and in this usual case

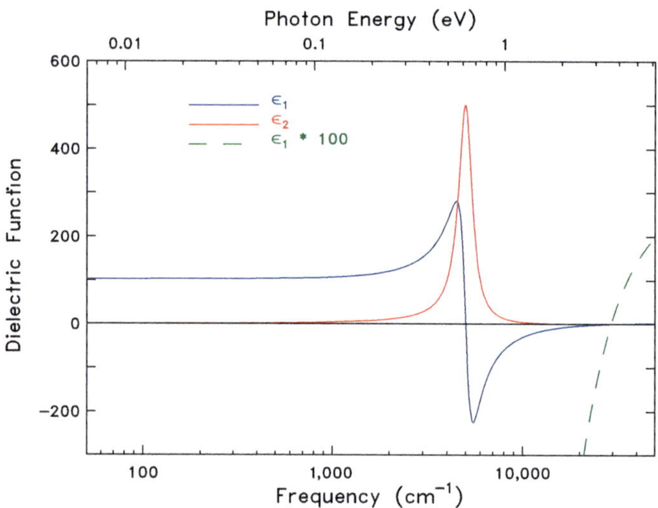

Fig. 4.10   Real and imaginary parts of the dielectric function of a Lorentzian oscillator.

* The mimicry is not very good. The linewidth is too narrow and the static dielectric constant too high.

$$\tilde{\omega}_p^2 = \frac{\omega_{pe}^2}{\epsilon_c} + \omega_e^2.$$

As a general statement, one can say that the zero is pushed up by ω_e and pulled down by γ.

4.4.5 Limiting Behavior

Low frequencies: The low frequency limit of Eq. 4.32a is $\epsilon_1 \approx \epsilon_c + \omega_p^2/\omega_e^2$. The imaginary part goes as ω at low frequencies; the conductivity σ_1 as ω^2.

Mid frequencies: At $\omega = \omega_e$, $\epsilon_1 = \epsilon_c$. The imaginary part is a maximum with $\epsilon_2 = \omega_{pe}^2/\omega\gamma$ and $\sigma_1 = \omega_{pe}^2/4\pi\gamma$. I have identified $\gamma = 1/\tau$ so that this combination of parameters is also: $\sigma_1 = \omega_{pe}^2\tau/4\pi$, the dc conductivity of a metal with plasma frequency ω_{pe} and mean free time $1/\tau$.

High frequencies: Above the resonance, ϵ_1 is negative, rising as the frequency increases, following $\epsilon_1 \approx \epsilon_c - \omega_{pe}^2/\omega^2$. It crosses through zero at the frequency given by Eq. 4.33. At these frequencies, $\epsilon_2 \sim 1/\omega^3$.

4.4.6 The Linewidth

The three parameters, ω_{pe}, ω_e and γ, control respectively the area, the frequency location, and the full width at half maximum of the resonance. The effect of changing γ is shown in Fig. 4.11. ϵ_2 is plotted in the main plot and ϵ_1 in the inset for three values of γ, 300, 1,000, and 3,000 cm^{-1}. The other parameters are as before: $\omega_{pe} = 50,000$ cm^{-1}, $\omega_e = 5,000$ cm^{-1}, and $\epsilon_c = 3.0$.

The resonance becomes broader and the maximum becomes lower as γ increases. The total area is nearly preserved.* A small movement of the peak value to lower frequencies can be discerned when γ is large.

The behavior for the real part is analogous. The maxima and minima, which are separated by $\sim \gamma$, become smaller and move further apart. It cannot be seen on the plot, but the zero of ϵ_1 moves to slightly lower frequencies as γ increases.

4.4.7 Conductivity

It is easy to write $\sigma_1(\omega)$; it is related to ϵ_2 by $\sigma_1(\omega) = \omega\epsilon_2/4\pi$. Then

$$\sigma_1(\omega) = \frac{\omega_{pe}^2\omega^2\gamma/4\pi}{(\omega_e^2 - \omega^2)^2 + \omega^2\gamma^2}. \tag{4.34}$$

The real and imaginary parts of σ are plotted in Fig. 4.12.

The shape of the Lorentzian is always the same: a peak in $\sigma_1(\omega)$ and a derivative-like structure[†] in $\sigma_2(\omega)$. Three parameters control the details: ω_e the location, ω_{pe}^2 (as I shall show) the area under the $\sigma_1(\omega)$ curve, and γ the full width at half maximum. The value of the conductivity at the maximum is $\sigma_1(\omega_e) = \omega_{pe}^2/4\pi\gamma$.

---

* Were I to plot the conductivity, the preservation would be perfect.
† The negative of the derivative actually.

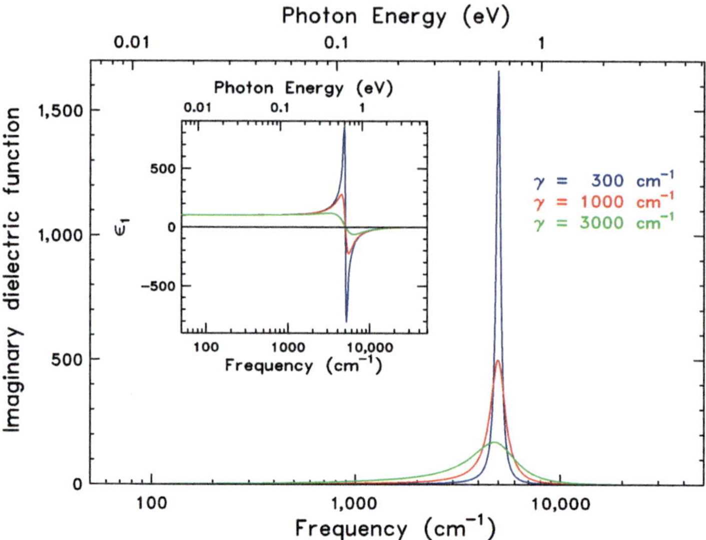

Fig. 4.11 The main plot shows the imaginary parts of the dielectric function of a Lorentzian oscillator for three values of the damping constant γ. The inset shows the real part.

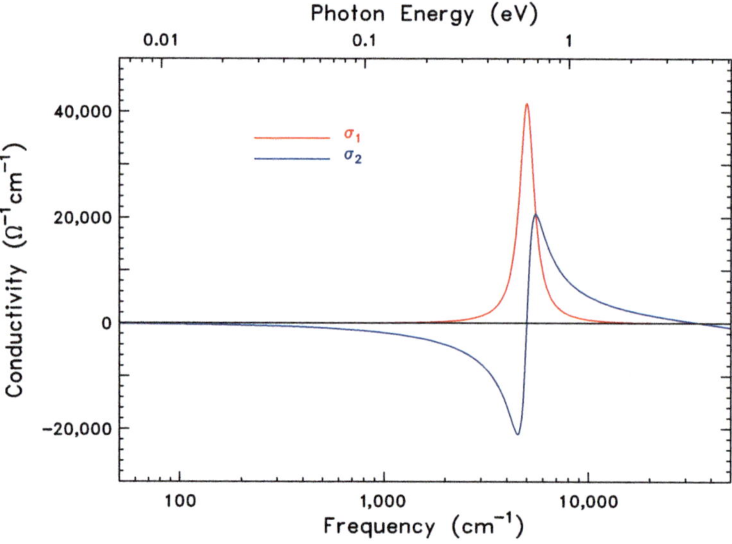

Fig. 4.12 Real and imaginary parts of the optical conductivity of a Lorentz oscillator.

4.4.8 The Refractive Index

Figure 4.13 shows the real and imaginary parts of the refractive index of the Lorentz model. At low frequencies, n is much bigger than κ. κ increases with frequency. At the resonance, $n \approx \kappa$ and both are relatively large. Above the resonance, n falls well below unity but κ stays large (because $\epsilon_1 < 0$). Then, at a certain frequency ϵ_1 changes sign and κ falls

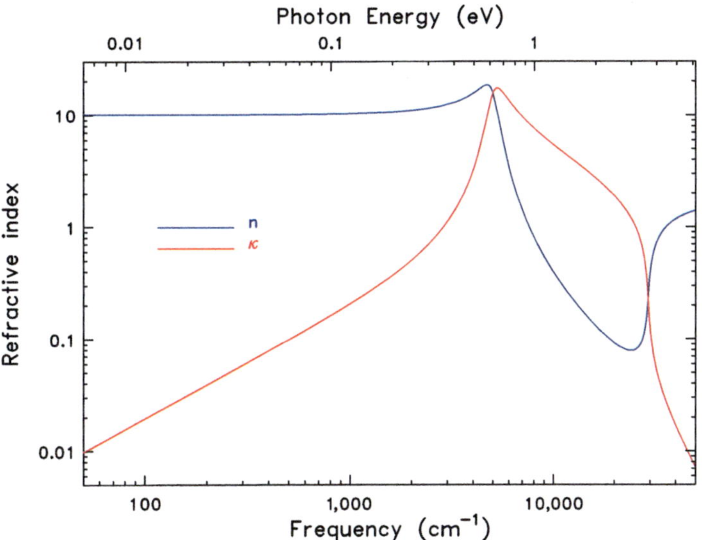

Fig. 4.13   Real and imaginary parts of the refractive index in the Lorentz model. Note the log-log scale.

rapidly while n jumps upward to a value bigger than 1. Note that both the lowest-frequency and the highest-frequency values of the refractive index n are just $n \approx \sqrt{\epsilon_1(\omega)}$ at that frequency.

4.4.9 Reflectance

Once I have N, I can calculate the reflectance from Eq. 3.33. The reflectance of the Lorentz model is shown in Fig. 4.14. At low frequencies, the real part of N dominates. As the frequency approaches ω_e, the reflectance increases, and above the resonance approaches that of a metal. Then, the reflectance falls rapidly from a value close to unity to a value close to zero. This drop occurs around 30,000 cm^{-1} (3.5 eV) for the parameters used for Fig. 4.14. This "plasma edge" occurs at the same frequencies where the rapid fall of κ and rise of n occur in Fig. 4.13. There is a minimum value of the reflectance, corresponding to the conditions* $n \approx 1$ and $\kappa \ll 1$.

4.4.10 Multiple Polarizable Electron Levels

Let's suppose that there are a number of different electrons[†] with different binding energies in the solid. I'll indicate the different levels with an index j, where j runs from the least

---

* The exact location of this "plasma minimum" is a balance of the frequency where $n - 1 \sim 0$ and, at the same time, κ has fallen to a relatively small value. It is *not* the same as the frequency where $\epsilon_1 = 0$ although these two frequencies are close to each other.

[†] Electrons are of course indistinguishable particles. However, in the multielectron atom and in solids made from such atoms, the electrons occupy different energy levels or energy bands, with the binding energy (or spring constant in the Lorentz model) different for each band.

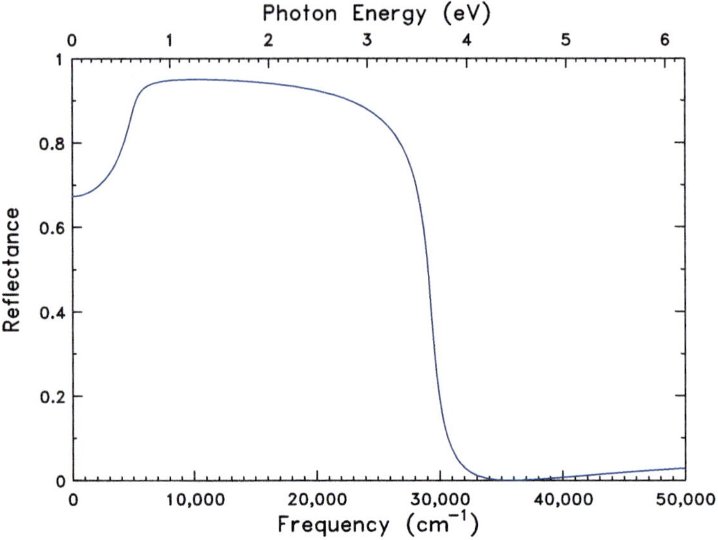

Fig. 4.14 Normal incidence reflectance in the Lorentz model.

tightly bound electron to the most tightly bound electron I want to consider. (Even more tightly bound electrons remain of course polarizable and will contribute to ϵ_c.)

Because I am ignoring electron–electron interactions, the equation of motion for the jth carrier remains unchanged from Eq. 4.20, except that it needs subscripts:

$$m_j\ddot{\mathbf{x}}_j = -m_j\omega_{0j}^2\mathbf{x}_j - m_j\gamma_j\dot{\mathbf{x}}_j - e\mathbf{E}_0 e^{-i\omega t},$$

where I've allowed for each level to have its own* spring constant ($m_j\omega_{0j}^2$) or resonant frequency ω_{0j}, dynamical mass m_j, and damping factor γ_j. The solution is unchanged from Eq. 4.21 except (still) for subscripts. The jth electron will have dipole moment $\mathbf{p}_j = -e\mathbf{x}_j$ and the total dipole moment per unit volume becomes

$$\mathbf{P} = \sum_j n_j\mathbf{p}_j$$

where n_j is the number density of type-j electrons.

Local field corrections are identical to those in Section 4.4.2. The dielectric function becomes a sum of Lorentzian oscillators of the form in Eq. 4.31.

$$\epsilon = \epsilon_c + \sum_j \frac{\omega_{pj}^2}{\omega_j^2 - \omega^2 - i\omega\gamma_j}, \tag{4.35}$$

where $\omega_{pj} = \sqrt{4\pi n_j e^2/m_j}$ is the plasma frequency associated with the jth kind of electrons and ω_j is the (local-field corrected) resonant frequency for that kind.

By assumption, each kind of electron has its own resonant frequency. If these frequencies are separated by amounts large compared to the linewidth, then ϵ_2 and σ_1 will have a peak

---

* But not its own charge!

Table 4.4 Parameters for five Lorentz oscillators.

Oscillator number j	ω_{pj}	ω_j	γ_j
1	5,000	500	30
2	10,000	1,000	60
3	50,000	5,000	300
4	55,000	5,500	330
5	200,000	20,000	1,200

$\epsilon_c = 3.0$

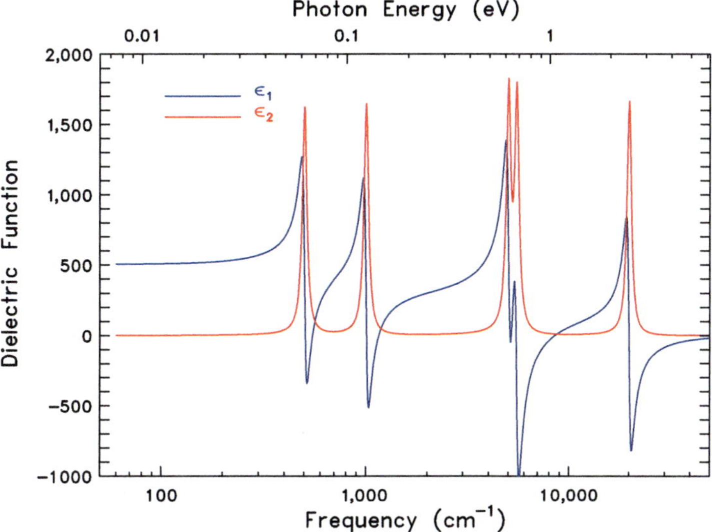

Fig. 4.15 Dielectric function for five Lorentz oscillators.

at each resonance. ϵ_1 will have a derivative-like structure and will step down to a lower value as the frequency is increased past the resonance. If the separation is small compared to the linewidth, there will be a single combined peak in σ_1, with corresponding feature in ϵ_1.

Figure 4.15 shows one such calculation. The parameters are given in Table 4.4. I chose the parameters so that ω_j/γ_j was the same for all and also so that ω_{pj}/ω_j was the same for all. This choice makes the "quality factor" Q of all resonances the same as well as their contributions to the static dielectric constant, $\epsilon_1(0)$.

4.5 Comments on Wave Propagation

I'll end this relatively long chapter with comments on how electromagnetic waves propagate in a material, using the Lorentz and the Drude models as a basis for my comments.

Electromagnetic wave propagation (and reflection/transmission at interfaces) is governed by the dielectric function, which in turn sets the refractive index. A wave traveling along x is written

$$\mathbf{E} = \mathbf{E}_0 e^{i\left(N\frac{\omega}{c}x - \omega t\right)} = \mathbf{E}_0 e^{i\left(n\frac{\omega}{c}x - \omega t\right)} e^{-\kappa\frac{\omega}{c}x}$$

with refractive index

$$N = \sqrt{\epsilon}$$

and optical absorption coefficient

$$\alpha = \frac{2\omega}{c}\kappa, \tag{4.36}$$

which I already wrote as Eq. 3.26.*

4.5.1 Insulators: Lorentz Model

I will discuss wave propagation in the Lorentz model first. In the discussion I'll take the hierarchy of frequencies as $\omega_p \gg \omega_e \gg \gamma = 1/\tau$. Even if the ratio of these quantities is only 3 or 5, I'll consider the inequality satisfied. The dielectric function and refractive index are plotted again in Figs. 4.16. These functions are the same as those in Figs. 4.10 and 4.13. I have added vertical bars to separate the diagram into 4 parts, labeled A–D. Each of these has its own flavor of wave propagation.

4.5.1.1 Mostly Transparent with Small Absorption Coefficient

In Region A, ϵ_1 is positive and in fact larger than unity. Moreover, $\epsilon_1 \gg \epsilon_2$ and $n \gg \kappa$. The solid is a transparent insulator with an absorption length (the inverse of α) that is longer than the wavelength, often much longer. The reflectance is governed largely by n; $\mathscr{R} \approx [(n-1)/(n+1)]^2$. Because n increases with frequency,† so does the reflectance. This behavior is termed "ordinary dispersion."

4.5.1.2 Strongly Absorbing, Near the Resonance

At the center of Region B, $\epsilon_2 \gg \epsilon_1$, so that

$$n \approx \kappa = \sqrt{\frac{\epsilon_2}{2}} \gg 1.$$

Waves are strongly damped, but currents and fields are more or less in phase. The damping length (δ or $1/\alpha$) and the wavelength differ by about 2π.

---

* Recall that the absorption coefficient describes the attenuation of the intensity, $I \sim EE^* \sim e^{-\alpha x}$, in the solid ($x$ is the distance the wave has traveled); the factor of 2 comes from the fact that the fields are squared to give the intensity.

† Believe me, it does. I know it is not obvious in the semilog plots of Fig. 4.16. I could make a better plot, or just consider the approximation $n^2 \approx \epsilon_1 \approx \epsilon_c + \omega_{pe}^2/(\omega_e^2 - \omega^2)$ as $\omega \to \omega_e$ from below.

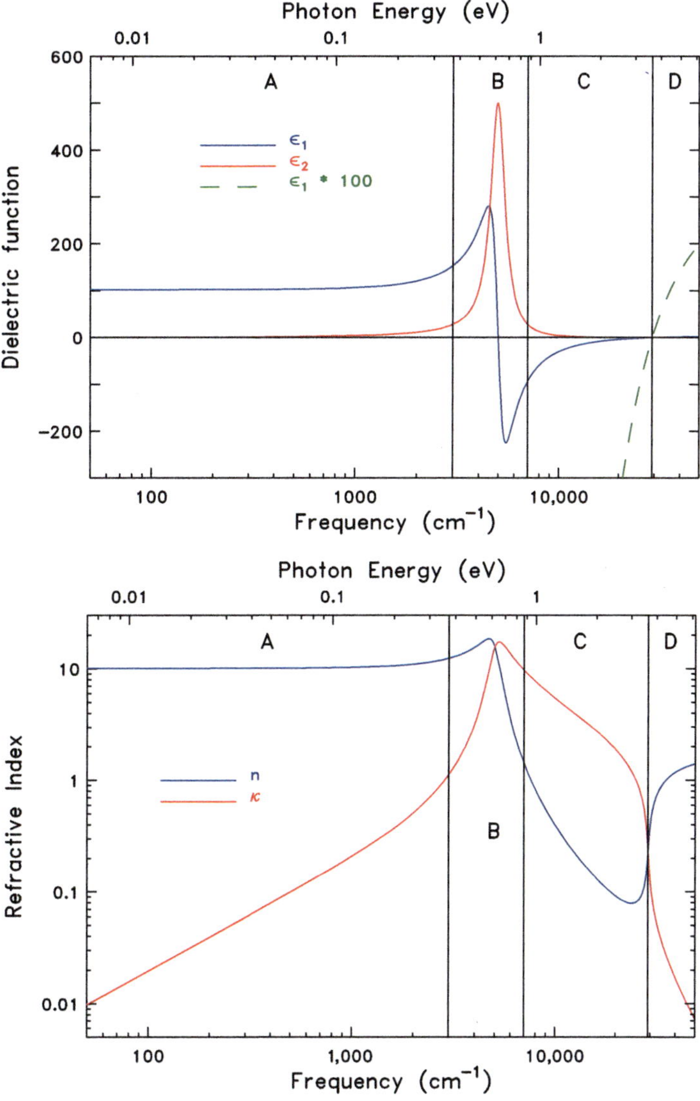

Fig. 4.16 (Top) dielectric function and (bottom) refractive index for the Lorentz oscillator. The four regions A–D are separated by vertical lines. The boundaries are fuzzy, and the exact location of the lines is not to be taken seriously.

Note that (as asserted in Chapter 1) "strongly absorbing" and "highly reflecting" go hand-in-hand here. To estimate the reflectance, I'll start with Eq. 4.17:

$$\mathcal{R} = 1 - \frac{4n}{(n+1)^2 + \kappa^2},$$

which becomes, near the resonance,

$$\mathcal{R} = 1 - \frac{2}{\kappa}.$$

Exactly on resonance, $\epsilon_2 = \omega_{pe}^2/\omega\gamma$ so that the reflectance becomes

$$\mathscr{R} = 1 - \sqrt{\frac{8\omega\gamma}{\omega_{pe}^2}}.$$

This behavior is the equivalent of Hagen Rubens, Eq. 4.18, derived originally for metals in the far infrared. A Drude metal may be considered a Lorentz oscillator with a zero frequency resonance.*

Around the middle of region B, the refractive index begins to decrease with increasing frequency. This behavior is termed "anomalous dispersion."

4.5.1.3 Mostly Damped, $\omega_e \ll \omega \ll$ the Zero of ϵ_1

In Region C, $\epsilon_1 < 0$ (negative) and $|\epsilon_1| \gg \epsilon_2$, so that n is a fraction of unity. and can be taken as close to zero. Then, $\kappa \approx \sqrt{|\epsilon_1|}$. So long as the frequency is not too close to either the resonance or the zero crossing, I may take $\epsilon_1 \approx -\omega_{pe}^2/\omega^2$ and $\kappa \approx \omega_{pe}/\omega$. The wave is strongly damped with

$$\alpha = \frac{2\omega}{c}\kappa = \frac{2\omega_{pe}}{c}$$

and a wavelength much larger than the vacuum wavelength.

To estimate the reflectance, I need to estimate n. I will start with $2n\kappa = \epsilon_2$ and use $\epsilon_2 \approx \omega_{pe}^2\gamma/\omega^3$ and $\kappa \approx \omega_{pe}/\omega$ (see above), so that $n = \omega_{pe}\gamma/2\omega^2$. Then taking $\mathscr{R}$ from Eq. 4.17 in the case where $\kappa \gg n$, I arrive at

$$\mathscr{R} = 1 - \frac{4n}{\kappa^2} = 1 - \frac{2\gamma}{\omega_{pe}}.$$

The reflectance is independent of frequency. It is also high. Here the response is inductive, so the dissipation is not high; instead, most of the incident energy is reflected.

4.5.1.4 High Frequencies, Above the Zero of ϵ_1

Here ϵ_1 is positive and bigger than ϵ_2; κ becomes small and $n \approx \sqrt{\epsilon_c - \omega_{pe}^2/\omega^2}$. The material is transparent again. The reflectance has a minimum very close[†] to the frequency where $n = 1$. Ordinary dispersion is recovered. The limiting high-frequency value of the reflectance is

$$\mathscr{R} = \frac{(\sqrt{\epsilon_c} - 1)^2}{(\sqrt{\epsilon_c} + 1)^2}. \tag{4.37}$$

4.5.1.5 Really High Frequencies

Equation 4.37 suggests that the reflectance is constant and remains that way as $\omega \rightarrow \infty$. But this suggestion is wrong. The high frequency "limiting" value of ϵ_c is a crude

---

* A resonance at zero implies of course zero restoring force.
[†] It would be exact except for the frequency dependence of κ.

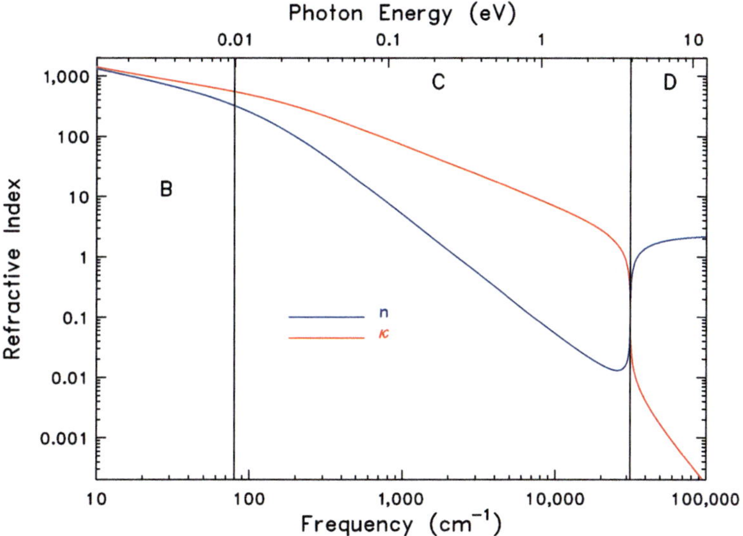

Fig. 4.17 Refractive index for the Drude metal. Three regions B–D are shown by the vertical lines. The boundaries are fuzzy, and exact location is not to be taken seriously.

approximation for the polarizability of the ion cores of the atoms making up the solid. Atomic physics [22] tells me that there are a series of resonances with the dielectric function decreasing a bit above each one. At the very highest frequencies, $\epsilon \approx 1 - \omega_P^2/\omega^2$ where $\omega_P = \sqrt{4\pi n_P e^2/m_e}$ is a plasma frequency for every electron in the solid (whose density is n_P). As ω gets large, ϵ is just below unity and the refractive index is $n = \sqrt{\epsilon} \approx 1 - \omega_P^2/2\omega^2$. I then get $\mathcal{R} \approx \omega_P^4/4\omega^4$.

4.5.2 Metals: Drude Model

A plot analogous to Fig. 4.16 is shown in Fig. 4.17. Everything I said in sections B–D above applies here too. (There is no region analogous to section A in metals.) I can rewrite the relevant equations, replacing ω_{pe} with ω_p, γ with $1/\tau$, and setting $\omega_e = 0$.

4.5.2.1 Strongly Absorbing, $\omega \ll 1/\tau$

In metals, region B starts at zero frequency. At these low frequencies, $\epsilon_2 \gg \epsilon_1$ or $\sigma_1 \gg \sigma_2$. In this limit, $n \approx \kappa$. I can write

$$\epsilon_1 \approx -\omega_p^2 \tau^2$$

and

$$\epsilon_2 \approx \frac{\omega_p^2 \tau}{\omega}$$

and

$$n = \kappa \approx \sqrt{\frac{\omega_p^2 \tau}{2\omega}} = \sqrt{\frac{2\pi \sigma_{\mathrm{dc}}}{\omega}}$$

where the last equality comes from $\sigma_{\mathrm{dc}} = \omega_p^2 \tau / 4\pi$. The dispersion is "anomalous" everywhere from $\omega = 0$ to just below the screened plasma frequency.

The field decays as $e^{-x/\delta}$ with

$$\delta = \frac{c}{\omega \kappa} \approx \sqrt{\frac{c^2}{2\pi \omega \sigma_{\mathrm{dc}}}} = \sqrt{\frac{2c^2}{\omega_p^2 \omega \tau}}. \tag{4.38}$$

Eq. 4.38 is the classical skin depth of the metal. It correctly gives the penetration length in the low frequency limit, where $\sigma_1(\omega) \approx \sigma_{\mathrm{dc}}$. I'll return to the skin effect in Chapter 14.

I'll start with Eq. 4.17 but let $n = \kappa \gg 1$ and write $\mathcal{R} = 1 - 2/\kappa$. Then the reflectance becomes

$$\mathcal{R} = 1 - \sqrt{\frac{8\omega}{\omega_p^2 \tau}} = 1 - \sqrt{\frac{2\omega}{\pi \sigma_{\mathrm{dc}}}}.$$

This is what Hagen and Rubens found and what I wrote already in Eq. 4.18.

4.5.2.2 Mostly Damped, $1/\tau < \omega < \omega_p/\sqrt{\epsilon_c}$

In Region C, $\epsilon_1 \approx -\omega_p^2/\omega^2$ and $\kappa \approx \omega_p/\omega$. The wave has a wavelength much larger than the vacuum wavelength and is strongly damped with skin depth

$$\delta = \frac{c}{\omega \kappa} \approx \frac{c}{\omega_p} \tag{4.39}$$

and

$$\alpha = \frac{2\omega}{c} \kappa \approx \frac{2\omega_p}{c}.$$

Anomalous dispersion for n continues.

To estimate the reflectivity, I need to estimate n. I will start with $2n\kappa = \epsilon_2$ and use $\epsilon_2 \approx \omega_p^2/\omega^3 \tau$ and $\kappa = \omega_p/\omega$, so that $n = \omega_p/2\omega^2\tau$. Then, taking $\mathcal{R}$ from Eq. 4.17 in the case where $\kappa \gg n$, I arrive at

$$\mathcal{R} = 1 - \frac{4n}{\kappa^2} = 1 - \frac{2}{\omega_p \tau}, \tag{4.40}$$

a frequency-independent reflectance.

4.5.2.3 High Frequencies, Above the Zero of ϵ_1

Here ϵ_1 is positive and quickly becomes bigger than ϵ_2, i.e., $\epsilon_1 \approx \epsilon_c - \omega_p^2/\omega^2$ whereas $\epsilon_2 \approx \omega_p^2/\omega^3 \tau$. The reflectance has a minimum very close* to the frequency where $n = 1$.

---

* It would be exact except for the frequency dependence of κ. Note that if $\epsilon_1 \gg \epsilon_2$ and $n = 1$ then $\epsilon_1 \approx 1$ also.

See Fig. 4.7. The limiting high-frequency value of the reflectance is

$$\mathcal{R} = \left(\frac{\sqrt{\epsilon_c} - 1}{\sqrt{\epsilon_c} + 1} \right)^2. \tag{4.41}$$

The light that is not reflected is transmitted, as long as the sample is not too thick.*

4.5.2.4 Really High Frequencies

Equation 4.41 suggests that the reflectance is constant and remains that way as $\omega \to \infty$. But, see the discussion in Section 4.5.1; the high frequency reflectance is $\mathcal{R} \sim 1/\omega^4$.

4.6 The Absorption Coefficient and a Not Uncommon Mistake

Light traveling in an absorbing medium is attenuated; the amount of attenuation is governed by the extinction coefficient κ, by the penetration depth δ, or, almost equivalently, by the absorption coefficient α. Starting with Eqs. 3.26 and 4.36, these are related by $\alpha = 2/\delta = 2\omega\kappa/c$. κ has no dimensions, α is in cm^{-1} in cgs units.

After traveling a distance d, the intensity is reduced by a factor of $e^{-\alpha d}$. So it is tempting to say that the transmittance of a sample of thickness d is

$$\mathcal{T} \approx e^{-\alpha d}. \tag{4.42}$$

If the reflectance is not large, this equation is a reasonable approximation; see the discussion in Chapter 7, Section 7.1.1, for the complete equation.

But what I want to mention here is another case: strongly absorbing materials where $\kappa \gg n$. This case corresponds to the upper half of region B and all of region C in Figs. 4.16 and 4.17. The decay length (the penetration depth or skin depth) is much less than the wavelength inside the solid, so that the picture in Fig. 3.1 does not represent the situation. The field decays over lengths so short that the oscillating part of the field is almost constant; oscillations are not observed. Were I to use Eq. 4.42 and calculate the absorption coefficient from $-\ln(\mathcal{T})/d$, I would be wrong. The reflection from the two surfaces is much more likely to govern the transmission than the absorption in the interior. Moreover, the sample is probably thinner than the wavelength, so that coherence of the multiple internal reflections needs to be incorporated into the solution. See the discussion in Section 7.2.

---

* How thick is too thick? Well, the intensity decays as $e^{-\alpha x}$ so too thick is a few times larger than $1/\alpha = c/2\omega\kappa = nc/4\pi\sigma_1(\omega) = \delta/2$.

5

Phonons

The atoms in a solid may be set to vibrating either by temperature or by external forces. One such force is the electric field of light. For the field to do anything, there must be a charge on the atom, where the charge is the sum of the positive nuclear and negative electronic charges. There is no need for there to be a charge, but most compounds, materials made up from two or more types of atoms, do have a certain amount of charge transfer from one type to another.* In the extreme case of ionic solids, like NaCl, valence electrons are transferred from one atom to the other. In NaCl, the sodium atoms have lost an electron and are present as Na^+ while the chlorine atoms have gained an electron and are present as Cl^-. NaCl is face-centered cubic, with each atom having six nearest neighbors of the opposite kind. These nearest neighbors are arranged at the corners of an octahedron surrounding the central atom. The bonding arises from the Coulomb attraction of the ions for their opposite-sign partners. Even though there are charged ions throughout the solid, there is no net polarization ($\mathbf{P} = 0$) because of the symmetrical arrangement of nearest neighbors.[†] However, if an electric field is applied, the positive ions will be pushed in the direction of the field, and the negative ions will be pulled in the opposite direction, creating a net polarization. There is a natural vibrational frequency, the transverse optical phonon, at which these ions will oscillate and couple most strongly to an external electromagnetic field. Here I will calculate the dielectric function contribution of these phonons. I'll start with a completely oversimplified model and then work up to something a little better.

5.1 Harmonic Oscillator

I'll start with a simple harmonic oscillator model of a diatomic molecule. This system is shown in Fig. 5.1. The electrical force on the ions is $\mathbf{F}_\pm = \pm Z^* e \mathbf{E}_{\text{loc}}$, where $\pm Z^* e$ is the effective charge of the ions and $\mathbf{E}_{\text{loc}}$ is the local field at the ions. The field, varying as $e^{-i\omega t}$, affects the ion separation x.

---

* Even if the bonding is mostly covalent, there is usually a small charge difference across the bonds of dissimilar atoms. It is this charge difference that leads to the infrared spectra of organic materials, an extremely important tool in analytical chemistry.

† I know a counterexample: ferroelectrics. These solids lack inversion symmetry and have a spontaneous electric polarization below their Curie temperature.

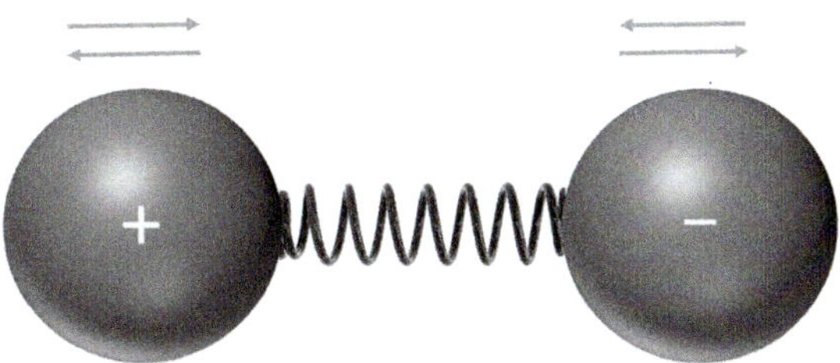

Fig. 5.1 The ionic force between a positive and a negative ion is represented by a spring. When light shines on the crystal, the spring is stretched and shrunk at the oscillation frequency of the light. The upper (lower) arrows show the force on the ions when the field is pointing right (left).

5.1.1 Equation of Motion

Newton's law is

$$\mu\ddot{\mathbf{x}} = -\mu\omega_0^2\mathbf{x} - \mu\gamma\dot{\mathbf{x}} + Z^*e\mathbf{E}_{\mathrm{loc}}, \tag{5.1}$$

where μ is the reduced mass of the ions,[*] $\mu\omega_0^2 = K$, the spring constant, and γ determines the strength of the damping force, which is proportional to the velocity. I'll take $\mathbf{E}_{\mathrm{loc}} = \mathbf{E} + 4\pi\mathbf{P}$ as before. Equation 5.1 is analogous to Eq. 4.20 for the Lorentz model, with a change of notation. The solution is then the analog to Eq. 4.21. I'll write the induced dipole as $\mathbf{p} = Z^*e\mathbf{x}$ and define an ionic plasma frequency as

$$\Omega_p = \sqrt{\frac{4\pi n_i (Z^*e)^2}{\mu}},$$

where n_i is the number density of ion pairs. The next step is to make the local field correction, finding eventually the dielectric function for the optical phonons, the analog of Eq. 4.31:

$$\epsilon = \epsilon_e + \frac{\Omega_p^2}{\omega_T^2 - \omega^2 - i\omega\gamma} \tag{5.2}$$

where ϵ_e is the electronic contribution, including *both* the core electrons and the valence electrons,[†] and ω_T is the transverse optical phonon frequency, related to the dilute-limit resonance by $\omega_T = \sqrt{\omega_0^2 - \Omega_p^2/3}$.

The dielectric function and other optical function for the Lorentz oscillator are plotted in Figs. 4.10–4.14. The functions for the vibrational absorption are the same Lorentzian, except that the frequency scale is different. The TO mode frequencies are typically at 40–600 cm^{-1} (6–100 meV or 1.6–25 THz).[‡]

---

[*] Relative motion!

[†] Here I assume – as typically is the case for insulators where phonon features are seen in the optical spectra – that the excitation energies for the electrons are much higher than the phonon energies, $\hbar\omega_T$.

[‡] The modes for very light atoms, such as C–H or O–H can be as high as 3,300 cm^{-1}. The frequencies for vibrations in organic solids, such as the C–C, C=C, and C=O bonds, are in 800–1,600 cm^{-1} range.

5.2 Lattice Dynamics

Like any mechanical object, a solid has a set of normal modes or vibrations of the lattice. If I want to describe these modes in terms of the motions of the individual atoms, there are an enormous number of them. Fortunately there are lots of simplifications I can make and the problem is quite tractable and much closer to real physics than the two-masses-on-a-spring problem I solved in the previous section. The most important is that the solid is *crystalline*, i.e., it has translation symmetry in the placement of the atoms as well as (sometimes) other symmetries. I'll discuss lattices and their symmetry in Chapter 15; for now I refer to solid-state physics texts [19, 23, 24].

I will still think in terms of springs and masses, but now the spring constant is not an actual spring but arises from the second derivative in a Taylor series expansion of the interatomic potential for the lattice. (The constant term can be suppressed by choosing the zero of energy and the linear term is zero because otherwise there would be an unbalanced force on the atoms and they would move to new equilibrium position where it was zero.)

When quantized, the lattice vibrations are called "phonons" by analogy to photons. In fact I'll seek plane wave solutions to the equations of motions. There are many physical effects due to the phonons, including a T^3 low-temperature specific heat, thermal conductivity in insulators, thermal expansion (due to anhamonicity), electron scattering in metals and semiconductors affecting the resistivity (and optical conductivity), and direct infrared absorption by certain vibrational modes. That the phonons are quantized is most easily proven by looking at neutron scattering and light scattering from them, processes that require them to have energy $\hbar\omega$ and momentum $\hbar\mathbf{k}$.

5.2.1 Equations of Motion

I'll now consider the problem of phonons in a diatomic crystal. The geometry is sketched in Fig. 5.2. The figure shows a single line or chain of atoms arranged with equal separation and connected by springs with force constant K. The charges and masses alternate along the chain, negative atoms with mass M_- and positive atoms with mass M_+. The springs are relaxed and each atom is at its equilibrium position, $\mathbf{r} = na\hat{x}$ for the negative ions and $\mathbf{r} = (na + a/2)\hat{x}$ for the positive ions. The lattice constant a is the length of the unit cell in the $\hat{x}$ direction and there are two atoms per unit cell. Clearly, I have restricted the problem to nearest-neighbor forces only, as there are no springs going to more distant atoms.

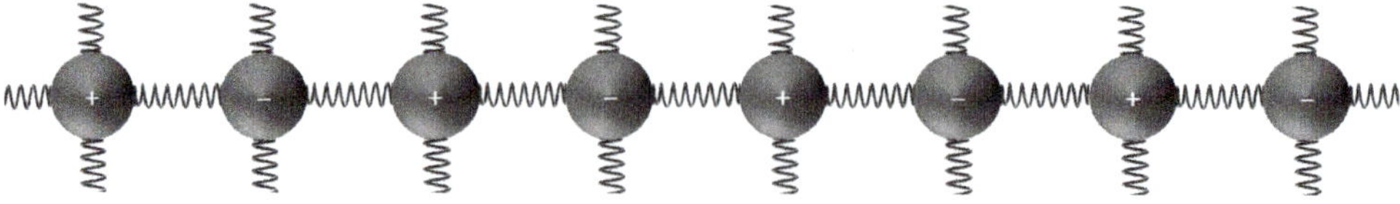

Fig. 5.2   A line or chain of atoms in a crystal. Positive and negative atoms alternate along the chain, linked by springs. Each atom can move from its equilibrium position by an amount **u**. The springs are stretched or compressed by the vector difference in the motions.

Note that this appears to be a one-dimensional problem but I show vestigial springs leading to neighboring atoms above and below the atoms shown. Other springs can connect atoms above and below the plane of the figure. As Kittel [24] has argued, if entire planes of atoms move together along $\hat{x}$ (parallel to the chain shown), or along $\hat{y}$ or $\hat{z}$ (perpendicular to the chain shown), one spring per atom is expanded and one compressed, just as in the one-dimensional case. (See the figures in his chapter on phonons.) Thus, especially for those diatomic crystals where each atom has six nearest neighbors (such as most of the alkali halides), the one dimensional model is a good representation of the three dimensional crystal.

So now, let me set the atoms in motion. Each atom has a displacement from equilibrium, which I'll denote as $u_{n,-}$ for the negative ion in the nth unit cell and $u_{n,+}$ for the positive ion in the nth unit cell. These are both vectors, but if I consider displacements only along x, y, or z (or if I do a one-dimensional problem) I only need to keep track of the signs. Now, I could write equations of motion for the atoms from inspection* but instead I'll use energy considerations. The reason is that writing the potential energy V and kinetic energy T and using Hamilton's equations is much more easily extended to longer-ranged forces and more complex structures [19]. The potential energy is stored in the springs and the kinetic energy in the motion of the atoms. All I need is to specify these energies in one unit cell and then I can sum over cells for the total.

The potential energy held by the nth cell is

$$V_n = \frac{1}{2}K(u_{n,-} - u_{n,+})^2 + \frac{1}{2}K(u_{n,+} - u_{n+1,-})^2. \tag{5.3}$$

The motion $u_{n,-}$ also extends and compresses the spring between it and the positive atom to its left but that energy is allocated to unit cell $n-1$. Note also that the energy (and the forces) depend on the difference in displacement of the neighbors. That suggests that long wavelength sinusoidal motion (where all neighbors undertake to move together) will have low frequencies and that motion where they move in opposite directions will have high frequencies.

The kinetic energy of the nth cell is just

$$T_n = \frac{p_{n,-}^2}{2M_-} + \frac{p_{n,+}^2}{2M_+}, \tag{5.4}$$

where $p_{n,j} = M_j \dot{u}_{n,j}$ is just the momentum of atom n, j with j counting atoms within the unit cell. (Here, $j = \pm$.) The Hamiltonian is the sum of total kinetic plus potential energies:

$$\mathcal{H} = \sum_n (T_n + V_n). \tag{5.5}$$

---

* I would start with

$$F_{n,-} = -K(u_{n,-} - u_{n-1,+}) - K(u_{n,-} - u_{n,+}),$$

and

$$F_{n,+} = -K(u_{n,+} - u_{n,-}) - K(u_{n,+} - u_{n+1,-}),$$

for the nearest-neighbor forces as in Fig. 5.2. And then use Newton's law.

One of Hamilton's equation (the one I need) applied to Eq. 5.5 is

$$-\dot{p}_{n,-} = \frac{\partial \mathcal{H}}{\partial u_{n,-}} \quad \text{and} \quad -\dot{p}_{n,+} = \frac{\partial \mathcal{H}}{\partial u_{n,+}}.$$

I have to be aware that there is a term containing $u_{n,-}$ in V_{n-1} and take that derivative too. I take the derivatives, collect terms, write $\dot{p}_{n,j} = M_j \ddot{u}_{n,j}$, and get the equation of motion of the negative atom to be

$$M_- \ddot{u}_{n,-} = K(u_{n,+} + u_{n-1,+} - 2u_{n,-}), \tag{5.6}$$

while the motion of the positive atom satisfies

$$M_+ \ddot{u}_{n,+} = K(u_{n,-} + u_{n+1,-} - 2u_{n,+}). \tag{5.7}$$

5.2.2 Plane-Wave Solutions, Range of k, and Boundary Conditions

I will seek plane-wave solutions to these equations. The reason for this choice* is that Eqs. 5.6 and 5.7 are those of a set of harmonic oscillators and I know that sines, cosines, and complex exponential functions solve the simple harmonic oscillator. I start with trial functions that are plane waves and which specify the displacement of the atoms from their equilibrium locations:

$$u_{n,j} = u_0 \epsilon_j e^{i(kna - \omega t)} \tag{5.8}$$

where u_0 is the amplitude of the wave, ϵ_j (with $j = -$ or $+$) is the relative amplitude of atom j, and $na = x$, the distance of the start of unit cell n from the origin. The wave is traveling along the $\hat{x}$ direction, but the motion ϵ_j could in principal be along any of the Cartesian axes.

I have started with a plane wave whose general form is $u \sim e^{i(kx - \omega t)}$. As I will show, this plane wave does the job and generates the relation between ω and k as well as the relative amplitudes and phases of the motion of the atoms. Initially, there is no obvious limit to the values that k could take on. But there is a well-justified upper limit to k that I should respect and it has to do with the finite spacing between the atoms. The motion of the atoms is all I can know about the vibrations in the crystal. In fact, it is the motion of the atomic nuclei that I can in principal measure and calculate. It makes little sense to talk about vibrations at places where no mass exists. That is why I restricted the values of x to $x = na$ right at the beginning.

If now I want to represent a sine wave by specifying the amplitude of the wave at a discrete set of points separated by distance a, I must satisfy the Nyquist criterion that at least two points per cycle of the wave are needed, so that $\lambda \geq 2a$. The wavelength can be longer than this but not shorter. Another way of saying this is that if there is a sine wave that passes through a set of points, then there are many higher frequency sine waves that also pass through the points but oscillate also in the space between the points. If the longest wavelength going through the points is $\lambda \geq 2a$, then the wave vector has magnitude $|k| = 2\pi/\lambda \leq \pi/a$. The other waves that pass through the points have wave vector $k \pm n \cdot 2\pi/a$

---

* Other than that it has been shown to work and is in all the textbooks.

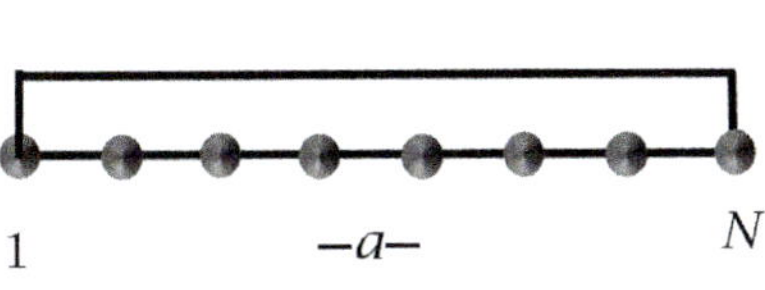

Fig. 5.3   A linear chain of N atoms is bent into a circle and the first and Nth atoms are connected by a bond. The system is periodic with period Na.

with n any integer. In the language of crystallography, the range $-\pi/a < k \leq \pi/a$ is the first Brillouin zone of the reciprocal lattice of the crystal [19, 23, 24]. The values $k = \pm\pi/a$ are the zone boundaries of the linear chain.

The range of k extends* over $\sim \pm 10^8$ cm^{-1}. In contrast, the vacuum q vector for an infrared electromagnetic wave of $\omega = 160$ cm^{-1} is $q = 2\pi/\lambda = 10^3$ cm^{-1}. Any lattice vibration with the same wavelength as the light wave is essentially at $k = 0$.

Now I specify the boundary conditions. If the linear chain has fixed ends, the solutions will be sines and cosines, i.e., standing waves. It is preferable to use "periodic boundary conditions." I'll write these mathematically (in one dimension) as

$$u_{n,j}(x + L) = u_{n,j}(x), \tag{5.9}$$

where L is the length of the chain.[†] To be concrete, think about a one-dimensional chain of N atoms. Such a chain is illustrated on the left side of Fig. 5.3. Then one connects the first and Nth atoms with a bond, bending the chain into a circle in order to make all bond lengths the same. Clearly, this loop satisfies Eq. 5.9 because the atomic displacements must be single valued and have the same value every time one goes once around the chain. This requirement means that k is quantized in steps of $2\pi/L$.

5.2.3 The Dispersion Relation

I put the function of Eq. 5.8 into the equations of motion (Eq. 5.6 and 5.7). The second derivative with respect to time brings down $-\omega^2$. There are common factors of u_0 and $e^{i(kna-\omega t)}$ that can be canceled. After doing these steps and collecting the coefficients of ϵ_j in a convenient way, I get

$$(2K - M_-\omega^2)\epsilon_- - K(1 + e^{-ika})\epsilon_+ = 0, \tag{5.10}$$

and

$$-K(1 + e^{+ika})\epsilon_- + (2K - M_+\omega^2)\epsilon_+ = 0. \tag{5.11}$$

---

* For NaCl, the repeat distance along the <100> directions (the cube edges) is $a = 5.64$ Å, so $\pi/a = 5.57 \times 10^7$ cm^{-1}.

[†] In three dimensions, I would make the displacements periodic along all three directions ($\hat{x}$, $\hat{y}$, or $\hat{z}$).

I have two equations and two unknowns. For there to be a solution, the determinant of the coefficients has to be zero. Calculating the determinant gives:

$$(2K - M_-\omega^2)(2K - M_+\omega^2) - K^2(1 + e^{-ika})(1 + e^{+ika}) = 0$$

Working on the algebra of this equation, I first use $(1 + e^{-ika})(1 + e^{+ika}) = 2 + 2\cos ka$. A bit more algebra and the trigonometric identity $1 - \cos ka = 2\sin^2(ka/2)$ leaves

$$\omega^4 - 2K\frac{M_- + M_+}{M_-M_+}\omega^2 + \frac{4K^2}{M_-M_+}\sin^2(ka/2).$$

The reduced mass μ is $\mu = M_-M_+/(M_- + M_+)$. $(1/\mu = 1/M_- + 1/M_+.)$ The total mass (in the unit cell) is $M = M_- + M_+$ so that $M\mu = M_-M_+$. These definitions simplify the equation above to

$$\omega^4 - \frac{2K}{\mu}\omega^2 + \frac{4K^2}{M\mu}\sin^2(ka/2) = 0,$$

which is quadratic in ω^2. The solution, after a bit of factoring, is

$$\omega^2 = \frac{K}{\mu}\left[1 \pm \sqrt{1 - \frac{4\mu}{M}\sin^2(ka/2)}\right]. \tag{5.12}$$

The equation for $\omega(k)$, called the "dispersion relation" is found by taking the square root of both sides.

Next, I can return to Eqs. 5.10 or 5.11 to find the ratio of ϵ_+ to ϵ_-. I choose Eq. 5.11:

$$\frac{\epsilon_+}{\epsilon_-} = \frac{1 + e^{ika}}{2 - \frac{M_+}{K}\omega^2}. \tag{5.13}$$

Note that the ratio is complex, indicating a phase difference between the motion of positive and negative atoms. In principal, I can substitute Eq. 5.12 to find the ratio as a function only of k or invert Eq. 5.12 to find $k(\omega)$ and substitute that to get the ratio as a function only of ω. I'll do this where it helps but only in limiting cases. Moreover, because I have a free parameter u_0, I should have defined a normalization for ϵ_+ and ϵ_- that makes, say, the sum of their squares equal to one. Because the system is perfectly linear, it does not matter very much unless I need to constrain the total energy of the oscillations to some value. In several pages, I'll write an equation relating the displacement to the electric field of the light and this will set the values of the displacements.

5.2.4 Acoustic and Optical Branches

For any value of k there are two values (called "branches") of ω, one where I chose the minus sign in Eq. 5.12 and one where I choose the plus sign. Figure 5.4 shows the dispersion relation, ω vs. k. I've chosen parameters to give a frequency scale appropriate for sodium chloride. Cl is heavier than Na by a little more than a factor of 1.5 so that the ratio $\mu/M = 0.24$.

The two branches are quite different. The lower branch has zero frequency at $k = 0$, rises linearly, and then bends over to have zero slope at the zone boundary $\pm\pi/a$.

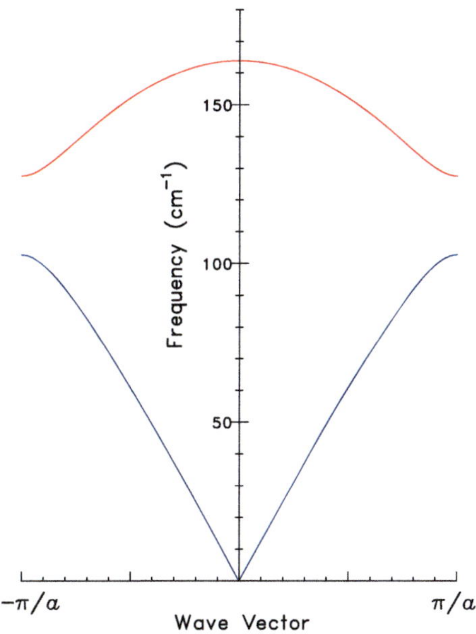

Fig. 5.4 Dispersion relation for the phonons in a diatomic crystal. The acoustic branch is shown in blue and the optic branch in red. The $k = 0$ optical phonon frequency was chosen to match NaCl. The range of k shown is within the first Brillouin zone, which also is one full cycle of $\sin^2(ka/2)$.

The upper branch has a high frequency (the highest in the figure) when $k = 0$. It decreases as k increases. The slope is zero both at $k = 0$ and at the zone boundary. There is a gap between the lower branch and the upper branch. No normal modes of the chain exist for frequencies in this gap.

5.2.5 Limiting Behavior

Let me look at limiting behavior of each branch individually. I'll start with the minus sign. The dispersion relation becomes

$$\omega_a = \sqrt{\frac{K}{\mu}}\left[1 - \sqrt{1 - \frac{4\mu}{M}\sin^2(ka/2)}\right]^{1/2}. \tag{5.14}$$

The subscript a stands for "acoustic" and this branch carries sound waves. As $k \to 0$ the sine can be replaced by its argument; the second term in the inner square root is small and I can expand it to give

$$\omega_a = \sqrt{\frac{K}{\mu}}\left[1 - \left\{1 - \frac{2\mu}{M}\frac{k^2a^2}{4}\right\}\right]^{1/2},$$
$$= \sqrt{\frac{2K}{M}}\frac{ka}{2}. \tag{5.15}$$

There is a linear relation between the normal mode frequency and the wave vector, making the product of the frequency and the wavelength ($f\lambda$) of the vibrations a constant. The constant is the speed v_s of a sound wave in the crystal: $v_s = \sqrt{2K/M}\,a/2$, equivalent to the Debye model for the solid.*

In the same limit, the displacement ratio becomes

$$\frac{\epsilon_+}{\epsilon_-} = \frac{1+1}{2-0} = 1, \tag{5.16}$$

showing that both atoms move together, in phase.

When $k = \pi/a$, the sine is unity; this zone boundary frequency is the maximum possible for this branch. Equation 5.14 becomes

$$\omega_a = \sqrt{\frac{K}{\mu}}\left[1 - \sqrt{1 - \frac{4\mu}{M}}\right]^{1/2}.$$

If I substitute for μ and M and do a little simplification, the term in square brackets becomes $[\,] = [1 - \Delta M/M]$ where ΔM is the mass difference between the two atoms. I now must specify which is heavier because ΔM must be positive. (I made the choice that the square root itself is a positive quantity when I solved the quadratic equation.) Because NaCl is the prototype for phonons in an ionic crystal, I'll take $M_- > M_+$. With this choice, the zone boundary acoustic branch frequency is

$$\omega_a = \sqrt{\frac{2K}{M_-}}. \tag{5.17}$$

Now I look at the case when I choose the plus sign. The dispersion relation becomes

$$\omega_o = \sqrt{\frac{K}{\mu}}\left[1 + \sqrt{1 - \frac{4\mu}{M}\sin^2(ka/2)}\right]^{1/2}. \tag{5.18}$$

The subscript o stands for "optic." This branch interacts with electromagnetic waves. As $k \to 0$ the sine can be replaced by its argument; the second term in the inner square root is small and I can expand it to give

$$\omega_o = \sqrt{\frac{K}{\mu}}\left[1 + \left\{1 - \frac{2\mu}{M}\frac{k^2 a^2}{4}\right\}\right]^{1/2},$$

$$= \sqrt{\frac{2K}{\mu}}\left[1 - \frac{\mu}{8M}k^2 a^2\right]. \tag{5.19}$$

The optic mode has a very high frequency (the highest of all the normal modes) when $k = 0$, which I'll call the transverse optic (TO) frequency

$$\omega_T = \sqrt{\frac{2K}{\mu}}. \tag{5.20}$$

---

* The various factors of 2 are unfamiliar if I only know the solution for the monatomic chain. But of course $M = M_- + M_+$ and $a = 2d$ where d is the distance between neighboring atoms. If the masses are identical, the monatomic chain result is returned.

The reason for this frequency being highest can be seen by looking at the displacement ratio. At $k = 0$, Eq. 5.13 is

$$\frac{\epsilon_+}{\epsilon_-} = \frac{1+1}{2 - \frac{M_+}{K}\omega_o^2}.$$

I substitute $\omega_o = \sqrt{2K/\mu}$ and after some algebra obtain

$$\frac{\epsilon_+}{\epsilon_-} = -\frac{M_-}{M_+}. \tag{5.21}$$

The atoms vibrate in opposite directions with the lighter atom having the larger amplitude. This opposing motion gives maximal compression and expansion amplitude to the springs and therefore has the highest possible frequency. It also moves positive charge in one direction and at the same time negative charge in the opposite direction. A time-dependent dipole moment is generated in the chain, analogous to what is sketched in Fig. 5.1. I'll therefore expect this $k = 0$ resonance to give rise to the Lorentzian dielectric function of Eq. 5.2.

When $k = \pi/a$, the sine is equal to unity and

$$\omega_o = \sqrt{\frac{K}{\mu}} \left[1 + \sqrt{1 - \frac{4\mu}{M}} \right]^{1/2}.$$

Here the square bracket is $[] = [1 + \Delta M/M]$. Then (with, as before, taking negative heavier) I find that the frequency at $k = \pi/a$,

$$\omega_o = \sqrt{\frac{2K}{M_+}}. \tag{5.22}$$

The optic-mode frequency is higher than the acoustic-mode frequency at any wave vector.

5.2.6 Phonon Notes

Here I'll just make a few random remarks about lattice dynamics and phonons. First, suppose the mass difference gets larger and larger, so $M_+ \ll \mathbf{M_-}$. When the masses are quite different the gap between the acoustic and optic branches becomes quite large. In addition, the reduced mass $\mu \sim M_+$, so the dispersion (the variation of ω with k) is small; the optical branch in a plot like Fig. 5.4 appears almost flat.

There are cases where the masses are all the same but the spring constants take two values. I could have a linear chain of carbon atoms with alternating single and double bonds. This example occurs as the polyacetylene chain, the first highly conducting polymer. It can be schematically represented as $-C=C-C=C-C=C-$. (Each carbon has a hydrogen attached to it as well; from the side the chain takes a zig-zag form.) The double bonds are stronger than the single bonds and the lattice dynamics have an acoustic and an optic branch.

Many solids have more than one phonon branch. As the chemical formula becomes more complex, the numbers of degrees of freedom increase. For the phonons, the key quantity is the number N of distinct atoms in the unit cell.* There is one for Ag, two for NaCl, and perhaps five for Al_2O_3. With N atoms in the unit cell, there are $3N$ degrees of freedom for the normal modes. Three of these are always the acoustic phonons, in which the entire unit cell moves as single unit, leaving $3N - 3$ optical modes. The optical modes involve relative motion of the atoms within the unit cell. Although not all of these modes are infrared active, many are.

For crystals with high symmetry, there are degeneracies. For example, the two transverse modes for a given $\mathbf{k}$ direction are degenerate in a cubic crystal but not degenerate in orthorhombic crystals.

The effective charge on the ions, Z^*e in Eq. 5.1 and to appear again in the next section, might naively be taken to be the charge of the closed shell ion. Then Na would have $Z^*e = +1$, Ca would have $+2$, O -2, and Cl -1. This guess would not be too far off but it would not be very accurate. Instead, modern theories of electronic structure/lattice dynamics theories [25–27] can calculate the "Born effective charge tensor,"

$$Z_{ij}^* = \frac{\Omega}{e} \frac{\partial P_i}{\partial d_j}$$

where Ω is the primitive cell volume, P_i is the ith component of the dipole moment/unit volume, and d_j is the jth component of the ion's displacement vector. Many lattice dynamical computational programs will calculate the Born tensor.[†]

The phonons are actually quantum mechanical objects. Phonons have no spin, so they obey Bose–Einstein statistics. It is straightforward to take the Hamiltonian of Eq. 5.5, view p and u as operators, Fourier transform them, and then replace with phonon creation and annihilation operators a^+ and a to arrive at $\mathcal{H} = \sum_{kv} \hbar\omega_{skv}(a_{skv}^+ a_{skv} + 1/2)$ where v is the polarization (two transverse and one longitudinal) and s is the branch (acoustic, optic). The analogy to photons is good: there is a thermal spectrum, transport of energy, Compton scattering, Bremsstrahlung, and photoelectric-effect-like interactions.

5.3 Interaction with Electromagnetic Waves

So now, I will add electromagnetic fields to the equation of motion for the lattice. I'll take the electric field to be the friendly plane wave with wave vector q and frequency ω: $\mathbf{E} = \mathbf{E}_0 e^{i(qz-\omega t)}$ with z the direction of travel. Let $\mathbf{E}$ be parallel to $\hat{x}$. I add this to Eqs. 5.7 and 5.6

$$M_+ \ddot{u}_{n,+} = K(u_{n,-} + u_{n+1,-} - 2u_{n,+}) + Z^*eE,$$

---

* By "distinct" I mean atoms that occupy distinct crystallographic positions. Atoms can be distinct even if they are the same element. For example, silicon, a monatomic solid, has two atoms per unit cell. Si has even got optical phonon branches, though not infrared active ones.
[†] Google "Born charge."

and

$$M_- \ddot{u}n, - = K(u_{n,+} + u_{n-1,+} - 2u_{n,-}) - Z^*eE,$$

where Z^*e is the Born charge of the ion.

If I need to solve these equations for all values of phonon wave vector k, then the index n is critical. But if (as is the case here) the wavelength of the electromagnetic field is much larger than that of the lattice, I can ignore the e^{ika} variation from site to site and therefore drop the index n. So these equations become

$$\ddot{u}_+ = \frac{2K}{M_+}(u_- - u_+) + \frac{Z^*e}{M_+}\mathbf{E},$$

and

$$-\ddot{u}_- = \frac{2K}{M_-}(u_- - u_+) + \frac{Z^*e}{M_-}\mathbf{E},$$

where I've divided through by the mass and changed signs in the second of these. Now add these, define $1/\mu = 1/M_+ + 1/M_-$ and the equation for $x \equiv u_+ - u_-$ is

$$\ddot{x} = -\frac{2K}{\mu}x + \frac{Z^*e}{\mu}\mathbf{E}. \qquad (5.23)$$

Using Eq. 5.20 makes it evident that Eq. 5.23 is identical to Eq. 5.1 (for the $\gamma = 0$ case). I want some damping and do not need to go through the solution again. The dielectric function is just the one given by Eq. 5.2, a Lorentz oscillator with damping. The reflectance of the Lorentz oscillator may be seen in Figs. 4.10–4.14. Some data for a few ionic solids will be shown in Chapter 6, Sections 6.5 and 6.6.

5.4 Transverse and Longitudinal Modes

The phonon modes of a three-dimensional crystal may be separated into transverse and longitudinal modes [24]. Some modes are optically active but not all. Some are Raman active and some are "silent." The selection rules can be determined by group theoretical analysis [28–30]. The infrared active modes are those that transform like the momentum (p_x, p_y, or p_z) or position (x, y, or z). In general, these are modes that are odd under inversion. (In the lingo of group theory, the infrared active modes are *ungerade* or uneven and carry the subscript u. Other modes are *gerade* or even with subscript g; these will be Raman active or inactive.)

The infrared-active transverse optical (TO) modes are the resonant frequencies of the harmonic oscillator; for each of these there is a corresponding longitudinal optical (LO) mode. In the language of transfer functions in control theory [31], the transverse modes are the poles and the longitudinal modes are the zeros of the dielectric function.*

---

* Just as the screened plasma frequency, $\omega_p/\sqrt{\epsilon_c}$, is the zero of the Drude dielectric function.

I discussed the plasma response of a metal in Section 4.3. The electrodynamics of the longitudinal mode of an insulator are similar. If the mode is longitudinal, then $\mathbf{q}$ is parallel to $\mathbf{E}$. Thus, $\nabla \cdot \mathbf{E} \neq 0$. In the insulating solid, there is no free charge, so $\nabla \cdot \mathbf{D} = 0$. For linear materials, $\mathbf{D} = \epsilon \mathbf{E}$, so $\nabla \cdot \mathbf{D} = \epsilon \nabla \cdot \mathbf{E} = 0$. Because $\nabla \cdot \mathbf{E} \neq 0$, $\epsilon = 0$. A further implication is: $\mathbf{D} = \epsilon \mathbf{E} = 0$. But $\mathbf{D} = \mathbf{E} + 4\pi \mathbf{P}$ so that $\mathbf{P} = -\mathbf{E}/4\pi$.

The frequency of the longitudinal mode is higher than that of the corresponding transverse mode. The reason for this difference is the long range Coulomb force that results from this polarization. I return to Eq. 4.26 (writing $\mathbf{E}_{\mathrm{app}}$ as just $\mathbf{E}$) and put in the value of $\mathbf{P}$ from above to find

$$\mathbf{E}_{\mathrm{loc}} = -\frac{8\pi}{3}\mathbf{P}.$$

So if the positive atoms move to the right and the negative to the left along the x axis, $\mathbf{P}$ is along $+\hat{x}$ and $\mathbf{E}_{\mathrm{loc}}$ is along $-\hat{x}$, opposing the motion. In fact, because $\mathbf{P}$ is linear in the displacement, this local field is just like a spring. The result is an additional restoring force and a stiffening of the mode. One way to think about it is to realize that on the surfaces of a slab of ionic crystal, a polarization charge density $\sigma_p = \mathbf{P} \cdot \hat{n}$ will appear. These surface charges are the sources of the electric field that raise the LO frequency. I note that there is also a local field correction of the TO frequency. This correction acts to reduce the TO frequency, as I discussed beginning in the context of the depolarizing field, Section 4.4.2.

5.4.1 The Longitudinal Frequency

The frequency of the zero of the dielectric function is easy to estimate. First, I'll write Eq. 5.2 on a common denominator:

$$\epsilon = \epsilon_e \frac{\Omega_p^2/\epsilon_e + \omega_T^2 - \omega^2 - i\omega\gamma}{\omega_T^2 - \omega^2 - i\omega\gamma},$$

and next I define

$$\omega_L^2 = \Omega_p^2/\epsilon_e + \omega_T^2, \tag{5.24}$$

so that

$$\epsilon = \epsilon_e \frac{\omega_L^2 - \omega^2 - i\omega\gamma_L}{\omega_T^2 - \omega^2 - i\omega\gamma_T}, \tag{5.25}$$

where I've put subscripts on the damping factors (relaxation rates) to suggest that different damping factors may affect the transverse and longitudinal modes. The real part of the dielectric function in Eq. 5.25 is zero at a frequency near ω_L. The frequency is exactly ω_L when the damping is negligible.

5.5 The Lyddane–Sachs–Teller Relation

In the limit as $\omega \to 0$, Eq. 5.25 becomes purely real:

$$\epsilon_1(0) = \epsilon_e \left(\frac{\omega_L}{\omega_T}\right)^2. \tag{5.26}$$

Table 5.1 Transverse and longitudinal frequencies in solids, and the static and high-frequency dielectric constants.

Mat'l	Transverse mode			Longitudinal mode			$\epsilon_1(0)$	ϵ_e
	cm^{-1}	THz	meV	cm^{-1}	THz	meV		
LiH	584	17.5	72.3	1114	33.4	138.1	12.9	3.6
LiF	308	9.2	38.1	637	19.1	78.9	8.9	1.9
LiCl	191	5.7	23.7	398	11.9	49.3	12.0	2.7
LiBr	159	4.8	19.7	324	9.7	40.1	13.2	3.2
NaF	239	7.2	29.6	414	12.4	51.3	3.1	1.7
NaCl	164	4.9	20.4	265	8.0	32.9	5.9	2.25
NaBr	133	4.0	16.4	207	6.2	25.7	6.4	2.6
KF	191	5.7	23.7	324	9.7	40.1	5.5	1.5
KCl	143	4.3	17.8	212	6.4	26.3	4.85	2.1
KI	101	3.0	12.5	138	4.1	17.1	5.1	2.7
RbF	154	4.6	19.1	286	8.6	35.5	6.5	1.9
RbI	74	2.2	9.2	101	3.0	12.5	5.5	2.6
CsCl	101	3.0	12.5	164	4.9	20.4	7.2	2.6
CsI	64	1.9	7.9	85	2.5	10.5	5.65	3.0
TlCl	64	1.9	7.9	159	4.8	19.7	31.9	5.1
TlBr	43	1.3	5.3	101	3.0	12.5	29.8	5.4
AgCl	101	3.0	12.5	180	5.4	22.4	12.3	4.0
AgBr	80	2.4	9.9	133	4.0	16.4	13.1	4.6
MgO	398	11.9	49.3	743	22.3	92.1	9.8	2.95
GaP	366	11.0	45.4	403	12.1	50.0	10.7	8.5
GaAs	271	8.1	33.5	292	8.8	36.2	13.13	10.9
GaSb	228	6.8	28.3	244	7.3	30.3	15.69	14.4
InP	302	9.1	37.5	345	10.3	42.8	12.37	9.6
InAs	218	6.5	27.0	239	7.2	29.6	14.55	12.3
InSb	186	5.6	23.0	196	5.9	24.3	17.88	15.6
SiC	790	23.7	98.0	950	28.5	117.7	9.6	6.7
C	1332	39.9	165.1	1332	39.9	165.1	5.5	5.5
Si	525	15.8	65.1	525	15.8	65.1	11.7	11.7
Ge	302	9.1	37.5	302	9.1	37.5	15.8	15.8
CaF$_2$	257	7.7	31.9	460	13.8	57.0	6.63	2.04
SrF$_2$	217	6.5	26.9	375	11.3	46.5	6.20	2.07
BaF$_2$	184	5.5	22.8	328	9.8	40.7	6.94	2.15

This is the Lyddane–Sachs–Teller relation [32–35]. The static dielectric constant is larger than the high-frequency (electronic) dielectric constant by the square of the ratio of longitudinal to transverse frequencies. Data for a large number of insulating (and mostly ionic) solids are in Table 5.1, adapted from Kittel [24].

5.6 Other Notations

The zero-frequency limit of Eq. 5.2 is

$$\epsilon_1(0) = \epsilon_e + \frac{\Omega_P^2}{\omega_T^2}.$$

This expression for the static dielectric constant is of course equivalent to Eq. 5.26 given the definition of ω_L in Eq. 5.24. Sometimes it is of interest to highlight the fact that Ω_P^2/ω_T^2 is the (additive) contribution to the static dielectric constant. Then one defines either $\Delta\epsilon = \epsilon_1(0) - \epsilon_e$ or the so-called oscillator strength S as

$$\Delta\epsilon \equiv S \equiv \frac{\Omega_P^2}{\omega_T^2}. \tag{5.27}$$

The dielectric function Eq. 5.2 in terms of S is then

$$\epsilon = \epsilon_e + \frac{S\omega_T^2}{\omega_T^2 - \omega^2 - i\omega\gamma}. \tag{5.28}$$

5.7 Multiple Modes

In solids with more than 2 atoms per unit cell, each infrared-active mode will contribute to the dielectric function. I'll model it as a sum of Lorentz oscillators just as in the case of multiple electronic orbitals. (See Chapter 4; Section 4.4.10.) Equation 5.2 becomes

$$\epsilon = \epsilon_e + \sum_j \frac{\Omega_{pj}^2}{\omega_{Tj}^2 - \omega^2 - i\omega\gamma_j}, \tag{5.29}$$

while the equivalent Eq. 5.28 is

$$\epsilon = \epsilon_e + \sum_j \frac{S_j\omega_{Tj}^2}{\omega_{Tj}^2 - \omega^2 - i\omega\gamma_j}. \tag{5.30}$$

In Eqs. 5.29 and 5.30, I've written the resonant frequency as ω_{Tj} rather than ω_j. Sometimes one sees one such notation and sometimes the other.

By analogy, Eq. 5.25 is written

$$\epsilon = \epsilon_e \prod_j \frac{\omega_{Lj}^2 - \omega^2 - i\omega\gamma_{Lj}}{\omega_{Tj}^2 - \omega^2 - i\omega\gamma_{Tj}}, \tag{5.31}$$

and the Lyddane–Sachs–Teller relation, Eq. 5.26, generalizes to

$$\epsilon(0) = \epsilon_e \prod_j \left(\frac{\omega_{Lj}}{\omega_{Tj}}\right)^2. \tag{5.32}$$

Each oscillator contributes to the static dielectric constant, with the ones at the lowest frequencies dominant. If a mode goes "soft" (i.e., its frequency goes to zero as temperature or some other parameter is varied), the dielectric constant will diverge; such soft modes may be found in ferroelectric and almost-ferroelectric crystals.

There are interesting interactions among the poles and zeros of the dielectric function when there are two or more modes. The effect is illustrated in Fig. 5.5. The four panels show the real dielectric function, real conductivity, the loss function, and the reflectance calculated starting with the dielectric function give by Eq. 5.29. The solid has two modes, a strong optical phonon with $\omega_{T1} = 300$ cm^{-1}, $\Omega_{p1} = 1{,}800$ cm^{-1} ($S_1 = 36$), and $\gamma_1 = 4$ cm^{-1}. There is a weak mode that I have given one of two TO frequencies. For the left column of plots it is at $\omega_{T2} = 150$ cm^{-1} and for the right column at $\omega_{T2} = 600$ cm^{-1}. The other parameters are $\Omega_{p2} = 200$ cm^{-1} (so $S_2 = 1.8$ when the mode is at 150 cm^{-1} and 0.11 when it is at 600 cm^{-1}), $\gamma_2 = 2$ cm^{-1}, and $\epsilon_e = 1.5$.

The real part of the dielectric function shows two of the usual derivative-like excursions. In the left panel, the weak mode has a small range of frequencies where it is negative, from 150 to 152 cm^{-1}, while the strong mode has a wide range, from 300 to 1,509 cm^{-1}. (The LO frequencies are 152 and 1,509 cm^{-1}.) The two modes have poles at the TO resonance frequencies, seen as peaks in σ_1, and zeros at the LO frequencies, seen as peaks in $-\,\mathrm{Im}(1/\epsilon)$. Finally the reflectance is high for frequencies where $\epsilon_1(\omega) < 0$ with dips toward zero near the LO frequencies.*

It is natural to associate the first pole and first zero with the phonon having the lower TO frequency and the second pole and second zero with the phonon with the higher TO frequency. For the case on the left in Fig. 5.5 this natural interpretation is correct. But for the case on the right, the first zero is really associated with the weaker mode, and occurs at a *lower* frequency than the pole. The behavior of $\epsilon_1(\omega)$ near the TO$_2$ frequency is shown in the little cartoon at the right of Fig. 5.5. The dielectric function ϵ_1 is negative and disperses upward as the TO$_2$ frequency is approached. It crosses through zero just below the TO$_2$ frequency and then, at the resonance, steps down to a minimum and then eventually returns to approach the curve that would be found if the mode did not occur. So it is correct to say that the zero of the weaker mode is below the pole and to continue to assign the zero at 1509 cm^{-1} to the stronger mode. The consequence of the narrow region where $\epsilon_1(\omega) > 0$ is a sharp dip in the reflectance.

---

* The dips are at frequencies a bit higher than the zeros of $\epsilon_1(\omega)$ because they occur where $n = 1$ rather than where $\epsilon_1 = 0$. If there is no damping, the dips reach zero but if there is some damping, the reflectance is always above zero.

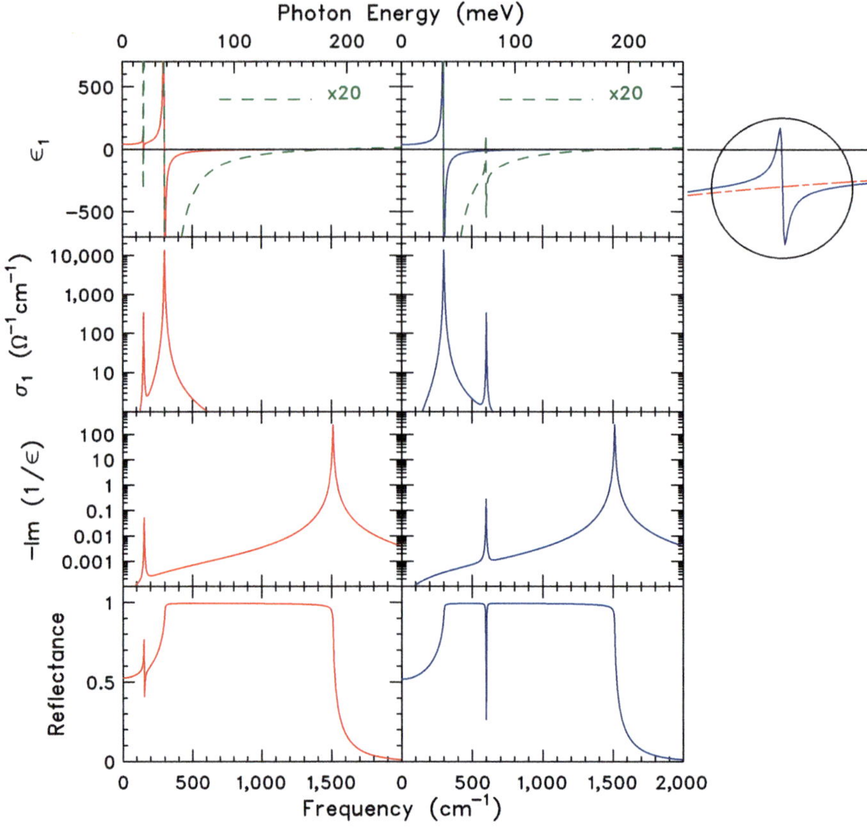

Fig. 5.5 Optical functions for a phonon system with both a strong and a weak vibrational mode. Top to bottom, the panels show $\epsilon_1(\omega)$, $\sigma_1(\omega)$, $-\mathrm{Im}(1/\epsilon)$, and $\mathcal{R}$. $\epsilon_1(\omega)$ is also shown magnified by 20 as the dashed lines. The left panels are for the case where the TO frequency of the weak mode is below the TO frequency of the strong mode and the right panels are for the case where the weaker mode is above the strong mode. The fragment on the right shows $\epsilon_1(\omega)$ around $\omega = 600$ cm^{-1} for the case where the weak mode is above the strong mode (blue solid line) and where it is below it (red dashed line). The horizontal line is the level where $\epsilon_1 = 0$.

This behavior, where there are narrow dips in the highly reflecting reststrahlen region, is seen in many perovskite materials [36–38], some cuprates [39], and other highly polar materials. The same effect occurs in doped polar semiconductors when the plasma frequency of the free carriers is a bit above the LO frequency of an optical phonon [40].

5.8 Polaritons

An electromagnetic wave traveling in a material with $\epsilon_1 \neq 1$ induces a polarization $\mathbf{P}$ in the material. In the case where the polarization is due to the optical phonons, there is physical motion of the atoms. The wave is a coupled combination of electromagnetic waves and lattice vibrations, or the coupled motion of photons and phonons. This coupled mode is called a "polariton." Polaritons actually occur for any wave traveling in a polarizable medium, but the relatively narrow linewidth of the phonon resonances (a few cm^{-1}) makes them a good vehicle for discussing these effects.

It is fair to ask about the condition required to make these effects important. There are a number of criteria I could construct. The simplest one is to think for a moment about the electric displacement vector $\mathbf{D}$. In general, $\mathbf{D} = \mathbf{E} + 4\pi\mathbf{P}$ and for linear, homogeneous, and isotropic materials, with $\mathbf{P} = X\mathbf{E}$ and $X = (\epsilon - 1)/4\pi$ (the dielectric susceptibility). So then $\mathbf{D} = \mathbf{E} + (\epsilon - 1)\mathbf{E}$ and the polarization is the larger component of $\mathbf{D}$ once $\epsilon > 2$.

The analysis is clearer (and the math easier) if I neglect the damping. So I'll use Eq. 5.25, setting the damping terms to zero:

$$\epsilon = \epsilon_e \frac{\omega_L^2 - \omega^2}{\omega_T^2 - \omega^2}. \tag{5.33}$$

This function is purely real.* In turn, the refractive index $N = \sqrt{\epsilon}$ is

$$N = n + i\kappa = \sqrt{\epsilon_e \frac{\omega_L^2 - \omega^2}{\omega_T^2 - \omega^2}}. \tag{5.34}$$

When $\epsilon > 0$, N is real and waves propagate without attenuation at a phase velocity given by $v = c/n$. When $\epsilon < 0$, N is purely imaginary, and the fields in the crystal are exponentially damped with attenuation length $\delta = c/\omega\kappa$.

What I want is the dispersion relation, the dependence of the frequency ω upon the wave vector q for *light* propagating in the polar material. I already know the dispersion relation for the phonons; it is given by Eq. 5.12 and plotted in Fig. 5.4 The last couple of sentences in the previous paragraphs signal what the physics is: the wave propagation is governed by n and n grows without bound as $\omega \to \omega_T$ from below. Wave speed will become slower and slower. The region of highly damped waves is between ω_T and ω_L. Above ω_L, there are again propagating waves; these are anomalous in that the phase velocity is greater than c until the frequency is high enough that $n \geq 1$.

To find the dispersion relation, I'll return to Eq. 3.3, Maxwell's equations for local, nonmagnetic, linear, isotropic, and homogeneous media, and consider plane wave functions for the fields. The answer is in Eqs. 3.9 and 3.10: $q = (\omega/c)\sqrt{\epsilon} = (\omega/c)N$. I square both sides and use Eq. 5.33 to obtain

$$q^2 = \frac{\omega^2 \epsilon_e}{c^2} \frac{\omega_L^2 - \omega^2}{\omega_T^2 - \omega^2}, \tag{5.35}$$

as an equation relating q to ω. But what I want is $\omega(q)$. Equation 5.35 is quadratic in ω^2 and the solution is

$$\omega^2 = \frac{\omega_L^2}{2} + \frac{c^2 q^2}{2\epsilon_e} \pm \frac{1}{2}\sqrt{\left(\omega_L^2 + \frac{c^2 q^2}{\epsilon_e}\right)^2 - 4\omega_T^2 \frac{c^2 q^2}{\epsilon_e}}.$$

---

* What happens at $\omega = \omega_T$ is implicit in discussions in Chapter 10. For the moment, I'll just avoid that single frequency. (The answer is that there is a delta function in the imaginary part $\epsilon_2(\omega)$ at $\omega = \omega_T$.)

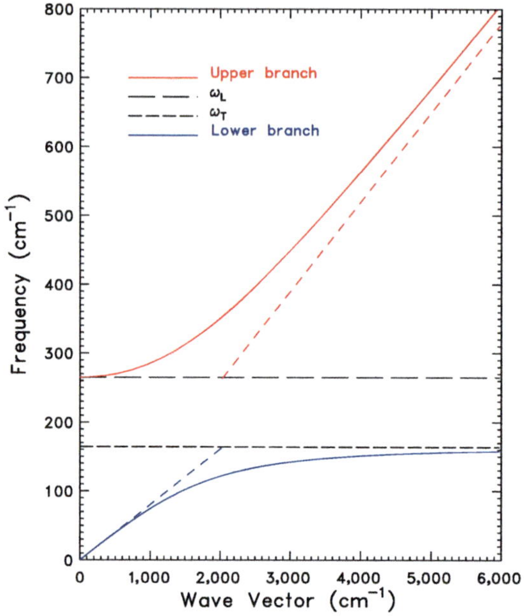

Fig. 5.6 Dispersion relation for light traveling in an ionic crystal with a single transverse optic mode. The TO and LO optical phonon frequencies, as well as the high-frequency dielectric constant were chosen to match NaCl. The range of q and k shown is very small compared to the width of the first Brillouin zone. The dispersion relations for light in a medium with $n = \sqrt{\epsilon_1(0)}$ and with $n = \sqrt{\epsilon_c}$ and the dispersion relations for LO and TO phonons are also shown.

For each value of q, there are two frequencies of propagating waves. One (the lower; choose $-$) is between zero and ω_T and the other (the upper; choose $+$) is between ω_L and ∞. A plot of the dispersion relations for photons (electromagnetic waves) and for phonons (lattice waves) is shown in Fig. 5.6. Note that the wave vector range shown is much smaller than the range in Fig. 5.4. If I wanted to plot the dispersion over the entire Brillouin zone but with the same detail as in Fig. 5.6, the diagram would be almost 10,000 times wider.

The lower branch starts just as would the case where there was no coupling of photons and phonons, with a slope governed by the refractive index as $\omega \to 0$, $n = \sqrt{\epsilon_1(0)}$. The dispersion in this uncoupled case is shown as a dashed line. But as the frequency approaches the transverse optical phonon frequency, ω_T, the polariton dispersion curve bends over and becomes eventually tangent to the TO phonon curve shown.

There are no propagating modes between ω_T and ω_L. Above the LO frequency, the polariton dispersion starts again. The dielectric function is only a little bit positive so that $n \ll 1$. The dispersion is tangent to the LO phonon curve for a while and then bends upward, eventually rising linearly with the same slope as the case when there was no coupling of photons and phonons. The high frequency propagation is governed by the high frequency refractive index, $n = \sqrt{\epsilon_c}$.

6

A Look at Real Solids

Chapters 4 and 5 have given me a framework to discuss optical spectra. I show in this chapter spectra for some metals, a semiconductor, and some ionic insulators, discussing them within this framework. Some of the concepts in Chapters 8–12 will be foreshadowed in this chapter. I think it is helpful to see examples of actual metals, semiconductors, and insulators before diving into the quantum-based theories and that's why this chapter is here and not later in the text.

6.1 Silver

Silver is an elemental metal. The atomic configuration of the 47 electrons in the silver atom is $1s^2 2s^2 2sp^6 3s^2 3p^6 3d^{10} 4s^2 4p^6 4d^{10} 5s^1$. The $5s$ electron is the "free" electron of the metal, so I expect one mobile electron per atom. The $4d$ levels are 4 eV below the $5s$ level in the atom [41, 42].

Silver has high reflectance in the visible. Indeed, it is the standard for a shiny metal; other things with high reflexivity are called "silvery." An example of "typical" silver reflectance data is shown in Fig. 6.1. The data are those collected by Palik [43–53]. The low frequencies are supplemented by a Drude reflectance, based on resistivity $\rho = 1.63$ $\mu\Omega$-cm at 300 K. The main panel shows the reflectance from 20 to 80,000 cm^{-1} (2.5 meV–10 eV). One can see the high metallic reflectance from far-infrared to near ultraviolet, a sharp and deep plasma edge around 32,000 cm^{-1} (4 eV), and the transitions from the d-bands to the conduction band above the plasma edge [45, 54].

The reflectance at higher photon energies, extending to 1,000 eV, is shown in the inset of Fig. 6.1. Here, one can see additional interband transitions followed by sharp core-level atomic transitions. Note that the reflectance above 100 eV ($\sim 800,000$ cm^{-1}) is pretty close to the expected ω^{-4} power law.

I'll use the Drude model to describe the optical properties of the free-electron metal silver (and many other metals). The high "metallic" reflectance and plasma edge are similar to the Drude model reflectance in Fig. 4.7. A fit to the silver data gives a plasma frequency of $\omega_p = 71,600$ cm^{-1} (8.88 eV) and a scattering rate of $1/\tau = 138$ cm^{-1} ($\tau = 38$ fs).*

The real and imaginary parts of the optical conductivity,† $\sigma_1(\omega)$ and $\sigma_2(\omega)$, are shown in Fig. 6.2. The real part, σ_1, equals the DC conductivity (610,000 Ω^{-1} cm^{-1}) at zero

---

* I must remember to multiply frequencies in cm^{-1} by $2\pi c$ to get radians/s before inverting.
† The conductivities and other optical functions have been calculated using Kramers–Kronig analysis of the reflectance to obtain the reflection phase. See Chapter 10.

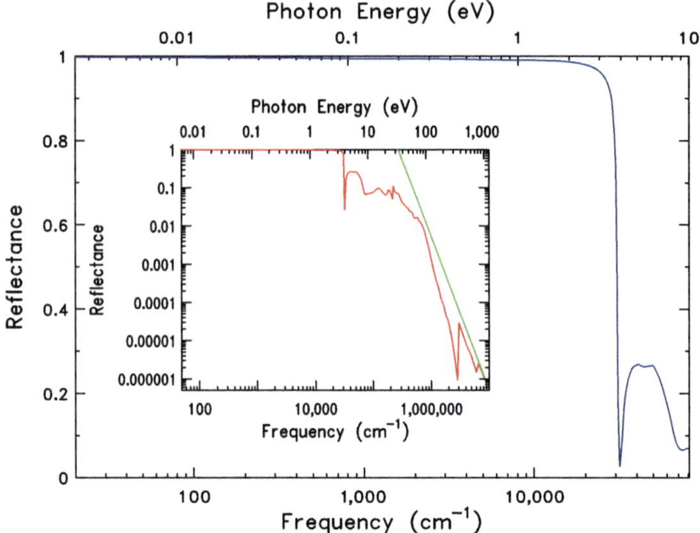

Fig. 6.1 Reflectance of silver [43–53]. The main panel shows the reflectance from far infrared to near ultraviolet on a logarithmic frequency scale. The inset shows the vacuum-ultraviolet and soft x-ray reflectance. The straight green line follows ω^{-4}.

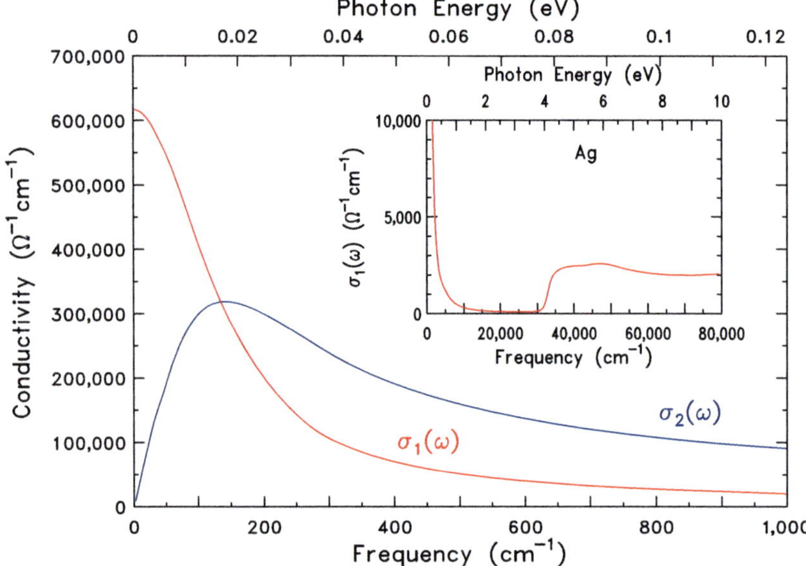

Fig. 6.2 Real and imaginary parts of the optical conductivity of silver. Inset: real part of the conductivity continued up to frequencies where the interband transitions occur.

frequency. It rolls off as frequency increases and becomes quite small above about $1{,}000\,\mathrm{cm}^{-1}$. Looking at the inset, where the scale is expanded, I can see that it falls to values of a few hundred $\Omega^{-1}\mathrm{cm}^{-1}$ (probably the limit of accuracy of the measurement). A set of absorption bands begins at $32{,}000\,\mathrm{cm}^{-1}$ (4 eV) and extend to above $80{,}000\,\mathrm{cm}^{-1}$ (10 eV).

The imaginary part, σ_2, is initially smaller than the real part. It rises (while the real part is falling) and the two functions are equal near $\omega = 140$ cm^{-1}, the frequency where σ_1 has fallen to half its DC value and where σ_2 has its maximum. Above this frequency, σ_2 falls, but is larger than σ_1. The peak of σ_2 and the half power point of σ_1 booth occur at $\omega = 1/\tau$ in the Drude model. I can see by comparison with Fig. 4.1 that the Drude model describes silver metal well. Indeed, the data and fits to the data give $1/\tau = 138$–140 cm^{-1}, essentially the same value I calculated in Chapter 4, Section 4.2.4 from the DC conductivity and carrier density.

All solids have electronic bands and interband transitions from filled to empty states are part of the optical spectrum. These interband transitions in silver are responsible for the features that begin at 4 eV (32,000 cm^{-1}) and are seen in Fig. 6.1 (inset) up to perhaps 300 eV. I'd use one or more Lorentz oscillators (Eq. 4.31) to represent these interband transitions. In silver the lowest energy interband transitions involve transitions of d electrons to the conduction band.

The inset of Fig. 6.2 illustrates one issue with using Lorentz oscillators for the interband transitions. The conductivity spectrum shows an almost step-like onset at 4 eV. In contrast, as can be seen from Eq. 4.34 and Fig. 4.12, the Lorentz dielectric function starts at zero when $\omega = 0$ and increases as ω^2 well below the resonance. The sudden onset of interband transitions is easy to understand. There is a minimum energy spacing of the band, equal to the distance between the top of the occupied band and the bottom of the empty band.* Consequently the contribution of the interband transitions to the conductivity spectrum is zero until the photon energy exceeds this spacing. The Lorentz oscillator does not represent this onset well at all. I can use many narrow oscillators if I want to fit data; it is not easy to see what the parameters of the fit tell me about the solid.

If a Drude model is correct for silver below the interband transitions, I expect large and negative values for the real part of the dielectric function. Silver does not disappoint in this regard; ϵ_1, shown in Fig. 6.3, is indeed large and negative. The limiting low-frequency value is $-\omega_p^2\tau^2 = -275,000$. As it should, it begins to roll up as ω approaches $1/\tau$. At higher frequencies, it follows $\epsilon_1 = \epsilon_c - \omega_p^2/\omega^2$, as appropriate for free carriers. It crosses zero around $31,000$ cm^{-1} (just under 4 eV). (A $20,000\times$ magnified view of the zero crossing is in Fig. 6.3.) There is considerable structure above the zero crossing qualitatively in accord with multiple Lorentz oscillators from the interband transitions.

The real and imaginary parts of the refractive index are plotted in Fig. 6.4. The refractive index of silver resembles closely the Drude model calculation show in Figs. 4.5 and 4.6. At low frequencies, n and κ are almost the same, in accord with Eq. 4.16. With increasing frequency, both become smaller and n becomes quite a bit smaller than κ. $n < 1$ between 3,000 and 30,000 cm^{-1}, where ϵ_1 changes sign. Unlike the model of Figs. 4.5 and 4.6, κ does not fall above the zero crossing. There is additional absorption due to the interband transitions.

Now, I look at the area under the curve of $\sigma_1(\omega)$. I can guess from the inset to Fig. 6.2 that there will be two contributions to this area: the Drude peak that starts at zero frequency

---

* Or the highest occupied level in the partially-occupied band, as is the case in silver.

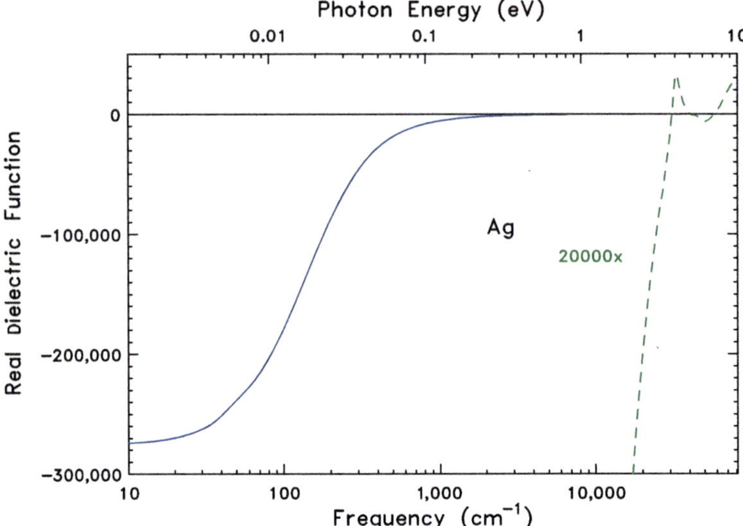

Fig. 6.3   Real part of the dielectric function of silver.

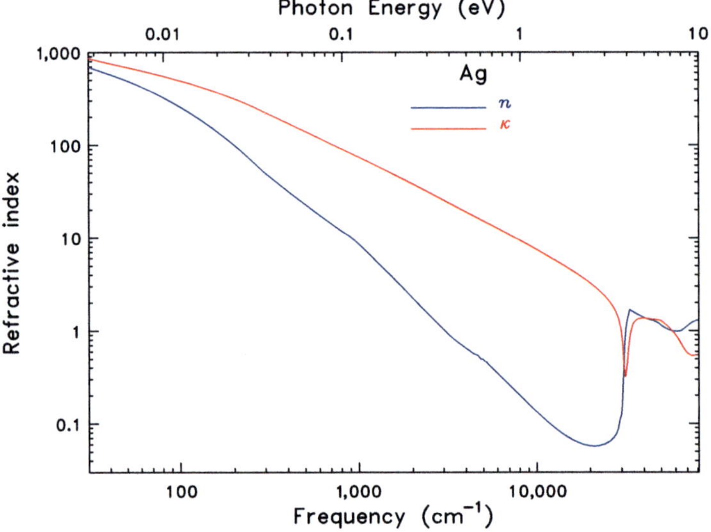

Fig. 6.4   Real (n) and imaginary (κ) parts of silver's refractive index.

and falls off to a small value by 5,000 cm^{-1} (0.6 eV) and the interband transitions that begin at 32,000 cm^{-1} (4 eV) and extend above 100,000 cm^{-1}.

This integral, known as the sum rule, will be discussed in Chapter 10. For the moment, let me integrate Eq. 4.10a:

$$\int_0^\infty \sigma_1(\omega')\,d\omega' = \int_0^\infty \frac{ne^2\tau/m}{1+\omega'^2\tau^2}\,d\omega' = \frac{\pi ne^2}{2m}.$$

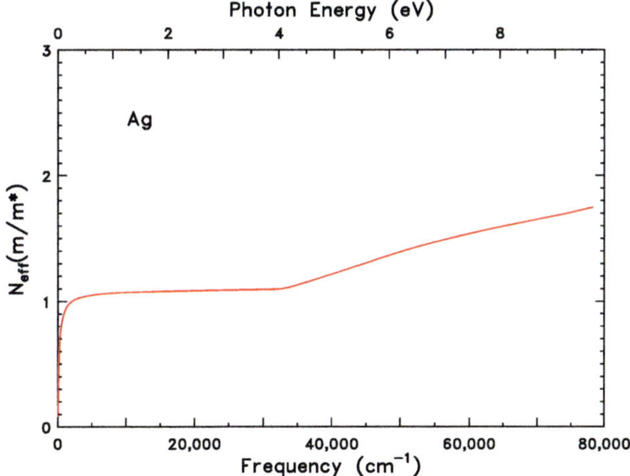

Fig. 6.5 $(m/m^*)N_{eff}(\omega)$ of silver.

I can turn this into an equation for the number N_{eff} of electrons per copper atom with effective mass ratio m/m^* participating in optical transitions at frequencies below ω by writing

$$\frac{m}{m^*}N_{eff}(\omega) = \frac{2mV_c}{\pi e^2}\int_0^\omega \sigma_1(\omega')\,d\omega', \tag{6.1}$$

where to allow for the effective mass being different from the electronic mass, I have included the factor m^*/m, the ratio of the effective mass of the charge carriers to the free electron mass. The quantity V_c is the unit cell volume (or formula volume); the carrier density is $n = N_{eff}/V_c$. Finally, I put the upper limit to ω rather than ∞, allowing me to plot the integral versus the upper limit ω. If the conductivity spectrum consists of a number of isolated bands, I will be able to see a series of steps in $(m/m^*)N_{eff}(\omega)$ as the upper limit passes through the bands, with flat segments between.

Figure 6.5 shows the result of evaluating Eq. 6.1 over 0–80,000 cm^{-1} (5 mev–10 eV) for the conductivity data in Fig. 6.2. For a simple metal like silver, the free carrier spectral weight is exhausted quickly, in the midinfrared, and the function saturates at the number of conduction electrons/atom (1 in the case of Ag) until the interband transitions set in [45, 54]. Note that the integral saturates at $(m/m^*)N_{eff}(\omega) = 1.04$, a quite satisfactory result [45].

6.2 Other Noble Metals

The reflectance of copper [43, 50, 55, 56] and gold [43, 57, 58] are compared to silver in Fig. 6.6. The three metals share a column in the periodic table. Copper and gold have electronic configurations similar to silver, except that the d levels are not so deep.* These metals also have high infrared reflectance and similar free-carrier optical conductivity and dielectric function spectra. The absorption from the d levels has onset in the red (copper) or

---

* And, of course, the outermost electron is in a $4s$ orbital for copper and in a $6s$ orbital for gold.

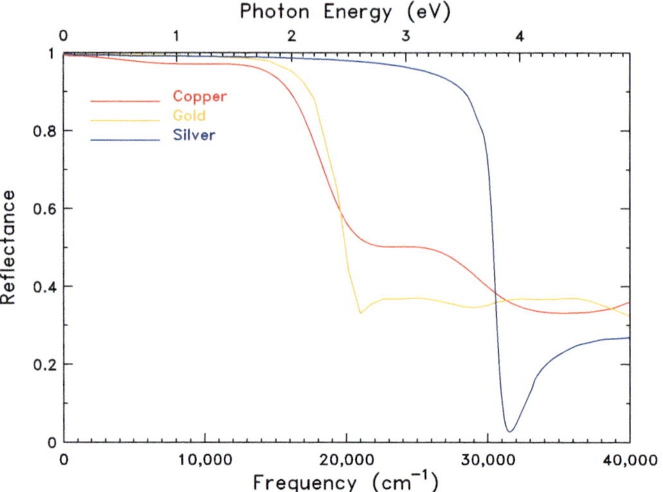

Fig. 6.6 Reflectance of copper [43, 50, 55, 56], gold [43, 57, 58], and silver [43–53].

yellow (gold) regions of the visible, and the reflectance falls from the high metallic value seen in the infrared to a lower value at this onset. This drop is responsible for the characteristic colors of these metals. Moreover, the "plasma minimum" is not so deep in copper and gold as in silver. A Drude–Lorentz model can describe their optical properties, with several oscillators fitting the interband transitions and a Drude model for the free carriers.

6.3 Aluminum

Aluminum is a trivalent atom ($1s^2 2s^2 2p^6 3s^2 3p^1$) and, as a metal, all of the electrons in the $n = 3$ level are free carriers, making the carrier density (and the plasma frequency) high. Now, I know that each Brillouin zone of a crystal can hold two electrons per atom, so the free carriers occupy two zones. This fact complicates the electronic properties somewhat, but a Drude model with three electrons per atom is a good starting point for describing the optical properties.

The far-infrared–ultraviolet reflectance spectrum [43, 50, 59–61] of Al is shown in Fig. 6.7. The main panel shows data from 70 to 50,000 cm^{-1} (0.01–6.2 eV) whereas the inset shows the vacuum ultraviolet and X-ray reflectance up to 6×10^6 cm^{-1} (800 eV). A trivalent metal, Al has its plasma edge deep in the vacuum ultraviolet, around 140,000 cm^{-1}/17 eV. In addition, aluminum has a weak interband transition in the infrared [61, 62].

The optical conductivity in the infrared is Drude-like. The far-infrared value of the conductivity is $\sigma_1(\omega) = 340,000 \ \Omega^{-1}\text{cm}^{-1}$. It rolls off as frequency increases. A fit gives $\omega_p = 139,000$ cm^{-1} (17.3 eV) and $1/\tau = 930$ cm^{-1} ($\tau = 5.7$ ps). Figure 6.8 shows the conductivity of Al, plotted only up to $\sigma_1 = 30,000$ cm^{-1} (about 9% of the low frequency conductivity). Using this scale emphasizes the $1/\omega^2$ fall off when $\omega\tau > 1$ and shows the interband transition at 13,600 cm^{-1} (1.7 eV). This band is reasonably well fit by a Lorentz oscillator.

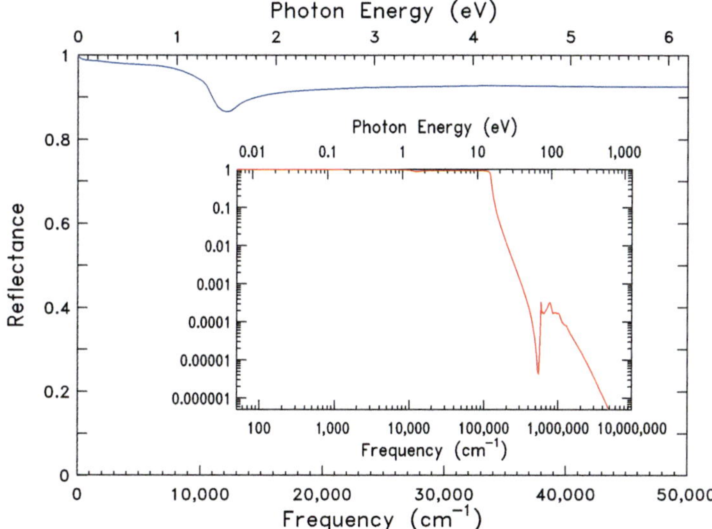

Fig. 6.7 Reflectance of aluminum [43, 50, 59–61]. The main panel shows the reflectance from far infrared to near ultraviolet whereas the inset shows the vacuum-ultraviolet and soft x-ray reflectance.

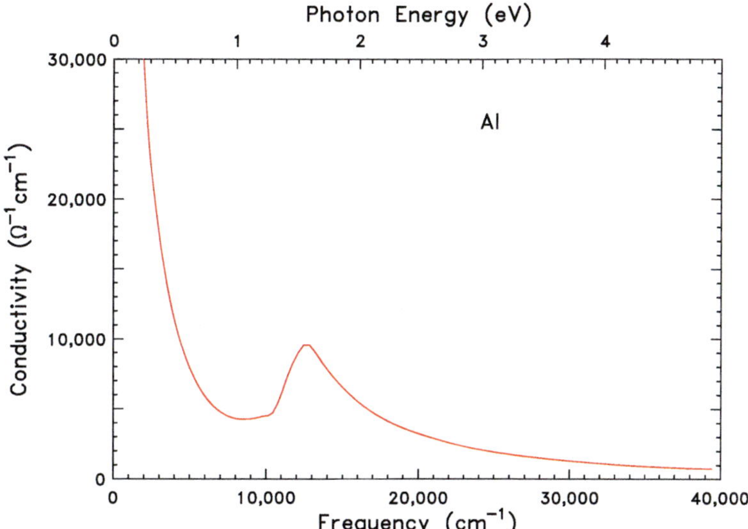

Fig. 6.8 Optical conductivity of aluminum.

6.4 Silicon

Silicon is an elemental semiconductor. The atomic configuration of the 14 electrons in the Si atom is $1s^2 2s^2 2p^6 3s^2 3p^2$ but in the crystal the outermost four electrons are hybridized to form equivalent covalent bonds to four neighboring Si atoms; this tetrahedral coordination gives crystalline Si the same face-centered cubic (fcc) structure as diamond, with two Si atoms per unit cell.

A semiconductor, such as silicon, germanium, GaAs, and many other materials, is to my mind just an insulator with a smallish energy "gap" between the highest occupied and the lowest unoccupied electronic orbitals in the crystal. In Si, this gap at 300 K ("room temperature")* is $E_g = 1.11$ eV $= 8950$ cm^{-1}. A consequence of the smallish gap is that the Boltzmann factor $e^{-E_g/kT}$ governing the probability of a thermal excitation to the unoccupied energy levels is small but not zero. The consequence is that very pure[†] semiconducting silicon has a finite resistivity, better than $\rho = 2,000$ Ω-cm at room temperature. This value is a billion times less conductive than good metals like silver but a trillion times better than really good insulators like Teflon. From the refractive index at long wavelengths [63, 64], I see that the static dielectric constant is positive: $\epsilon_1(0) = 11.66$ at 300 K.

Silicon's value as the basis of electronic materials comes from the ability to control its electrical properties by the incorporation of atoms such as phosphorous, arsenic, or boron in the crystal. Each of these "dopant" atoms adds to the mobile charge density; at maximum dopant atom concentration of about 2×10^{20} cm^{-3} (about 0.4%) one finds about 500 $\mu\Omega^{-1}$cm^{-1} for the resistivity [65]. This heavily doped material has electrical and optical properties of a metal.[‡]

I'll discuss here the optical properties of pure silicon. Even if the impurity concentration is well above that of the most pure materials, it is not high enough to affect the far-infrared–ultraviolet reflectance. Figure 6.9 shows the 300 K reflectance of silicon [43, 66–71]. The main panel shows the reflectance from far infrared to near ultraviolet. The long wavelength reflectance is $R = 0.3$ and is almost constant with frequency, rising slightly as frequency increases. This is the reflectance expected for a Lorentz model well below the resonance. See Fig. 4.14. The reflectance is completely determined by the real part of the dielectric function, consistent with Eq. 4.32a and Fig. 4.10. See also Fig. 4.13.

As frequency increases, the reflectance also increases, rising to a series of maxima in the ultraviolet, at 27,500, 37,000, and (broad) 60,000 cm^{-1} (3.4, 4.6, and 7.4 eV). At still higher frequencies the reflectance falls, eventually following the ω^{-4} behavior of free electrons. (See the inset.)

The optical conductivity is shown in Fig. 6.10 and the real part of the dielectric function in Fig. 6.11. The conductivity is zero (or very small) at low frequencies but rises to give a series of peaks in the ultraviolet as in the reflectance. The maximum value of the conductivity (27,000 Ω^{-1}cm^{-1}) is below the DC conductivity of silver or other good metals (600,000 Ω^{-1}cm^{-1}) but considerably above the maximum in the interband region of silver (3,000 Ω^{-1}cm^{-1}) or aluminum (9,000 Ω^{-1}cm^{-1}). The details of the conductivity spectrum are determined by the electronic band structure of silicon [72–76]. See Chapter 12 for some more discussion and also see the chapters on semiconductors in solid-state

---

* The gap is slightly temperature dependent, increasing to 1.17 eV at $T = 0$.
† Nothing is perfectly pure, but Si crystals can be had with impurity concentrations of 1×10^{12} cm^{-3} or 20 impurity atoms for every trillion silicon atoms.
‡ More common doping levels are in the range 10^{13}–10^{18} cm^{-3}. At the higher values metallic properties already are observed.

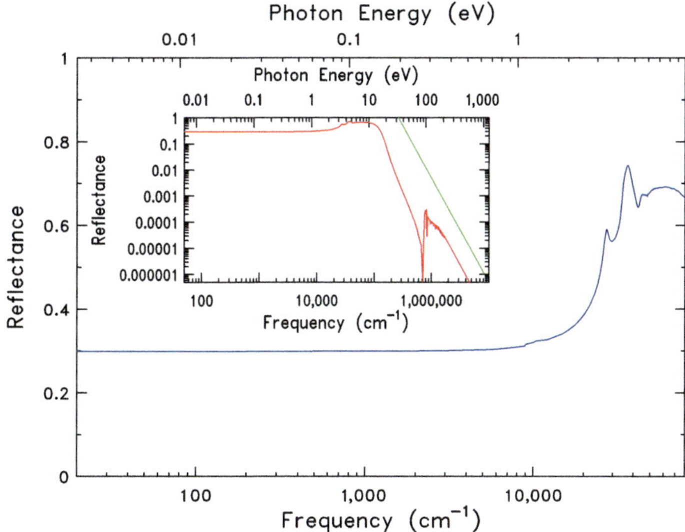

Fig. 6.9 Reflectance of silicon [43, 66–71]. The main panel shows the reflectance from far infrared to near ultraviolet whereas the inset shows the vacuum-ultraviolet and soft x-ray reflectance. The straight line in the inset follows ω^{-4}.

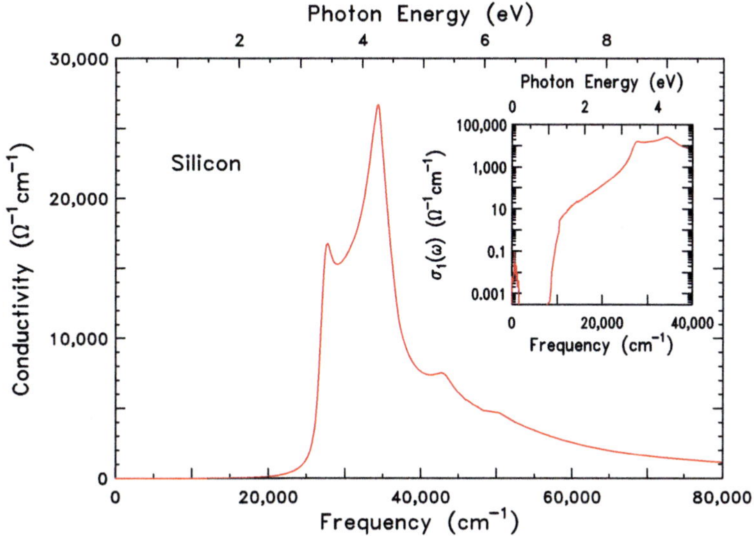

Fig. 6.10 Optical conductivity of silicon.

textbooks [19, 23, 24]. The peaks correspond to high-spectral weight transitions from the ground state (the valence band) to the excited state (the conduction band).

The conductivity spectrum resembles in some ways that of a Lorentz oscillator (Fig. 4.12) – or perhaps as a sum of several overlapping oscillator dielectric functions. It is small at low frequencies, is large in the visible and ultraviolet, and then falls off at higher frequencies. There are differences as well. Most notably, the low frequency edge of

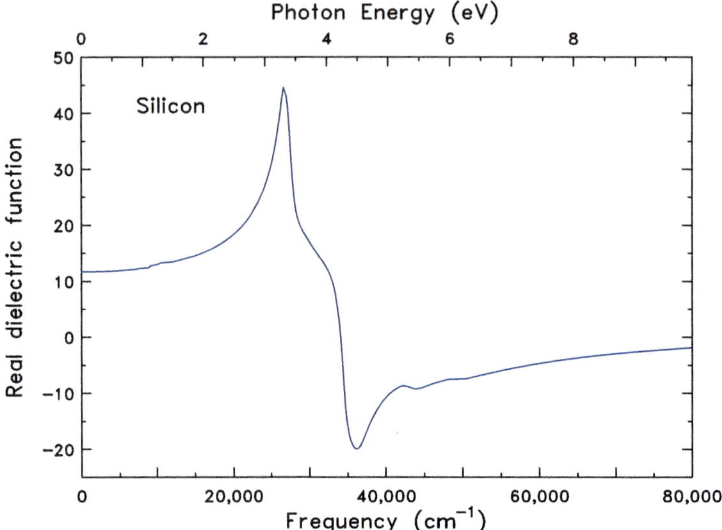

Fig. 6.11 Real part of the dielectric function of silicon.

the spectrum is very steep whereas the high frequency falloff is rather gentle. The Lorentz-model conductivity rises as ω^2 while the conductivity of Si rises much more rapidly. This edge is seen more clearly in the inset, where the conductivity is plotted using a logarithmic frequency scale. The conductivity is offscale low in the midinfrared region and rises very steeply at 9,000 cm^{-1} (1.1 eV). This energy corresponds to the energy band gap of Si.

A weak low-frequency peak in the conductivity can be seen in the inset. This absorption occurs between 400 and 1,200 cm^{-1} (50–150 meV) with considerable structure and is attributed to vibrational overtones, i.e., multiphonon processes. At still lower frequencies a free carrier (Drude) conductivity contribution appears, with the conductivity in the mΩ^{-1}cm^{-1} range.

That the transitions have high spectral weight is confirmed by the real part of the dielectric function (Fig. 6.11). At low frequencies, $\epsilon \approx \epsilon(0) = 11.66$. The dielectric function increases monotonically to a maximum near 27,000 cm^{-1} (3.3 eV), falls with some structure, and becomes negative at 34,000 cm^{-1} (4.2 eV). There is a minimum a bit above this frequency, and $\epsilon(\omega)$ remains negative over a wide range of frequencies, finally crossing zero at 132,000 cm^{-1} (16.4 eV).

The negative region of ϵ_1 is similar to that of a metal below its plasma frequency. The extinction coefficient κ is large and there is a region where $n < 1$. But there is high absorption in the region where ϵ is positive when frequencies are above the energy band gap, on account of the large value of $\sigma_1(\omega)$.

6.5 Sodium Chloride

Sodium chloride (NaCl) is an insulating ionic compound composed of ions of Na$^+$ and Cl$^-$. NaCl is representative of about 20 alkali halides, ranging from LiF to CsI. The alkaline earth halides, like BaF_2, have quite similar optical properties. The cohesive energy of these crystals is due to the Coulomb attraction of the oppositely charged ions. Both ions are closed shell and the lowest electronic interband transitions are in the ultraviolet region of the spectrum. Most (including NaCl) crystallize in the fcc structure; a few (such as CsI) are body-centered cubic (bcc).

The reflectance of NaCl [43] is shown over a wide frequency range in Fig. 6.12. The spectrum has electronic transitions across the band gap in the ultraviolet as well as a vibrational peak (transverse-optical or TO phonon) in the infrared. The maximum reflectance in the infrared reststrahlen region exceeds 90%, a value that is in the realm of metallic reflectance. The minimum near the longitudinal optical (LO) phonon frequency is very close to zero. A Lorentz model fits this feature well, although the structure in the region where the reflectance is falling quickly with increasing frequency indicates some deviations from the simple harmonic oscillator.

The reflectance spectrum in the ultraviolet is more intricate. There are both sharp peaks and broader features. The energy scale is quite high, 50,000–160,000 cm^{-1}(6–20 eV).

These feature are seen better in the optical conductivity spectrum, shown in Fig. 6.13, and in the dielectric function, Fig. 6.14. The lowest-energy feature in the electronic absorption conductivity shows a very strong and narrow peak at 62,500 cm^{-1} (7.75 eV), with additional broad and sharp features at higher energies. The sharp peak leads to the usual derivative-like signature in the dielectric function.

This peak is attributed to an exciton. Excitons are described in Chapter 12, Section 12.10. The exciton is a localized bound state of an electron–hole pair. Because there is a finite binding energy, it requires less energy to produce the exciton than to promote an electron to the conduction band. I can think of the electron in the conduction band and the hole left

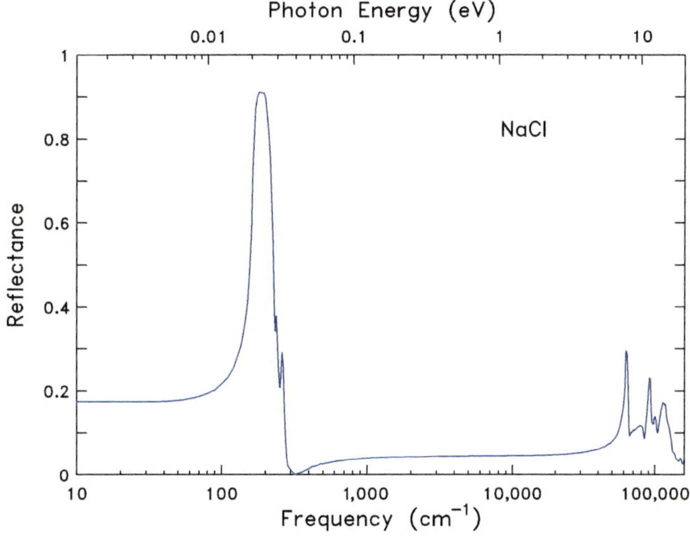

Fig. 6.12 Reflectance of NaCl [43]. The reflectance is shown from far infrared to the ultraviolet.

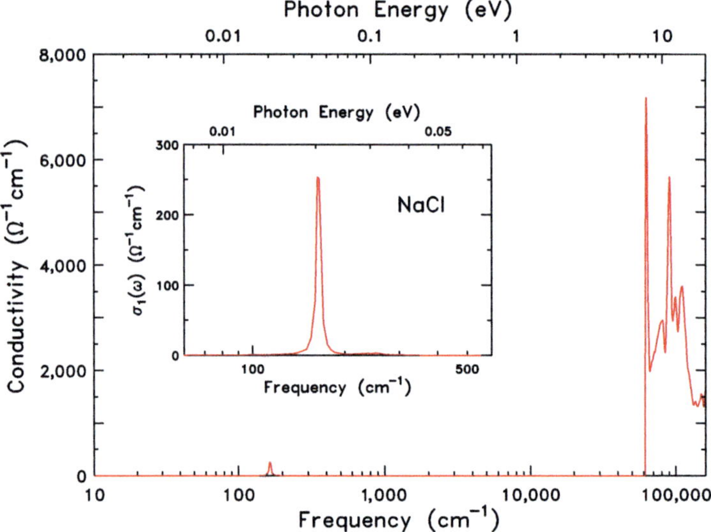

Fig. 6.13 Optical conductivity of NaCl. The inset shows the infrared region on an expanded scale.

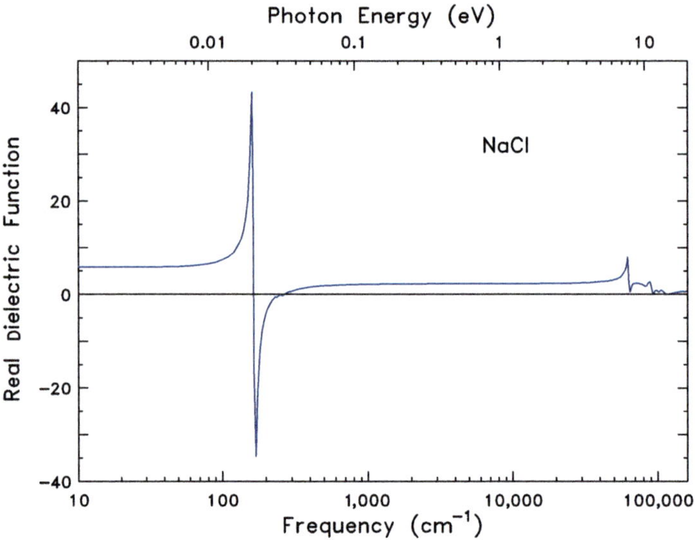

Fig. 6.14 Real part of the dielectric function of NaCl.

behind in the valence band as free carriers. These interband transitions appear at just higher energies than the exciton energy in Fig. 6.13.

The optical phonon feature that was so prominent in the reflectance is a modest feature in the optical conductivity.* The phonon spectrum is shown expanded in the inset to

---

* Had I plotted ϵ_2, then the phonon feature would reach $\epsilon_2 \approx 95$ and would have dwarfed the ultraviolet electronic peaks. $\epsilon_2 = 4\pi\sigma_1/\omega$ and the factor of almost 500 between the frequencies of the phonon and of the electronic transitions has a big effect.

Fig. 6.13; the peak at the transverse optical phonon frequency, 164 cm^{-1}, is quite narrow, with a full width at half maximum of about 5 cm^{-1} and a peak height of about 260 Ω^{-1}cm^{-1}. The phonon is quite evident in the dielectric function, with a strong derivative-shape and a rapid change from a large positive value to a large negative value through the resonance (the TO frequency) and a zero crossing back to positive at a higher frequency (the LO frequency).

The additional dipole moment available by polarizing the lattice is evident from the larger value of ϵ_1 below the resonance than the value above the resonance. The same thing happens across the electronic transitions.

It is fair to ask why a material with a conductivity of 260 Ω^{-1}cm^{-1} has a reflectance not much smaller than that of a metal, for which $\sigma_1 \approx 10^5$ Ω^{-1}cm^{-1}. The answer is that it is not the value of the conductivity that matters. Instead, it is the fact that ϵ_1 is negative, combined with a small value of ϵ_2 over much of the range where $\mathscr{R}$ is large. A negative ϵ_1 (and small ϵ_2) means that the refractive index n is small but the extinction coefficient κ is large, making the reflectance close to unity. See Eq. 3.33 or, maybe better, Eq. 4.17.

6.6 Other Alkali Halides

The optical properties of other alkali halides are very similar to those of NaCl, although of course there are different quantitative values for the parameters. Figure 6.15 shows the optical conductivity spectra of CsI, KBr, NaCl, and LiF and Fig. 6.16 shows their dielectric functions.

The frequency of the TO phonon is lowest for CsI and highest for LiF. This behavior is completely understandable, because the solution to Eq. 5.1 has a resonance at $\omega_0 = \sqrt{K/\mu}$

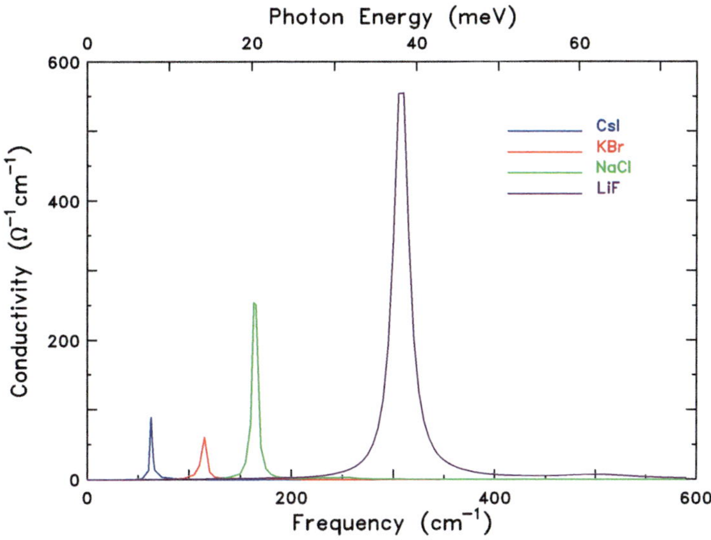

Fig. 6.15 Optical conductivity of CsI, KBr, NaCl, and LiF.

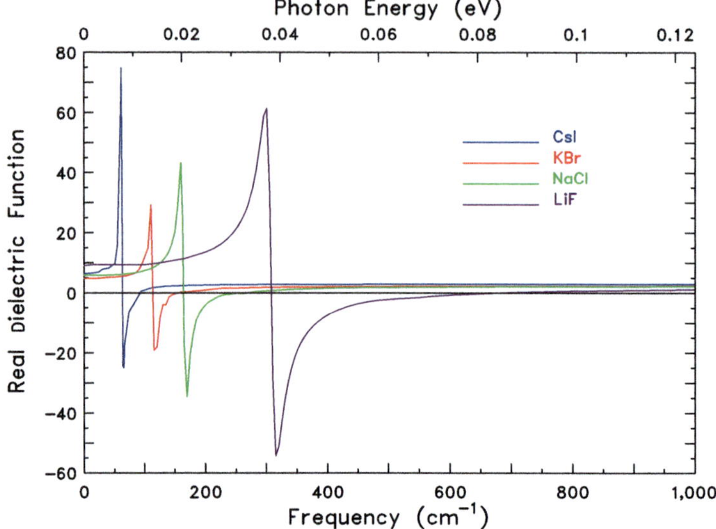

Fig. 6.16 Real part of the dielectric function of CsI, KBr, NaCl, and LiF.

Table 6.1 Transverse and longitudinal frequencies in four alkali halides, and the static and high-frequency dielectric constants.

Material	ω_T cm^{-1}	ω_L cm^{-1}	γ cm^{-1}	ϵ_e	$\epsilon_1(0)$	LST $\epsilon_1(0)$
CsI	62.5	90	3	3.04	5.65	6.30
KBr	115	159	5	2.36	4.58	4.55
NaCl	164	267	5	2.25	5.90	5.96
LiF	307	665	17	1.92	9.04	9.01

with K the spring constant and μ the reduced mass. The higher the reduced mass, the lower the frequency. Of course, the force constant K also plays a role, and CsI is softer than LiF, with a lower sound velocity.

The behavior of ϵ_1 is not monotonic with the mass. The high frequency value is in the order LiF:NaCl:KBr:CsI, consistent with the band gaps of these compounds. (I'm thinking that for ω well below the resonance ω_e, the dielectric function of the Lorentzian is $\epsilon_1(\omega) = 1 + \omega_{pe}^2/\omega_e^2$. I'm ignoring any variation of plasma frequencies—which there is no doubt.) But the low frequency (as $\omega \to 0$) ϵ_1 is in the order KBr:CsI:NaCl:LiF. I look at Eq. 5.26 and know that it must be a very large value for ω_L that makes the LiF have such a large $\epsilon_1(0)$.

It is straightforward to read off the relevant parameters from the data in Figs. 6.15 and 6.16. The peak in σ_1 occurs at ω_T. The zero above the resonance* of ϵ_1 is at ω_L. The

---

* There is another zero right at the resonance as ϵ_1 changes sign from positive to negative.

value of ϵ_1 at the lowest measured frequency gives $\epsilon_1(0)$. The value of ϵ_1 well above the resonance* gives ϵ_e.

I did what I just described for the data in Figs. 6.15 and 6.16. The results are shown in Table 6.1. The last column shows the static dielectric constant calculated from the Lyddane–Sachs–Teller relation, Eq. 5.26. The agreement is poor for CsI but for the other three compounds, the LST static dielectric constant is within 1% of the measured value.

---

* This value is not as easy to pick as the other parameters. The dielectric function is monotonically increasing between resonances. See Fig. 4.15; note that it changes from concave down to concave up between the resonances. This inflection point is a good place to define the high frequency limiting value.

7

Transmission and Reflection

Now that I've gone through semiclassical models of metals and insulators and have looked at data for some common materials, I'll discuss an experiment to learn about their optical properties. I want to know how light of different frequencies propagates in the material. I've discussed the propagation of light in a solid, starting with Section 3.4 in Chapter 3. The propagation is governed by the complex refractive index or, equivalently, the complex dielectric function.

A cartoon of typical experimental setups was shown in Fig. 1.3. To learn about my sample, I need to send light at wavelength λ (or frequency ω) from some optical device (laser, spectrometer, interferometer) to the sample, where some if it will be reflected (see Section 3.6) and some will cross the surface and begin to travel in the solid. If the sample is very thick, so that the light has little chance to make it to the back surface,* the reflectance (Eq. 3.33) is all one can measure. See Chapter 10 for discussion of Kramers–Kronig analysis of this "single-bounce" reflectance.

Here, I want to consider the right-hand side of Fig. 1.3. A significant fraction of the light makes it all the way through the sample, where its intensity may be measured to find the transmittance $\mathscr{T}(\omega)$. Of course some light also is reflected (from both surfaces), so I can move the detector to the other side and measure also the reflectance $\mathscr{R}(\omega)$ of the sample.[†]

7.1 Incoherent Light

I'll begin by considering the case of a slab that is thick compared to the wavelength. Moreover, it has some slight variation in thickness across the beam. The thickness variation is a few wavelengths, so that I can combine intensities and not consider coherence (interference) until later. Figure 7.1 shows a diagram of the paths taken by light as it passes

---

* I know the scale for "very thick": the electric field of light is attenuated as $\mathbf{E} \sim e^{-\omega\kappa x/c}$ at a distance x into the sample; the intensity goes as $I = I_0 e^{-\alpha x}$. So if the thickness d is large compared to $c/\omega\kappa$ or $1/\alpha$, the field or intensity is $e^{-\text{large}}$ when the light reaches the back surface. In this case, the reflected light comes entirely from the first surface. In the other limit, where $d < 1/\alpha$, the light that reaches the back surface will be in part reflected. Some of that light will pass through the front surface and become a component of the reflected field.

[†] The reflectance of the partially transmissive sample contains contributions from both front and back surfaces. In contrast the opaque sample returns only the single-bounce reflectance. The distinction is important for analysis if reflectance using the Kramers–Kronig relations: one must analyze the sample and correct for the extra light coming from the back surface before beginning the Kramers–Kronig analysis.

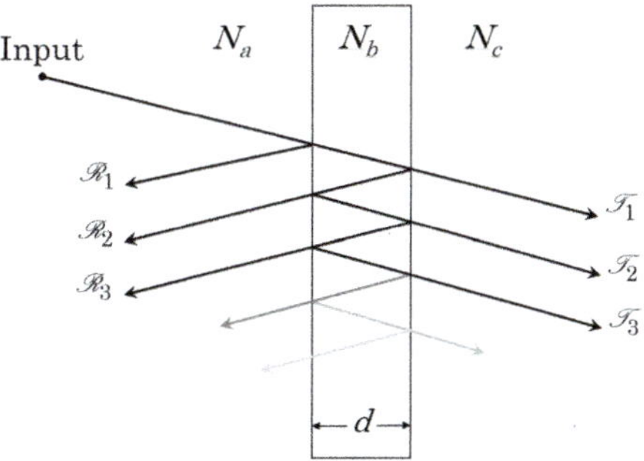

Fig. 7.1 Multiple internal reflections in a slab with refractive index N_b and thickness d set between semi-infinite media with refractive indices N_a and N_c. The intensity in each successive pass is smaller than the one before it, indicated in cartoon form by the fainter lines of the last few rays.

through a slab of thickness d. The light is incident from the left.* At the first surface, some if the incident light is reflected, and some transmitted. This transmitted light passes through and some exits from the rear side. The rest is reflected, and then bounces back and forth between front and back, with a fraction leaking out of each surface on each pass.

7.1.1 The Transmittance

To set up the problem, I'll follow the paths of the individual rays shown in Fig. 7.1. I will take the refractive index of the semi-infinite media on both sides to be the same, $N_a = N_c$. Let the incident ray have unit intensity. The reflected intensity at the front surface (sometimes called the prompt reflectance) is $\mathscr{R}_{ab}$, where the subscripts mean "from a onto b." The amplitude reflectivity r_{ab} is given by Eq. 3.30 and the reflectance is $|r_{ab}|^2$

$$\mathscr{R}_{ab} = \left| \frac{N_a - N_b}{N_a + N_b} \right|^2 = \frac{(n_a - n_b)^2 + (\kappa_a - \kappa_b)^2}{(n_a + n_b)^2 + (\kappa_a + \kappa_b)^2}. \tag{7.1}$$

The intensity just inside the slab at the front surface is $1 - \mathscr{R}_{ab}$ because there is no absorption at the interface. There is, however, absorption as the light travels through the slab, following Eq. 3.25, $I = I_0 e^{-\alpha x}$. When the light reaches the rear surface, some is reflected. The fraction reflected is $\mathscr{R}_{bc} = \mathscr{R}_{ab}$. Some, $= 1 - \mathscr{R}_{ab}$, is transmitted. Thus, the intensity of the first ray exiting from the slab is

$$\mathscr{T}_1 = (1 - \mathscr{R}_{ab})e^{-\alpha x}(1 - \mathscr{R}_{ab}),$$

consisting of reflection losses at both surfaces and bulk absorption.

I will now follow the other rays that lead to transmission. The second ray travels three times through the sample and has intensity

$$\mathscr{T}_2 = (1 - \mathscr{R}_{ab})e^{-\alpha x}\mathscr{R}_{ab}e^{-\alpha x}\mathscr{R}_{ab}e^{-\alpha x}(1 - \mathscr{R}_{ab}).$$

---

* The diagram shows the rays inside the slab at an angle to the slab normal; I'll work out the case for normal incidence. I drew the oblique incidence to allow me to follow the various partial waves as they travel back and forth in the sample.

It is transmitted at the front, passes through the slab, is reflected, passes through, is reflected, passes through, and is transmitted. The third ray is

$$\mathscr{T}_3 = (1 - \mathscr{R}_{ab})e^{-\alpha x}\mathscr{R}_{ab}e^{-\alpha x}\mathscr{R}_{ab}e^{-\alpha x}\mathscr{R}_{ab}e^{-\alpha x}\mathscr{R}_{ab}e^{-\alpha x}(1 - \mathscr{R}_{ab}).$$

It is transmitted at the front, passes through the slab, is reflected, passes through, is reflected, passes through, is reflected, passes through, is reflected, passes through, and is transmitted.

Now note that $\mathscr{T}_2$ and $\mathscr{T}_3$ both contain all the factors in $\mathscr{T}_1$ multiplied by $\mathscr{R}_{ab}^2 e^{-2\alpha x}$ and $\mathscr{R}_{ab}^4 e^{-4\alpha x}$ respectively. The next term has of course the 6th power of $\mathscr{R}_{ab}e^{-\alpha x}$, and the total transmittance is an infinite sum of the partial rays. The transmittance is then

$$\mathscr{T} = (1 - \mathscr{R}_{ab})^2 e^{-\alpha x}[1 + \mathscr{R}_{ab}^2 e^{-2\alpha x} + \mathscr{R}_{ab}^4 e^{-4\alpha x} + \cdots].$$

Now, I remember the expansion of $1/(1 - \varepsilon) = 1 + \varepsilon + \varepsilon^2 + \cdots$, which converges as long as $\varepsilon < 1$. Clearly $\mathscr{R}_{ab} < 1$ as is $e^{-2\alpha x}$, so the series is convergent. I sum it to get

$$\mathscr{T} = \frac{(1 - \mathscr{R}_{ab})^2 e^{-\alpha x}}{1 - \mathscr{R}_{ab}^2 e^{-2\alpha x}}. \tag{7.2}$$

7.1.2 The Reflectance

I can use the same approach for the reflected rays:

$$\mathscr{R}_1 = \mathscr{R}_{ab},$$

and

$$\mathscr{R}_2 = (1 - \mathscr{R}_{ab})e^{-\alpha x}\mathscr{R}_{ab}e^{-\alpha x}(1 - \mathscr{R}_{ab}),$$

and

$$\mathscr{R}_3 = (1 - \mathscr{R}_{ab})e^{-\alpha x}\mathscr{R}_{ab}e^{-\alpha x}\mathscr{R}_{ab}e^{-\alpha x}\mathscr{R}_{ab}e^{-\alpha x}(1 - \mathscr{R}_{ab}).$$

Now I keep $\mathscr{R}_1$ separate and add all the partial rays for $\mathscr{R}_2, \mathscr{R}_3\ldots\mathscr{R}_\infty$ together, factor out common terms, and find an infinite series almost the same as in $\mathscr{T}$. The reflectance is:

$$\mathscr{R} = \mathscr{R}_{ab} + \mathscr{R}_{ab}\frac{(1 - \mathscr{R}_{ab})^2 e^{-2\alpha x}}{1 - \mathscr{R}_{ab}^2 e^{-2\alpha x}}. \tag{7.3}$$

The second term is positive, so that the slab reflectance is larger than the single-bounce reflectance $\mathscr{R}_{ab}$.

7.1.3 Some Approximations

A common case is the one where $\kappa \ll 1$. Why should I say that? Well, the absorption coefficient is $\alpha = 2\omega\kappa/c = 4\pi\kappa/\lambda$. Suppose the slab thickness is $d = 10\lambda$. This is about as thin as it can be and still be considered "thick." In this case, $\alpha d \approx 120\kappa$, where I've used the old fashioned approximation $\pi \approx 3$. The transmittance is (neglecting reflection at the surfaces)

$$\mathcal{T} \approx e^{-120\kappa}.$$

Now I want to get light through the slab, at least at the 1% level. So I take $\mathcal{T} = 0.01$ and find $-120\kappa = \ln(0.01)$. Thus $\kappa = 0.037$, and is indeed rather small.

The advantage of having small κ is that I can neglect it* in Eq. 7.1. Assuming vacuum before and behind the slab, I take $N_a = N_c = 1$ and write $N_b = n$, making $\mathcal{R}_{ab} = (n-1)^2/(n+1)^2$. Substituting this expression into Eq. 7.2 and doing a little algebra, I find

$$\mathcal{T} = \frac{16n^2}{(n+1)^4 e^{\alpha d} - (n-1)^4 e^{-\alpha d}}. \tag{7.4}$$

Eq. 7.4 simplifies further if instead of having small κ, I have zero κ. Then $\alpha = 0$ and the transmittance simplifies to

$$\mathcal{T} = \frac{2n}{n^2 + 1}. \tag{7.5}$$

Eq. 7.5 has the correct limit when $n = 1$. Note that it is different from the transmission across the first surface, $\mathcal{T}_1 = 1 - \mathcal{R}_1 = 4n/(n+1)^2$, which represents the intensity just inside the layer if the back surface is not there. More importantly, Eq. 7.5 is larger than $\mathcal{T}_1\mathcal{T}_2 = \mathcal{T}_1^2 = 16n^2/(n+1)^4$, the intensity of the ray transmitted without reflections through the slab. The multiple internal reflections augment the transmitted intensity by about 7% if $n = 3$ and by 80% if $n = 10$. They should not be neglected.

7.2 Coherent Light

If the material is thin, or if it is thick but the two surfaces are parallel to within a fraction of a wavelength, then the multiple internal reflections will add coherently; the consequences are that there are interference patterns (or standing wave effects) in the transmitted and reflected light [8, 77]. Layers with thicknesses corresponding to the wavelength will usually show these effects; they are responsible for the iridescent colors in an oil film on water [78], for the light scattered off of the wings of certain insects [79], and for many other examples in nature. But it is not necessary for the layer to be comparable to the wavelength. The arm cavities of LIGO [80] are 4 km in length whereas the laser wavelength is 1 μm; nevertheless, the mirrors are held so parallel by a complex and effective control system that there is constructive interference of the light bouncing back and forth between the mirrors even though they are separated by 4×10^9 wavelengths.

In the coherent case I must add the amplitudes of the fields[†] and, after finding the field resulting from superposing the components, I calculate the intensity or the Poynting vector

---

[*]   Actually, I will be neglecting κ^2, which is about 0.0014 in this example.
[†]   One typically adds the electric field and lets the magnetic field come along for the ride.

in order to find out about the energy.* One must keep track of the phase of the fields as the light propagates; the phase advances as $e^{iq\ell}$ for travel along the path ℓ. There are also phase changes on reflection or transmission across an interface.

7.2.1 The Geometry

I'll start by defining the geometry, illustrated in Fig. 7.2. This diagram is not much different from Fig. 7.1; I show the amplitude (not the intensity) of reflected and transmitted partial waves. The layer has thickness d and refractive index N_b. It is sandwiched between media with indices N_a and N_c. I'll take the origin of the x-axis (along which the wave travels) at the front surface; the back is at $x = d$. A plane wave of unit amplitude is incident at normal incidence from the right medium: $\mathbf{E}_{\text{inc}} = \hat{e}e^{iq_a x}$ with $q_a = N_a\omega/c$. As long as the media are isotropic, I can ignore the field direction. I'll evaluate the fields at $x = 0$ and $x = d$, either inside or outside the layer. The first surface has reflection coefficient r_{ab}, as derived in Chapter 3, Eq. 3.30,

$$r_{ab} = \frac{N_a - N_b}{N_a + N_b},$$

and transmission coefficient t_{ab}, Eq. 3.31,

$$t_{ab} = \frac{2N_a}{N_a + N_b}.$$

There are equivalent equations for r_{bc}, r_{ba}, t_{bc}, and t_{ba}. In each of these, subscripts like ab mean "incident from a into b."

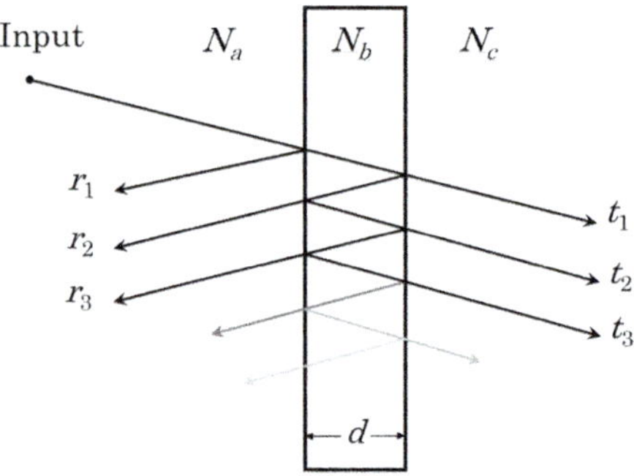

Amplitude transmission and reflection of a layer with thickness d and refractive index N_b between media with indices N_a and N_c.

---

* Actually, one should *always* combine the fields coherently. The incoherent case arises because the light is in a beam wide compared to the wavelength. If the sample thickness varies over this beam dimension, then the phase differences among the combined waves will vary in the same way, with some parts adding constructively and some destructively. An average over these phases gives the resultant transmittance and reflectance.

7.2.2 Transmission Coefficient

Now, I will follow the various partial rays to find their sum, the transmitted field. I start in medium a, at the surface ($x = 0$). For ray t_1 the light is transmitted across the ab interface, travels a distance d, and is transmitted across the bc interface. The field is then

$$t_1 = t_{ab}e^{iq_bd}t_{bc}.$$

Some of the light is reflected at the back surface and travels to the front surface where some of it is reflected and travels to the back surface where some if it is transmitted. Starting with the incident field, I find that t_2 is

$$t_2 = t_{ab}e^{iq_bd}r_{bc}e^{iq_bd}r_{ba}e^{iq_bd}t_{bc} = t_{ab}r_{bc}r_{ba}t_{bc}e^{3iq_bd}$$
$$= t_1 r_{bc}r_{ba}e^{2iq_bd}.$$

The third partial ray is

$$t_3 = t_1(r_{bc}r_{ba}e^{2iq_bd})^2,$$

because the ray has made five passes across medium b, internally reflecting twice at each surface. Thus, just as in the discussion leading to Eq. 7.2, the multiple internal reflections lead to an infinite series, which I can sum to get t, the amplitude transmission coefficient of the layer.

$$t = \frac{t_{ab}t_{bc}e^{iq_bd}}{1 - r_{bc}r_{ba}e^{2iq_bd}}. \tag{7.6}$$

7.2.3 Reflection Coefficient

An identical analysis (see Eq. 7.3) gives the amplitude reflection coefficient of the layer, r

$$r = r_{ab} + \frac{t_{ab}r_{bc}t_{ba}e^{iq_bd}}{1 - r_{bc}r_{ba}e^{2iq_bd}}. \tag{7.7}$$

7.2.4 Intensities

The transmitted and reflected intensities (transmittance and reflectance) are just the absolute value squared of these amplitude coefficients:

$$\mathcal{T} = tt^* \quad \text{and} \quad \mathcal{R} = rr^*. \tag{7.8}$$

The phase term in the numerator of the transmitted amplitude disappears when multiplied by its complex conjugate,* but the denominator contains terms like $\cos(2n\frac{\omega}{c}d)$. The transmittance therefore has oscillations as a function of the frequency or the thickness, with period $2n\omega d/c = 2\pi(2d/\lambda_{in})$. Here $2d$ is the round-trip path length and λ_{in} is the wavelength of the light in the slab.

---

* The exponential function containing κ does not disappear of course, and the quantity $e^{-\alpha d}$ appears in the numerator of $\mathcal{T}$.

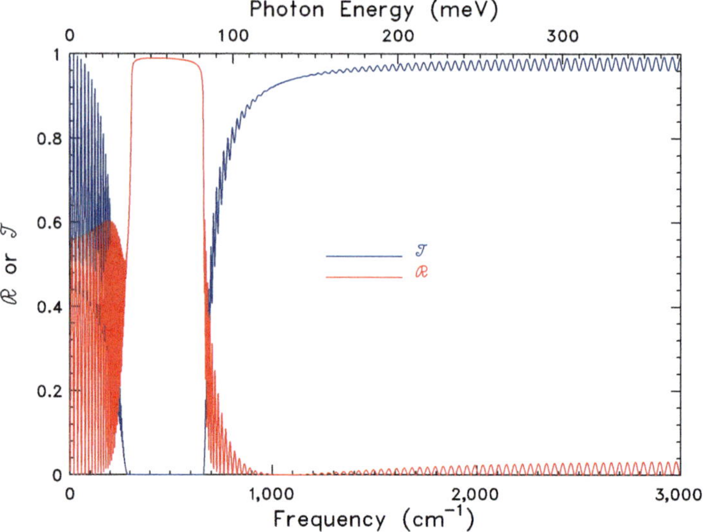

Fig. 7.3   Transmittance and reflectance of a slab of "LiF" using a linear frequency scale.

7.2.5 An Example: An Ionic Insulator

As an illustration Figs. 7.3 and 7.4 show the calculated transmittance of a material with a strong harmonic oscillator, such as an optical phonon, in its dielectric function. The media on either side of the material are both taken to have $N_a = N_c = 1$. The parameters for layer b are chosen to resemble those of LiF. The dielectric function is the product dielectric function of Eq. 5.25,

$$\epsilon = \epsilon_e \frac{\omega_L^2 - \omega^2 - i\omega\gamma_L}{\omega_T^2 - \omega^2 - i\omega\gamma_T}$$

with $\omega_L = 662$ cm^{-1}, $\omega_T = 307$ cm^{-1}, $\gamma_L = \gamma_T = 2$ cm^{-1}, $\epsilon_c = 1.5$ and $d = 0.01$ cm.

The oscillations in $\mathcal{T}$ and $\mathcal{R}$ are obvious. They occur below ω_T and above ω_L. Because the refractive index is larger below ω_T than above ω_L[*] the period is smaller and the amplitude larger at low frequencies than at high. In the transparent regions the maxima in reflectance correspond to minima in transmittance.[†]

It is not obvious in the figure, but the period of the oscillations becomes shorter as the frequency is increased from zero to ω_T because the refractive index is increasing as the resonance is approached. The transmittance is zero and reflectance high in the reststrahlen region where $\epsilon_1(\omega) < 0$. The dielectric function becomes positive at $\omega_L = 662$ cm^{-1}, but the reflectance has its minimum at about 1,000 cm^{-1}, the frequency where $n = 1$. Because the damping in the model is low, one can see interference oscillations in the transmittance and reflectance at frequencies below the reflectance minimum.

---

[*] $n \approx 2.6$ at low frequencies and ≈ 1.2 at high frequencies.
[†] Energy is conserved; some is absorbed at frequencies near ω_T.

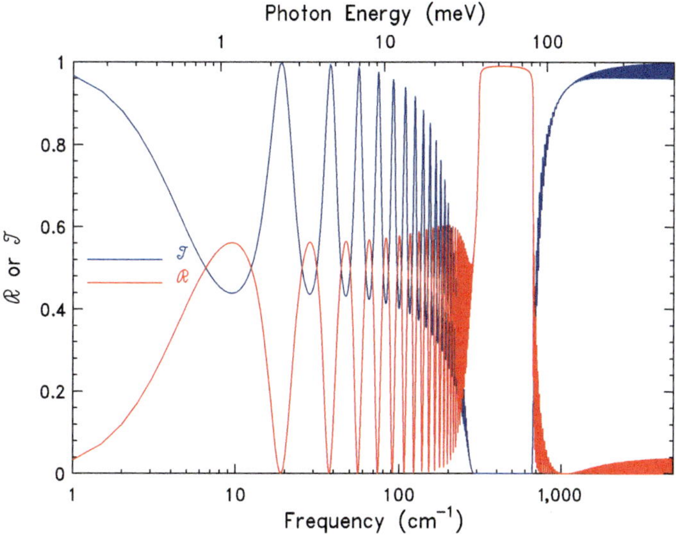

Fig. 7.4 Transmittance and reflectance of a slab of "LiF" using a logarithmic frequency scale.

7.2.6 A Second Example: A Metal Film

As a second example, I'll calculate the transmittance of a thin metal film. The film will have to be very thin for any light to get through. It is likely that the film is thin enough that electron scattering from the surface becomes important, decreasing the conductivity. In the calculations here, I will ignore this likelihood; the optical properties will be those shown in Figs. 4.1 (conductivity) and 4.3 (dielectric function). The parameters for the Drude model are given in Chapter 4, Section 4.2.7; they are meant to mimic silver metal at room temperature. The results of the calculation are shown in Fig. 7.5 for thicknesses of 10, 30, and 100 Å.

The transmittance is very low at low frequencies and decreases as the square of the film thickness. The flat spectrum continues until $\omega \approx 1/\tau$. Above this frequency the transmittance increases, following approximately $\mathcal{T} \sim \omega^2$. At high frequencies the transmittance approaches 100%. The thinner films reach high transmittance at lower frequencies than the thicker films do.

7.2.7 Layer Transmittance and Reflectance

It is certainly possible to find analytic equations for the transmittance and reflectance – what one measures – from Eqs. 7.6, 7.7, and 7.8 but it may not be profitable to slog through the algebra involved. Instead, let me make a number of simplifications. First, I'll take $N_a = N_c = 1$, i.e., put vacuum on both sides of the layer.* I can then write $N_b = N$, without subscript. Moreover, by symmetry $r_{ba} = r_{bc}$ and $t_{ba} = t_{bc}$. Note that these

---

* This simplification is not really justified: many samples consist of a thin film on a thick substrate. That case is additionally complicated in practice by the fact that the film is thin and hence coherent whereas the substrate is thick, and incoherent.

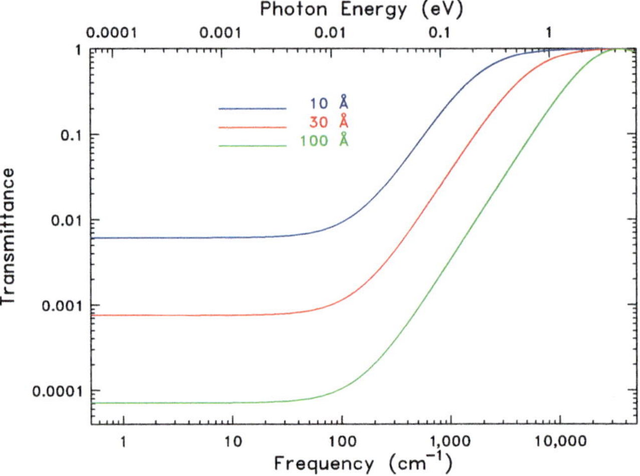

Fig. 7.5   Transmittance of a metal meant to represent silver. The calculated transmittance curves for three thicknesses are shown versus frequency. Both axes use a logarithmic scale.

coefficients represent what happens when the light is incident on the vacuum from within the material. Then I have from Eqs. 3.30 and 3.31

$$r_{ba} = \frac{N-1}{N+1} = -r_{ab}$$

$$t_{ba} = \frac{2N}{N+1} = 1 + r_{ba}$$

$$t_{ab} = \frac{2}{N+1} = 1 - r_{ba}$$

and Eq. 7.6 becomes (after substituting and multiplying top and bottom by $(N+1)^2 e^{-iqd}$),

$$t = \frac{4N}{(N+1)^2 e^{-iqd} - (N-1)^2 e^{iqd}} \tag{7.9}$$

whereas Eq. 7.7 becomes (after substituting, multiplying top and bottom by $(N+1)^2$, combining both terms on a common denominator, doing some algebra, and then multiplying top and bottom by e^{-iqd},

$$r = \frac{2i(1-N^2)\sin(qd)}{(N+1)^2 e^{-iqd} - (N-1)^2 e^{iqd}}. \tag{7.10}$$

7.2.8 Nonabsorbing Layer

If now the layer is nonabsorbing ($\kappa = 0$), the transmittance, tt^*, becomes (starting with Eq. 7.9, manipulating the denominator, and making use of the double-angle formula)

$$\mathscr{T} = \frac{4n^2}{4n^2 + (n^2 - 1)^2 \sin^2(n\omega d/c)}.$$

This Airy function has minimum transmittance when $\sin^2(n\omega d/c) = 1$

$$\mathscr{T}_{\min} = \left(\frac{2n}{n^2+1}\right)^2$$

and maximum transmittance when $\sin^2(n\omega d/c) = 0$

$$\mathscr{T}_{\max} = 1.$$

The reflectance is

$$\mathscr{R} = \frac{(n^2-1)^2 \sin^2(n\omega d/c)}{4n^2 + (n^2-1)^2 \sin^2(n\omega d/c)}$$

with maximum

$$\mathscr{R}_{\max} = \left(\frac{n^2-1}{n^2+1}\right)^2$$

and minimum

$$\mathscr{R}_{\min} = 0.$$

The medium is nonabsorbing; consequently energy conservation tells me that $\mathscr{T} + \mathscr{R} = 1$. The period of the oscillations in $\mathscr{T}$ and $\mathscr{R}$ is the distance between the zeros of $\sin^2(n\omega d/c)$, which occur when $n\omega d/c = J\pi$ where J is an integer. Thus

$$\Delta\omega = \frac{\pi c}{nd}. \tag{7.11}$$

Here the frequency is in radians/s, $\omega = 2\pi f$, with f the frequency in Hz. Eq. 7.11 then becomes $\Delta f = c/(2nd)$. Finally, if the frequency is in cm^{-1}, $\nu = f/c = 1/\lambda$, and

$$\Delta\nu = \frac{1}{2nd}, \tag{7.12}$$

with d in cm. Each period corresponds to adding one more cycle to the standing wave inside the layer.

Observation of these multiple-internal-reflection in a measurement allows for a simple, if coarse, determination of the refractive index. One picks out the transmittance maxima and calculates

$$n(\nu) = \frac{1}{2\Delta\nu d}, \tag{7.13}$$

where $\Delta\nu$ is the spacing between two adjacent maxima and ν is the average of their frequencies. The resolution can be improved by a factor of two by carrying out the same procedure for the minima.

Figure 7.6 shows a calculation of the transmittance of a nonabsorbing layer with variation of the refractive index (top to bottom) from 1.4, 2, 4, 8, 16, to 32. The thicknesses were adjusted to keep nd constant: 70.7, 50, 25, 12.5, 6.25, and 3.125 µm.*

---

* I note that a refractive index of 32 (dielectric constant of 1024) is uncommon. But the same effect – a high reflectance at the interfaces – can be achieved by applying multilayer dielectric coatings to the surfaces of a low index spacer.

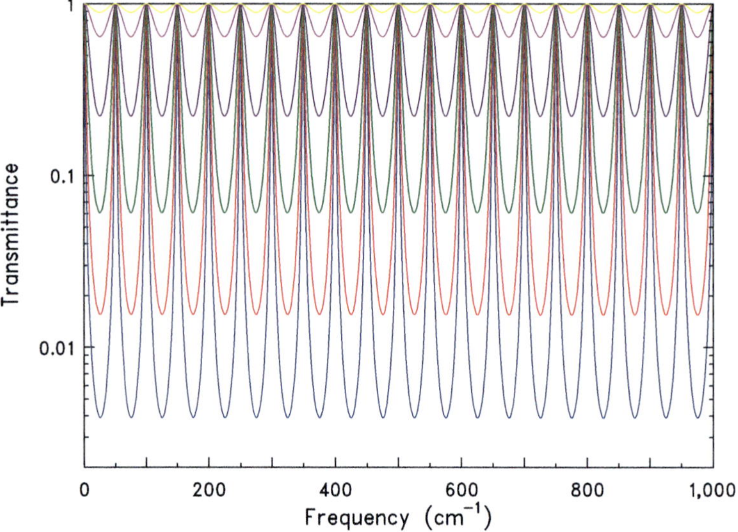

Fig. 7.6 Transmittance of a layer having a refractive index (top to bottom) of 1.4, 2, 4, 8, 16, and 32 and thickness of 70.7, 50, 25, 12.5, 6.25, and 3.125 μm respectively.

Note that the maximum transmittance is independent of the refractive index or reflection coefficient of the vacuum-material interface; it is always unity* when the partial waves for the multiply-internally-reflected light all add constructively (see Fig. 7.2.). As the reflection coefficient of the surfaces increase, the minima between the 100% transmittance peaks become deeper and deeper, and the shape of the transmittance spectra, which is close to sinusoidal for low reflection surfaces, becomes more and more a set of narrow peaks between low transmittance regions. This spectrum is the Airy function transmittance of a Fabry–Perot interferometer.

I'll not plot the reflectance, but I can imagine it by looking at Fig. 7.6 because I know (energy conservation!) that $\mathscr{R} = 1 - \mathscr{T}$. So for $n = 1.4$, the reflectance would oscillate between zero and about 10%; the reflectance maxima correspond to transmittance minima. For $n = 32$, the reflectance exceeds 99% over much of the spectrum, punctuated by narrow dips to zero when the partial waves inside interfere constructively.

The material with $n = 32$ has a single-surface reflectance $(n - 1)^2/(n + 1)^2 = 0.88$. So it is reasonable to ask: how can a slab with two such highly reflecting surfaces in series have transmittance equal to 1 and zero reflectance? The answer of course is "For most frequencies it does not." But in the special case where all the partial waves interfere constructively, the energy stored in the slab grows enormously. Some of this energy leaks out of the front surface with a phase that is 180° out of phase with the light promptly reflected; the two sum to the zero reflectance. Some light also leaks out of the back and energy conservation tells me that this leakage is equal to the incident intensity, making $\mathscr{T} = 1$.

* Assuming no absorption!

7.3 The Matrix Method

The use of "ABCD" matrices to calculate wave propagation appears in a variety of guises, such as to follow Gaussian laser beams through lenses, mirrors, and other optical components [81, 82]. Here, I'll discuss how it is done for multilayer thin films [8, 77, 83]. I will do this for normal incidence; the extension to oblique incidence is relatively straightforward.

The method resembles the way one solves the barrier penetration problem in quantum mechanics. Instead of following each partial ray as I did in Fig. 7.2 and in the analysis leading to Eqs. 7.6 and 7.7, one considers a forward-going wave and a backwards-going wave in the barrier and in the medium from which the waves come. Matching boundary conditions then gives the solution. So I will (conceptually at least) combine all the forward-going waves in Fig. 7.2 into one and will do the same thing with all the backwards-going waves.

The geometry is shown in Fig. 7.7. There are L layers in the film, each with refractive index N_ℓ and thickness d_ℓ. The total number of media is then $L + 2$ because the light enters the film from a semi-infinite space on the left and leaves into a semi-infinite space on the right. The total number of interfaces is $L + 1$, and these are numbered $0 \ldots L$ according to the medium on the left of the interface. Thus, interface 2 (between medium 2 and 3) is at $x = d_1 + d_2$ when the origin is in interface 0 and the light is propagating along the x axis.

The fields are given by Eqs. 3.11 and 3.12, which I rewrite below:

$$\mathbf{E} = \mathbf{E}_0 e^{iN\frac{\omega}{c}\hat{q}\cdot\mathbf{r}} = \begin{cases} \mathbf{E}_0 e^{in\frac{\omega}{c}x} e^{-\kappa\frac{\omega}{c}x} & \equiv E_f \quad \text{(forward)}, \\ \mathbf{E}_0 e^{-in\frac{\omega}{c}x} e^{+\kappa\frac{\omega}{c}x} & \equiv E_b \quad \text{(backward)}, \end{cases} \tag{7.14}$$

where $\hat{q} = \hat{x}$ for forward and $\hat{q} = -\hat{x}$ for backward. I have incorporated the $e^{-i\omega t}$ time dependence into $\mathbf{E}_0$ (or evaluated everything at $t = 0$).

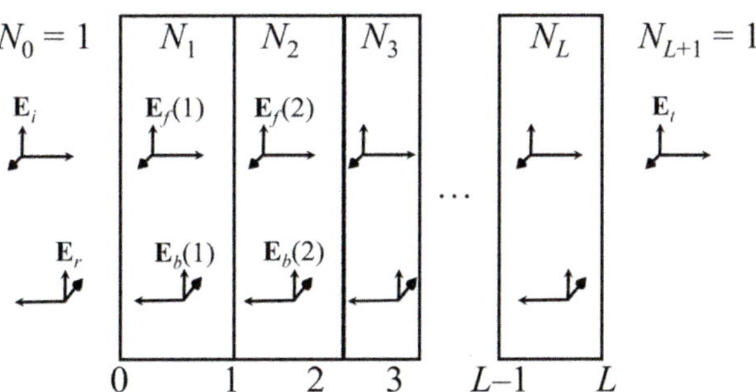

A multilayer optical system. There are L layers in the film, each with complex refractive index N_ℓ. The triplets of arrows indicate the **q**, **E**, and **H** vectors of forward-going light (to the right) and backward going light (to the left). Labels are on only some of the electric field vectors. All electric field vectors are along $+\hat{y}$ although in reality there will be phase reversals at some (but probably not all) reflections.

Some of the light is reflected and some transmitted at each interface. The coefficients are given in Eqs. 3.30 and 3.31. I rewrite these here, substituting the interface number for the letters used previously:

$$r_{\ell,\ell+1} = \frac{N_\ell - N_{\ell+1}}{N_\ell + N_{\ell+1}} = -r_{\ell+1,\ell},$$

and

$$t_{\ell,\ell+1} = \frac{2N_\ell}{N_\ell + N_{\ell+1}} \qquad t_{\ell+1,\ell} = \frac{2N_{\ell+1}}{N_\ell + N_{\ell+1}},$$

where a subscript like $\ell, \ell + 1$ means "incident from ℓ into $\ell + 1$."

Now, let me focus on one interface, say interface 1, which separates medium 1 from 2. Light approaches the interface as forward-going in medium 1, which I will write as $E_f(1)$, and as backward-going in medium 2, $E_b(2)$. Some of each incoming beam is transmitted and some reflected, so there are four outgoing waves, which superimpose into two fields for light leaving the interface:

$$E_f(2) = t_{1,2}E_f(1) + r_{2,1}E_b(2),$$
$$E_b(1) = r_{1,2}E_f(1) + t_{2,1}E_b(2).$$

As the light travels from one side of a single layer to the other it picks up a phase (and a change of amplitude) according to Eq. 7.14, which I will write as

$$E_f(x = d) = E_f(x = 0)e^{iN\frac{\omega}{c}d} \equiv E_f(x = 0)e^{i\phi},$$

and

$$E_b(x = 0) = E_f(x = d)e^{-iN\frac{\omega}{c}d} \equiv E_f(x = d)e^{-i\phi},$$

where for each layer, $\phi_\ell = N_\ell(\omega/c)d_\ell$.

So now I'll combine both of these factors into equations for the ℓth interface:

$$E_f(\ell + 1) = t_{\ell,\ell+1}E_f(\ell)e^{i\phi_\ell} + r_{\ell+1,\ell}E_b(\ell + 1),$$
$$E_b(\ell)e^{-i\phi_\ell} = r_{\ell,\ell+1}E_f(\ell)e^{i\phi_\ell} + t_{\ell+1,\ell}E_b(\ell + 1). \tag{7.15}$$

The left sides of these equations give the amplitudes of the waves leaving interface ℓ in terms of the waves approaching it, including the phase. If I use $r_{\ell,\ell+1} = -r_{\ell+1,\ell}$ and $t_{\ell,\ell+1}t_{\ell+1,\ell} = 1 - r_{\ell,\ell+1}^2$ which I can prove from Eq. 3.30 and 3.31, I can write Eq. 7.15 in matrix form,

$$\begin{pmatrix} E_f(\ell) \\ E_b(\ell) \end{pmatrix} = \mathbf{M}_\ell \begin{pmatrix} E_f(\ell + 1) \\ E_b(\ell + 1) \end{pmatrix}, \tag{7.16}$$

where the transfer matrix $\mathbf{M}_\ell$ is a 2×2 matrix that takes me from one layer to another. In turn, $\mathbf{M}_\ell$ is conveniently written as a product of a propagation matrix $\mathbf{P}_\ell$ and an interface matrix $\mathbf{I}_\ell$,

$$\mathbf{M}_\ell = \mathbf{P}_\ell \mathbf{I}_\ell, \tag{7.17}$$

where

$$\mathbf{P}_\ell = \begin{pmatrix} e^{-i\phi_l} & 0 \\ 0 & e^{i\phi_l} \end{pmatrix},$$ (7.18)

and

$$\mathbf{I}_\ell = \frac{1}{t_{\ell,\ell+1}} \begin{pmatrix} 1 & r_{\ell,\ell+1} \\ r_{\ell,\ell+1} & 1 \end{pmatrix}.$$ (7.19)

The hard part is done. To calculate the transfer matrix for the entire structure, I just multiply the matrices for each interface and layer, starting with interface 0 and ending with interface L,

$$\mathbf{M}_{0-L} = \mathbf{I}_0 \mathbf{P}_1 \mathbf{I}_1 \mathbf{P}_2 \mathbf{I}_2 \ldots \mathbf{P}_L \mathbf{I}_L.$$ (7.20)

The forward and backward waves in medium 0 are respectively the incident and reflected fields. (The incident field has unit amplitude.) The forward-going wave in medium $L+1$ is the transmitted light; there is no backward-going wave. Thus the entire multilayer structure satisfies an equation of the form of Eq. 7.16

$$\begin{pmatrix} 1 \\ r \end{pmatrix} = \mathbf{M}_{0-L} \begin{pmatrix} t \\ 0 \end{pmatrix}.$$ (7.21)

Multiplying out the matrix equation allows me to find r and t for the multilayer structure. After computing $\mathbf{M}_{0-L}$, I will find it to be of the form

$$\mathbf{M}_{0-L} = \begin{pmatrix} A & B \\ C & D \end{pmatrix}.$$ (7.22)

Then $1 = At$ and $r = Ct$ so that

$$t = \frac{1}{A} \qquad r = \frac{C}{A}.$$ (7.23)

Once I have the amplitudes, the transmittance and reflectance are $\mathscr{T} = tt^*$ and $\mathscr{R} = rr^*$. (Medium 0 and medium $L+1$ both have $n = 1$.) Computation is quick and relatively easy to program. The curves in Figs. 7.3, 7.4, and 7.6 were calculated using the matrix method.

7.4 Inverting $\mathscr{R}$ and $\mathscr{T}$ to Find ϵ

If my goal is to use experimental measurements of $\mathscr{R}$ and $\mathscr{T}$ to find out about the optical constants $(n, \kappa, \epsilon, \ldots)$, I'll need to figure out how to invert the transcendental equations for the measured quantities to extract the things in which I am interested. The inversion is not trivial, and often involves some approximations or computation.

The computational approach is to do some sort of least-square minimization of the difference between measurements and calculation. The trick is to get on the right cycle of the fringe pattern; when that is done, the method can work well.

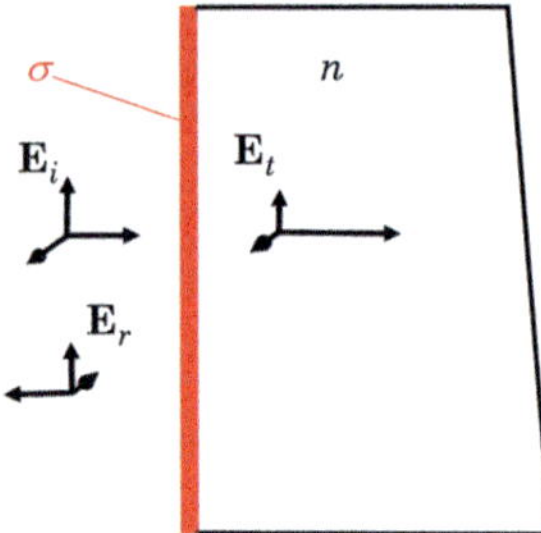

Fig. 7.8 Geometry of a thin film on a thick substrate. The film (red) has complex optical conductivity σ while the substrate has refractive index n. Light is incident from the left at normal incidence with amplitude E_i. Some is reflected with amplitude E_r and some is transmitted with amplitude E_t. The wave vectors and electric and magnetic field vectors are shown but only the electric fields are labeled.

7.4.1 The Glover-Tinkham Formula

I can derive a useful approximation to the exact formulae, useful for the case where the film is very thin. This case is not very restrictive; it is only for very thin layers that any light at all is transmitted through a highly conducting (or superconducting) layer. The formula was derived by Glover and Tinkham [84, 85] many years ago, and probably has origins even earlier. This approximation is easy to invert in order to extract the optical conductivity from the measured transmittance and reflectance.

The basic physics of the situation is that we know that the electric and magnetic field amplitude inside the film goes as $e^{-x/\delta}$ where x is the distance the wave has propagated into the film and δ is the optical penetration depth.* If the thickness d of the film satisfies $d \ll \delta$ (and also is small compared to the wavelength inside the film), then the field does not change measurably within the film: $\mathbf{E}(d) = \mathbf{E}(0)$. I am then permitted to treat the film as a sheet of surface current with surface current density $\mathbf{K}$. The geometry is shown in Fig. 7.8 and the boundary conditions [2] are

$$E_{\|1} - E_{\|2} = 0 \tag{7.24}$$

$$H_{\|1} - H_{\|2} = \frac{4\pi}{c} K, \tag{7.25}$$

where the subscript $\|$ specifies the tangential or parallel components of the field, 1 refers to the side of the film where the incident and reflected fields are, and 2 to the side where the transmitted field is.

The geometry is as follows. The film defines the y–z plane. The incident and transmitted $\mathbf{q}$ are along $\hat{x}$ while the reflected $\mathbf{q}$ is along $-\hat{x}$. The electric fields are all taken as along $\hat{y}$. Eq. 3.3c then tells me that the incident and transmitted magnetic fields are along $\hat{z}$ and the

---

* To be explicit, I can always write at time $t = 0$

$$\mathbf{E}(x) = \mathbf{E}_0 e^{in\omega x/c} e^{-\kappa\omega x/c}$$

with n and κ the real and imaginary parts of the refractive index N. Then the penetration depth is $\delta = c/\omega\kappa$ and the wavelength inside the film is $\lambda_{in} = 2\pi c/\omega n = \lambda_0/n$ with λ_0 the vacuum wavelength.

reflected one along $-\hat{z}$. The coefficients I need are the amplitude reflection and transmission coefficients r and t, defined via $\mathbf{E}_r = r\mathbf{E}_i$ and $\mathbf{E}_t = t\mathbf{E}_i$. Moreover, $H = nE$ in the substrate and $H = E$ in the $n = 1$ vacuum on the left side of the film. I might as well take $E_i = 1$ and write Eqs. 7.24 and 7.25 as

$$1 + r - t = 0 \tag{7.26}$$

$$1 - r - nt = \frac{4\pi}{c}K. \tag{7.27}$$

Compare to Eqs. 3.28 and 3.29, where there was no surface current, but where I did allow both sides of the interface to have a complex refractive index.

What should I use for K? Well, it is the surface current density, which for a thin layer with nonzero thickness d is $K = jd = \sigma Ed$, where I used Ohm's law to write it in terms of the complex conductivity σ. This answer only raises another question: What should I use for E? Well, the field is the same on both sides of the film and continuous across the interface (Eq. 7.24) so I can use either t or $1 + r$. (Recall that I took the incident field to have unit amplitude.) The former is easier, so I have from the magnetic-field boundary condition:

$$1 - r = \left(n + \frac{4\pi}{c}\sigma d\right)t.$$

The electric-field boundary condition allows me to eliminate $r = t - 1$ so that

$$t = \frac{2}{n + 1 + Z_0\sigma d} \tag{7.28}$$

where I replaced $4\pi/c$ with Z_0, the impedance of free space, 377 Ω. (See Appendix A.) The units are correct. Recall the DC resistance of a rectangular slab, $R = \rho L/Wd$ where ρ is the resistivity, L the length, W the width, and d the thickness. If the slab is a square ($L = W$), then the square resistance or sheet resistance is $R_\square = \rho/d = 1/\sigma d$. I can consider that at finite frequencies $\sigma(\omega)d = 1/Z_\square$ and write $t = 2/(n + 1 + Z_0/Z_\square)$. Either works, and both indicate that the transmission of light by a conducting thin film is controlled by the film's impedance.

The amplitude reflection coefficient is easy to calculate once I have t:

$$r = \frac{1 - n - Z_0\sigma d}{1 + n + Z_0\sigma d}, \tag{7.29}$$

The amplitude reflection coefficient is negative; the reflected electric field points opposite to the incident field (and opposite to how it is drawn in Fig. 7.8).

The transmittance into the substrate is the ratio of intensity behind the film to incident intensity. The ratio of the Poynting vector amplitudes, using Eq. 3.23, gives $\mathscr{T}_f = ntt^*$ and

$$\mathscr{T}_f = \frac{4n}{(n + 1 + Z_0\sigma_1 d)^2 + (Z_0\sigma_2 d)^2}. \tag{7.30}$$

The reflectance is rr^*

$$\mathscr{R}_f = \frac{(n - 1 + Z_0\sigma_1 d)^2 + (Z_0\sigma_2 d)^2}{(n + 1 + Z_0\sigma_1 d)^2 + (Z_0\sigma_2 d)^2}. \tag{7.31}$$

The subscript f for "film" is there to remind me that this is the transmittance across the film and into the substrate. When the light makes it to the back surface, some will be reflected, return to the film, be reflected again, and continue to produce within the substrate either coherent or incoherent multiple internal reflections. See the discussion above that leads to Eq. 7.8 and to Eqs. 7.2 and 7.3. The derivations need to be modified to include the fact that the two surfaces of the thin-film-on-substrate sample are not the same. The film surface is governed by $\mathcal{T}_f$ and $\mathcal{R}_f$ and the back surface by the usual single-surface coefficients, $\mathcal{R}_{bc} = (n-1)^2/(n+1)^2$ and $\mathcal{T}_{bc} = 1 - \mathcal{R}_{bc} = 4n/(n+1)^2$. Eq. 7.2 becomes

$$\mathcal{T} = \frac{\mathcal{T}_f(1 - \mathcal{R}_{bc})e^{-\alpha x}}{1 - \mathcal{R}'_f \, \mathcal{R}_{bc} e^{-2\alpha x}}, \tag{7.32}$$

where I've allowed for some absorption in the substrate, with absorption coefficient α. Note carefully the prime on $\mathcal{R}'_f$. It indicates that the multiple-internal-reflection terms have the light incident on the film from within the substrate rather than from refractive index unity vacuum. To write $\mathcal{R}'_f$, I exchange n and 1 in Eq. 7.31.

Similarly, the full reflectance of the film-on-substrate, Eq. 7.3, becomes

$$\mathcal{R} = \mathcal{R}_f + \mathcal{R}_{bc} \frac{\mathcal{T}_f^2 e^{-2\alpha x}}{1 - \mathcal{R}'_f \, \mathcal{R}_{bc} e^{-2\alpha x}}, \tag{7.33}$$

where again I note the presence of $\mathcal{R}'_f$.

It is of course possible to have coherent multiple internal reflections in the substrate. Standing waves will occur, like those shown in Figs. 7.3 and 7.4. The film typically increases the reflectance of the surface on which is is deposited, increasing the amplitude of the standing waves over the uncoated case.

Figure 7.9 shows the film thickness dependence of the transmittance, reflectance, and absorptance ($\mathcal{A} \equiv 1 - \mathcal{R} - \mathcal{T}$, the fraction of the incident energy that is absorbed) of a metal

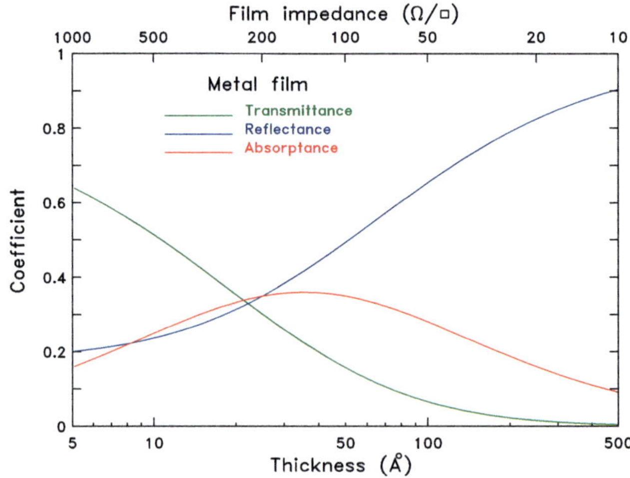

Fig. 7.9 Thickness dependence of the transmittance, reflectance, and absorptance of a metal film on a transparent substrate. The sheet impedance of the metal is shown on the top axis.

film on a transparent substrate. The sheet impedance, $1/\sigma d$ of the film is shown across the top. The metal was taken to have a conductivity $\sigma = 20{,}000\ \Omega^{-1}\ \mathrm{cm}^{-1}$, independent of film thickness. Its resistivity is then $\rho = 50\ \mu\Omega\text{-cm}$. This value is typical of a dirty metal such as Nichrome, a nickel–chromium alloy used for resistive heaters. The optical conductivity of such metals is nearly constant from DC to the midinfrared. The substrate index of refraction was taken as $n = 1.9$, appropriate for crystal quartz and other substrates in the far-infrared region.

For very thin films ($\sim$1Å, a not really realistic thickness, because the ionic radii of Ni and Cr are about 2 Å) the transmittance and reflectance are close to those of the substrate. As thickness increases, the transmittance falls and the reflectance and absorptance rise, with the three becoming nearly equal, 1/3 each, for $d \approx 24$ Å. At this thickness and for the conductivity chosen, $R_\square = 250\ \Omega$. With increasing thickness, the transmittance continues to fall, the reflectance to rise. The absorptance shows a broad maximum around 33 Å, where $R_\square = 150\ \Omega$, and decreases as thickness continues to increase.

What must I keep in mind and what did I learn by making this calculation? (1) The $100\ \mathrm{cm}^{-1}$ skin depth is 2,000 Å for this conductivity. Therefore, the approximation I made, $d \ll \delta$, is adequate below about 200 Å thickness but not so good for thicker films. If frequency is lowered, the approximation is better. (2) It is the sheet impedance $R_\square$ that matters and transmittance becomes quite small when $R_\square$ falls below 10–20 Ω. (3) $R_\square$ cannot be too large either; the transmittance is pretty insensitive to conductivity changes when $R_\square$ is above 1,000–2,000 Ω. (4) I cannot make a broadband absorber out of a thin metal film. The maximum absorptance is not more than 35%, far below the 99% that might be the requirement for a good absorber.[*]

7.4.2 Inverting the Glover-Tinkham Formula

If I measure the reflectance and the transmittance of the film on the substrate and if I know also the refractive index and absorption of the substrate, I can use these measurements to determine the real and imaginary parts of the conductivity. The substrate adds complexity without adding clarity, so let me do this inversion for the free-standing film, with refractive index $n = 1$ on both sides. Eqs. 7.30 and 7.31 become

$$\mathscr{T} = \frac{4}{(2 + Z_0 \sigma_1 d)^2 + (Z_0 \sigma_1 d)^2} \tag{7.34}$$

and

$$\mathscr{R} = \frac{(Z_0 \sigma_1 d)^2 + (Z_0 \sigma_1 d)^2}{(2 + Z_0 \sigma_1 d)^2 + (Z_0 \sigma_1 d)^2}. \tag{7.35}$$

This pair may be combined to find

$$\sigma_1 = \frac{1}{Z_0 d}\frac{1 - \mathscr{R} - \mathscr{T}}{\mathscr{T}} = \frac{1}{Z_0 d}\frac{\mathscr{A}}{\mathscr{T}}$$

---

[*] In some applications the $\sim$33% that is transmitted could be considered as lost too. In others, where for example I want to turn electromagnetic energy into heat, transmitted light does not contribute to generating this heat.

where $\mathscr{A} = 1 - \mathscr{R} - \mathscr{T}$ is the absorptance. Then, because I now know σ_1, I can calculate

$$\sigma_2 = \frac{1}{Z_0 d}\sqrt{\frac{4}{\mathscr{T}} - (2 + Z_0\sigma_1 d)^2}.$$

Inversion of transmittance and reflectance spectra has been done a number of times. In most cases, the experiments were aimed at superconductors [85–88]. I will show a few examples in Chapter 11.

The free-electron theory of metals was developed in the late 1920s and early 1930s by Sommerfeld [89] and others [90, 91]. The basic idea is that one or more electrons of each atom in the solid are detached from their atomic location and free to move throughout the crystal as a free gas. The electrons are fermions, subject to the Pauli principle, and therefore follow the Fermi–Dirac distribution function. Each state is occupied by at most one electron. The electron density is high enough that the Fermi temperature is much higher than the physical temperature, so the Fermi–Dirac function is relatively sharp. The electrons are said to form a "Fermi gas" with a Fermi surface that is the constant-energy surface of the most energetic electrons.

The Fermi–Dirac statistics overwhelm other forces. Thus, the free-electron model treats the potential energy in which the electron moves (the ion potential) as a constant. (Including the potential can be done of course; this is the realm of electronic band structure. Band-structure calculations have been done for many materials systems – not just elements.) For many metals, the effect is mostly to modify slightly the effective mass of the charge carriers, to adjust the numerical value of the density of states at the Fermi surface, and to find bands above and below the one where the free carriers live. I discussed in the look at real solids just completed that the band structure – in particular the presence of the filled d-electron levels below the Fermi level – is the reason why gold, silver and copper appear different. The basic low-energy physics is however very close to what the free-electron model predicts.*)

The electron–electron interaction is also left out of the free-electron model, so that one individual electron state is independent of the other electron states. Here, I can take refuge in Landau's Fermi-liquid theory. This theory of interacting fermions describes the properties of metals in cases where the interaction between the particles of the many-body system is not small. There is a one-to-one correspondence between the elementary excitations of a Fermi gas system and a Fermi liquid system. These excitations are called "quasiparticles;" the quasiparticle states have the same quantum numbers (momentum spin, charge) as the electrons of the Fermi gas.

8.1 Schrödinger Equation for Free Electrons

The simple free-electron model is surprisingly successful in explaining many experimental phenomena, including electrical conductivity, heat conductivity, the Wiedemann–Franz law

---

* In contrast, band structure is essential for understanding semiconductors and insulators.

(a relation of the ratio of electrical conductivity to thermal conductivity), the electronic contribution to the heat capacity, Hall effect, and optical properties. I'll start with this Hamiltonian, a sum of kinetic and potential energies:

$$\mathcal{H} = \frac{p^2}{2m} + \mathcal{V},$$

where p is the momentum, m the mass, and $\mathcal{V}$ the potential. In general I would have $\mathcal{V} = \mathcal{V}(\mathbf{r})$ where $\mathbf{r}$ is the location of the electron and the potential would be periodic in the lattice structure and account for the interaction of the electron with atoms and ions in the crystal. But in the free electron model I take $\mathcal{V} = $ constant and I might as well take it to be zero.*

My quantum mechanics class told me that the momentum operator is $\mathbf{p} = -i\hbar\nabla$, so that the free electron Hamiltonian is

$$\mathcal{H} = -\frac{\hbar^2\nabla^2}{2m} \tag{8.1}$$

and the time independent Schrödinger equation is

$$\mathcal{H}\psi = \mathcal{E}\psi, \tag{8.2}$$

where $\mathcal{E}$ is the energy eigenvalue and ψ is the wave function of the single electron I am considering.

8.2 Wave Function

Consider a plane-wave wave function:

$$\psi = \psi_0 e^{i\mathbf{k}\cdot\mathbf{r}}, \tag{8.3}$$

with ψ_0 a complex amplitude and $\mathbf{k}$ a wave vector to be determined by solving the Schrödinger equation.

Of course (Appendix C) $\nabla^2\psi = -k^2\psi$. I need to normalize the wave function:

$$1 = \langle\psi|\psi\rangle = \int_V dV\,\psi^*(\mathbf{r})\psi(\mathbf{r}) = \int_V dV\,|\psi_0|^2 = |\psi_0|^2 V$$

where V is the entire volume of the crystal. Then,

$$\psi_0 = \frac{e^{i\phi}}{\sqrt{V}}$$

with ϕ an arbitrary phase. (The time-dependent Schrödinger equation leads to a phasor $e^{-i\omega t}$ where the frequency ω is related to the energy $\mathcal{E}$ by Planck's constant: $\mathcal{E} = \hbar\omega$.)

Substituting Eq. 8.3 into Eq. 8.2 with the Hamiltonian of Eq. 8.1 yields

$$+\frac{\hbar^2 k^2}{2m}\psi = \mathcal{E}\psi.$$

---

* I need a wall at the surface to keep the electrons in the crystal, but the choice of $\mathcal{V}$'s constant value has no effect. I take the wall height to be infinite. (And then I eliminate the walls by using periodic boundary conditions.)

Multiply from the left by ψ^* and integrate to find

$$\mathcal{E} = \frac{\hbar^2 k^2}{2m}. \tag{8.4}$$

Eq. 8.4 is just the kinetic energy of a free particle. The momentum operator is $\mathbf{p} = -i\hbar\nabla$ and so $\langle\psi|\mathbf{p}|\psi\rangle = \hbar\mathbf{k}$ and so I can identify $\hbar\mathbf{k}$ as the momentum. The plane-wave wave functions are eigenfunctions of both the Hamiltonian and of momentum. The energy is quadratic in k, with zero energy* at $k = 0$. The probability density to find the electron at any position in the crystal is constant, $P(\mathbf{r}) = \psi^*\psi = 1/V$. For conceptual clarity, let me assume a cube of edge length $L = Ja$ where a is the lattice constant of the crystal and J is a (large) integer. There is one free electron per atom. The volume is $V = L^3 \approx J^3 a^3$. If L is 1 cm, and $a \approx 4.6$ Å, then the solid contains about 10^{22} electrons and $J \approx 2.2 \times 10^7$.

8.3 Exclusion Principle and Boundary Conditions

The electrons are fermions; only one electron can have a particular set of quantum numbers. The quantum numbers here are the three components of the wave vector $\mathbf{k}$ and the spin: $\{k_x, k_y, k_z, \sigma\}$ where $\sigma = \uparrow$ or $\downarrow$. The allowed values of the wave vector components are set by the boundary conditions. If I consider particle-in-a-box conditions, then the wave function will be zero outside the solid, and the solutions are sines and/or cosines depending on the choice of coordinate-system origin.

As in Chapter 5, Section 5.2.2, for phonons, it is much better† to use periodic boundary conditions. I'll write these mathematically as

$$\psi(\mathbf{r} + \mathbf{L}) = \psi(\mathbf{r}), \tag{8.5}$$

where $\mathbf{L}$ is oriented along one or another of the $\hat{x}$, $\hat{y}$, or $\hat{z}$ directions. To begin, look at Fig. 5.3 where one takes a linear chain and bends it into a circle, connecting the first and Nth sites with a bond. Clearly, this loop satisfies Eq. 8.5 because the wave function must be single valued and give the same value every time one goes once around the chain. One may also think in a slightly different way, illustrated in Fig. 8.1. I take the block at the center and make many copies of it and lay them all next to each other, filling space with replicas of the center block. The system boundary is moved off to infinity, but the fact that the copies are exact means that Eq. 8.5 is obeyed. For example the electron colored gray that is about to leave the block on its right-hand side will reappear from the neighbor on the left.

I will apply periodic boundary conditions to the free-electron wave function of Eq. 8.3 for $\mathbf{L} = Ja\hat{x}$:

$$\psi(\mathbf{r}) = \psi_0 e^{i\mathbf{k}\cdot\mathbf{r}} = \psi(\mathbf{r} + \mathbf{L}) = \psi_0 e^{i\mathbf{k}\cdot(\mathbf{r} + Ja\hat{x})}$$

---

* Recall: the potential energy is zero.

† The sine/cosine solutions yield standing waves in the block of material, just as the case of waves on a string with fixed ends. These waves carry no momentum and do not contribute to the current until one forms the proper linear combinations.

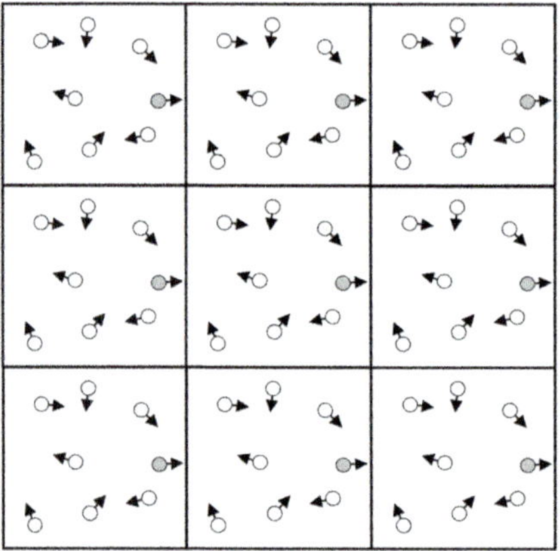

A square system, with edge L, containing N electrons is repeated many times and the copies laid as tiles to cover the plane. The system is periodic in both directions with period L.

with a the lattice constant. Now, $\mathbf{k} \cdot \hat{\mathbf{x}} = k_x$. The periodic boundary conditions for this specific value of $\mathbf{L}$ then yield

$$e^{ik_x Ja} = 1. \tag{8.6}$$

This equation has a large number of solutions; the phase of the complex exponential $k_x Ja$ must be an integer (positive or negative) times 2π radians. Hence

$$k_x = \frac{2\pi}{Ja} j_x \qquad \text{where } j_x = 0, \ \pm 1, \ \pm 2, \ldots$$

Applying periodic boundary conditions in the $\hat{\mathbf{y}}$ and $\hat{\mathbf{z}}$ directions give equivalent quantization of k_y and k_z. Because the components of the wave vector are set by the integer values of $\{j_x, j_y, j_z\}$ and are no longer continuous variables, the energy, Eq. 8.4, is also quantized:

$$\mathcal{E}(\mathbf{k}) = \frac{\hbar^2 k^2}{2m} = \frac{\hbar^2}{2m} \left(\frac{2\pi}{Ja} \right)^2 (j_x^2 + j_y^2 + j_z^2)$$

I note that there are many degeneracies in the energy spectrum of the free electrons. The two values of spin (up and down) are degenerate as well as many permutations of integers (positive and negative). These permutations are distinguishable because the momentum is distinguishable:

$$\mathbf{k} = (2\pi/Ja)(j_x\hat{\mathbf{x}} + j_y\hat{\mathbf{y}} + j_z\hat{\mathbf{z}}). \tag{8.7}$$

The Pauli principle says that no two electrons can have the same quantum numbers. I can call the set $\{j_x, j_y, j_z, \sigma\}$ the quantum numbers. I now pour electrons into the metal, stopping when I reach electrical neutrality.* The first few energies are

$$\mathcal{E}_0 = 0 \qquad \{0,0,0, \uparrow\} \quad \{0,0,0, \downarrow\}$$
$$\mathcal{E}_1 = \delta \qquad \{1,0,0, \uparrow\} \quad \{0,1,0, \uparrow\} \quad \ldots \quad \{0,0,1, \downarrow\}$$
$$\mathcal{E}_2 = 2\delta \qquad \{1,1,0, \uparrow\} \quad \{1,0,1, \uparrow\} \quad \ldots \quad \{0,1,1, \downarrow\}$$

where $\delta = (2\pi\hbar)^2/2mJ^2a^2$ is a typical spacing between levels.

* One to a few electrons per atom in simple metals; one to a few per formula unit in metallic compounds.

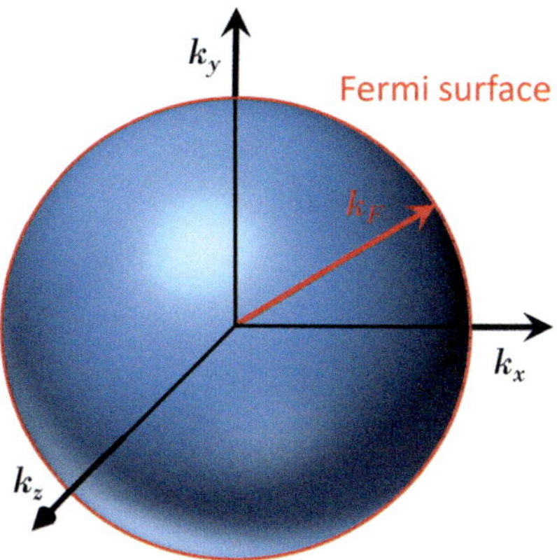

Fig. 8.2   The Fermi sphere of the free-electron metal. The surface of this sphere is the Fermi surface.

8.4 The Fermi Energy

The values of **k** allowed by the boundary conditions form a grid of points when plotted in three-dimensional space* with axes k_x, k_y, k_z. Because the energy $\mathcal{E}$ is quadratic in the wave vector k, the surfaces of constant energy are spheres of radius k. Indeed,

$$\mathcal{E}(\mathbf{k}) = \frac{\hbar^2}{2m} \left(\frac{2\pi}{Ja} \right)^2 j^2$$

where $j^2 = j_x^2 + j_y^2 + j_z^2$ is an integer.

The highest energy of any of these electrons (at $T = 0$) is called the Fermi energy:

$$\mathcal{E}_F = \frac{\hbar^2}{2m} k_F^2 = \frac{\hbar^2}{2m} \left(\frac{2\pi}{Ja} \right)^2 j_F^2$$

where k_F is the radius of the sphere corresponding to the highest energy electrons and j_F is the radius of an equivalent sphere in coordinates of the integers j_x, j_y, j_z. A diagram of this Fermi sphere is shown in Fig. 8.2.

Now let me calculate the radius k_F of the Fermi surface. There are N electrons in the metal. Because of the spin degeneracy, two of these electrons occupy each point on the grid of points indexed by j_x, j_y, and j_z. The total number of points inside the sphere is just the volume of the sphere in "j-space." Hence,

$$\frac{N}{2} = \frac{4\pi}{3} j_F^3 \tag{8.8}$$

or

$$j_F = \left(\frac{3N}{8\pi} \right)^{1/3}.$$

The Fermi wave vector is

$$k_F = \frac{2\pi}{Ja} j_F = \frac{1}{Ja} \left(3\pi^2 N \right)^{1/3}.$$

---

* This space is called "momentum space," "k-space," or "reciprocal space." Its axes have units of inverse length.

Table 8.1 Free-electron Fermi surface parameters for simple metals.

Valence	Metal	n 10^{22} cm^{-3}	k_F 10^8 cm^{-1}	v_F 10^8 cm/s	$\mathcal{E}_F$ eV	T_F 10^4 K
1	Li	4.70	1.11	1.29	4.72	5.48
	Na	2.65	0.92	1.07	3.23	3.75
	K	1.40	0.75	0.86	2.12	2.46
	Rb	1.15	0.70	0.81	1.85	2.15
	Cs	0.91	0.64	0.75	1.58	1.83
	Cu	8.45	1.36	1.57	7.00	8.12
	Ag	5.85	1.20	1.39	5.48	6.36
	Au	5.90	1.20	1.39	5.51	6.39
2	Be	24.2	1.93	2.23	14.14	16.41
	Mg	8.60	1.37	1.58	7.13	8.27
	Ca	4.60	1.11	1.28	4.68	5.43
	Sr	3.56	1.02	1.18	3.95	4.58
	Ba	3.20	0.98	1.13	3.65	4.24
	Zn	13.10	1.57	1.82	9.39	10.90
	Cd	9.28	1.40	1.62	7.46	8.66
3	Al	18.06	1.75	2.02	11.63	13.49
	Ga	15.30	1.65	1.91	10.35	12.01
	In	11.49	1.50	1.74	8.60	9.98
4	Pb	13.20	1.57	1.82	9.37	10.87
	Sn(w)	14.48	1.62	1.88	10.03	11.64

But $(Ja)^3 = V$ or $Ja = V^{1/3}$. Hence*

$$k_F = \left(\frac{3\pi^2 N}{V}\right)^{1/3} = \left(3\pi^2 n\right)^{1/3} \tag{8.9}$$

where $n = N/V$ is the number density of electrons.

The Fermi energy is the energy of an electron with wave vector k_F.

$$\mathcal{E}_F = \frac{\hbar^2}{2m}\left(3\pi^2 n\right)^{2/3}. \tag{8.10}$$

Related to the Fermi energy is the Fermi temperature T_F, defined as $\mathcal{E}_F/k_B$, where k_B is Boltzmann's constant. I also define the Fermi momentum p_F and Fermi velocity v_F.

$$p_F = \hbar k_F = \sqrt{2m\mathcal{E}_F}$$

and

$$v_F = \frac{p_F}{m}.$$

These quantities correspond respectively to the momentum and group velocity of an electron at the Fermi surface. Fermi surface parameters for many metals are shown in Table 8.1, after Ref. 24.

---

* I can now forget j_i and J. It was helpful for me to introduce them; it was not necessary. In fact, the periodic boundary conditions may also be left behind at this point.

Fermi energies are in the range of 1.5–15 eV; Fermi temperatures are thus 18,000–180,000 K. These temperatures are much above 300 K, the temperature of the room. I did not make a bad mistake by considering (as I have done since this chapter began) $T = 0$. The Fermi velocities are surprisingly large; they are in the range 0.8–2×10^8 cm/s. The average is about $c/200$. Our quintessential metal, silver has $\mathcal{E}_F = 5.5$ eV, $T_F = 64{,}000$ K and $v_F = 1.4 \times 10^8$ cm/s. I'll use these numbers when numbers are needed.

8.5 The Effect of Temperature

At zero temperature the Fermi–Dirac distribution is perfectly sharp because every electron is in the lowest energy state that it can access. At finite temperatures, the probability of occupancy is given by the Fermi–Dirac distribution function $f(\mathcal{E})$,

$$f(\mathcal{E}) = \frac{1}{1 + e^{\frac{\mathcal{E}-\mu}{k_B T}}} \tag{8.11}$$

where μ is the chemical potential or Fermi level. At zero temperature $\mu = \mathcal{E}_F$; it has a weak temperature dependence until T becomes a significant fraction of T_F. A plot of the Fermi–Dirac distribution function at four temperatures is shown in Fig. 8.3. The calculation was done for silver, with $\mathcal{E}_F = 5.5$ eV. The effect of temperature is to broaden the distribution. Some states below the Fermi level are empty; some above it are occupied. The "width" of the distribution, the energy range where it drops from 90% to 10% of the $T = 0$ value, is $\Delta\mathcal{E} \approx 4.4 k_B T$, 38 meV at 100 K.

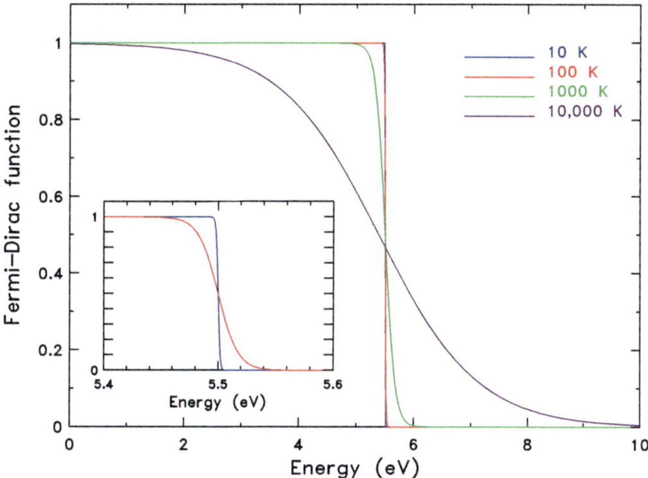

The Fermi–Dirac distribution function $f(\mathcal{E})$ at four temperatures, 10, 100, 1,000, and 10,000 K. The inset shows the detail around $\mathcal{E}_F$ at 10 and 100 K.

8.6 The Density of States

I may need the density of states, so let me work it out now. I can write an equation for $N_\mathcal{E}$, the number of electrons with energy between the zero of energy and energy $\mathcal{E}$, by noting that Eq. 8.10 is such an equation, so

$$\mathcal{E} = \frac{\hbar^2}{2m}\left(\frac{3\pi^2 N_\mathcal{E}}{V}\right)^{2/3}$$

so that

$$N_\mathcal{E} = \left(\frac{2m\mathcal{E}}{\hbar^2}\right)^{3/2}\frac{V}{3\pi^2}.$$

I now raise $\mathcal{E}$ by just a little bit $d\mathcal{E}$ and find that the number goes up by dN.

$$N_\mathcal{E} + dN = \left(\frac{2m(\mathcal{E}+d\mathcal{E})}{\hbar^2}\right)^{3/2}\frac{V}{3\pi^2}.$$

Because $d\mathcal{E}$ is infinitesimal, I can expand $(\mathcal{E}+d\mathcal{E})^{3/2} = \mathcal{E}^{3/2}(1 + 3d\mathcal{E}/2\mathcal{E} + \cdots)$ to find

$$N_\mathcal{E} + dN = \left(\frac{2m\mathcal{E}}{\hbar^2}\right)^{3/2}\frac{V}{3\pi^2}\left(1 + \frac{3d\mathcal{E}}{2\mathcal{E}}\right).$$

The first term on the right is just $N_\mathcal{E}$ and the second is dN. The density of states (the number of states between $\mathcal{E}$ and $\mathcal{E}+d\mathcal{E}$) is

$$\frac{dN}{d\mathcal{E}} = \mathscr{D}(\mathcal{E}) = \left(\frac{2m}{\hbar^2}\right)^{3/2}\frac{V}{2\pi^2}\sqrt{\mathcal{E}}.$$

The density of states* grows as the energy increases, as the square root of the energy.[†]

8.7 Electrical Conductivity

In the absence of an external field, the metal does not carry current. The electrons are moving in all directions and the average or net motion is zero.

Consider now the effect of an electric field on these electrons. To be definite, I will specify the field as our plane wave traveling along $\hat{x}$ with electric field oriented in the $\hat{y}$ direction.

$$\mathbf{E} = \hat{y}E_0 e^{i(qx-\omega t)}. \tag{8.12}$$

The field exerts a force on each and every charged electron:

$$\mathbf{F}_{\text{ext}} = -e\mathbf{E} = -e\hat{y}E_0 e^{iqx}e^{-i\omega t}.$$

---

* Sometimes the symbol $N(\mathcal{E})$ is used for the density of states.

[†] The $\sqrt{\mathcal{E}}$ behavior is correct for three-dimensional metals, where the energy depends on three—x, y, and z—components of $\mathbf{k}$. In two dimensions $\mathscr{D}(\mathcal{E})$ is constant and in one dimension $\mathscr{D}(\mathcal{E}) \sim 1/\sqrt{\mathcal{E}}$.

I can put the space variation into $\mathbf{E}_0$ and ignore the fact that the field varies in space with a length scale $\lambda = 2\pi/q$. The reason is that the wavelength λ is long compared to the effective range of the electron (the mean free path ℓ) and also large compared to the lattice constant. I also have not made a local-field correction. The electron is completely delocalized.* Hence the local field is the average field, which is the applied field.

Newton's second law, $F = ma$, is best written

$$\sum \mathbf{F} = \frac{d\mathbf{p}}{dt} = \hbar \frac{d\mathbf{k}}{dt}. \tag{8.13}$$

So the external force and any internal forces will change the k-state of the electron. The resultant response is not hindered by the Pauli exclusion principle: as the electron in state k evolves to k', the one at k' has gotten out of the way by shifting to k''.

What other forces act on the electrons? Well, the electrons are not bound; indeed, the potential energy has been set to zero. But the electrons can suffer collisions with impurities, other defects, lattice vibrations (phonons), and the surface. These collisions have a mean free time τ and relax the system back towards equilibrium. In the spirit of Drude's model, I will consider that the relaxation generates an impulse $\mathbf{F}_{\mathrm{coll}}\tau$ which changes the momentum $\hbar\mathbf{k}$ back to the equilibrium value[†] $\hbar\mathbf{k}_0$. Thus,

$$\mathbf{F}_{\mathrm{coll}} = -\hbar \frac{\delta\mathbf{k}}{\tau} = -\hbar \frac{\mathbf{k} - \mathbf{k}_0}{\tau}. \tag{8.14}$$

The collisions enter the picture as abrupt changes in momentum that occur every τ seconds. Note that $\hbar\delta\mathbf{k} = m\delta\mathbf{v} = m\mathbf{v}_{\mathrm{drift}}$ where $\mathbf{v}_{\mathrm{drift}}$ is the average velocity acquired by the collection of electrons, known as the drift velocity.[‡]

Let me think for a moment about how the applied electric field, with its $e^{-i\omega t}$ time dependence, affects the motion of an electron. The electron has wave vector $\mathbf{k}_0$ or velocity $\mathbf{v}_0 = \hbar\mathbf{k}_0/m$; this velocity will be changed (in magnitude, direction, or both) by the field. I will write the $\mathbf{k}$ of my electron in the presence of the field as

$$\mathbf{k} = \mathbf{k}_0 + \delta\mathbf{k}\, e^{-i\omega t}, \tag{8.15}$$

with $\delta\mathbf{k}$ in the direction of the field, $\hat{\mathbf{y}}$. $\delta\mathbf{k}$ represents the linear response of the electron to the external field. $\mathbf{k}_0$ is the wave vector in the absence of the field; it does not depend on the time. δk contains the amplitude and phase of the driven motion of the electron. Newton's second law becomes

$$\hbar \frac{d\mathbf{k}}{dt} = -\hbar \frac{\mathbf{k} - \mathbf{k}_0}{\tau} - e\mathbf{E}. \tag{8.16}$$

On substituting for $\mathbf{k}$ and $\mathbf{E}$, taking derivatives, and canceling $e^{-i\omega t}$, I get

$$-i\hbar\omega\delta\mathbf{k} = \frac{-\hbar\delta\mathbf{k}}{\tau} - e\mathbf{E}_0 \tag{8.17}$$

---

* ψ has amplitude everywhere in the crystal.
† Drude considered the equilibrium value to be zero, but I know that it is close to the Fermi momentum.
‡ Without the electric field of the light, the average velocity is *zero*. The average speed is some fraction of the Fermi velocity, but speed and velocity are different things.

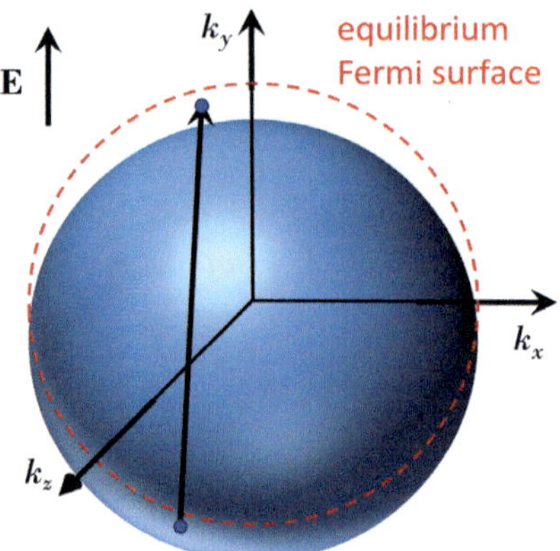

  The momentum-space location of the Fermi sphere in equilibrium is within the dashed line that indicates the Fermi surface. When an electric field is applied along $\hat{y}$, the electrons are accelerated downward. (They are negatively charged.) The displaced sphere is shown in blue. There is a current, because more electrons are going down than are going up. The acceleration continues until it is balanced by the scattering of electrons; one such event is shown.

A little algebra then yields this relation between the shift of $\mathbf{k}$ and the electric field:

$$\delta k = -\frac{e\tau/\hbar}{1 - i\omega\tau}E_0, \tag{8.18}$$

where all vectors are in the $\hat{y}$ direction. Equation 8.18 looks a lot like the Drude velocity, Eq. 4.7, except that $\hbar$ appears. The appearance will be fleeting, however, and the theory is at best semiclassical.

The picture of what happen in the metal due to the external electric field is not intuitive. Every electron has changed its $\mathbf{k}$-vector under the influence of the electric field. This displacement occurs alike and by the same amount to the deeply buried electrons with small $\mathbf{k}$ values and low speeds and to the Fermi-surface electrons which are whizzing along at $c/200$. The Fermi surface, which is centered at $\mathbf{k} = 0$ in equilibrium, is moved as a whole by $\delta\mathbf{k}$, as illustrated in Fig. 8.4.

Were there no scattering, the displacement of the Fermi sphere would continue to grow; the electric field exerts a force on the electrons, causing acceleration. At zero frequency, both $\delta\mathbf{k}$ and the velocity of each electron would change linearly with the time, increasing for left-moving electrons and decreasing for right-moving electrons. With scattering, the relaxation eventually balances the acceleration and there is a fixed, steady-state* displacement. Because scattering events are uncorrelated and affect at random one or the other of the independent and noninteracting electrons, the relaxation process, shown schematically in Fig. 8.4, must respect the Pauli principal, taking the electron from a filled state to an empty state. If the scattering is elastic, then the energy of the initial and final states should be the same. So the relaxation process takes an electron from the leading edge of the displaced Fermi surface, where electrons occupy states that are empty in equilibrium, and deposits it just outside the trailing edge, where states that once were occupied have been emptied.

---

* But not equilibrium.

The electrons that are relaxed are therefore very close to the Fermi energy and move at the Fermi velocity.* Thus the relation between the mean free time τ and mean free path ℓ is

$$\ell = v_F \tau. \tag{8.19}$$

The electrical current is $\mathbf{j} = -ne\mathbf{v}_d$ where n is the electron density and $\mathbf{v}_d$ is an average or drift velocity. This velocity is *not* the Fermi velocity, but the average of all electrons' responses to the field. The averaging process goes like:

$$\mathbf{v}_d = \langle \mathbf{v_k} \rangle = \langle \mathbf{v}_0 + \delta\mathbf{v} \rangle = \langle \delta\mathbf{v} \rangle = \delta\mathbf{v}$$

where the last equality comes from the fact that every electron feels the same force. In the above, the velocity of an electron with wave vector $\mathbf{k}$ is $\mathbf{v_k} = \mathbf{p_k}/m = \hbar\mathbf{k}/m$ and $\mathbf{v}_d = \hbar\delta\mathbf{k}/m$, making

$$\mathbf{j} = -\frac{ne\hbar}{m}\delta\mathbf{k}. \tag{8.20}$$

I then substitute $\delta\mathbf{k}$ from Eq. 8.18 to obtain

$$\mathbf{j} = -\left(\frac{ne\hbar}{m}\right)\left(\frac{-e\tau/\hbar}{1 - i\omega\tau}\right)\hat{\mathbf{y}}E_0.$$

Using Ohm's law, $\mathbf{j} = \sigma\mathbf{E}$, I find finally

$$\sigma = \frac{ne^2\tau/m}{1 - i\omega\tau}, \tag{8.21}$$

an equation identical to Eq. 4.8. Equation 8.21 gives a complex conductivity. The dielectric function and other optical constants follow immediately as I have shown many times before. I write $\epsilon = \epsilon_c + 4\pi i\sigma/\omega$ to allow for core polarizability and have

$$\epsilon = \epsilon_c - \frac{\omega_p^2}{\omega^2 + i\omega/\tau} \tag{8.22}$$

with ω_p the plasma frequency,

$$\omega_p = \sqrt{\frac{4\pi ne^2}{m}}, \tag{8.23}$$

as before.

8.8 Discussion of the Drude Model

8.8.1 Low Frequencies and the Steady-State

To think a bit more about the conduction process in metals, let me first address zero frequency (or any frequency low with respect to $1/\tau$). In this case, I can take $dv/dt \approx 0$,

---

* The realization that electron relaxation involves the fast-moving Fermi-surface electrons is where quantum mechanics (and quantum statistical mechanics) leads to new concepts in the Drude theory.

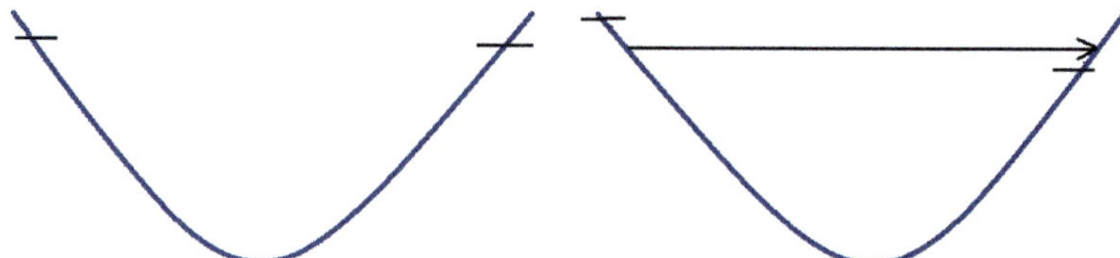

Energy plotted vs k_y in equilibrium (left) and with a steady current flow (right). An elastic scattering event is also shown.

which is the same as saying that $d\mathbf{k}/dt \ll (\mathbf{k} - \mathbf{k}_0)/\tau$ in Eq. 8.16. I have a steady state (not equilibrium!) solution; the acceleration is zero. (I ignore the initial transient when the field is turned on). The conductivity is

$$\sigma = \frac{ne^2\tau}{m},$$

and Eq. 8.18 becomes

$$\delta k = -\frac{e\tau}{\hbar}E_0.$$

The Fermi surface is displaced by an amount δk in the negative field direction as shown in Fig. 8.4. Now, the energy dispersion is still $\mathcal{E} = \hbar^2 k^2/2m$ so that the average energy of the electrons has increased. (The field did work on them.) Figure 8.5 shows the energy dispersion and Fermi energy as a function of k_y (for, say $k_x = k_z = 0$) in equilibrium (left) and with an electric field applied and corresponding steady-state current flow. Remember that $m\mathbf{v} = \hbar\mathbf{k}$; for electrons moving in the direction of electron flow, the most energetic ones have a bit higher energy than in equilibrium. In contrast, for electrons moving against the direction of electron flow, the most energetic ones have an energy that is a bit smaller than in equilibrium. Scattering can take electrons from the first to the second direction.

Scattering from impurities, other defects, and the surface generally is elastic. The direction of the electron momentum is changed but not the energy. Scattering by absorption of phonons is inelastic, but phonon energies are tens of meV whereas the Fermi energy is 1,000 times higher. So on the scale of Fig. 8.5 the distinction is not visible.*

8.8.2 More Numbers

Figures 8.4 and 8.5 suggest that δk is large. It is not. Let me estimate it from Eq. 8.20. I'll take the current density as 10^7 A/cm^2. (This is a huge current; the electrical code typically restricts current densities in copper house wiring [93] to 115 A/cm^2.) Given that e is 1.6×10^{-19} C, the current is carried by a flow of 0.6×10^{26} electrons/s cm^2. I invert Eq. 8.20: $\delta\mathbf{k} = -m\mathbf{j}/ne\hbar = 10^4$ cm^{-1}. Compare this to the Fermi wave vector (Table 8.1)

---

* There are phenomena where the distinction is important. Elastic vs. inelastic scattering plays a large role in localization phenomena where the phase of the wave function is critical. In the far-infrared response at low temperatures there is the Holstein effect [92], where phonon emission affects the optical conductivity at frequencies equal to or larger than the phonon frequencies. See Section 13.3.1.

of order 10^8 cm^{-1}. Very high current densities displace the Fermi surface by only 0.01% of its radius.*

The energy change is also small. The energy of the Fermi-surface electrons moving in the $-y$ direction is increased to $\hbar^2(k_F + \delta k)^2/2m$. Expanding, we see that the energy goes to

$$\mathcal{E}_F + \delta\mathcal{E} = \frac{\hbar^2 k_F^2}{2m} + \frac{\hbar^2 k_F \delta k}{m}$$

or

$$\frac{\delta\mathcal{E}}{\mathcal{E}_F} = \frac{2\delta k}{k_F} \approx 10^{-4}.$$

This value corresponds to the energy spread of the Fermi surface at temperatures of 1 K or so. So at most temperatures (and most currents) the thermal spread of the Fermi distribution is much larger than the displacement due to the current.

8.8.3 The Perfect Conductor

Now suppose there are no collisions; the metal is a perfect conductor.[†] The equation of motion (Eq. 8.16) is simplified.

$$\hbar\frac{d\mathbf{k}}{dt} = -e\mathbf{E} \tag{8.24}$$

and the solution is (at finite frequencies)[‡]

$$\delta k = -\frac{ie}{\hbar\omega}E_0.$$

The oscillations of k are 90° out of phase with the field. Equation 8.20 is unchanged and so, after a bit of algebra I find for the conductivity.

$$\sigma = i\frac{ne^2}{m\omega}. \tag{8.25}$$

This expression is the high frequency limit of Eq. 8.21. That makes sense because no scattering means that the mean free time for collisions is infinite and *any* finite frequency makes $\omega\tau$ big compared to 1. We have ($\omega > 0$)

$$\sigma_1 = 0$$

$$\sigma_2 = \frac{ne^2}{m\omega}.$$

---

* I would be more correct if I said that very high current densities increase the speed of the fastest electrons–those at the Fermi surface–by only 0.01% of their zero-current speed.
† But not a superconductor. The perfect conductor does not exhibit the Meissner effect, Josephson effect, or the energy gap in the optical conductivity. See Chapter 11.
‡ At DC, the solution is a δk that grows linearly with time: a constant acceleration.

The dielectric function of the perfect metal is purely real.*

$$\epsilon = \epsilon_c - \frac{4\pi n e^2/m}{\omega^2} = \epsilon_c - \frac{\omega_p^2}{\omega^2}.$$

So both conductivity and dielectric function tell us that the perfect metal does not dissipate electromagnetic energy.

But wait! ϵ_1 is negative for frequencies below $\omega_p/\sqrt{\epsilon_c}$, which I'll call the screened plasma frequency. The fields inside are damped, and does not that damping mean attenuation and energy loss? The answer is no. The refractive index below the screened plasma frequency is indeed imaginary

$$N = \sqrt{\epsilon_c - \frac{\omega_p^2}{\omega^2}},$$

making $n = 0$ and $\kappa \approx \omega_p/\omega$ (where I have neglected ϵ_c). But in this case the reflectance (Eq. 3.33) is

$$\mathscr{R} = \frac{1 + \kappa^2}{1 + \kappa^2} \equiv 1,$$

or 100%. All the incident energy is reflected. Electromagnetic energy is conserved! Moreover, if I remember that the reflected electric field direction is $180°$ from the incident field direction, then $\mathbf{E}_{\text{inc}} + \mathbf{E}_{\text{refl}} = 0 = \mathbf{E}_{\text{transm}}$, making the field in the perfect conductor zero, as it should be.

8.8.4 Intraband Transitions

Now, let me look qualitatively at the absorption process (σ_1 finite and frequency low). If I think of photons, the process is the absorption of a photon of energy $\hbar\omega$ and the promotion of the electron to a higher state. Because $\mathcal{E} \sim k^2$ and $p = \hbar k$, this process involves a change of momentum of the electron. The initial and final electron states are both on the same $\mathcal{E}$ vs. k curve, so this is an intraband – within the same band – transition.

I look at the energy and momentum conservation in the process here. The Pauli principle requires that initial state be full (below the Fermi level at $T = 0$) and the final state be empty (above the Fermi level at $T = 0$).[†] I write the initial energy as[‡]

$$\mathcal{E}_i = \mathcal{E}_F = \frac{\hbar^2 k_F^2}{2m}$$

and the final state energy as

$$\mathcal{E}_f = \mathcal{E}_F + \hbar\omega = \frac{\hbar^2(k_F^2 + 2k_F\Delta k)}{2m}$$

---

* Except at $\omega = 0$ where the imaginary part, or conductivity, is infinite. I'll calculate the DC conductivity of the perfect conductor in Chapter 10, Section 10.1.2.

† This statement is a zero temperature picture, but it is unchanged if T is finite. The electron still goes from occupied state to empty state.

‡ Instead of taking the initial state at $\mathcal{E}_F$, I could take at any energy between $\mathcal{E}_F - \hbar\omega$ and $\mathcal{E}_F$, but nevermind.

where Δk is the amount the final state wave vector is beyond k_F to account for the final state energy. I have neglected $(\Delta k)^2$. Taking the difference leads to

$$\hbar\omega = \frac{\hbar^2 k_F \Delta k}{m}$$

or

$$\Delta k = \frac{m\omega}{\hbar k_F} = \frac{\omega}{v_F}.$$

The photon starts with momentum $p_\gamma = \hbar q$ where $q = \omega/c$ and ends with zero. (It is absorbed.) But

$$q \ll \Delta k \quad \text{because} \quad \frac{\omega}{c} \ll \frac{\omega}{v_F}.$$

It is not possible to conserve momentum in this energy-conserving process. So how does the metal absorb energy? The collisions change the momentum of the electron and transfer momentum to the crystal as a whole. The process is efficient at low frequencies, when $\omega \ll 1/\tau$, and inefficient at high frequencies.

8.8.5 Uncertainty Principle

Uncertainty principle arguments are always worth exploring. An electron with quantum numbers $\mathbf{k}$ travels freely between collisions for a time τ. Whether the collision is elastic or inelastic, the length of time the electron occupies state $\psi(\mathbf{k})$ is τ. The consequence is that the energy is not known better than $\Delta\mathcal{E} = \hbar/\tau$. I use Eq. 8.4, $\mathcal{E} = \hbar^2 k^2/2m$, to find (by differentiation) that the energy blurring of $\Delta\mathcal{E}$ means a momentum spread of Δk. $\Delta\mathcal{E} = \hbar^2 k\,\Delta k/m$ or (because the relevant energy is the Fermi energy)

$$\Delta k = \frac{\Delta\mathcal{E} k_F}{2\mathcal{E}_F} = \frac{\hbar k_F}{2\mathcal{E}_F\tau} = \frac{m}{\hbar k_F\tau} = \frac{1}{v_F\tau}$$

But! $v_F\tau$ is the mean free path ℓ! So the position uncertainty is the mean free path and the momentum uncertainty is $\hbar\Delta k = \hbar/\ell$. This calculation should encourage you to stop thinking about the electron as a point-like particle skating around like a billiard ball on a table filled with obstacles and think of it as a quantum-mechanical wave packet that is delocalized over a range given by the mean free path.*

8.8.6 Yet More Numbers

I first specified paramaters for a standard metal in Chapter 4, Section 4.2.4: $\sigma \approx 6 \times 10^5\ \Omega^{-1}\text{cm}^{-1}$ or $\rho = 1/\sigma = 1.6\ \mu\Omega\text{-cm}$. The mean free time is $\tau = 2 \times 10^{-14}$ s. From this number and $v_F = 2 \times 10^8$ cm/s I get

$$\ell = 4 \times 10^{-6}\ \text{cm} = 40\ \text{nm}.$$

A mean free path of 40 nm–400 Å–is a characteristic value for the noble metals at room temperature.

---

* If the idea of the physical extent of the wave function being the mean free path bothers you, do not forget that the wave function of the plane-wave state $e^{i\mathbf{k}\cdot\mathbf{r}}$ has finite amplitude everywhere in the solid.

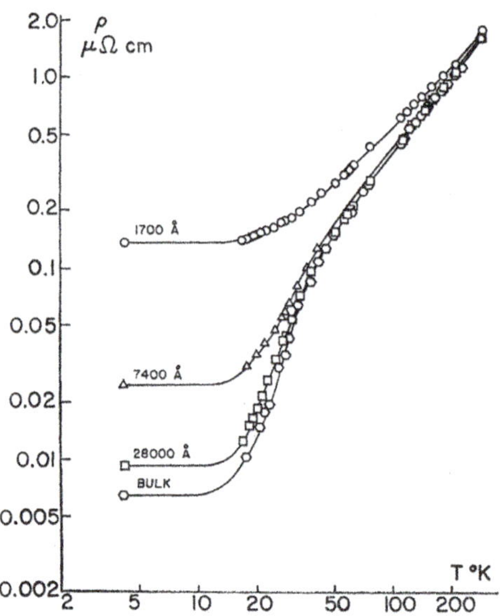

Fig. 8.6   Resistivity $\rho_{dc} = 1/\sigma_{dc}$ of bulk and thin-film silver at temperatures between 4 and 300 K. See Ref. 95.

The resistance of pure metals is strongly temperature dependent, decreasing by large factors as the temperature is reduced. The physical reason for this decrease is that the scattering is due to interactions of the electrons with lattice vibrations (phonons) and, as T decreases both the number and the mean energy of the phonons decreases [24, 94].

Figure 8.6 shows an example, the temperature dependence of the resistivity $\rho_{dc} = 1/\sigma_{dc}$ of silver films of various thicknesses and also that of a bulk (mm-sized) crystal [95]. The resistivity is linear at high temperatures, and then, below the Debye temperature of 220 K, falls rapidly as a power law, $\rho \sim T^n$, with n in the range 2–5, before settling down to the low temperature residual value. The bulk crystal has a residual resistivity ratio (RRR) of 250. The mean free path at low temperatures is very long, $\ell = 10\,\mu$m. This is a macroscopic length, about the diameter of a cotton fiber. It is by no means a record: crystals of Ga have been produced with mean free paths of millimeters [96], so that electrons traveled the full thickness of the crystal before a collision; ℓ was in fact limited by the crystal size, and the collision was with its surface.

8.9 The Boltzmann Transport Equation

The Boltzmann equation was devised initially to work out the nonequilibrium behavior of a classical ideal gas [97]. It addresses the changes in the distribution function (the probability of finding a particle in a unit volume of phase space) arising from external forces, diffusion of particles, and collisions of the particles. The Boltzmann equation can be written also for quantum mechanical particles. It is extremely powerful, allowing the influence of electric fields, temperature gradients, magnetic fields, and chemical potential gradients (alone or in combination) to be included [19, 98–100]. I'll use it to calculate the optical conductivity and I'll find that it gives me a way to derive (once again) the Drude equation for the conductivity.

8.9.1 The Distribution Function

I know that the equilibrium distribution function of electrons in the solid follows the Fermi–Dirac distribution, Eq. 8.11:

$$f_{\mathbf{k}}^0 = \frac{1}{1 + e^{\frac{\mathcal{E}_{\mathbf{k}} - \mu}{k_B T}}}$$

where I've added a subscript $\mathbf{k}$ to indicate that the energies are a function of the wave vector and the superscript 0 to indicate that this is the *equilibrium* function. In equilibrium there are no electrical currents or flow of energy in the solid. (The cases where there are DC currents are called the *steady state*.) In the above equation, T is the temperature and μ is the chemical potential. μ is temperature dependent but at practical temperatures in simple metals, μ may be taken to be the Fermi energy. The chemical potential is set by

$$\sum_{\mathbf{k}, \sigma} f_{\mathbf{k}}^0 = N = \int d\mathcal{E}\, f_{\mathbf{k}}^0(\mathcal{E}) \mathscr{D}(\mathcal{E})$$

where the electrons have quantum numbers $\mathbf{k}$ and σ where σ is the spin. They can have a band index n also but I'll omit the band index; were I to include it, I'd need to sum over bands [19]. Finally, $\mathscr{D}(\mathcal{E})$ is the density of states.

The nonequilibrium distribution function differs from the equilibrium one in two ways. First, I allow for temperature variations within the sample so that

$$f_{\mathbf{k}}(\mathbf{r}) = \frac{1}{1 + e^{\frac{\mathcal{E}_{\mathbf{k}} - \mu(\mathbf{r})}{k_B T(\mathbf{r})}}},$$

where the varying temperature causes (small) variations in the chemical potential. Second, I'll impose an external force $\mathbf{F}_{\text{ext}}$. Newton's law gives $\hbar \dot{\mathbf{k}} = \mathbf{F}_{\text{ext}}$ because $\mathbf{k} = m v/\hbar$. The velocity can be calculated from

$$\mathbf{v}_{\mathbf{k}} = \frac{1}{\hbar} \nabla_{\mathbf{k}} \mathcal{E}_{\mathbf{k}}. \tag{8.26}$$

The external force will cause the occupancy of some values of $\mathbf{k}$ to increase and of others to decrease.

The nonequilibrium distribution $f_{\mathbf{k}}(\mathbf{r}, t)$ gives the probability that there is an electron at $\mathbf{r}$ at time t with wave vector $\mathbf{k}$. Because of spin, I'll have to say that it is the probability for a specific spin orientation and there is an equal probability for the other orientation. I'll take care of the spin degeneracy by multiplying by two at the appropriate points. Moreover, I have to avoid violation of the uncertainty principle by thinking that $\mathbf{r}$ points to a volume large enough to accommodate the position uncertainty of the charge carriers.* So a correct approach is to say that the number of electrons in a small volume $d\mathbf{r}d\mathbf{k}$ of phase space is

$$dN = \frac{2}{(2\pi)^3} f_{\mathbf{k}}(\mathbf{r}, t) d\mathbf{r}d\mathbf{k}$$

---

* I made an uncertainly principle argument in Section 8.8.5.

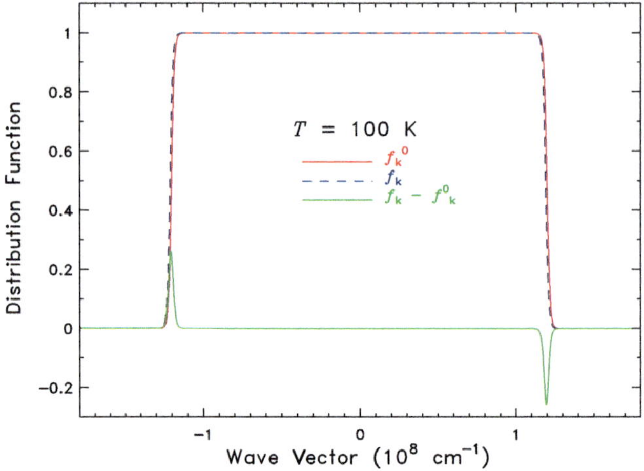

Fig. 8.7   Equilibrium Fermi function, $f_{\mathbf{k}}^0$, a current-carrying nonequilibrium distribution, $f_{\mathbf{k}}$, and the difference, $f_{\mathbf{k}} - f_{\mathbf{k}}^0$.

where the 2 in the numerator is for spin. I want the number of particles dN to be large enough to be able to consider the statistics of the particles in $d\mathbf{r}$.

Figure 8.7 shows the equilibrium Fermi function, $f_{\mathbf{k}}^0$, a current-carrying nonequilibrium distribution, $f_{\mathbf{k}}$, and the difference, $f_{\mathbf{k}} - f_{\mathbf{k}}^0$. All three are plotted against $\mathbf{k}$, which for free electrons is $\mathbf{k} = \pm\sqrt{2m\mathcal{E}}/\hbar$. The calculation was done at 100 K, for a metal (silver!) with $\mathcal{E}_F = 5.5$ eV and $k_F = 1.2 \times 10^8$ cm^{-1}. This plot is a cut along the k_y axis of the equilibrium and non-equilibrium Fermi spheres shown in Fig. 8.4. The current is upward, so because the electron is negative, $f_{\mathbf{k}}$ is shifted downward. In order to make the shift barely visible, I had to shift by 1×10^6 cm^{-1}, about 100× the amount of the estimate of δk I made earlier in this chapter (Section 8.8.2), which did itself correspond to a huge current.

The difference displays a pair of spikes at k_F. The current can truly be said to be carried by Fermi-surface electrons.

Sums or integrals over the distribution function give physical quantities. The one I need is the current,

$$\mathbf{j}(\mathbf{r},t) = -e \sum_{\mathbf{k}} f_{\mathbf{k}} \mathbf{v}_{\mathbf{k}}$$

$$= -e \frac{2}{(2\pi)^3} \int d^3 k f_{\mathbf{k}}(\mathbf{r},t) \mathbf{v}_{\mathbf{k}}. \tag{8.27}$$

The next step in the development is to let time advance by dt. The electron moves according to the equation of motion to a new place in phase space, at $\mathbf{r}' = \mathbf{r} + d\mathbf{r}$ at time $t + dt$ with wave vector $\mathbf{k}' = \mathbf{k} + d\mathbf{k}$. Now, conservation of charge tells me that the product of the probability of occupation and the volume of phase space, $f_{\mathbf{k}}(\mathbf{r},t)d\mathbf{r}d\mathbf{k}$, is preserved over time. Liouville's theorem* tells me that the volume of phase space is preserved. Hence, if there are no collisions to take electrons out of the phase space volume or to bring them into it, $f_{\mathbf{k}'}(\mathbf{r}',t+dt) = f_{\mathbf{k}}(\mathbf{r},t)$. (If dt is small enough, collisions will be unlikely.)

---

* See Appendix H of Ref. 19.

8.9.2 Boltzmann's Equations

When there are collisions, the change in the distribution function in time dt is due to collisions, because otherwise there is no change. So

$$f(\mathbf{k}+d\mathbf{k},\mathbf{r}+d\mathbf{r},t+dt) - f(\mathbf{k},\mathbf{r},t) = \left(\frac{\partial f_{\mathbf{k}}}{\partial t}\right)_{\text{coll}} dt \tag{8.28}$$

where I have changed notation, moving the subscript $\mathbf{k}$ into the argument of f. (I'll change back in a minute.) I note that

$$d\mathbf{k} = \frac{d\mathbf{k}}{dt}dt = \frac{\mathbf{F}_{\text{ext}}}{\hbar}dt$$

where $\mathbf{F}_{\text{ext}}$ is the external force, typically $\mathbf{F}_{\text{ext}} = -e\mathbf{E} + (\mathbf{v} \times \mathbf{B})/c$ (or $\mathbf{F}_{\text{ext}} = -e\mathbf{E}$ if there is no external magnetic field) and

$$d\mathbf{r} = \frac{d\mathbf{r}}{dt}dt = \mathbf{v}_{\mathbf{k}}dt,$$

where $\mathbf{v}_{\mathbf{k}}$ is the velocity of the electron with wave vector $\mathbf{k}$. For linear response, I will do a Taylor expansion of the first term in Eq. 8.28:

$$\begin{aligned} f(\mathbf{k}+d\mathbf{k},\mathbf{r}+d\mathbf{r},t+dt) = &\, f(\mathbf{k},\mathbf{r},t) \\ &+ dt\left(\frac{\mathbf{F}_{\text{ext}}}{\hbar}\cdot\nabla_{\mathbf{k}}f(\mathbf{k},\mathbf{r},t) + \mathbf{v}_{\mathbf{k}}\cdot\nabla_{\mathbf{r}}f(\mathbf{k},\mathbf{r},t) + \frac{\partial f(\mathbf{k},\mathbf{r},t)}{\partial t}\right) \end{aligned} \tag{8.29}$$

I substitute Eq. 8.29 into Eq. 8.28, cancel the two identical f terms, divide by dt and obtain the Boltzmann equation:

$$\frac{\mathbf{F}_{\text{ext}}}{\hbar}\cdot\nabla_{\mathbf{k}}f_{\mathbf{k}} + \mathbf{v}_{\mathbf{k}}\cdot\nabla_{\mathbf{r}}f_{\mathbf{k}} + \frac{\partial f_{\mathbf{k}}}{\partial t} = \left(\frac{\partial f_{\mathbf{k}}}{\partial t}\right)_{\text{coll}}. \tag{8.30}$$

Note that I have returned to writing $\mathbf{k}$ as a subscript.

There are many approaches to the collision term, including detailed calculations of scattering processes. For examples of some of these approaches, see the Boltzmann equation discussion in Ref. 19. The simplest approach however is to say that the nonequilibrium distribution relaxes to equilibrium with a time scale $\tau_{\mathbf{k}}$. This relaxation-time approximation then sets

$$\left(\frac{\partial f_{\mathbf{k}}}{\partial t}\right)_{\text{coll}} = -\frac{f_{\mathbf{k}} - f_{\mathbf{k}}^0}{\tau_{\mathbf{k}}}. \tag{8.31}$$

If I define $\delta f_{\mathbf{k}} = f_{\mathbf{k}} - f_{\mathbf{k}}^0$ and note that the partial of $f_{\mathbf{k}}^0$ with respect to time is zero (the definition of equilibrium), then the solution of Eq. 8.31 is

$$\delta f_{\mathbf{k}}(t) = \delta f_{\mathbf{k}}(0)e^{-t/\tau_{\mathbf{k}}}.$$

If the external conditions that lead to the nonequilibrium state are removed, it relaxes to the equilibrium state with time constant $\tau_{\mathbf{k}}$.

That the partial of $f_{\mathbf{k}}^0$ with respect to time is zero does not mean that scattering does not occur at equilibrium. What does happen is that the events that remove electrons from the element of phase space that I am tracking are exactly balanced by events which put other electrons back in. In the nonequilibrium state, the balance is removed.

Even in the relaxation-time approximation, the Boltzmann equation needs further simplification. I'll assume that $\delta f_\mathbf{k} \ll f_\mathbf{k}^0$ and write

$$\nabla_\mathbf{k} f_\mathbf{k} = \nabla_\mathbf{k}(f_\mathbf{k}^0 + \delta f_\mathbf{k}) \approx \nabla_\mathbf{k} f_\mathbf{k}^0 = \frac{\partial f_\mathbf{k}^0}{\partial \mathcal{E}} \nabla_k \mathcal{E} = \frac{\partial f_\mathbf{k}^0}{\partial \mathcal{E}} \hbar \mathbf{v_k} \tag{8.32}$$

where I used the chain rule to get fourth term and Eq. 8.26 for the last. Also

$$\nabla_\mathbf{r} f_\mathbf{k} = \nabla_\mathbf{r}(f_\mathbf{k}^0 + \delta f_\mathbf{k}) \approx \nabla_\mathbf{r} f_\mathbf{k}^0 = \frac{\partial f_\mathbf{k}^0}{\partial T} \nabla_\mathbf{r} T + \frac{\partial f_\mathbf{k}^0}{\partial \mu} \nabla_\mathbf{r} \mu \tag{8.33}$$

where I used again the chain rule to get last term.

So now I collect everything, rearrange a few terms, and obtain the linearized Boltzmann equation in the relaxation-time approximation

$$\mathbf{F}_{\text{ext}} \cdot \mathbf{v_k} \frac{\partial f_\mathbf{k}^0}{\partial \mathcal{E}} + \mathbf{v_k} \cdot \left(\nabla_\mathbf{r} T \frac{\partial f_\mathbf{k}^0}{\partial T} + \nabla_\mathbf{r} \mu \frac{\partial f_\mathbf{k}^0}{\partial \mu} \right) + \frac{\partial \delta f_\mathbf{k}}{\partial t} = -\frac{\delta f_\mathbf{k}}{\tau_\mathbf{k}}. \tag{8.34}$$

I note that each of the three derivatives is peaked at the Fermi level and will be quite narrow, essentially delta functions if $kT \ll E_F$. I show the energy derivative at two temperatures in Fig. 8.8. (The derivative at 10 K is a spike 10 times higher than the 100 K curve and at 10,000 K is quite wide and 10 times smaller than the 1,000 K curve; neither display well on this plot.) Note that the horizontal scale is expanded compared to the one in Fig. 8.3. Even on the expanded scale, the derivatives are sharp at $\mathcal{E}_F$.

The peaky behavior of the derivatives tells me that all the transport quantities which the Boltzmann equation allows me to investigate are determined by the Fermi-surface properties of the solid. There are many of these quantities if I impose at the same time electric fields, magnetic fields, and temperature gradients and measure electrical currents (or voltages) and heat transfer. For some of the details, take a look at Ref. 101.

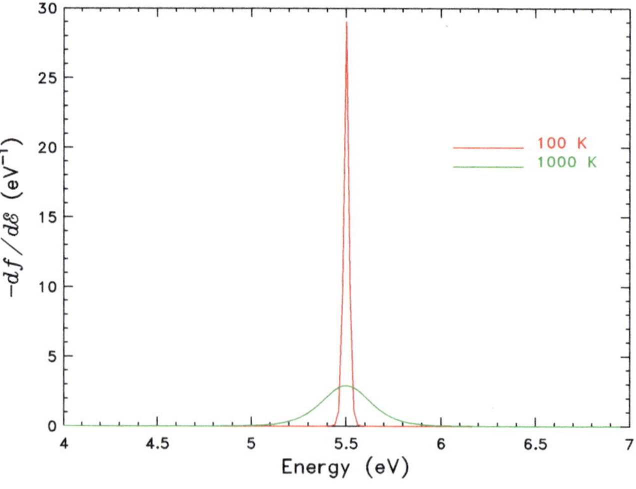

Fig. 8.8 The energy derivative of the Fermi–Dirac distribution function at two temperatures.

8.9.3 The Optical Conductivity

For optics I want the temperature the same everywhere in the sample, so that the gradients of temperature and electrochemical potential are both zero. Moreover, the external force is due to the electric field $\mathbf{E}$ of the light, the magnetic field being as inconsequential as ever. The field will be the plane-wave field I always use, Eqs. 2.5 or 8.12, but in the local approximation I can ignore the space part and write $\mathbf{F}_{\text{ext}} = -e\mathbf{E}(\mathbf{r}, t) = -e\mathbf{E}_0 e^{-i\omega t}$. This field oscillates and the current and the distribution function will also oscillate, although possibly with different phase. So I write $\delta f_{\mathbf{k}}(t) = \delta_{\mathbf{k}}(0)e^{-i\omega t}$. The partial derivative with respect to time in Eq. 8.34 then can be replaced with $-i\omega \delta f_{\mathbf{k}}$, leading to

$$-e\mathbf{E} \cdot \mathbf{v}_{\mathbf{k}} \frac{\partial f_{\mathbf{k}}^0}{\partial \mathcal{E}} - i\omega \delta f_{\mathbf{k}} = -\frac{\delta f_{\mathbf{k}}}{\tau_{\mathbf{k}}}.$$

Now it is just a little algebra to find the deviation of the distribution function from equilibrium in response to the electric field:

$$\delta f_{\mathbf{k}} = -\frac{e\tau_{\mathbf{k}}\mathbf{E} \cdot \mathbf{v}_{\mathbf{k}}}{1 - i\omega\tau_{\mathbf{k}}} \left(-\frac{\partial f^0}{\partial \mathcal{E}} \right) \tag{8.35}$$

where I fiddled the minus signs to have one in front of the partial of the distribution function because that derivative, shown in Fig. 8.8, is by itself negative.

Now use Eq. 8.27 to find the current as a function of the field

$$\mathbf{j} = -e\frac{2}{(2\pi)^3} \int d^3k \left[f^0 - \frac{e\tau_{\mathbf{k}}\mathbf{E} \cdot \mathbf{v}_{\mathbf{k}}}{1 - i\omega\tau_{\mathbf{k}}} \left(-\frac{\partial f^0}{\partial \mathcal{E}} \right) \right] \mathbf{v}_{\mathbf{k}}.$$

The integral of the first term in the square brace is clearly zero. (There is no current if $f_{\mathbf{k}} = f_{\mathbf{k}}^0$!) I simplify the rest and get

$$\mathbf{j} = \frac{e^2}{4\pi^3} \int d^3k \frac{\mathbf{v}_{\mathbf{k}}\tau_{\mathbf{k}}(\mathbf{v}_{\mathbf{k}} \cdot \mathbf{E})}{1 - i\omega\tau_{\mathbf{k}}} \left(-\frac{\partial f^0}{\partial \mathcal{E}} \right). \tag{8.36}$$

To proceed towards Ohm's law, I need to make some assumptions about the system I am studying. If I were interested in very high temperatures, I probably should not treat the energy derivative as a delta function, but evaluate the integral over the entire occupied volume in k space. But for now, I will restrict myself to the low temperature condition, by using

$$\left(-\frac{\partial f^0}{\partial \mathcal{E}} \right) = \delta(\mathcal{E} - \mathcal{E}_F).$$

I will also convert the integral over d^3k to an energy integral times an integral over a constant energy surface.

$$d^3k = \frac{d\mathcal{E} \, dS_{\mathcal{E}}}{|\nabla_k \mathcal{E}|} = \frac{d\mathcal{E} \, dS_{\mathcal{E}}}{\hbar |\mathbf{v}_{\mathbf{k}}|}$$

The surface $S_{\mathcal{E}}$ becomes the Fermi surface when I apply the delta function and $\mathbf{v}_{\mathbf{k}} \rightarrow \mathbf{v}(\mathbf{k} = \mathbf{k}_F)$, with $\mathbf{k}_F$ the Fermi wave vector. This quantity clearly can vary with direction. I note also that Eq. 8.36 gives a tensor value for the conductivity: $\mathbf{j}$ is in general not parallel to $\mathbf{E}$.

A final note is that I've allowed for **k**-dependent scattering by writing the scattering time as $\tau_\mathbf{k}$. I'll simplify things a lot in a minute, but first, I write Ohm's law in components, $j_i = \sigma_{ij} E_j$. Clearly i and j can each be x, y, or z. (I will discuss anisotropy and the tensor dielectric function/conductivity in Chapter 15.) With all of this, Eq. 8.36 then allows me to write one of the elements of the conductivity tensor (after doing the trivial energy integration) as

$$\sigma_{ij} = \frac{e^2}{4\pi^3\hbar} \int_{S_F} dS_{\mathcal{E}} \frac{\tau_\mathbf{k} v_{\mathbf{k}i} v_{\mathbf{k}j}}{|\mathbf{v_k}|(1 - i\omega\tau_\mathbf{k})}, \tag{8.37}$$

where the constant energy surface is the Fermi surface and all the wave vectors **k** become $\mathbf{k}_F$.

To finish this section, I'll use the free electron model with a constant (energy- and wave-vector-independent) scattering time. In this case, the velocity vectors whose components add to the Fermi velocity, $v_x^2 + v_y^2 + v_z^2 = v_F^2$, are all equal. Hence $v_x = v_y = v_z = v_F/\sqrt{3}$. Thus in Eq. 8.37 the ratio $v_{\mathbf{k}i} v_{\mathbf{k}j}/|\mathbf{v_k}| = v_F/3$. Thus, the conductivity, now a scalar, can be written.

$$\sigma = \frac{e^2 v_F}{12\pi^3\hbar} \left(\frac{\tau}{1 - i\omega\tau} \right) \int_{S_F} dS_{\mathcal{E}}, \tag{8.38}$$

It remains to do the surface integral, which is $\int dS_{\mathcal{E}} = 4\pi k_F^2$, and to remember that $v_F = \hbar k_F/m$ to find $\sigma = (e^2 k_F^3/3\pi m)(\tau/1 - i\omega\tau)$. The last step is to look back at Eq. 8.9 which gives the Fermi wave vector in terms of the electron density, $k_F = (3\pi^2 n)^{1/3}$. The conductivity becomes at last

$$\sigma = \frac{ne^2\tau/m}{1 - i\omega\tau}, \tag{8.39}$$

identical to Eqs. 4.8 and 8.21.

So after about five pages of labor, I get the same conductivity that I found either classically in Chapter 4 or in the simple free-electron model earlier in this chapter. Let me list the assumptions (or simplifications) I made in order to get here:

- Collisions are represented by a scattering time $\tau_\mathbf{k}$.
- The scattering time is the same for all electrons, independent of their energy or wave vector.
- The deviations from equilibrium are small.
- The temperature is uniform in the sample.
- The temperature is small.

Once I begin to consider the atomic and electronic structure of a solid, I have to move from the comfortable confines of the free-electron model and allow for there to be a potential energy. For many important solids (silicon!), the electronic structure and optical properties are given by band theory; for others (those called highly correlated or strongly interacting materials), one must go beyond band structure and consider the strong Coulomb forces and other interactions among the electrons.

I'll discuss interband transitions and strongly interacting materials in subsequent chapters. Here, I want to look at the interaction of electromagnetic fields with a quantum-mechanical solid.

I will introduce the issue by considering that the solid has at least two "bands," each containing a fixed number of k states. Allowing for spin, I'll take this number as being equal to two states per atom (or per chemical formula – such as GaAs – in compound materials) in the solid. Moreover, I'll posit that there are enough electrons to fill completely the lower-energy band, called the valence band. The higher energy band – the conduction band – consists of empty states.* How is the optical response affected by this electronic structure? Well, the Pauli principle remains dominant, so there are no intraband transitions in a full band; all states are occupied.† Thus, the lowest-energy optically allowed excitations are from the highest (occupied) state in the valence band to the lowest (unoccupied) state in the conduction band. More important, *all* transitions are from valence to conduction band.

These optical transitions (lowest or not) must satisfy energy conservation,

$$\mathcal{E}_c = \mathcal{E}_v + \hbar\omega \tag{9.1}$$

where $\mathcal{E}_c$ is the unoccupied state in the conduction band, $\mathcal{E}_v$ is the occupied state in the valence band, and $\hbar\omega$ is the photon energy. The transition must also satisfy momentum conservation,

$$\mathbf{k}_c = \mathbf{k}_v + \mathbf{q}_\gamma \tag{9.2}$$

where $\mathbf{k}$ represents the crystal momentum of the electrons and $\mathbf{q}_\gamma$ that of the photon. The scales are quite different. The range of $\mathbf{k}$ in each band is $\pm\pi/a \sim \pm 5 \times 10^8$ cm^{-1}. But, if $\mathcal{E}_c - \mathcal{E}_v$ is of order 1 eV, then the photon wave vector is $\mathbf{q} = 2\pi/\lambda \sim 5 \times 10^4$ cm^{-1}. The ratio of photon momentum to electron momentum is 1×10^{-4}. Hence, $\mathbf{k}_c = \mathbf{k}_v$; optical transitions are *vertical*, because on a plot of energies versus $\mathcal{E}_v$ and $\mathcal{E}_c$, the energy of the final state is directly above the energy of the initial state.

---

* These names are historical, and a little misleading. In typical semiconductors, conduction can be accomplished by either electrons in the conduction band or holes in the valence band.

† And none in an empty band either.

There is one loophole. If the valence band to conduction band transition can take advantage of some other excitation, then energy conservation becomes

$$\mathcal{E}_c = \mathcal{E}_v + \hbar\omega \pm \hbar\Omega \tag{9.3}$$

where $\hbar\Omega$ is the energy of the excitation* (e.g., a phonon). Momentum conservation must also be satisfied:

$$\mathbf{k}_c = \mathbf{k}_v + \mathbf{q}_\gamma \pm \mathbf{k}_\Omega \tag{9.4}$$

where $\mathbf{k}_\Omega$, the crystal momentum of the excitation, is typically in the range $\pm\pi/a$ (like that of the electrons) so the initial and final state can be widely separated on the dispersion relation.

The selection rules given by Eqs. 9.1 and 9.2 are for direct optical transitions. Those in Eqs. 9.3 and 9.4 are for indirect optical transitions.

9.1　The Solid with an Electromagnetic Field

I start with a crystalline solid and the electromagnetic wave. The Hamiltonian is a sum of kinetic and potential energies:

$$\mathcal{H} = T + V.$$

Here, T is the kinetic energy,

$$T = \frac{1}{2m}\left(\mathbf{p} + \frac{e}{c}\mathbf{A}\right)^2,$$

with $\mathbf{p} = -i\hbar\nabla$ the momentum and $\mathbf{A}$ the vector potential.[†] The potential energy is periodic in $\mathbf{T}$ the translation vector of the lattice;

$$V(\mathbf{r}) = V(\mathbf{r} + \mathbf{T}).$$

* The $\pm$ is there because the excitation may either be created or (if excited by thermal processes) destroyed.
† There is a certain mysterious character to the vector potential. It is a function from which the magnetic field and the electric field both can be calculated: In nonmagnetic materials,

$$\mathbf{B} = \mathbf{H} = \nabla \times \mathbf{A} \qquad \text{and} \qquad \mathbf{E} = -\frac{1}{c}\frac{\partial \mathbf{A}}{\partial t}, \tag{9.5}$$

when the scalar potential is equal to zero. (As it should be for transverse fields: if Φ is of plane-wave form then the gradient, $\mathbf{E}$, is parallel to $\mathbf{k}$.) Let me construct a vector potential consistent with our plane-wave fields:

$$\mathbf{E} = \hat{e}E_0 e^{i(\mathbf{q}\cdot\mathbf{r}-\omega t)}$$

from which Maxwell's equations lead to $q = N\omega/c$ and

$$\mathbf{H} = \frac{c}{\omega}\mathbf{q} \times \mathbf{E} = N(\hat{q} \times \hat{e})E_0 e^{i(\mathbf{q}\cdot\mathbf{r}-\omega t)}.$$

I write an $\mathbf{A}$ that gives $\mathbf{E}$ via Eq. 9.5

$$\mathbf{A} = -i\hat{e}E_0\frac{c}{\omega}e^{i(\mathbf{q}\cdot\mathbf{r}-\omega t)}.$$

(Continued on next page.)

The kinetic energy is proportional to the square of a vector, the scalar product of the vector with itself. I must preserve the order of the terms. Thus

$$T = \frac{1}{2m}\left(p^2 + \frac{e}{c}\mathbf{p}\cdot\mathbf{A} + \frac{e}{c}\mathbf{A}\cdot\mathbf{p} + \frac{e^2}{c^2}A^2\right).$$

Now, look at the commutators

$$[x, p_x] = xp_x - p_x x = i\hbar$$

and*

$$[f(\mathbf{r}), p_x] = f(\mathbf{r})p_x - p_x f(\mathbf{r}) = i\hbar\frac{\partial}{\partial x}f(\mathbf{r})$$

so

$$[A_x, p_x] = A_x p_x - p_x A_x = i\hbar\frac{\partial}{\partial x}A_x$$

and the same for $[A_y, p_y]$ and $[A_z, p_z]$. I then can reform the vectors, with the derivative turning into a divergence

$$\mathbf{p}\cdot\mathbf{A} = \mathbf{A}\cdot\mathbf{p} - i\hbar\nabla\cdot\mathbf{A}.$$

Thus

$$T = \frac{1}{2m}\left(p^2 + \frac{2e}{c}\mathbf{A}\cdot\mathbf{p} - \frac{ie\hbar}{c}\nabla\cdot\mathbf{A} + \frac{e^2}{c^2}A^2\right). \tag{9.6}$$

Now, the external field is weak, so I will keep only the first-order terms,[†] dropping the one in A^2.[‡] I use the Coulomb gauge, $\nabla\cdot\mathbf{A} = 0$, eliminating the third term. I also use $\mathbf{p} = -i\hbar\nabla$ to find

$$\mathcal{H} = \mathcal{H}_0 + \mathcal{H}_1 = -\frac{\hbar^2}{2m}\nabla^2 - \frac{ie\hbar}{mc}\mathbf{A}\cdot\nabla + V(\mathbf{r}).$$

9.2 Perturbation Expansion

Perturbation theory is widely used in quantum mechanics and many other areas of physics. The basic idea is to break a problem that cannot be solved exactly into two parts. There is a part with a known solution, described by a zero-order or unperturbed Hamiltonian, and

(Check it. The time derivative brings down $-i\omega$.) Then

$$\nabla\times\mathbf{A} = i\mathbf{q}\times\left(-i\hat{e}E_0\frac{c}{\omega}\right)e^{i(\mathbf{q}\cdot\mathbf{r}-\omega t)}$$

$$= N(\hat{q}\times\hat{e})E_0 e^{i(\mathbf{q}\cdot\mathbf{r}-\omega t)}$$

$$= \mathbf{H}$$

with $H = NE$.

* The first equation is obvious from $p_x = -i\hbar\frac{\partial}{\partial x}$ and $xp_x C = 0$ and $p_x x C = -i\hbar C$.

† In a minute I will start a perturbation theory calculation.

‡ Were it to be kept, there would be nonlinear behavior, as $(e^{-i\omega t})^2$ becomes $e^{-2i\omega t}$, a doubled frequency.

a second part, assumed weak, that is the perturbation Hamiltonian. The wave functions and energy levels of the zero-order system are corrected, usually in a power series in the strength of the perturbation. It is not uncommon to truncate the series after the first nonzero term.

I'll take the zero-order Hamiltonian to be the Hamiltonian for the electrons in the solid

$$\mathcal{H}_0 = -\frac{\hbar^2}{2m}\nabla^2 + V,$$

which I will *not* solve. Instead, it is solved by others, who provide me and the world with the energy-band structure of the solid. I therefore suppose that I know the eigenfunctions of $\mathcal{H}_0$,

$$\mathcal{H}_0\,|n\rangle = \mathcal{E}_n\,|n\rangle \Rightarrow \langle n|\mathcal{H}_0|n\rangle = \mathcal{E}_n,$$

where the $\mathcal{E}_n$ are a set of energies in the various bands and the vectors $|n\rangle$ are a complete orthonormal set, satisfying $\langle n|n'\rangle = \delta_{nn'}$. This zero-order Hamiltonian $\mathcal{H}_0$ is independent of the time. Hence I know that the time-dependent wave function associated with $|n\rangle$ is

$$\psi_n(t) = |n\rangle\,e^{-i\mathcal{E}_n t/\hbar}.$$

The perturbation comes from the electromagnetic field

$$\mathcal{H}_1 = -i\frac{e\hbar}{mc}\mathbf{A}\cdot\nabla.$$

This perturbation contains the vector potential $\mathbf{A}(t)$ and so is time dependent. I use time-dependent perturbation theory, writing the perturbed wave function as a linear combination of the eigenfunctions of $\mathcal{H}_0$,

$$\psi = \sum_{n'} a_{n'}(t)\,|n'\rangle\,e^{-i\mathcal{E}_{n'}t/\hbar}.$$

(If the energy is in the continuum, the sum becomes an integral.)

Now, I calculate using the time-dependent Schrödinger equation

$$\mathcal{H}\psi = (\mathcal{H}_0 + \mathcal{H}_1)\psi = i\hbar\frac{\partial\psi}{\partial t}.$$

Let me do this in pieces. First $\mathcal{H}_0$

$$\mathcal{H}_0\psi = \sum_{n'} a_{n'}(t)\mathcal{H}_0\,|n'\rangle\,e^{-i\mathcal{E}_{n'}t/\hbar},$$

$$= \sum_{n'} a_{n'}(t)\mathcal{E}_{n'}\,|n'\rangle\,e^{-i\mathcal{E}_{n'}t/\hbar},$$

then the perturbation

$$\mathcal{H}_1\psi = \sum_{n'} a_{n'}(t)e^{-i\mathcal{E}_{n'}t/\hbar}\mathcal{H}_1\,|n'\rangle,$$

and then the time derivative

$$i\hbar\frac{\partial\psi}{\partial t} = i\hbar\sum_{n'}\frac{\partial a_{n'}(t)}{\partial t}\,|n'\rangle\,e^{-i\mathcal{E}_{n'}t/\hbar} + i\hbar\sum_{n'} a_{n'}(t)\,|n'\rangle\left(-i\frac{\mathcal{E}_{n'}}{\hbar}\right)e^{-i\mathcal{E}_{n'}t/\hbar}.$$

I put this all together to get

$$\sum_{n'} a_{n'}(t)e^{-i\mathcal{E}_{n'}t/\hbar}\mathcal{E}_{n'}\left|n'\right\rangle + \sum_{n'} a_{n'}(t)e^{-i\mathcal{E}_{n'}t/\hbar}\mathcal{H}_1\left|n'\right\rangle$$

$$= i\hbar \sum_{n'} \frac{\partial a_{n'}(t)}{\partial t}\left|n'\right\rangle e^{-i\mathcal{E}_{n'}t/\hbar} \;+ \sum_{n'} a_{n'}(t)\left|n'\right\rangle \mathcal{E}_{n'}e^{-i\mathcal{E}_{n'}t/\hbar}.$$

The first and last terms are identical and cancel. I am left with

$$\sum_{n'} a_{n'}(t)e^{-i\mathcal{E}_{n'}t/\hbar}\mathcal{H}_1\left|n'\right\rangle = i\hbar \sum_{n'} \dot{a}_{n'}(t)\left|n'\right\rangle e^{-i\mathcal{E}_{n'}t/\hbar}$$

where the dot over $a_{n'}(t)$ on the right-hand side signifies a time derivative. Multiply from the left by $\langle n|$

$$\sum_{n'} a_{n'}(t)e^{-i\mathcal{E}_{n'}t/\hbar}\left\langle n\left|\mathcal{H}_1\right|n'\right\rangle = i\hbar \sum_{n'} \dot{a}_{n'}(t)\left\langle n\middle|n'\right\rangle e^{-i\mathcal{E}_{n'}t/\hbar}.$$

Because $\left\langle n\middle|n'\right\rangle = \delta_{nn'}$, only one term on the right, $n' = n$, survives, so

$$i\hbar\dot{a}_n(t)e^{-i\mathcal{E}_n t/\hbar} = \sum_{n'} a_{n'}(t)e^{-i\mathcal{E}_{n'}t/\hbar}\left\langle n\left|\mathcal{H}_1\right|n'\right\rangle.$$

Now I multiply by $e^{i\mathcal{E}_n t/\hbar}$, define the matrix element as $\mathcal{H}_{1nn'} = \left\langle n\left|\mathcal{H}_1\right|n'\right\rangle$, and the frequency difference as $\omega_{nn'} = (\mathcal{E}_n - \mathcal{E}_{n'})/\hbar$, and I get

$$i\hbar\dot{a}_n(t) = \sum_{n'} a_{n'}(t)\mathcal{H}_{1nn'}e^{i\omega_{nn'}t}. \tag{9.7}$$

Other than the omission of the A^2 term I have yet to make an approximation. Now is the time to do so. First-order perturbation theory consists of considering the field to have been turned on at time $t = 0$ and using $a_{n'}(0)$ in Eq. 9.7 rather than $a_{n'}(t)$. Moreover, I will assume that the system at $t = 0$ is in its ground state ($n = 0$) so that $a_0(0) = 1$ and all the others are zero. Only one term on the right of Eq. 9.7 remains:

$$i\hbar\dot{a}_n(t) = \mathcal{H}_{1n0}e^{i\omega_{n0}t}. \tag{9.8}$$

9.3　The Matrix Element of the Perturbation

Let me take a look again at $\mathcal{H}_1$:

$$\mathcal{H}_1 = -i\frac{e\hbar}{mc}\mathbf{A}\cdot\nabla \tag{9.9}$$

with

$$\mathbf{A} = -i\frac{c}{\omega}\mathbf{E}.$$

I'll take the electric field to be a real (observable) field:

$$\mathbf{E} = \mathbf{E}_0 \cos(\mathbf{q} \cdot \mathbf{r} - \omega t) = \frac{\mathbf{E}_0}{2} \left(e^{i(\mathbf{q} \cdot \mathbf{r} - \omega t)} + e^{-i(\mathbf{q} \cdot \mathbf{r} - \omega t)} \right).$$

Hence

$$\mathcal{H}_1 = -\frac{e\hbar}{m\omega} \cos(\mathbf{q} \cdot \mathbf{r} - \omega t)\mathbf{E}_0 \cdot \nabla$$

and the matrix element I need is

$$\mathcal{H}_{1n0} = -\frac{e\hbar}{m\omega} \langle n|\cos(\mathbf{q} \cdot \mathbf{r} - \omega t)\mathbf{E}_0 \cdot \nabla|0\rangle.$$

I can make a number of simplifications. First, the matrix element contains $\cos(\mathbf{q} \cdot \mathbf{r} - \omega t)$, but I have argued that $\mathbf{q} = 0$. Essentially, the light field varies slowly on the scale of the unit cell or the size of an electron wave packet so that I may take it as constant in space on that scale.* This long-wavelength approximation allows me to take the time-dependent parts as well as the field amplitude vector out of the matrix element. I get

$$\mathcal{H}_{1n0} = -\frac{e\hbar}{2m\omega} \left(e^{-i\omega t} + e^{i\omega t} \right) \langle n|\nabla|0\rangle \cdot \mathbf{E}_0. \tag{9.10}$$

I will return to momentum by using $-i\hbar\nabla = \mathbf{p}$ so that $\langle n|\nabla|0\rangle = (i/\hbar)\langle n|\mathbf{p}|0\rangle$. Note that it is the component of the momentum (i.e., velocity) in the direction of the field that is important. Now, I'll manipulate the momentum matrix element,

$$\langle n|\mathbf{p}|0\rangle = m\frac{d}{dt}\langle n|\mathbf{r}|0\rangle .$$

To take the derivative, I need to insert the time dependence of $\langle n|$ and $|0\rangle$. These are $e^{i\omega_n t}$ and $e^{-i\omega_0 t}$ respectively.† The derivative with respect to time then works on

$$\langle n|\mathbf{p}|0\rangle = m\frac{d}{dt}\left(e^{i\omega_n t}\langle n|\mathbf{r}|0\rangle\, e^{-i\omega_0 t} \right) = im(\omega_n - \omega_0)e^{i\omega_n t}\langle n|\mathbf{r}|0\rangle\, e^{-i\omega_0 t}.$$

Now $\omega_n = \mathcal{E}_n/\hbar$ for each n, so $\hbar(\omega_n - \omega_0)$ is the energy difference between the initial and final states, which in turn equals (by energy conservation, Eq. 9.1) the photon energy $\hbar\omega$. So the momentum matrix element is equal to

$$\langle n|\mathbf{p}|0\rangle = im\omega \langle n|\mathbf{r}|0\rangle .$$

I put everything together and obtain

$$\mathcal{H}_{1n0} = \frac{e}{2}\left(e^{-i\omega t} + e^{i\omega t} \right) \langle n|\mathbf{r}|0\rangle \cdot \mathbf{E}_0. \tag{9.11}$$

The quantity $-e\mathbf{r}$ is the classical value of the electric dipole moment associated with the electron's optical transition. At this point it is convenient to let the electric field be polarized in a specific direction, which I will take to be the $\hat{x}$ direction, making $\langle n|\mathbf{r}|0\rangle \cdot \mathbf{E}_0 = \langle n|x|0\rangle\, E_0$. Later I may need to do an angular average.

---

* The plane wave states $e^{i\mathbf{k} \cdot \mathbf{r}}$ of course have equal amplitude everywhere in the crystal. But the appropriate length scale is not the crystal size, but instead is some length between the lattice constant and the mean free path.

† Remember that $\langle n|$ is the complex conjugate of $|n\rangle$.

9.4 Fermi's "Golden Rule"

I put the matrix element of Eq. 9.11 into Eq. 9.8:

$$i\hbar \dot{a}_n = \frac{e}{2}\left(e^{i(\omega_{n0}-\omega)t} + e^{i(\omega_{n0}+\omega)t}\right)\langle n|x|0\rangle \cdot E_0. \tag{9.12}$$

The next step is to integrate Eq. 9.12. I turn on the perturbation at time $t' = 0$ and integrate to time t. The left side integrates to

$$i\hbar \int_0^t dt'\,\dot{a}_n = i\hbar a_n(t)$$

because $a_n(0)$ is by hypothesis zero for all times $t \leq 0$. The right side contains terms like

$$\int_0^t dt'\, e^{i(\Omega)t'} = \frac{e^{i(\Omega)t} - 1}{i\Omega}.$$

Then

$$a_n(t) = \frac{e}{2\hbar}\left(\frac{1 - e^{i(\omega_{n0}-\omega)t}}{\omega_{n0} - \omega} + \frac{1 - e^{i(\omega_{n0}+\omega)t}}{\omega_{n0} + \omega}\right)\langle n|x|0\rangle\, E_0. \tag{9.13}$$

The first term inside the parenthesis is resonant when $\omega \approx \omega_{n0}$ and corresponds to a photon absorption process. The energy balance is $\hbar\omega = \hbar\omega_{n0} = \mathcal{E}_n - \mathcal{E}_0$: the electron is promoted from state $|0\rangle$ to state $|n\rangle$ and a photon disappears. The second term corresponds to a photon emission process. It is resonant when $\omega = -\omega_{n0}$: the electron drops from state $|n\rangle$ to state $|0\rangle$ and a photon appears. The second term is negligible for absorption and I'll discard it at this time.*

I notice that if a common term of $e^{i(\omega_{no}-\omega)t/2}$ is factored from the absorption term, then it can be cast as

$$a_n(t) = -\frac{ie}{\hbar}\left(\frac{\sin\frac{(\omega_{n0}-\omega)}{2}t}{\omega_{n0} - \omega}\right)\langle n|x|0\rangle\, E_0 e^{i\frac{(\omega_{n0}-\omega)}{2}t}. \tag{9.14}$$

What I want to know is not the amplitude to be in state $|n\rangle$ but the rate at which transitions to that state occur. So first I'll write the probability P_n to be in that state:

$$P_n(t) = a_n^* a_n = \frac{e^2}{\hbar^2}\left(\frac{\sin^2\frac{(\omega_{n0}-\omega)}{2}t}{(\omega_{n0} - \omega)^2}\right)|\langle n|x|0\rangle|^2|E_0|^2. \tag{9.15}$$

The transition rate W_n is the derivative of the probability with respect to time. The time-dependent factor is the quantity $\sin^2(\omega_{n0} - \omega)t/2$ and (using a trigonometric identity in the last step)

---

* The numerators are at most two. The denominator of the absorption term approaches zero at the resonance, making that term arbitrarily large. The denominator of the emission term is always big and so that term can be neglected.

$$\frac{d}{dt}\sin^2\frac{(\omega_{n0}-\omega)}{2}t = 2\sin\left[\frac{(\omega_{n0}-\omega)}{2}t\right]\cos\left[\frac{(\omega_{n0}-\omega)}{2}t\right]\left(\frac{\omega_{n0}-\omega}{2}\right)$$

$$= \left(\frac{\omega_{n0}-\omega}{2}\right)\sin(\omega_{n0}-\omega)t$$

and so

$$W_n = \frac{e^2}{2\hbar^2}|\langle n|x|0\rangle|^2|E_0|^2\frac{\sin(\omega_{n0}-\omega)t}{\omega_{n0}-\omega} \tag{9.16}$$

is the rate at which transitions from the ground state to the excited state occur. The rate is proportional to the square of the electric field, or to the *intensity*, $I \sim |E_0|^2$.

That the transition rate is proportional to the intensity should not be a surprise. I've calculated the response of the semiconductor to photons, and the intensity I is $I = \Phi_\gamma \hbar\omega$, where Φ_γ is the photon flux (photons/unit area·unit time) and $\hbar\omega$ is the photon energy.

The quantity at the end of Eq. 9.16, $F(\omega) \equiv \sin(\omega_{n0}-\omega)t/(\omega_{n0}-\omega)$, oscillates with time with period $T = 2\pi/|\omega_{n0}-\omega|$. The period is small and oscillations are fast when $|\omega_{n0}-\omega|$ is large. In the other limit, when $\omega \approx \omega_{n0}$, the period is long; if $t \ll T$ then I can expand the sine function and find that $F(\omega) \approx t$.

Figure 9.1 shows the function $F(\omega)$ at two times t. The function has a maximum centered at w_{n0} that grows with t and a width that shrinks with t. As the time becomes larger, the function becomes increasingly narrow. In fact, this function is one representation of the delta function. By integrating it, I find that (independent of the time)

$$\int_{-\infty}^{\infty}d\omega\frac{\sin(\omega_{n0}-\omega)t}{(\omega_{n0}-\omega)} = \pi.$$

I can therefore write $\sin(\omega_{n0}-\omega)t/(\omega_{n0}-\omega) = \pi\delta(\omega_{n0}-\omega)$. I do that now and get

$$W_n = \frac{\pi e^2}{2\hbar^2}|\langle n|x|0\rangle|^2|E_0|^2\delta(\omega_{n0}-\omega). \tag{9.17}$$

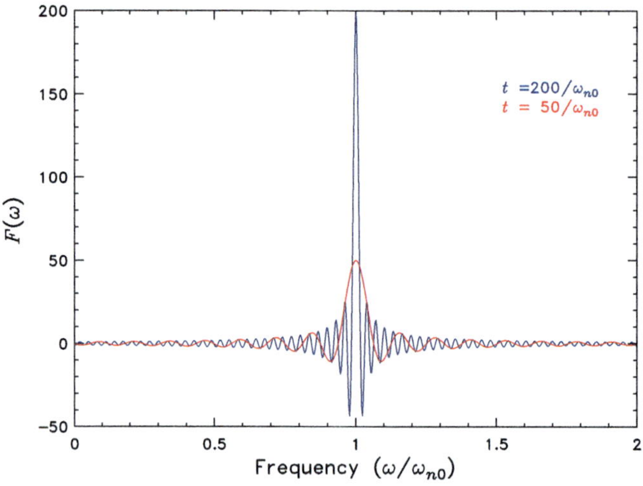

Fig. 9.1   Frequency dependence of $F(\omega) = \sin(\omega_{n0}-\omega)t/(\omega_{n0}-\omega)$ at two times t.

This equation is a version of Fermi's "golden rule" of quantum mechanics. It says that the transition rate is proportional to the square of a matrix element between ground and excited state with an energy-conserving delta function.* If I integrate it over a density of states $\rho(\omega)$ for the excited states, then the delta function picks out $\rho(\omega_{n0})$, giving another version of the golden rule.

$$W_n = \frac{\pi e^2}{2\hbar^2}\rho(\omega_{n0})|\langle n|x|0\rangle|^2|E_0|^2. \tag{9.18}$$

One generally takes $\langle n|x|0\rangle$ as independent of frequency and removes it from integrals. It is then the frequency dependence of the density of states that governs the transition rate.

The physics is contained in W_n, but it would be good to convert to the conductivity or dielectric function. I'll do so by considering the relation between the absorption of light and the transition rate. If light is absorbed in a small thickness dx in the solid then Eq. 3.25 tells me that the change of intensity is $dI(x) = -\alpha I(x)dx$; the energy lost per unit area per second is $\alpha I(x)dx$. Equation 9.16 gives the energy lost per unit area per second as $\hbar\omega W_n dx$. Equating these, I get

$$\alpha = \frac{\hbar\omega W_n}{I}.$$

If now I use Eq. 3.23 to write the intensity in terms of the field amplitude and Eq. 3.27 for the relation between α and σ_1, I can write

$$\sigma_1 = \frac{2\hbar\omega W_n}{|E_0|^2}. \tag{9.19}$$

9.5 Electric Dipole Transitions

Another approach to the dielectric function goes by way of the dipole moment. The dipole moment[†] is

$$\mathfrak{p} = \langle\psi_f|-e\mathbf{r}|\psi_i\rangle + \langle\psi_i|-e\mathbf{r}|\psi_f\rangle$$

where $\psi_i = |0\rangle e^{-i\omega_0 t}$ is the initial (ground) state wave function ($\hbar\omega_0 = \mathcal{E}_0$) and $\psi_f = \sum a_n|n\rangle e^{-i\omega_n t}$ ($\hbar\omega_n = \mathcal{E}_n$) is the excited state wave function. Note that I am letting the excited state be a linear combination of all the unoccupied states $|n\rangle$. Energy conservation will determine which value of n is important when the frequency is ω.

The induced dipole moment will be parallel to the electric field, the $\hat{x}$ direction, leading to

$$\mathfrak{p}_x = -e\sum\left\{a_n^* e^{i\omega_{n0}t}\langle n|x|0\rangle + a_n e^{-i\omega_{n0}t}\langle 0|x|n\rangle\right\}$$

with the energy difference $\omega_{n0} = \omega_n - \omega_0$.

---

* I can show the energy conservation explicitly by writing $\delta(\omega_{n0} - \omega) = \hbar\delta(\mathcal{E}_n - \mathcal{E}_0 - \hbar\omega)$.
[†] I'll use, for the moment, $\mathfrak{p}$ for the dipole moment because $\mathbf{p}$ is used for momentum.

Now substitute for $a_n(t)$ from Eq. 9.13 and calculate and substitute for $a_n^*(t)$ also. Note that $\mathfrak{p}_x$ has one factor of $\langle n|x|0\rangle$ as does a_n, so there will be terms like $|\langle n|x|0\rangle|^2$ in the dipole moment. After some algebra, I get

$$\mathfrak{p}_x = \frac{2e^2}{\hbar} \sum_n \frac{\omega_{n0}}{\omega_{n0}^2 - \omega^2} |\langle n|x|0\rangle|^2 E_0 \cos \omega t = \alpha_e E$$

where α_e is the polarizability and $E = E_0 \cos \omega t$ is the electric field.

I can add a (simulated) damping by letting $\omega_{n0} \to \omega_{n0} - i\gamma_n/2$, with γ_n a small imaginary part. If I wanted to think about it, I would say that γ_n is related to the inverse of the excited state lifetime, and so to the width of the transition. I could go further and say that lifetime effects stop the narrowing of $F(\omega)$ in Fig. 9.1 when the time reaches this lifetime. With this included, the polarizability is

$$\alpha_e = \frac{2e^2}{\hbar} \sum_n \frac{\omega_{n0}}{\tilde{\omega}_{n0}^2 - \omega^2 - i\omega\gamma_n} |\langle n|x|0\rangle|^2. \tag{9.20}$$

The tilde on $\tilde{\omega}_{n0}$ is meant to indicate that the damping pulls down the resonant frequency. (Expanding a square root, I get $\tilde{\omega}_{n0} \approx \omega_{n0} - \gamma_n^2/8\omega_{n0}$.) But local field corrections, introduced in Chapter 4, Sections 4.4.2 and 4.4.3, will lead to larger shifts. Also, I have cheated a bit and written $\omega_{n0}\gamma_n = \omega\gamma_n$; I am comfortable doing this because the delta function in Eq. 9.17 enforces $\omega_{n0} = \omega$.

Equation 9.20 gives the polarizability of the site where the excitation lives. All that I have used here is that there is a ground state $|0\rangle$ and a set of excited states $|n\rangle$ at some location in the crystal. The details of the location are not specified, nor is the path taken by the electron going from $|0\rangle$ to $|n\rangle$. I may reasonably assert that the transitions occur within an atom or pair of atoms, or perhaps a molecule in the solid. They involve one electron leaving the ground state (where its absence is called a hole) and entering one or another of the excited states. There is a total of $\mathfrak{n}$ such electrons in each unit volume of the solid.* Now, in the ground state of the solid there are many electrons, and these electrons occupy many energy levels. The ground state, $|0\rangle$, represents the initial energy of *any* of these electrons. So in one case it might be an electron near the top of the valence band and in another case a core electron in the band made from the $1s$ orbitals of the atoms in the crystal.

With these considerations, I'll just add up all possible states, make local-field corrections (by dropping the tilde!), and obtain the dielectric function:

$$\epsilon = 1 + \frac{8\pi \mathfrak{n} e^2}{\hbar} \sum_n \frac{\omega_{n0}}{\omega_{n0}^2 - \omega^2 - i\omega\gamma_n} |\langle n|x|0\rangle|^2. \tag{9.21}$$

9.6 The Oscillator Strength

Equation 9.21 is of Lorentzian form, just like Eq. 4.31. I can make it look a bit more familiar by defining

---

* I used n as the index of the states in the unperturbed Hamiltonian because I wanted to suggest these were the principal quantum numbers of the atomic energy levels. So I will use this Gothic $\mathfrak{n}$ for the number density.

$$f_{n0} \equiv \frac{2m\omega_{n0}}{\hbar} \, | \langle n|x|0\rangle |^2 = \frac{2}{\hbar\omega_{n0}m} \, | \langle n|p_x|0\rangle |^2 \tag{9.22}$$

where m is the mass of the electron.* The quantity f_{n0} is known as the "oscillator strength" of the transition. Using it, I can write ϵ as

$$\epsilon(\omega) = 1 + \frac{4\pi n e^2}{m} \sum_n \frac{f_{n0}}{\omega_{n0}^2 - \omega^2 - i\omega\gamma_n} \tag{9.23}$$

and it is clear that n is the total number density of electrons in the solid, $n = ZN_{\text{atoms}}/V$, with Z the atomic number (or the average atomic number in a compound). The imaginary part, ϵ_2, has a series of peaks at the resonant frequencies $\{\omega_{n0}\}$. The widths of these peaks are set by the values of $\{\gamma_n\}$; as $\gamma_n \to 0$ the width of that peak becomes narrower and narrower until, eventually, it becomes a delta function,[†] $\epsilon_2 \sim \delta(\omega - \omega_{n0})$.

9.6.1 Limiting values of ϵ

I learn something by thinking about Eq. 9.23 in both low- and high-frequency limits. First, when $\omega \to 0$,

$$\epsilon_1(0) = 1 + \frac{4\pi n e^2}{m} \sum_n \frac{f_{n0}}{\omega_{n0}^2}. \tag{9.24}$$

Note that $\epsilon_2(0) = 0$; this approach does not consider free carriers. Note further that if f_{n0} is large, ω_{n0} is small, or both, that value of n contributes a lot to the static dielectric constant.

Next, I evaluate ϵ at high frequencies.

$$\epsilon_1(\infty) = 1 - \frac{4\pi n e^2}{m\omega^2} \sum_n f_{n0}. \tag{9.25}$$

Eq. 9.25 resembles the response of perfectly free carriers.[‡]

9.6.2 Another Definition of Oscillator Strength

Note that in many papers, especially papers about lattice dynamics (phonons), the dielectric function is written as I did in Eq. 5.28, with a factor $S_n\omega_n^2$ in the numerator:

$$\epsilon(\omega) = 1 + \sum_n \frac{S_n\omega_n^2}{\omega_n^2 - \omega^2 - i\omega\gamma_n}.$$

Clearly

$$S_n\omega_n^2 = \frac{4\pi n e^2}{m} f_{n0}.$$

---

* The actual mass, $9.10938291 \times 10^{-28}$ grams, and not some effective mass.
[†] I can prove this from Kramers–Kronig; see Chapter 10.
[‡] As it should if the frequency is high enough. Note that $\sum_n f_{n0}$ is just a number. I'll compute it in a couple of pages.

9.7 Oscillator Strength Sum Rule

I turn to the sum that appears in Eq. 9.25, a sum over all the states n of f_{n0}.

$$\sum_n f_{n0} = \frac{2m}{\hbar} \sum_n \omega_{n0} |\langle n|x|0\rangle|^2$$

Now $|\langle n|x|0\rangle|^2 = (\langle n|x|0\rangle)^*(\langle n|x|0\rangle) = \langle 0|x|n\rangle \langle n|x|0\rangle$ and $\omega_{n0} = (\mathcal{E}_n - \mathcal{E}_0)/\hbar$. Finally, I'll use up the factor of 2 by writing the same thing twice:

$$\sum_n f_{n0} = \frac{m}{\hbar^2} \sum_n \left\{ \langle 0|x|n\rangle (\mathcal{E}_n - \mathcal{E}_0) \langle n|x|0\rangle + \langle 0|x|n\rangle (\mathcal{E}_n - \mathcal{E}_0) \langle n|x|0\rangle \right\}. \qquad (9.26)$$

Of course I did not know that I wanted to write Eq. 9.26 as I just wrote it until I had worked the following little problems: First I look at the matrix element of the commutator of $\mathcal{H}_0$, the unperturbed Hamiltonian with x:

$$\langle n|[\mathcal{H}_0, x]|0\rangle = \langle n|\mathcal{H}_0 x - x\mathcal{H}_0|0\rangle$$
$$= \langle n|\mathcal{E}_n x - x\mathcal{E}_0|0\rangle$$
$$= (\mathcal{E}_n - \mathcal{E}_0) \langle n|x|0\rangle$$

where I used the fact that $\mathcal{H}_0 |n\rangle = \mathcal{E}_n |n\rangle$ and went leftward in the first term of the commutator and rightward in the second. An identical process yields

$$\langle 0|[\mathcal{H}_0, x]|n\rangle = -(\mathcal{E}_n - \mathcal{E}_0) \langle 0|x|n\rangle.$$

I inspect Eq. 9.26 and see that I can include both commutators neatly into it:

$$\sum_n f_{n0} = \frac{m}{\hbar^2} \sum_n \left\{ \langle 0|x|n\rangle \langle n|[\mathcal{H}_0, x]|0\rangle - \langle 0|[\mathcal{H}_0, x]|n\rangle \langle n|x|0\rangle \right\}. \qquad (9.27)$$

Next, I recall the trick from quantum mechanics [15] where the teacher "inserts a complete set of states"[*]

$$\sum_n \langle 0|A|n\rangle \langle n|B|0\rangle = \langle 0|A \left\{ \sum_n |n\rangle \langle n| \right\} |B\rangle 0 = \langle 0|AB|0\rangle$$

where A and B are arbitrary operators and $\sum_n |n\rangle \langle n| = 1$.

This trick simplifies Eq. 9.27 a lot,

$$\sum_n f_{n0} = \frac{m}{\hbar^2} (\langle 0|x [\mathcal{H}_0, x]|0\rangle - \langle 0|[\mathcal{H}_0, x] x|0\rangle)$$

$$= \frac{m}{\hbar^2} \langle 0|(x [\mathcal{H}_0, x] - [\mathcal{H}_0, x] x)|0\rangle \qquad (9.28)$$

$$= -\frac{m}{\hbar^2} \langle 0|[[\mathcal{H}_0, x], x]|0\rangle$$

---

[*] Or, rather, the way I have derived it, de-inserts them.

where I chose to write a minus sign in front with the double commutator in the last line because I liked its looks more than the other possibility $([x, [\mathcal{H}_0, x]],)$.

Equation 9.28 is a remarkable result: The sum over all the f_{n0} would seem to require knowledge of all the excited states $\{|n\rangle\}$ but instead it requires only the *ground state* wave function $|0\rangle$. However, the analysis can continue a bit further. I recall that

$$\mathcal{H}_0 = T + V$$

with T the kinetic energy and V the potential energy. The latter is a complicated function of location in the crystal, $V = V(\mathbf{r})$. But! Any function of $\mathbf{r}$ commutes with x. Thus $[V, x] = 0$ and

$$\sum_n f_{n0} = -\frac{m}{\hbar^2} \langle 0 |[[T, x], x]|0 \rangle.$$

So, in addition to the ground-state wave function being the only one that matters, only the kinetic energy in the ground state, along with the dipole moment (x), appears in the sum. Of course the ground state wave function depends on the potential; the calculation must at some time find the eigenfunctions and eigenstates of $(T + V)|n\rangle = \mathcal{E}_n |n\rangle$.

9.7.1 Kinetic Energy

In the Schrödinger picture, the kinetic energy is just $p^2/2m$ or

$$T = -\frac{\hbar^2}{2m}\nabla^2$$

but this is not the only possibility. In the tight-binding approach to band structure, one often includes a second-quantized term in the Hamiltonian that allows electrons to hop between two atomic orbitals:

$$\mathcal{H} = -t \sum_{\langle i,j\rangle,\sigma} (c_{i,\sigma}^{\dagger} c_{j,\sigma} + h.c.),$$

where t is a hopping integral, $c_{i\sigma}^{\dagger}$ and $c_{j\sigma}$ are creation and annihilation operators, $\langle i, j\rangle$ are site indices for the two orbitals, and σ is a spin index.

The Schrödinger picture is particularly useful in the evaluation of $\sum_n f_{n0}$. I write out the commutator:

$$[[T, x], x] = [(Tx - xT), x]$$
$$= Tx^2 - 2xTx + x^2T$$
$$= -\frac{\hbar^2}{2m}(\nabla^2 x^2 - 2x\nabla^2 x + x^2\nabla^2).$$

Now, as I did earlier,* I write

$$[[T,x],x]\,C = -\frac{\hbar^2}{2m}(\nabla^2 x^2 - 2x\nabla^2 x + x^2\nabla^2)C = -\frac{\hbar^2}{m}C.$$

Thus,

$$\sum_n f_{n0} = -\frac{m}{\hbar^2}\langle 0|\left(-\frac{\hbar^2}{m}\right)|0\rangle,$$

$$= \langle 0|0\rangle$$

$$= 1.$$

This extremely simple result is called the Thomas-Reiche-Kuhn sum rule. It sets a limit on the values of all the oscillator strengths in the dielectric function. Moreover, the sum is independent of lots of details about what is happening in the solid.

9.7.2 Sum Rule for the Conductivity

I apply this sum rule to the expression I wrote for the conductivity.

$$\sigma_1(\omega) = \frac{ne^2}{m}\sum_n \frac{f_{n0}\omega^2\gamma_n}{(\omega_{n0}^2 - \omega^2)^2 + \omega^2\gamma_n^2}.$$

I integrate both sides of this

$$\int_0^\infty d\omega\,\sigma_1(\omega) = \frac{ne^2}{m}\sum_n f_{n0}\int_0^\infty d\omega\,\frac{\omega^2\gamma_n}{(\omega_{n0}^2 - \omega^2)^2 + \omega^2\gamma_n^2}.$$

I can do some changes of variables to simplify the integral on the right, and then look it up. Or I can just look it up. This definite integral equals $\pi/2$. Therefore,

$$\int_0^\infty d\omega\,\sigma_1(\omega) = \frac{\pi}{2}\frac{ne^2}{m}\sum_n f_{n0} = \frac{\pi}{2}\frac{ne^2}{m} \tag{9.29}$$

because the sum on f is unity. The integral is independent of the value of the resonant frequency or of the damping.

---

* If you do not like this, then I will try it on an arbitrary function $f(x)$. ($f(x) = C$ is one such function.) I do it in one dimension, so $\nabla^2 = \frac{\partial^2}{\partial x^2}$ and f' means derivative with respect to x.

$$[[T,x],x]\,f(x) = -\frac{\hbar^2}{2m}(\nabla^2 x^2 - 2x\nabla^2 x + x^2\nabla^2)f(x)$$

$$= -\frac{\hbar^2}{2m}\left\{\frac{\partial}{\partial x}[2xf(x) + x^2 f'(x)] - 2x\frac{\partial}{\partial x}[f(x) + xf'(x)] + x^2 f''(x)\right\}$$

$$= -\frac{\hbar^2}{2m}\{2f(x) + 2xf'(x) + x^2 f''(x) + 2xf'(x) - 2xf'(x)$$

$$\qquad\qquad - 2xf'f'(x) - 2x^2 f''(x) + x^2 f''(x)\}$$

$$= -\frac{\hbar^2}{m}f(x)$$

Most of the terms cancel, leaving only the result of taking two derivatives of x^2 multiplied by $f(x)$. Just as in the case where $f(x) = C$.

I will check this by considering the Drude conductivity. The only charge carriers in this model are those in the Fermi sphere, and

$$\sigma_1(\omega) = \frac{\omega_p^2 \tau/4\pi}{1 + \omega^2 \tau^2}$$

with $\omega_p^2 = 4\pi n e^2/m$.

Do the integral:

$$\int_0^\infty d\omega\, \sigma_1(\omega) = \int_0^\infty d\omega\, \frac{\omega_p^2 \tau/4\pi}{1 + \omega^2 \tau^2}$$
$$= \frac{\omega_p^2}{4\pi} \int_0^\infty dz\, \frac{1}{1 + z^2}$$

where $z = \omega \tau$. The integral is $\pi/2$ so that

$$\int_0^\infty d\omega\, \sigma_1(\omega) = \frac{1}{8}\omega_p^2 = \frac{\pi}{2}\frac{ne^2}{m}. \tag{9.30}$$

The integral is independent of the mean free time τ. It says that if, for example, the metal is cooled to low temperatures, and the conductivity increases by the residual resistivity ratio, then the increase in the area under $\sigma_1(\omega)$ at very low frequencies is offset by a decrease at higher frequencies. The Drude function becomes taller and narrower and the sum rule explains how this happens.

The Kramers–Kronig relation [102, 103] is a consequence of our experience that observable effects are *causal*, i.e., that the cause precedes the effect.* This notion seems sensible and it is a component of most parts of physics.

This chapter will start with a couple of examples where I show that there must be some relationship between the real and imaginary parts of the complex dielectric function or of the complex conductivity. Then, I'll derive the relation for a number of our favorite response functions and follow with the use of Kramers–Kronig analysis for the reflectance, a common method for estimating response functions from experiment. The chapter ends with the derivation of some sum rules from the Kramers–Kronig relations.

10.1 Absorption and Dispersion Must Be Related

10.1.1 A Notch Filter

To illustrate what causality implies for the optical constants that I have defined and used, let me do a *gedankenexperiment*. I imagine the following situation: I need an optical filter that is opaque for one wavelength and transparent for a wide band of wavelengths around the stop-wavelength. The transmission of such a filter should be 100% except at the specific wavelength of opacity, where it is zero. This filter, and others like it, differencing only in the specific wavelength of operation, could be part of some wavelength multiplexing scheme where the data streams of many receivers are put onto an optical transmission line; the filter is used to select out the wavelength of a single receiver and to pass the others onward.[†] The design transmission of the filter is shown in Fig. 10.1, plotted versus frequency rather than versus wavelength.

Now let me imagine that my signal is a pulse such as the one in the upper panel of Fig. 10.2. The pulse has several oscillations over a period of about 1 ps; outside this time window, there is very small pulse amplitude. The power spectrum of the pulse is shown in the inset; the bandwidth is about 4 THz wide, centered at 5 THz. Clearly the pulse is not a sine or cosine with a single frequency. Indeed, Fourier teaches that I can write the function $f(t)$ as an integral:

$$f(t) = \frac{1}{2\pi} \int_{-\infty}^{\infty} d\omega \, F(\omega) \, e^{i\omega t},$$

---

* Noncausal effects are part of what is called *magic*.
[†] I know; there are many holes in this concept. It is, after all, only a gedankenexperiment.

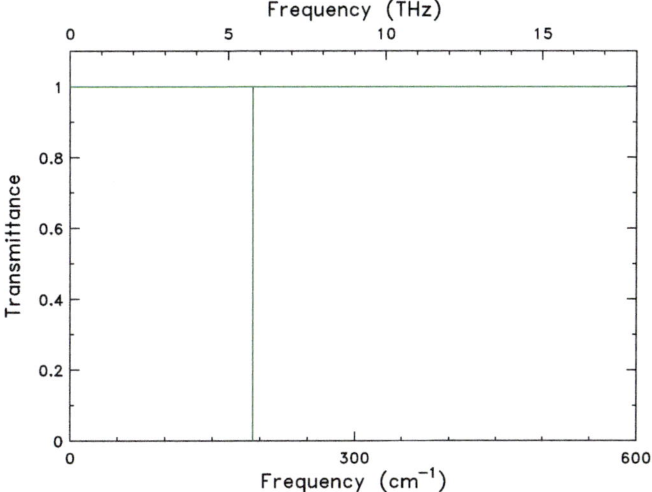

Fig. 10.1   A hypothetical filter that transmits 0–600 cm^{-1} (0–18 THz) except for a band 0.001 cm^{-1} (30 MHz) wide at $\omega_0 = 194$ cm^{-1} (5.8 THz).

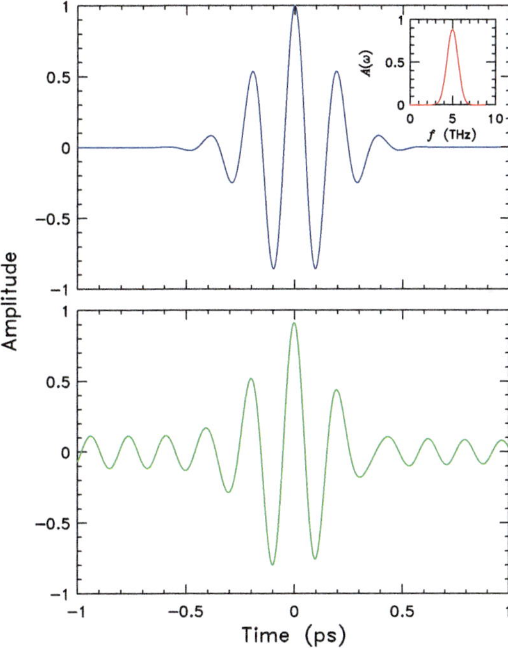

Fig. 10.2   Upper panel: A pulse with frequency components between 3 and 7 THz. The zero of time is chosen to be at the peak of the pulse. The actual time when the signal arrives at the detector is determined by when it was transmitted and how long it took to get there. The inset shows the power spectrum of the pulse. Lower panel: The pulse after the amplitude at 5.8 THz was set to zero.

where $F(\omega)$ is the spectrum needed* to give $f(t)$. To give a good representation of the pulse, many frequencies will enter $F(\omega)$. Now, I will let the pulse be incident on my filter. The filter removes one frequency, $\omega_0 = 5.8$ THz, from the spectrum. Suppose the amplitude of the this component is A; the frequency spectrum behind the filter will be something like[†]

$$G(\omega) = F(\omega) - A e^{i\phi}\delta(\omega - \omega_0)$$

where ϕ is the phase of the ω_0 component. I have subtracted this contribution from the spectrum to represent the effect of the filter.

The pulse will be changed of course, now its spectrum is

$$\begin{aligned}
g(t) &= \frac{1}{2\pi}\int_{-\infty}^{\infty} d\omega\, G(\omega)\, e^{i\omega t}, \\
&= \frac{1}{2\pi}\int_{-\infty}^{\infty} d\omega\, [F(\omega) - A e^{i\phi}\delta(\omega - \omega_0)]\, e^{i\omega t}, \\
&= f(t) - A\cos(\omega_0 t + \phi).
\end{aligned}$$

This result looks quite reasonable at first. My filter has done what I asked it to do: remove the ω_0 part of the pulse and transmit the rest. But then I realize that $\cos(\omega_0 t + \phi)$ is finite at all times,[‡] including large negative times! The bottom panel of Fig. 10.2 shows the amplitude after one frequency component was removed. The effect of the filter is to produce a signal at the detector long before the pulse arrived at the filter. This early appearance is a violation of causality and of course cannot happen.

The fix is exactly the mechanism used to generate the pulse in the first place: the phases of nearby Fourier components are adjusted so as to interfere destructively – indeed, perfectly so – and to cancel everything outside the times where the signal is nonzero. The filter can (and will) produce amplitudes at later times than the end of the original pulse; such ringing is an unavoidable consequence of having a narrow-band filter response. (Think of a struck bell.) But no signal will appear before the leading edge of the pulse arrives.

The filter must have a dispersive behavior, with a refractive index different from unity, especially near the absorption frequency. The refractive index of the Lorentzian oscillator, shown in Fig. 4.13, displays this behavior. The frequency-dependent refractive index will give frequency dependence to the velocity of the waves in the medium, accomplishing the required phase shifting. The amount of phase shift depends on the absorption properties of the filter, meaning that there must be a relation between the refractive index and extinction coefficient or between the real and imaginary parts of the conductivity.

10.1.2 The Perfect Conductor

The notch filter example is striking, but gives no easy route to finding the relation between real and imaginary parts of the response function. Here is a second example, said to be due

---

* I calculate $F(\omega)$ by taking the inverse transform of $f(t)$.

[†] I think of the pulse as being measured as a voltage or field; I will have to worry about specifying the real and imaginary parts correctly so that the measured signal will be real. But this is *my* gedankenexperiment and I won't think too much about this detail.

[‡] Except for its zeros, of course.

to Tinkham [104]. If there are no collisions, the conductivity of the Drude metal, Eqs. 4.8 or 8.21, is

$$\sigma = i\frac{ne^2}{m\omega} \tag{10.1}$$

i.e., zero real part and an imaginary part that diverges as $1/\omega$.

This form for σ is easy to derive. The equation of motion, Eq. 4.4, with no damping ($\Gamma = 0$) is

$$m\ddot{\mathbf{r}} = -e\mathbf{E}_{\text{ext}}.$$

I take $e^{-i\omega t}$ time dependences for $\mathbf{E}_{\text{ext}}$ and $\mathbf{v} = \dot{\mathbf{r}}$ to find that

$$\mathbf{v} = -i\frac{e}{m\omega}\mathbf{E}$$

and then write $\mathbf{j} = -ne\mathbf{v} = \sigma\mathbf{E}$ to obtain Eq. 10.1.

This derivation is fine for monochromatic electric fields, and there can be no causality issues for them because the field amplitude is constant* for $-\infty < t < \infty$. But here, I will do the opposite. Let the applied field be an *impulse,* a delta function in time[†] and of unit amplitude

$$\mathbf{E}(t) = \hat{e}\delta(t).$$

This impulsive field has a frequency spectrum that is infinitely wide. I know this because I know that a representation of the delta function is

$$\delta(t) = \frac{1}{2\pi}\int_{-\infty}^{\infty} d\omega\, e^{i\omega t},$$

with frequency spectrum

$$D(\omega) \equiv \int_{-\infty}^{\infty} dt\, \delta(t)e^{-i\omega t} = 1,$$

making the spectrum $\mathbf{E}(\omega)$ have unit amplitude at all frequencies: $\mathbf{E}(\omega) = \hat{e}$.

The time dependence of the current is also the Fourier transform of its frequency spectrum. Using $\mathbf{j}(\omega) = \sigma(\omega)\mathbf{E}(\omega)$, I get

$$\mathbf{j}(t) = \frac{1}{2\pi}\int_{-\infty}^{\infty} d\omega\, \sigma(\omega)\mathbf{E}(\omega)e^{i\omega t}.$$

Now, I substitute $\mathbf{E}(\omega) = \hat{e}$ and Eq. 10.1 for the conductivity:

$$\mathbf{j}(t) = i\frac{ne^2}{2\pi m}\hat{e}\int_{-\infty}^{\infty} d\omega\, \frac{e^{i\omega t}}{\omega}.$$

---

* When I took classical mechanics and the instructor covered the forced harmonic oscillator, there was discussion of the *turn-on transient.* The oscillator responded at its resonant frequency when the force was initially applied, but then the internal damping reduced the amplitude of oscillations at this frequency to a negligible value after a few relaxation times had passed. If I turn on the field at $t = -\infty$, there is not a problem with the transient.

[†] Which might as well be centered at $t = 0$.

I look up the integral and find

$$\mathbf{j}(t) = \begin{cases} +\dfrac{ne^2}{2m}\hat{e} & \text{if } t > 0 \\[2mm] -\dfrac{ne^2}{2m}\hat{e} & \text{if } t < 0 \end{cases}. \tag{10.2}$$

Here is another noncausal result. The current flows in the direction of the electric field polarization after the pulse arrives (as I would expect) but flows in the *opposite* direction beforehand.

What is missing? Well, I can get Eq. 10.1 by taking the Drude conductivity, Eq. 4.8, and letting $\tau \rightarrow \infty$: $\sigma = (ne^2\tau/m)/(1 - i\omega\tau) \rightarrow (ne^2\tau/m)/(-i\omega\tau) = ine^2/m\omega$. Taking this limit is reasonable as long as I do not think about the $\omega = 0$ case; if I do, I would start with $\sigma = (ne^2\tau/m) \rightarrow \infty$. The Drude DC conductivity is real; $\sigma_1(\omega)$ has a maximum at zero frequency and a width of $1/\tau$ (see Fig. 4.1). As τ increases, the peak becomes higher and higher and the width narrower and narrower. In the limit, the real part becomes a delta function in frequency, located at the origin. Let me add this delta function to σ:

$$\sigma = \mathcal{A}\delta(\omega) + i\frac{ne^2}{m\omega}, \tag{10.3}$$

where $\mathcal{A}$ is the weight (unknown at the moment) of the delta function.* The delta function term in σ is responsible for a current responding to the impulsive electric field as

$$\mathbf{j}(t) = \frac{1}{2\pi}\int_{-\infty}^{\infty} d\omega\, \mathcal{A}\delta(\omega)\hat{e}e^{i\omega t} = \frac{\mathcal{A}}{2\pi}\hat{e}.$$

Note that this current is constant in time and also violates causality! However, if I take

$$\mathcal{A} = \frac{ne^2\pi}{m}.$$

I can make this contribution be $\mathbf{j} = \hat{e}ne^2/2m$. The delta-function contribution exactly cancels the current from the $-ne^2/m\omega$ term for negative times and doubles it for positive times. The total current is now

$$\mathbf{j}(t) = \begin{cases} 0 & \text{if } t < 0 \\[2mm] \dfrac{ne^2}{m}\hat{e} & \text{if } t > 0 \end{cases}, \tag{10.4}$$

a step function. The impulse starts a current which, in the absence of damping, continues indefinitely.[†] Equation 10.4 makes me much happier than Eq. 10.2.

---

* Although I might guess that it is related to the area under $\sigma_1(\omega)$ in Eq. 4.10a. I calculated this area already; see Eq. 9.30.

[†] The result in Eq. 10.4 is easy to obtain from the equation of motion: I write $\mathbf{j} = -ne\mathbf{v}$ and $m\mathbf{a} = -e\mathbf{E}(t)$ to find

$$\frac{d\mathbf{j}}{dt} = -ne\frac{d\mathbf{v}}{dt} = \frac{ne^2}{m}\mathbf{E}(t) = \frac{ne^2}{m}\delta(t)\hat{e},$$

where $\hat{e}$ is a unit vector in the $\mathbf{E}$ direction. I can integrate this equation (from $-\infty$ to t) to get

$$\mathbf{j}(t) = \begin{cases} 0 & \text{if } t < 0 \\[2mm] \dfrac{ne^2}{m}\hat{e} & \text{if } t > 0 \end{cases}.$$

The current is zero for $t < 0$ and then jumps to a final value when the impulse arrives.

10.2 Kramers–Kronig Integrals

The Kramers–Kronig integrals are derived in a number of textbooks [2, 3, 54, 105] and have been discussed by many authors [60, 106–114]. The subject is mostly approached by considering integrals on the complex frequency plane, although the original derivations by Kramers and Kronig relied on model dielectric functions [114].

10.2.1 Time Domain Response and Causality

To begin, I discuss the way in which the dipole moment per unit volume $\mathbf{P}$ depends on the electric field $\mathbf{E}$. These vectors are related by X, a susceptibility,*

$$\mathbf{P} = X\mathbf{E}$$

where X is the electric susceptibility and $X = (\epsilon - 1)/4\pi$. More generally, I can write the dipole moment per unit volume at location $\mathbf{r}$ and time t as an integral of the electric field over nearby locations and recent times,

$$\mathbf{P}(\mathbf{r},t) = \int d^3r' \int dt'\, X(\mathbf{r},\mathbf{r}',t,t')\mathbf{E}(\mathbf{r}',t').$$

Now, I want local response, so that I will be able to use local relations like $\mathbf{j} = \sigma\mathbf{E}$ and $\mathbf{D} = \epsilon\mathbf{E}$; to obtain this, I write the space part of X as a delta function. Moreover, the time part should depend only on the difference between t and t':

$$X(\mathbf{r},\mathbf{r}',t,t') = \delta(\mathbf{r} - \mathbf{r}')X(t - t').$$

Then

$$\mathbf{P}(\mathbf{r},t) = \int_{-\infty}^{\infty} dt'\, X(t - t')\mathbf{E}(\mathbf{r},t'), \tag{10.5}$$

where the polarization at $\mathbf{r}$ depends only on the field at $\mathbf{r}$. Of course, I do not make the response local in time. I know that a response can be retarded in time, with a phase difference between the sinusoidal field and the sinusoidal polarization or a ringing after an impulsive field.

Causality sets a requirement on the form of $X(t - t')$: there can be no response before the stimulus. Suppose $\mathbf{E} = \mathbf{A}\delta(t' - t_0)$, an impulse of amplitude A at time t_0. Then $\mathbf{P}(t) = X(t - t_0)\mathbf{A}$. Causality requires that $X(t - t_0) = 0$ for $t < t_0$ so that $\mathbf{P}(t)$ is likewise zero for $t < t_0$. Equation 10.5 becomes

$$\mathbf{P}(\mathbf{r},t) = \int_{-\infty}^{t} dt'\, X(t - t')\mathbf{E}(\mathbf{r},t'). \tag{10.6}$$

---

* I am going to use X to represent the relation between dipole moment per unit volume and electric field for linear (and isotropic and homogeneous) media. The fields are functions of space and time and X will depend on the time difference between when the field had a specific value and when the dipole moment per unit volume response exists. But later, $X = X(\omega)$, the relation between field at ω and dipole moment per unit volume at ω. I will therefore state the arguments of X explicitly here. Elsewhere, it will almost always be $X(\omega)$.

Eq. 10.6 is a statement that the dipole moment at time t can be affected by the electric field values at all times *before t* but none after.

Equation 10.6 may be rewritten if I define $\tau = t - t'$. Then $dt' = -d\tau$ and

$$\mathbf{P}(\mathbf{r},t) = \int_0^\infty d\tau\, X(\tau)\mathbf{E}(\mathbf{r},t-\tau), \tag{10.7}$$

where the lower limit is at zero to force the response to be causal.

10.2.2 Fourier Transformation into the Frequency Domain

The Kramers–Kronig integral takes place in the frequency domain. I know that the time response and frequency response are Fourier transform pairs.

$$E(\omega) = \int_{-\infty}^\infty dt\, E(t)e^{-i\omega t}, \qquad E(t) = \frac{1}{2\pi}\int_{-\infty}^\infty d\omega\, E(\omega)e^{i\omega t}. \tag{10.8}$$

Similarly,

$$P(\omega) = \int_{-\infty}^\infty dt\, P(t)e^{i\omega t}, \tag{10.9}$$

and (using the shift theorem)

$$\begin{aligned} X(\omega) &= \int_{-\infty}^\infty dt\, X(t-t')e^{i\omega(t-t')}, \\ &= \int_0^\infty d\tau\, X(\tau)e^{i\omega\tau}, \end{aligned} \tag{10.10}$$

where the second line uses the same substitution I used in going from Eqs. 10.5 and 10.6 to Eq. 10.7. This version is important because it shows two things.

First, I can take $\mathbf{P}$ and $\mathbf{E}$ to be real in Eq. 10.7. Hence $X(\tau)$ is also real. Of course $X(\omega)$ is complex on account of the complex exponential in Eq. 10.10. But the reality of $X(\tau)$ along with Eq. 10.10 means that $X(-\omega) = X^*(\omega)$ (ω real) or that $X_1(-\omega) = X_1(\omega)$ and $X_2(-\omega) = -X_2(\omega)$.

Second, in the next few pages I am going to allow ω to be a complex quantity, $\omega_1 + i\omega_2$, so that I can consider the properties of $X(\omega)$ as a complex function of a complex variable. Equation 10.10 then shows that, as long as $X(\tau)$ is finite for all τ, as it should be because I am restricting myself to linear materials, $X(\omega)$ has no poles in the upper half plane,* even as $\mathrm{Im}(\omega) \to +\infty$.

Now, I substitute Eq. 10.5 into Eq. 10.9, noting that the directions of $\mathbf{P}$ and $\mathbf{E}$ are the same, to get

$$P(\omega) = \int_{-\infty}^\infty dt \int_{-\infty}^\infty dt'\, X(t-t')E(t')e^{i\omega t},$$

I use the usual trick, exchanging the order of integration. I will also insert $e^{-i\omega t'}e^{i\omega t'} = 1$

$$P(\omega) = \int_{-\infty}^\infty dt'\, E(t')\left[\int_{-\infty}^\infty dt\, X(t-t')e^{i\omega t}e^{-i\omega t'}\right]e^{i\omega t'}.$$

---

* There are of course divergences in the *lower* half plane, where $e^{-\omega_2 t}$ rapidly diverges as $-\omega_2$ or t become large.

The term in square braces is $X(\omega)$. After moving it out of the integral, what remains is $E(\omega)$. Thus, I get a linear relation between polarization and electric field:

$$P(\omega) = X(\omega)E(\omega) \tag{10.11}$$

Note that the convolution theorem of Fourier analysis would allow me to go from Eq. 10.5 to Eq. 10.11 directly; the former is a convolution in the time domain whereas the latter is a product in the frequency domain.

10.2.3 The Complex ω Plane

I now consider $X(\omega)$ to be a complex function of a complex frequency, $\omega = \omega_1 + i\omega_2$. The range of both real and imaginary parts is $-\infty \leftrightarrow \infty$. The function has these properties:

1. $X(\omega)$ is analytic in the upper half plane. (This arises—as discussed above—because $X(\tau)$ is zero for $\tau < 0$.)
2. $X(\omega) \to 0$ as $|\omega| \to \infty$ at least as fast as $1/\omega$.
3. $X(\omega)$ is Hermitian: $X(-\omega) = X^*(\omega)$.

Are these reasonable? Well, the Lorentz oscillator satisfies these conditions. It is:

$$X(\omega) = \frac{ne^2/m}{\omega_0^2 - \omega^2 - i\omega\gamma}.$$

This function has poles when the denominator is zero. The poles are at frequencies

$$\omega = \pm\sqrt{\omega_0^2 - \gamma^2/4} - i\gamma/2$$

and lie in the lower half plane. As the frequencies become very large $X(\omega) = -ne^2/m\omega^2$ and so is zero as $|\omega| \to \infty$. Finally, I write it as

$$X(\omega) = \frac{ne^2/m}{(\omega_0^2 - \omega^2)^2 + \omega^2\gamma^2}\left[(\omega_0^2 - \omega^2) + i\omega\gamma\right].$$

By inspection, I can see that $X(-\omega) = X^*(\omega)$; the only term that is affected either by complex conjugation or by negation—and in the same way—is $i\omega\gamma$.

Free carriers have the same high-frequency limiting behavior. Indeed, all particles are free at high enough frequencies, so the $X(\omega){\sim}1/\omega^2$ behavior is quite general. See also Eq. 9.25.

Electromagnetism is unchanged as $\omega \to -\omega$ or, equivalently, whether I take an $e^{-i\omega t}$ or an $e^{i\omega t}$ form for the electric field. Let me start with the former,

$$E = E_0 e^{-i\omega t}$$

Eq. 10.11 can be written

$$P(\omega) = [X_1(\omega) + iX_2(\omega)][E_0\cos(\omega t) - iE_0\sin(\omega t)]$$

but the observable is the real part of this equation,

$$\mathrm{Re}(P) = X_1(\omega)E_0\cos(\omega t) + X_2(\omega)E_0\sin(\omega t).$$

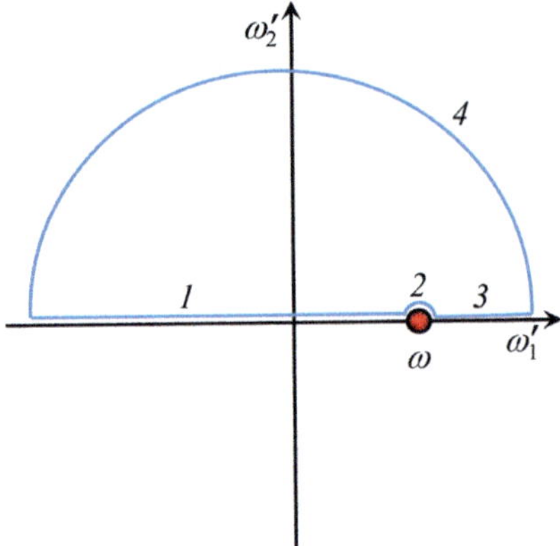

Fig. 10.3   Contour used for deriving the Kramers–Kronig relation in the complex plane, where $\omega' = \omega_1' + i\omega_2'$. The four segments that make up the closed contour are shown.

I want this equation to be unchanged when ω is replaced by $-\omega$. This condition does hold, because

$$X_1(\omega) = X_1(-\omega)$$

and

$$X_2(\omega) = -X_2(-\omega).$$

$X_1(\omega)$ is said to be an *even* function while $X_2(\omega)$ is an *odd* function of ω. This statement is equivalent to saying that X is Hermitian, so that

$$X(-\omega) = X^*(\omega).$$

10.2.4   Use of Cauchy's Theorem

I will use Cauchy's integral theorem [115] to derive the Kramers–Kronig relation [102, 103]. This theorem states that the line integral along a closed path of an analytic function is zero. My susceptibility $X(\omega')$ is such a function if I restrict myself to the upper half plane in the complex ω' space.* In addition, the function $X(\omega')/(\omega' - \omega)$, with ω exactly on the real axis, is also analytic in the upper half plane. Cauchy's theorem then states that

$$\oint d\omega' \, \frac{X(\omega')}{\omega' - \omega} = 0 \tag{10.12}$$

for any closed contour in the upper half plane. I will use the contour shown in Fig. 10.3. It starts (*1*) at $(-\infty, 0)$, runs along the real axis until it is close to the point $(\omega, 0)$. It then (*2*) makes a semicircle around $(\omega, 0)$ in the upper half plane, and (*3*) continues along the real axis to $(\infty, 0)$. The path is closed by making a large[†] semicircle (*4*) of radius infinity in the upper half plane, joining the start of segment (*1*).

---

* I write the frequency with a prime because I want to use ω' as the dummy variable of integration.
[†] Very large!

Now, I break the integral up into its four segments

$$0 = \int_{(1)} d\omega' \frac{X(\omega')}{\omega' - \omega} + \int_{(2)} d\omega' \frac{X(\omega')}{\omega' - \omega} + \int_{(3)} d\omega' \frac{X(\omega')}{\omega' - \omega} + \int_{(4)} d\omega' \frac{X(\omega')}{\omega' - \omega},$$

and look at each one of these. On segment (4), the length of the semicircle increases proportional to $|\omega|$, but the integrand vanishes. The integral is zero so long as $X(\omega)$ falls at least as fast as $1/|\omega|$. And it does, according to our assumptions above. The line of segment (2) may be parameterized by writing $\omega' = \omega + ue^{i\phi}$ where ϕ is the angle the radius vector makes to the Re ω' axis. As the path moves around the semicircle, the angle goes from π to 0. The quantity u is the (constant) radius of the semicircle. If u is small enough, the value of $X(\omega')$ is the same for all points on the little semicircle.* I can replace $X(\omega')$ with $X(\omega)$. Finally, $d\omega' = ue^{i\phi}i\,d\phi$. Then

$$\int_{(2)} d\omega' \frac{X(\omega')}{\omega' - \omega} = iuX(\omega) \int_{\pi}^{0} d\phi\, e^{i\phi} \frac{1}{ue^{i\phi}} = iX(\omega)\phi\Big[_{\pi}^{0} = -i\pi X(\omega).$$

Segments (1) and (3) combine to make a principal-value integral,

$$\int_{(1)} d\omega' \frac{X(\omega')}{\omega' - \omega} + \int_{(3)} d\omega' \frac{X(\omega')}{\omega' - \omega} = \mathcal{P} \int_{-\infty}^{\infty} \frac{X(\omega')}{\omega' - \omega}\, d\omega'.$$

I combine all four segments to obtain

$$\oint d\omega' \frac{X(\omega')}{\omega' - \omega} = \mathcal{P} \int_{-\infty}^{\infty} d\omega' \frac{X(\omega')}{\omega' - \omega} - i\pi X(\omega) = 0.$$

This equation can be rearranged to give the Kramers–Kronig relation,

$$X(\omega) = \frac{1}{i\pi}\mathcal{P} \int_{-\infty}^{\infty} d\omega' \frac{X(\omega')}{\omega' - \omega}. \tag{10.13}$$

There is an i in the denominator, so that the real part of X is given by an integral of the imaginary part of X and vice versa. I substitute $X(\omega) = X_1(\omega) + iX_2(\omega)$ on both sides; Equation 10.13 must hold for both its real and its imaginary parts, giving

$$X_1(\omega) = \frac{1}{\pi}\mathcal{P} \int_{-\infty}^{\infty} d\omega' \frac{X_2(\omega')}{\omega' - \omega} \tag{10.14}$$

and

$$X_2(\omega) = -\frac{1}{\pi}\mathcal{P} \int_{-\infty}^{\infty} d\omega' \frac{X_1(\omega')}{\omega' - \omega}, \tag{10.15}$$

where $\mathcal{P}$ denotes the Cauchy principal value. The real and imaginary parts of X are not independent: the complex function can be constructed given just one of its parts.

If I had chosen segment (2) to be a semi-circle below instead of above the point $(\omega, 0)$, I would have had[†] $\int_{(2)} = +i\pi X(\omega)$ but the pole at ω would be *inside* the closed contour, so the contour integral will pick up the residue of the pole, $2i\pi X(\omega)$, making the result (of course) be the same.

---

* How small is small enough? In the case of the Lorentz oscillator, I must have $u \ll \gamma$.
[†] Because then $\omega' = \omega - ue^{i\phi}$.

10.2.5 Eliminating Negative Frequencies

Equations 10.14 and 10.15 are not as useful as they could be, because the integrals cover $-\infty \rightarrow \infty$, implying that I must know the susceptibility at negative frequencies.[*] However, as discussed earlier in this chapter, near the end of Section 10.2.3, $\chi_1(\omega)$ is an even function of frequency and $\chi_2(\omega)$ is an odd function of frequency. I can use these properties to reduce the integration range to $0 \rightarrow \infty$. I start with Eq. 10.14, which gives the real part, $\chi_1(\omega)$, and transform the integral into two of specific parities by multiplying both the numerator and the denominator of the integrand by $\omega' + \omega$. Separating into two integrals yields

$$\chi_1(\omega) = \frac{1}{\pi}\mathcal{P}\int_{-\infty}^{\infty} d\omega' \frac{\omega'\chi_2(\omega')}{\omega'^2 - \omega^2} + \frac{\omega}{\pi}\mathcal{P}\int_{-\infty}^{\infty} d\omega' \frac{\chi_2(\omega')}{\omega'^2 - \omega^2}$$

Now, $\chi_2(\omega)$ is odd so that the second integral vanishes; the part from $-\infty \rightarrow 0$ exactly cancels the part from $0 \rightarrow \infty$. Further, $\omega\chi_2(\omega)$ is even, so the part of the first integral from $-\infty \rightarrow 0$ exactly equals the part from $0 \rightarrow \infty$. So I only need to do the integral from $0 \rightarrow \infty$ and multiply by two, obtaining[†]

$$\chi_1(\omega) = \frac{2}{\pi}\mathcal{P}\int_{0}^{\infty} d\omega' \frac{\omega'\chi_2(\omega')}{\omega'^2 - \omega^2}. \tag{10.16}$$

I carry out the same derivation for the imaginary part and get

$$\chi_2(\omega) = -\frac{2\omega}{\pi}\mathcal{P}\int_{0}^{\infty} d\omega' \frac{\chi_1(\omega')}{\omega'^2 - \omega^2}. \tag{10.17}$$

Eqs. 10.16 and 10.17 show the Kramers–Kronig relations in the form usually used for analysis of optical functions.

10.2.6 Case When the DC Conductivity Is Finite

I have made a mistake (or, perhaps, have omitted something) in the previous section. Suppose the solid has a finite DC conductivity. Many solids do, including metals, superconductors, doped semiconductors, and so forth. Then at low frequencies, $\chi(\omega) = i\sigma_{dc}/\omega$, which diverges at $\omega = 0$. There is a pole at the origin and the contour of Fig. 10.3 runs right through it. To handle this, I will write the susceptibility for the conductor as an analytic part χ_a plus the divergent term:

$$\chi = \chi_a + i\sigma_{dc}/\omega.$$

I can then assert that $\chi_a \equiv \chi - i\sigma_{dc}/\omega$ is analytic in the upper half plane and so satisfies the Kramers–Kronig relation. The integral of Eq. 10.17 becomes

---

[*] My signal generator or spectrometer only has positive frequencies on its dial.

[†] If you want to work through this in detail, break the single integral over $-\infty \rightarrow \infty$ (Eq. 10.14) into two, one from $-\infty \rightarrow 0$ and a second from $0 \rightarrow \infty$. In the first, define $\omega'' = -\omega'$ and substitute for ω' everywhere, including $d\omega'$ and the limits. You will arrive a the results given here.

$$\chi_2(\omega) = \frac{\sigma_{dc}}{\omega} - \frac{2\omega}{\pi}\mathcal{P}\int_0^\infty d\omega'\,\frac{\chi_1(\omega')}{\omega'^2 - \omega^2}. \tag{10.18}$$

(Eq. 10.16 is unchanged.)

10.2.7 Kramers–Kronig Analysis of Other Optical Functions

There are many complex optical functions in addition to the susceptibility: dielectric function, conductivity, refractive index, etc. and there are Kramers–Kronig relations among all of them. Many can be obtained by substitution. For example, $\epsilon = 1 + 4\pi\chi$ and $\sigma_1 = \omega\chi_2 = \omega\epsilon_2/4\pi$ so that

$$\epsilon_1(\omega) = 1 + \frac{2}{\pi}\mathcal{P}\int_0^\infty d\omega'\,\frac{\omega'\epsilon_2(\omega')}{\omega'^2 - \omega^2}, \tag{10.19}$$

or

$$\epsilon_1(\omega) = 1 + 8\mathcal{P}\int_0^\infty d\omega'\,\frac{\sigma_1(\omega')}{\omega'^2 - \omega^2}. \tag{10.20}$$

Also,

$$\epsilon_2(\omega) = \frac{4\pi\sigma_{dc}}{\omega} - \frac{2\omega}{\pi}\mathcal{P}\int_0^\infty d\omega'\,\frac{[\epsilon_1(\omega') - 1]}{\omega'^2 - \omega^2}, \tag{10.21}$$

or, equivalently,

$$\sigma_1(\omega) = \sigma_{dc} - \frac{2\omega^2}{\pi}\mathcal{P}\int_0^\infty d\omega'\,\frac{\sigma_2(\omega')}{\omega'(\omega'^2 - \omega^2)}. \tag{10.22}$$

Kramers–Kronig relations are commonly used for estimating many optical functions from experiments that give one member of a complex pair. Among these are $\sigma_1 \rightarrow \sigma_2$, $\mathrm{Im}(1/\epsilon) \rightarrow \mathrm{Re}(1/\epsilon)$, $n \rightarrow \kappa$, κ or $\alpha = 2\omega\kappa/c \rightarrow n$, and $\mathscr{R} \rightarrow \phi$. Of these, the calculation of the phase shift on reflectance from the reflectance itself is by far the most commonly used and will be discussed in detail in the following subsection.

Table 10.1 shows a number of functions used in Kramers–Kronig integrals. But note also that moments of these functions also are consistent with the Kramers–Kronig relations as long as the moments satisfy the convergence condition specified at the beginning of Section 10.2.3. I write the general integral as

$$\mathcal{I}(\omega) = \frac{2}{\pi}\mathcal{P}\int_0^\infty dx\,\frac{f(x)}{x^2 - \omega^2} \tag{10.23}$$

and then show in the left column the form $f(x)$ must take to be put in to Eq. 10.23. The right column shows the function that would be obtained and its dependence on $\mathcal{I}(\omega)$. The complex quantities here are the dielectric function, $\epsilon = \epsilon_1 + i\epsilon_2$, the conductivity, $\sigma = \sigma_1 + i\sigma_2$, the susceptibility, $\chi = \chi_1 + i\chi_2$, the refractive index, $N = n + i\kappa$, the loss function, $(1/\epsilon) = \epsilon_1/(\epsilon_1^2 + \epsilon_2^2) - i\,\epsilon_2/(\epsilon_1^2 + \epsilon_2^2)$, and the surface impedance, $Z_s = R_s + iX_s$.

Table 10.1 A function $f(x)$ is integrated to give $\mathcal{I}(\omega)$. The possibilities for $f(x)$ and the result in terms of $\mathcal{I}(\omega)$ are given here.

$f(x)$	$\rightarrow$	KK transform
$x\epsilon_2(x)$	$\rightarrow$	$\epsilon_1(\omega) = 1 + \mathcal{I}$
$\epsilon_1(x) - 1$	$\rightarrow$	$\epsilon_2(\omega) = (4\pi\sigma_{\rm dc}/\omega) - \omega\mathcal{I}$
$\sigma_1(x)$	$\rightarrow$	$\sigma_2(\omega) = -\omega\mathcal{I}$
$\sigma_2(x)/x$	$\rightarrow$	$\sigma_1(\omega) = \sigma_{\rm dc} + \omega^2\mathcal{I}$
$x\chi_2(x)$	$\rightarrow$	$\chi_1(\omega) = \mathcal{I}$
$\chi_1(x)$	$\rightarrow$	$\chi_2(\omega) = (\sigma_{\rm dc}/\omega) - \omega\mathcal{I}$
$x\kappa(x)$	$\rightarrow$	$n(\omega) = 1 + \mathcal{I}$
$n - 1$	$\rightarrow$	$\kappa(\omega) = -\omega\mathcal{I}$
$x[-\,{\rm Im}(1/\epsilon(x)]$	$\rightarrow$	${\rm Re}[1/\epsilon(\omega)] = 1 - \mathcal{I}$
${\rm Re}[1/\epsilon(x)] - 1$	$\rightarrow$	$-\,{\rm Im}(1/\epsilon(\omega)] = \omega\mathcal{I}$
$xX_s(x)$	$\rightarrow$	$R_s = Z_0 + \mathcal{I}$
$R_s - Z_0$	$\rightarrow$	$X_s = -\omega\mathcal{I}$

$Z_0 = 377\ \Omega = 4\pi/c$ is the impedance of free space.

10.3 Kramers–Kronig Analysis of Reflectance

Kramers–Kronig analysis, of – for the most part – reflectance data, is commonly used to estimate the optical conductivity, dielectric function, sum rules, and other optical functions for new materials. Many reports of Kramers–Kronig analysis of reflectance have appeared, spanning more than a 50 years [116], with studies of metals [44, 45, 49–52, 59, 60], pure and doped elemental solids [117–119], organic conductors [120–122], charge-density-wave materials [123–125], conducting polymers [126–128], cuprate superconductors [129–139], manganites [140–142], pnictides [143–149], heavy-Fermions [150], multiferroics [151–154], topological insulators [155–157], and many others. In addition, a number of methods papers have appeared [109, 111, 158–163].

When I measure the reflectance,* I am taking the ratio of the reflected intensity or power to the incident intensity or power. Phase information is not available.† The amplitude reflectivity, Eq. 3.32, does have a phase; indeed I can write it as

$$r = \rho e^{i\phi} = \frac{1 - N}{1 + N},$$

where $\rho = \sqrt{\mathcal{R}}$ is the magnitude of the reflectivity, $\mathcal{R}$ is the single-bounce reflectance, ϕ is the phase shift on reflection, and $N = n + i\kappa$ is the complex refractive index.

---

* Reflectance here means the *single-bounce* reflectance, $\mathcal{R}$, as discussed in Chapter 3 (Section 3.2) and given in Eq. 3.33.

† Measuring the phase shift on reflection is difficult. The phase of the electromagnetic field changes by 2π for every wavelength of travel of the light, so that a phase-sensitive measurement requires control of lengths to a small fraction of the wavelength.

It would be nice to know ϕ, because I could invert Eq. 3.32 to get

$$N = \frac{1 - \sqrt{\mathscr{R}e^{i\phi}}}{1 + \sqrt{\mathscr{R}e^{i\phi}}},\tag{10.24}$$

using the known ϕ and the measured reflectance.*

Kramers–Kronig analysis is one way of estimating this phase [54, 158, 164, 165]. Consider

$$\ln r = \ln \rho + i\phi.$$

Here, $\ln \rho$ is the real part and ϕ is the imaginary part. The reflectance must be causal, and hence so must be the log of the reflectance. Equations 10.14 and 10.15 become

$$\ln \rho(\omega) = \frac{1}{\pi} P \int_{-\infty}^{\infty} d\omega' \, \frac{\phi(\omega')}{\omega' - \omega},\tag{10.25}$$

and

$$\phi(\omega) = -\frac{1}{\pi} P \int_{-\infty}^{\infty} d\omega' \, \frac{\ln \rho(\omega')}{\omega' - \omega}.\tag{10.26}$$

The hermiticity of r, $r(-\omega) = [r(\omega)]^*$, proves that ρ is even and ϕ is odd. I use these properties as I did earlier for X to obtain

$$\phi(\omega) = -\frac{2\omega}{\pi} P \int_{0}^{\infty} d\omega' \, \frac{\ln \rho(\omega')}{\omega'^2 - \omega^2}.\tag{10.27}$$

Equation 10.27 is perfectly usable for numerical analysis, but there is one improvement that can be made [54]. I write

$$P \int_{0}^{\infty} d\omega' \, \frac{1}{\omega'^2 - \omega^2} = 0.\tag{10.28}$$

The integral is zero because the negative area for $\omega' < \omega$ cancels the positive area for $\omega' > \omega$. Thus, I can add

$$+\frac{2\omega}{\pi} \ln \rho(\omega) P \int_{0}^{\infty} d\omega' \, \frac{1}{\omega'^2 - \omega^2}$$

---

* Every time I encounter the formula in Eq. 10.24 I think that it contains an algebra error and that the minus and plus signs are transposed. But then I recall that the phase shift for nonabsorbing materials with $n > 1$ is $-\pi$ so that r is negative. The range of ϕ is $-\pi \le \phi \le 0$. If I want to have a positive phase, I could define $\theta = \phi + \pi$ so that $e^{i\theta} = -e^{i\phi}$. In this case,

$$N = \frac{1 + \sqrt{\mathscr{R}e^{i\theta}}}{1 - \sqrt{\mathscr{R}e^{i\theta}}}$$

and, of course,

$$\rho e^{i\theta} = \frac{N - 1}{N + 1}.$$

The range of θ is $0 \le \theta \le \pi$. The real part of $\rho e^{i\theta}$ becomes negative when $n < 1$ and κ is small, requiring $\theta > \pi/2$.

to the right hand side of Eq. 10.27 without affecting the phase. Collecting terms, replacing ρ with $\sqrt{\mathscr{R}}$, and using the properties of the log,* I get

$$\phi(\omega) = -\frac{\omega}{\pi} \mathcal{P} \int_0^\infty d\omega' \, \frac{\ln[\mathscr{R}(\omega')/\mathscr{R}(\omega)]}{\omega'^2 - \omega^2}. \tag{10.29}$$

This modification has two advantages. First, if there are errors in the calibration of the reflectance measurements, so that the data for $\mathscr{R}$ are in error by a constant factor, the results for ϕ are unaffected. Second, both numerator and denominator of the integrand are zero when $\omega' = \omega$. L'Hôpital's rule shows that the ratio does not diverge; hence the pole has been removed. The second consequence is more important than the first. Removing the pole is important for simple evaluation of the integral. Although the scale error does not affect the phase, it *does* affect $\mathscr{R}$, and Eq. 10.24 shows that N depends on both phase and reflectance.

As a final note, Eq. 10.28 shows that the factor $[\epsilon(\omega') - 1]$ may be replaced by $\epsilon(\omega')$ in Eqs. 10.21 and 10.22. However, if I want to do a numerical integral, I will prefer to keep it as it is, so that $[\epsilon(\omega') - 1] \to 0$ as $\omega \to \infty$.

10.3.1 Extrapolations

The alert reader will have noticed that the range of the integrals in Eqs. 10.16–10.22 and 10.29 is 0 to ∞ and may wonder how one acquires data over that entire range. The answer of course is that data are always limited to a finite range of frequencies. The experimenter typically has data from far infrared through near ultraviolet, covering, say, 5 meV–5 eV (40–40,000 cm^{-1}). This is a reasonably wide bandwidth, but extrapolations must always be made outside the measured range. The high frequency extrapolation is especially problematic and can cause significant distortions to the conductivity over the entire measured range, with consequences for sum rules as well. I will discuss these extrapolations in the context of Kramers–Kronig analysis of reflectance, but the same approaches can be used for other functions.

One must estimate the reflectance between zero and the lowest measured frequency. The best approach is to employ a model that reasonably describes the low-frequency data. Such models include Drude for metals, Lorentz for insulators, often a sum of several Lorentzians, sometimes a Drude plus Lorentzians. Many other functions exist. When a good fit of the model to the data is obtained, a set of reflectance points may be calculated between zero and the lowest measured frequency, using a spacing between points similar to that of the lowest-frequency data, and combined with the measured data.

Other approaches to the low-frequency extrapolations include an assumption that the reflectance is constant to DC or the use of the Hagen Rubens formula for a conductor, $\mathscr{R} = 1 - A\omega^{1/2}$. (Other funcitons, such as $\mathscr{R} = 1 - A\omega^s$, with s in the range 1–4, are used as well.) The constant A is adjusted so the extrapolation goes through the first few points and then, using a spacing between points similar to that of the lowest-frequency data, a

---

* $\ln \sqrt{\mathscr{R}} = (^1/_2) \ln \mathscr{R}$ and $\ln A - \ln B = \ln(A/B)$.

set of extrapolated reflectance points is calculated between zero and the lowest measured frequency and combined with the measured data.

I have already mentioned that the high-frequency extrapolation can be a source of major error [166]. It is good to use data from other experiments on identical or similar samples if these exist. Or, one can measure far into the ultraviolet with synchrotron radiation if one has access to such a facility. Moving to the highest frequencies, one knows that in the limit as $\omega \to \infty$, the dielectric function is mostly real and slightly smaller than unity, following $\epsilon = 1 - \omega_p^2/\omega^2$, where ω_p is the plasma frequency of all the electrons in the solid and $\omega \gg \omega_p$. Then $n = 1 - \omega_p^2/2\omega^2$ and $\mathscr{R} = \omega_p^4/8\omega^4 \equiv C/\omega^4$. It has been usual to fill the region between the highest-frequency data point and the transition to ω^{-4} with a weaker power law, $\mathscr{R} \sim B/\omega^s$ with $0 < s < 4$. (s does not have to be an integer.) This midrange power law is adjusted to match the slope of the data and to give pleasing curves, but the choice of power (typically between 1 and 3) is arbitrary. The value of B is chosen so that the power law joins smoothly to the data at the high frequency limit; the value of C is also chosen for a smooth transition between mid- and high-frequency extrapolations. The free parameters are the exponent s and the frequency for the crossover from ω^{-s} and ω^{-4}. The integrals in the extrapolation range may be done analytically [54] and their contributions added to the phase obtained by numerical integration of Eq. 10.29 over the low-frequency extrapolation and the measured data.

Although simple, this midrange power-law extrapolation is arbitrary and affects in very noticeable ways the conductivity spectrum, sum rules, etc. There is a better way. It is fairly simple to extrapolate using X-ray atomic scattering functions developed by Henke and co-workers [22, 167]. These extrapolations basically treat the solid as a linear combination of its atomic constituents and, knowing the chemical formula and the density, allow the computation of dielectric function, reflectivity, and other optical functions. The "Henke reflectivity" can be used over photon energies of 10 eV–34 keV, after which a $1/\omega^4$ continuation is perfectly fine.

Photoabsorption in the x-ray region has been considered by a number of authors [168–175]. It is described by an atomic scattering function, a complex quantity: $f = f_1 + if_2$. The approach used by these authors is to combine experiment and theory and determine the imaginary part of the scattering factor, f_2, for each atomic species. This quantity has peaks or discontinuities at the absorption thresholds for each electronic level and falls to zero as $\omega \to \infty$. The real part f_1 is obtained by Kramers–Kronig integration of f_2, It increases with frequency via a series of plateaus, each approximately equal to the number of "free" electrons at that photon energy. Electrons are free if their binding energies are much less than the photon energy. The limiting high frequency value is – except for relativistic corrections – then the atomic number Z.

The starting point is a set of tables of the scattering factors for each element devised by Henke, Gullikson, and Davis [22] in 1993. A related web site also exists [167] with the ability to calculate optical properties, including reflectance. I find however that better results are found if two adjustments are made, one to the scattering functions f and one to the procedure of using these function to calculate reflectance. The first is that the functions of Ref. 22 provide f_2 from 10 to 30,000 eV but only have f_1 from 30 to 30,000 eV. I have redone the Kramers–Kronig integrals of f_2 to provide f_1 also from 10 eV (80,000 cm^{-1}), extrapolating $f_2 \sim \omega^2$ at low frequencies and $f_2 \sim \omega^{-1}$ at high frequencies.

To estimate the high-frequency properties of a material, one makes the assumption that the solid consists of a linear combination of its component atomic constituents, with the dielectric function determined by the scattering factors and the number density of the constituents. The dielectric function is then

$$\epsilon = 1 - \sum_j \frac{4\pi n_j e^2}{m\omega^2}(f_1^j - i f_2^j), \tag{10.30}$$

where the sum runs over atoms j at number density n_j and with complex scattering factor f^j. Note that this has the right limiting high-frequency behavior, because $f_1^j \to Z^j$ (with Z^j here the atomic number) and $f_2 \to 0$, so that $\epsilon \to 1 - \sum_j 4\pi n_j Z^j e^2/m\omega^2$. The complex refractive index* is

$$N = \sqrt{\epsilon} \tag{10.31}$$

and the reflectance $\mathcal{R}$ is

$$\mathcal{R} = \left| \frac{1-N}{N+1} \right|^2. \tag{10.32}$$

The procedure is implemented in the following way. The user supplies the chemical formula, such as Ag or La$_{1.85}$Sr$_{0.15}$CuO$_4$ and either the appropriate† volume V_c or the density ρ. With this information, the reflectance $\mathcal{R}$ can be calculated at 340 logarithmically spaced points over 80,000–2.4$\times$10^8 cm^{-1} (10–30,000 eV) using Eqs. 10.30–10.32. A bridge needs to be placed over the gap between the highest experimental point (say, 40,000 cm^{-1} or 5 eV) and the beginning of the extrapolated reflectance at 80,000 cm^{-1}. The user has the option of a power series in ω, in $1/\omega \propto \lambda$ or a cubic spline.‡ As it turns out, the bridge has a modest effect in some cases, minimal in others. Then, the Kramers–Kronig integral, Eq. 10.29, is computed to obtain the phase.

---

* Note that Ref. 22 and the website Ref. 167 write an equation for the refractive index of a monatomic solid:

$$N = 1 - \frac{n r_0 \lambda^2}{2\pi}(f_1 - i f_2),$$

with $r_0 = e^2/mc^2$ the classical radius of the electron and $\lambda = 2\pi c/\omega$ the wavelength. This is clearly the first term in an expansion of $N = \sqrt{\epsilon}$. It is preferable to compute the dielectric function from f using Eq. 10.30, and then take the complex square root to obtain N.

† There is occasional confusion – both here and in the application of partial sum rules – about the value to use for the volume V_c. This volume is described variously as the "volume of the unit cell" or the "volume of one formula unit." Of course any carefully described volume will work, but generally one wants the number of charge carriers per atom or per primitive unit cell. And of course the conventional cell often contains several times more atoms than one would guess from the chemical formula. So the user must decide what she or he wants to determine. It may be the number of carriers per Ag atom in silver metal, the number per buckyball in C$_{60}$, the number per dopant in P-doped Si, or the number per copper atom in Bi$_2$Sr$_2$CaCu$_2$O$_8$. Once the decision is made, compute the volume in the crystal allocated to the desired quantity. One (almost) failsafe approach is to obtain the density ρ of the crystal and the mass M of the entity one is interested in, such as one Ag atom, 60 C atoms, one silicon atom divided by the dopant concentration, or 1/2 the mass of 2Bi + 2Sr + Ca + 2Cu + 8O. Then $V_c = M/\rho$.

‡ A cubic spline bridge occasionally causes excitement by taking the reflectance in the bridge region over unity or below zero.

Persons who wish to test the approach will find a Windows version of the code (and the Kramers–Kronig routine that uses the extrapolation and generates the bridge function) in the program suite associated with this book. See Appendix G. Use `xro.exe` to generate the file for extrapolation, `kk.exe` to do the Kramers–Kronig integral, and `op.exe` to calculate optical functions. The use of reflectance calculated from a dielectric function made up of a sum of atomic scattering functions for a material provides a reliable and reproducible method of extrapolating measured data. It removes a certain amount of arbitrariness in the use of Kramers–Kronig analysis of reflectance.

10.3.2 Kramers–Kronig Analysis of Transmittance

The (coherent) transmittance $\mathcal{T}$ of a thin* parallel-plate slab is related to the amplitude transmission coefficient by

$$\mathcal{T} = |te^{i\phi_t}|^2$$

where t is the amplitude and ϕ the phase change on passing through the sample, including the coherent addition of multiple internal reflections.

As in reflectance, $\ln t + i\phi_t$ are causal functions that satisfy causality [176–179],

$$\phi_t(\omega) = -\frac{\omega}{\pi}\mathcal{P}\int_0^\infty d\omega'\,\frac{\ln[\mathcal{T}(\omega')/\mathcal{T}(\omega)]}{\omega'^2 - \omega^2} + \frac{\omega}{c}d, \tag{10.33}$$

where d is the thickness and the last term is a correction for the phase gained on going through an identical thickness of vacuum. The equation for the transmission of a slab in terms of the refractive index or dielectric function is discussed in Chapter 7. Given $\mathcal{T}$ and ϕ_t, these may be inverted numerically for n and κ.

Like reflectance, Kramers–Kronig analysis of transmittance requires extrapolations. At low frequencies, I use one of these:

- Transmission is assumed constant to DC
- Metallic conductivity assumed: $\mathcal{T} = B + C\omega^2$
- Superconducting: $\mathcal{T} = A\omega^2$

Above the experimental regime, it is good to use data from other experiments on identical or similar samples if these exist. At the highest frequencies, in the x-ray range, the data can be extrapolated as $\mathcal{T} \sim 1 - C\omega^{-4}$ as appropriate for completely free electrons. Between the highest data point and the x-ray range atomic scattering factors or a slower power law can be used.

10.4 Another Look at the Conductivity

The conductivity σ, which appears in Ohm's law, $\mathbf{j} = \sigma\mathbf{E}$, is a well behaved function with no poles in the upper half plane. At very high frequencies, the conductivity is that of free

---

* The thickness scale is λ/n, the wavelength divided by the refractive index. If the slab is several wavelengths thick, the requirements on the variations in thickness and in the parallelism of the light passing through it become severe.

carriers, Eq. 10.1. Thus it falls like $1/\omega$ and therefore (barely) satisfies the requirements specified earlier in this chapter (Section 10.2.3). Moreover σ_1 is an even function and σ_2 is odd, so that the derivation from Eq. 10.12 to 10.17 goes through exactly, and I arrive at [105]

$$\sigma_1(\omega) = \frac{2}{\pi} \mathcal{P} \int_0^\infty d\omega' \, \frac{\omega' \sigma_2(\omega')}{\omega'^2 - \omega^2}, \tag{10.34}$$

and

$$\sigma_2(\omega) = -\frac{2\omega}{\pi} \mathcal{P} \int_0^\infty d\omega' \, \frac{\sigma_1(\omega')}{\omega'^2 - \omega^2}. \tag{10.35}$$

Eq. 10.35 is identical to Eq. 10.20 but Eq. 10.34 differs from Eq. 10.22. The two integrals employ different moments of σ_1.

10.5 Sum Rules

A sum rule is a statement that the integral of an optical function has a fixed value, is zero, or is related to a specific value of another optical function. I introduced the oscillator strength sum rule (or f sum rule) in Chapter 9, Section 9.7. Here I will briefly discuss what the Kramers–Kronig relation says about sum rules [60, 107, 180–186].

10.5.1 The f Sum Rule

I start by writing the high-frequency limiting value of the real part of the dielectric function. (I have written this several times before.)

$$\epsilon_1(\omega) = 1 - \frac{\omega_p^2}{\omega^2},$$

where ω_p is a plasma frequency for every electron in the solid. In the notation of Eqs. 9.21 and 9.29,

$$\omega_p = \sqrt{\frac{4\pi \mathfrak{n} e^2}{m}},$$

where $\mathfrak{n}$ counts *every* electron in the solid. I use this funny symbol to reduce confusion with the conduction or valence electron density n.

I equate this equation to the Kramers–Kronig integral of Eq. 10.20, keeping in mind that ω must be very large,

$$1 - \frac{\omega_p^2}{\omega^2} = 1 + 8\mathcal{P} \int_0^\infty d\omega' \, \frac{\sigma_1(\omega')}{\omega'^2 - \omega^2}.$$

I now split the integral into two integrals, with the boundary frequency, ω_f, less than ω but high enough that $\sigma_1(\omega_f)$ is essentially zero:

$$-\frac{\omega_p^2}{8\omega^2} = \mathcal{P} \int_0^{\omega_f} d\omega' \, \frac{\sigma_1(\omega')}{\omega'^2 - \omega^2} + \mathcal{P} \int_{\omega_f}^\infty d\omega' \, \frac{\sigma_1(\omega')}{\omega'^2 - \omega^2}.$$

Because $\sigma_1(\omega_f)$ is arbitrarily small, the second integral can be neglected. Because $\omega \gg \omega_f$, the term ω'^2 in the first integral can be neglected. Hence

$$-\frac{\omega_p^2}{8\omega^2} = \mathcal{P} \int_0^{\omega_f} d\omega' \, \frac{\sigma_1(\omega')}{-\omega^2}.$$

I can now change sign, cancel the factors of ω^{-2}, drop the principal value symbol (there are no divergences of the integrand), and extend the upper limit to ∞ (because by ω_f I have already gotten all the area under σ_1) to get an equation identical to Eqs. 9.29 and 9.30:

$$\int_0^\infty d\omega' \, \sigma_1(\omega') = \frac{\omega_p^2}{8} = \frac{\pi}{2} \frac{ne^2}{m}. \tag{10.36}$$

See Chapter 9 for an evaluation of Eq. 10.36 for the Drude conductivity.

Equation 10.36 is also a sum rule on $\omega\epsilon_2$ because $\omega\epsilon_2 = 4\pi\sigma_1$; the right hand side is 4π larger of course. A similar development of the energy loss function, $\mathrm{Im}(1/\epsilon)$ gives:

$$\int_0^\infty d\omega' \, \omega' \, \mathrm{Im}\left(-\frac{1}{\epsilon(\omega')}\right) = \frac{\pi\omega_p^2}{2} = \frac{2\pi^2 ne^2}{m}. \tag{10.37}$$

For a simple free-electron (Drude) metal, σ_1 has a maximum at zero frequency and rolls off with characteristic width $1/\tau$. The area under σ_1 mostly occurs at low frequencies. The loss function has a peak at ω_p, with width (in a simple model where τ is not a function of frequency) $1/\tau$. But the integrals of both functions are proportional to the plasma frequency squared.

10.5.2 The Static Dielectric Function Sum Rule

Equation 10.20 shows that the real part (the dispersive or reactive part) of the dielectric function is related to the real part of the optical conductivity. Evaluating this equation for $\omega = 0$ yields the static dielectric function $\epsilon_{dc} \equiv \epsilon_1(\omega = 0)$ as the ω^{-2} moment of σ_1,

$$\epsilon_{dc} = 1 + 8\mathcal{P} \int_0^\infty d\omega' \, \frac{\sigma_1(\omega')}{\omega'^2}. \tag{10.38}$$

This equation is not useful for conductors, because there is a divergence of the integrand at $\omega' = 0$ unless σ_1 is zero there. For insulators, one can see that conductivity peaks at low frequencies are the largest factors in determining the static dielectric constant. Peaks with high values of the optical conductivity and located at low frequencies are required for high values of the dielectric constant.* Moreover, there is a trend for materials with high dielectric constants or refractive indices in their transparent regions to have small band gaps; large-band-gap materials have low values of ϵ and n.

---

* The high-temperature paraelectric phase of many ferroelectric insulators has a "soft mode," a lattice vibration where the transverse-optic frequency goes to zero as the temperature is reduced towards the Curie temperature. The dielectric constant thus diverges at the transition.

10.5.3 DC Conductivity Sum Rules

A third sum rule uses Eq. 10.34, evaluated at $\omega = 0$:

$$\int_0^\infty d\omega' \, \frac{\sigma_2(\omega')}{\omega'} = \frac{\pi}{2}\sigma_{\text{dc}},$$

or, substituting for σ_2,

$$\int_0^\infty d\omega' \, [1 - \epsilon_1(\omega')] = 2\pi^2\sigma_{\text{dc}}. \tag{10.39}$$

Eq. 10.39 makes it evident that it is OK to convert the principal value integral into a regular integral. This equation says that the parts of ϵ_1 that are bigger than unity are balanced by parts below unity except for the free carrier pole. Note that all solids have real parts of their dielectric functions that are slightly smaller than unity at very high energies, becoming unity asymptotically as the frequency goes to infinity.

10.5.4 Sum Rules for the Refractive Index

The real and imaginary parts of the refractive index satisfy Kramers–Kronig relations. These are given in Table 10.1 (and Eq. 10.23). I see that

$$n = 1 + \frac{2}{\pi}\int_0^\infty d\omega' \, \frac{\omega'\kappa(\omega')}{\omega'^2 - \omega^2}.$$

As in the derivation for the integral of σ_1, I argue that $\kappa \to 0$ as $\omega \to \infty$ fast enough that I can stop the integral as some frequency $\omega_f \ll \omega$. I write for these large values of ω,

$$n = \sqrt{\epsilon} \approx 1 - \frac{\omega_p^2}{2\omega^2}$$

where the factor of 2 comes from expanding the square root. Then

$$-\frac{\omega_p^2}{2\omega^2} = \frac{2}{\pi}\int_0^\infty d\omega' \, \frac{\omega'\kappa(\omega')}{-\omega^2}$$

or

$$\int_0^\infty d\omega' \, \omega'\kappa(\omega') = \frac{\pi}{4}\omega_p^2. \tag{10.40}$$

Equation 10.40 is the f sum rule for the extinction coefficient. It could be written also in terms of the absorption coefficient α, because $\alpha = 2\omega\kappa/c$.

Now, I can write Eq. 10.36 in terms of the imaginary part of the dielectric function, where it reads

$$\int_0^\infty d\omega' \, \omega'\epsilon_2(\omega') = \frac{\pi}{2}\omega_p^2,$$

so that the result of the integrations is just twice that of Eq. 10.40. Therefore

$$\int_0^\infty d\omega' \, \omega'\epsilon_2(\omega') = 2\int_0^\infty d\omega' \, \omega'\kappa(\omega').$$

But $\epsilon_2 = 2n\kappa$! Hence, after minimal algebra, this equation becomes

$$\int_0^\infty d\omega'\,\omega'\kappa(\omega')[n(\omega') - 1] = 0. \tag{10.41}$$

Finally, I can start with the Kramers–Kronig integral for κ in terms of n from Table 10.1 and do the usual high frequency evaluation (where $\kappa = 0$) to obtain a sum rule for the refractive index alone,

$$\int_0^\infty d\omega'\,[n(\omega') - 1] = 0, \tag{10.42}$$

known as the inertia sum rule.

I think that it is remarkable that the integral of n, whether weighted by $\omega\kappa$ or not, comes out to be the same as the refractive index of the vacuum. At some frequencies n is larger than one and at others smaller than one. But (in two senses) the average value is unity.

10.6 Partial Sum Rules

The sum rule integrals extend from zero to infinity. It is often informative to compute a partial sum rule. I can carry out the integral from zero to some frequency ω, and then plot the result as a function of the upper limit of the integration. I'll formulate the problem as follows. The right-hand side of Eq. 10.36 can be written

$$\frac{\pi n e^2}{2m} = \frac{\pi e^2}{2m V_c} N_{\text{eff}} \frac{m}{m^*}$$

where m^* is the effective mass, m is the free electron mass, V_c is the unit cell volume (or formula volume), and N_{eff} is the number of effective electrons within the volume V_c participating in optical transitions. Then I can write

$$\frac{m}{m^*} N_{\text{eff}}(\omega) = \frac{2m V_c}{\pi e^2} \int_0^\omega d\omega'\,\sigma_1(\omega') \tag{10.43}$$

for the number of electrons with effective mass ratio m/m^* participating in optical transitions at frequencies below ω.

Figure 10.4 shows $N_{\text{eff}}(m/m^*)$ for a model of a metal similar to Ag. The model uses a Drude free carrier component, with $1/\tau \approx 130$ cm^{-1} and three Lorentz oscillators, at 40,000 cm^{-1} (5 eV), 54,000 cm^{-1} (5.7 eV), and 70,000 cm^{-1} (8.7 eV). The conductivity is shown in the inset. The main plot shows the partial sum rule. The integral of the Drude conductivity saturates at 1 electron/Ag atom. $N_{\text{eff}}(m/m^*)$ is flat over a wide frequency range and then steps up when the interband transitions begin. See section 6.1 and Fig. 6.5 for the partial sum rule analysis of Ag.

Similar partial sum rules may also be written for the loss function, for the extinction coefficient, and for the absorption coefficient.

I may also write a partial sum rule based on Eq. 10.38,

$$\Delta\epsilon_{\text{dc}}(\omega) = 1 + 8\mathcal{P}\int_0^\omega d\omega'\,\frac{\sigma_1(\omega')}{\omega'^2}, \tag{10.44}$$

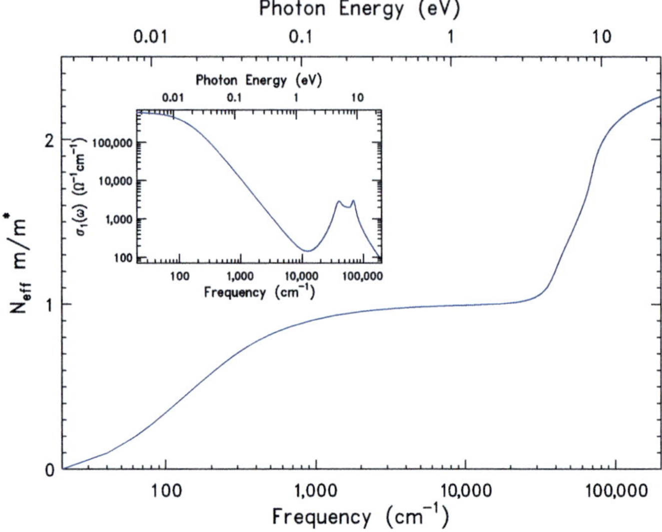

Fig. 10.4 $N_{\mathit{eff}}\,(m/m^*)$ for a model of a metal similar to Ag. The inset shows the conductivity whose integral is shown in the main plot. V_c is the volume allocated to one metal atom.

where I write $\Delta\epsilon_{\mathrm{dc}}(\omega)$ to indicate that the function shows the contributions to the static dielectric constant from successively occurring peaks in the conductivity. Note that Eq. 10.44 is equivalent in some ways to writing the static dielectric function as a sum of Lorentz oscillators, as in Eqs. 4.35 and 5.30, as a functon of the number of oscillators included in the sum:

$$\Delta\epsilon_{\mathrm{dc}}(N) = 1 + \sum_{j=1}^{N} \frac{\omega_{pj}^2}{\omega_j^2},$$

where I've set $\epsilon_c = 1$. If the oscillators are narrow compared to their separation, the integral in Eq. 10.44 will step up as the upper limit ω includes more and more resonances ω_j.

A third partial sum rule that I have seen used is based on Eq. 10.42. (One based in Eq. 10.41 is also possible.) I write

$$\Delta n(\omega) = \int_0^\omega d\omega'\,[n(\omega') - 1].$$

This function will start positive and increase as ω increases. Indeed, because the refractive index often looks like the derivative of the extinction coefficient, it will resemble the latter function. But when n becomes smaller than unity at high frequencies, $\Delta n(\omega)$ will begin to decrease with increasing ω and (if all has gone well) ramp down to zero.

Superconductors

11.1 Superconducting Phenomena

All discussions of superconductivity are supposed to start in 1911, in Leiden, when and where Kamerlingh Onnes [187] and his assistants discovered that the resistance of solid mercury abruptly falls to zero at a temperature of 4.2 K. Shown in Fig. 11.1, the resistance is falling with temperature, similar to other pure metals. (See Fig. 8.6, for example.) But then, the resistance takes a sudden drop to zero at 4.2 K. Onnes reported that the resistance was smaller than 10^{-5} Ω compared to 0.11 Ω just above T_c; it is now believed to be identically zero. Evidence for zero resistance includes experiments, done on rings of superconducting material, in which continuous and steady currents are sustained for years with no applied voltage.

In the slightly more than a century since its discovery, superconductivity has been found to be widely distributed, occurring in elements [187], alloys [188], intermetallic compounds [189], oxides [190], ionic compounds [191], organic charge-transfer compounds [192], polymers [193], and many other systems [194–207]. It seems that the only two prerequisites are (1) that the material be conductive, so that there are carriers to condense into the charged superfluid, and (2) that it not be a magnet; strong magnetic fields destroy superconductivity. But I should note that the conductivity does not need to be that of free-electron metals and that there are counterexamples to prerequisite 2.

Figure 11.2 shows the "records" for superconductivity over a 107 year period, 1911–2018. Several classes of materials are shown. The blue circles are metals, alloys and intermetallic compounds that are superconducting due to electron phonon interactions. The record as of 2018 is held by H_3S under enormous (155 GPa) pressure; it is thought to be an electron–phonon superconductor. The high-T_c cuprates are shown as red squares; organic conductors as magenta pincushions; fullerenes as purple triangles; heavy-Fermion metals as gold diamonds; finally, the Fe-based superconductors are in green oh-plus symbols. The diagram is probably neither complete nor accurate. It does however indicate the enormous range of material types where superconductivity is observed.

The second key property of superconductors was discovered in 1929, when Meissner [208, 209] observed that a superconductor expels magnetic flux, making the condition for the field inside the superconductor be that $B = 0$, where B is the magnetic field. The flux expulsion is sketched in Fig. 11.3. Start with the constitutive equations, Eqs. 2.3 and 2.14

$$0 = B = H + 4\pi M = (1 + 4\pi\chi)H.$$

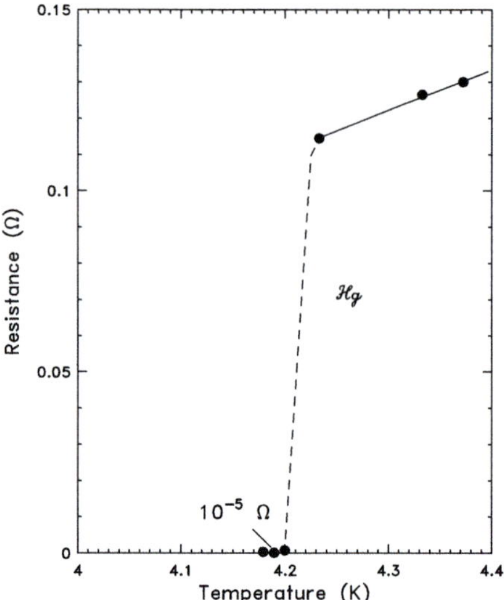

Fig. 11.1 Resistance *vs.* temperature for Hg. (After Onnes [187].)

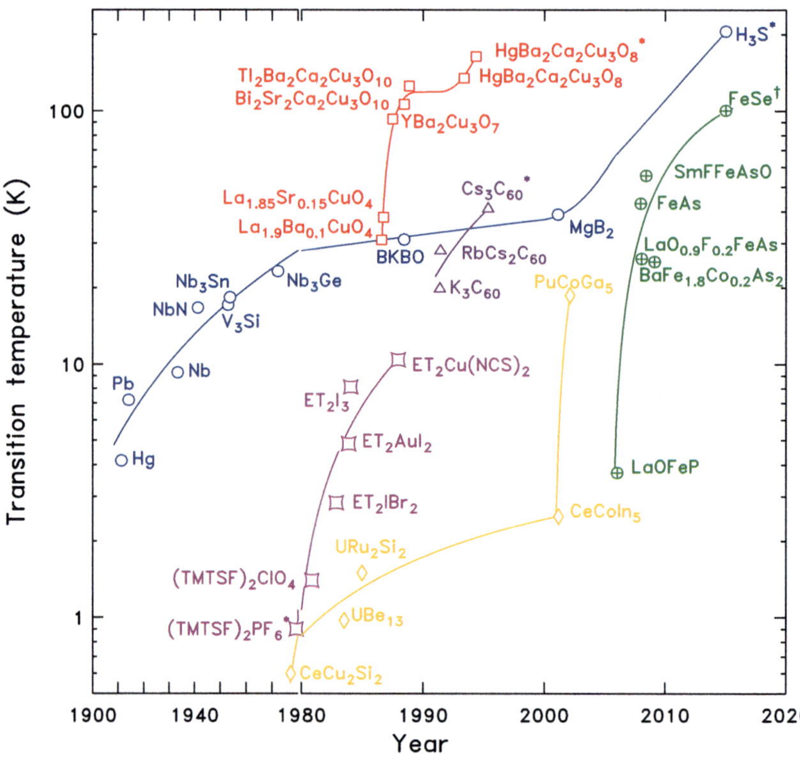

Fig. 11.2 Record transition temperatures in many classes of superconductors over a little more than a century. Note that there is a change of scale in the time axis at 1980; the first third covers 80 years and the second two-thirds covers 40 years. The materials with an asterisk attached require pressure to become superconducting (or to reach the T_c shown). The material with a dagger (FeSe) is a monolayer film.

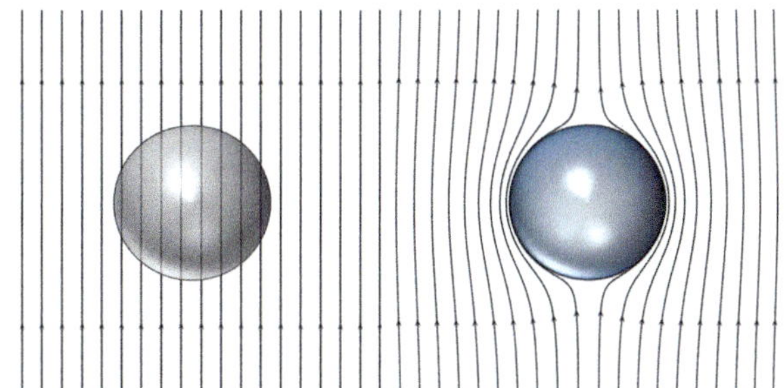

Fig. 11.3 Two spheres in a uniform external magnetic field **B**. The sphere on the left is a normal metal; the field penetrates uniformly. The sphere on the right is a superconductor (below T_c). The field is expelled, making **B** $= 0$ inside. At the same time, the sphere acquires a uniform magnetization.

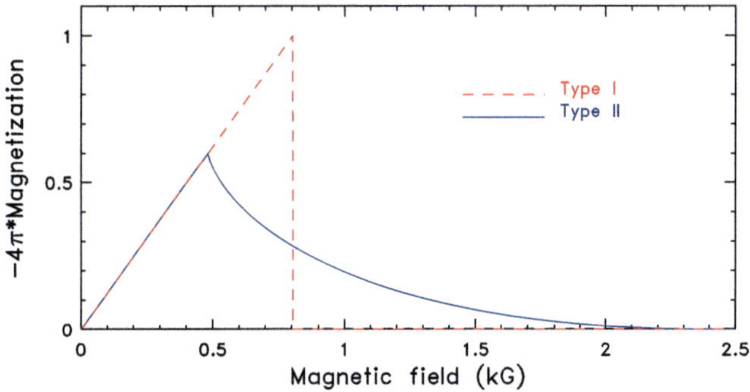

Fig. 11.4 Magnetization versus applied field for type I and type II superconductors. Parameters are chosen to represent Pb. Clean Pb is a type I superconductor with $H_c = 800$ G (0.08 T). Adding impurities to Pb renders it a type II superconductor.

B is zero but H is not zero. Hence,

$$\chi = -\frac{1}{4\pi}. \qquad (11.1)$$

In the Meissner state, the superconductor is a perfect diamagnet. The magnetic moment of the sphere is generated by currents which circulate on the surface of the sphere parallel to the equator.* The diamagnetic response continues in all or in part as the field increases, until an upper critical field is reached and superconductivity is destroyed.

Superconductors come in two flavors, illustrated in Fig. 11.4. Type I superconductors function as described above, with perfect diamagnetic response, following Eq. 11.1, up to the critical field H_c when superconductivity ceases to exist. Type II superconductors are perfect diamagnets up to a (typically relatively low) lower critical field called H_{c1}. Above this field, an inhomogeneous state occurs, with a magnetic flux lattice existing in the super-conductor. The density of flux in the material grows with increasing field. At a (typically relatively high) upper critical field, H_{c2}, the flux penetration is complete. Superconductivity (zero resistance) persists up to H_{c2}. The materials that remain superconducting at high

---

* Actually the currents are not on the surface but decay exponentially into the superconductor with a characteristic penetration depth. I will calculate this depth shortly.

magnetic fields are all type II and some of these (NbTi, Nb_3Sn, $YBa_2Cu_3O_7$, etc.) are used in superconducting magnets.

A list of superconductors is given in Table 11.1. The table lists the material, transition temperature, field required to destroy superconductivity, and the type (I or II). Transition temperatures can be seen to vary from 15 mK (α-W) to to 200–210 K. The current record* is held by H_3S at very high pressure [204–207].

Although you cannot see anything with your eyes, superconductivity affects optical properties in two important ways. First, the infinite DC conductivity (a delta function at $\omega = 0$) gives a large inductive response to the imaginary part of the conductivity, just as

Table 11.1 A list of superconducting materials. The materials are organized into elements, compounds, alloys, organics, cuprates, and pnictides. H_{ci} is the maximum field at which zero resistance is observed. The classification of I or II identifies the way magnetic flux is expelled.

Formula	T_c (K)	H_{ci} (T)	Type
	Metals		
α-Hg	4.15	0.04	I
Al	1.20	0.01	I
In	3.4	0.03	I
Sn	3.72	0.03	I
Ta	4.48	0.09	I
Pb	7.19	0.08	I
Nb	9.26	0.82	II
α-W	0.015	0.00012	I
α-U	0.68		I
Zn	0.855	0.005	I
V	5.03	1	II
	Compounds		
C_6K	1.5		II
$C_{60}K_3$	19.8	0.013	II
$C_{60}Rb_x$	28		II
In_2O_3	3.3		II
V_3Si	17.1	23	II
Nb_3Sn	18.3	24.5	II
Nb_3Ge	23.2	37	II
MgB_2	39	74	II
H_3S	203 (at 1,500 kbar)		
	Binary alloys		
NbO	1.38		II
TiN	5.6		
ZrN	10		
NbTi	10	15	II
NbN	16	15.3	II

* There are T_c's below 15 mK but none – so far – at 300 K.

Table 11.1 *continued.*

Formula	T_c (K)	H_{ci} (T)	Type
Charge-transfer salts			
$(TMTSF)_2PF_6$	1.1 (at 6.5 kbar)		II
$(TMTSF)_2ReO_4$	1.2 (at 9.5 kbar)		II
$(TMTTF)_2ClO_4$	1.4		II
β-$(ET)_2I_3$	1.5		II
β'-$(ET)_2SF_5CH_2CF_2SO_3$	5.3		II
κ-$(ET)_2Cu(NCS)2$	10.4	10	II
Cuprates (high-T_c)			
$La_{1.85}Sr_{0.15}CuO_4$	38	45	II
$La_2CuO_{4.11}$	43		II
$YBa_2Cu_3O_7$ (123)	92	140	II
$Bi_2Sr_2CuO_6$ (Bi-2201)	20		II
$Bi_2Sr_2CaCu_2O_8$ (Bi-2212)	92	107	II
$Bi_2Sr_2Ca_2Cu_3O_{10}$ (Bi-2223)	106		II
$Tl_2Ba_2CaCu_2O_8$ (Tl-2212)	108		II
$Tl_2Ba_2Ca_2Cu3O_{10}$ (Tl-2223)	125	75	II
$HgBa_2CuO_4$ (Hg-1201)	94	II	
$HgBa_2Ca_2Cu_3O_8$ (Hg-1223)	134	190	II
$HgBa_2Ca_2Cu_3O_8$ (Hg-1223)	164 (at 300 kbar)		II
$Nd_{1.7}Ce_{0.3}CuO_4$	24		II
Iron-based (pnictide)			
$LaO_{0.9}F_{0.2}FeAs$	28.5		II
$PrFeAsO_{0.89}F_{0.11}$	52		II
$GdFeAsO_{0.85}$	53.5		II
$BaFe_{1.8}Co_{0.2}As_2$	25.3		II
$SmFeAsO_{0.85}$	55		II
$Ba_{0.6}K_{0.4}Fe_2As_2$	38		II
$CaFe_{0.9}Co_{0.1}AsF$	22		II
$Sr_{0.5}Sm_{0.5}FeAsF$	56		II
$FeSe$	27		II

in the case of the perfect conductor, Eq. 10.3. Second there is a gap (called 2Δ for reasons that will become apparent) for excitations, which makes the real part of the conductivity zero for (finite) frequencies below 2Δ.*

---

* The gap energy is in the far infrared, which is why your eyes see no difference between normal and superconducting states.

11.2 Theoretical Background

This book is not intended for a course in superconductivity, so I will derive neither BCS theory [210, 211] nor the phenomenological Ginzburg-Landau theory [212–214]. Here I will just list some of the results. The book by Tinkham (Ref. 215) is a good place to start reading about these theories.

BCS theory [210, 211] is the microscopic theory of metallic superconductors. These superconductors have the following properties

- Below T_c, electrons form Cooper pairs. The Cooper state is a superposition of two electrons of the normal metal into a zero-momentum, spin-singlet state $|\mathbf{k} \uparrow, -\mathbf{k} \downarrow\rangle$.
- The superconducting condensate is a coherent superposition of Cooper pairs, involving all the conduction electrons in the metal.
- Because the pairs are made up of two electrons, the number density of Cooper pairs is $n_c = n/2$; the charge is $Q_c = 2e$; and the mass is $m_c = 2m$.
- The Cooper pairs exists because of an effective attractive interaction between the two electrons.
- In metallic superconductors, this attraction arises through the interaction with the lattice (the phonons).
- The attractive interaction reduces the energy of the electrons. In fact, the Fermi surface is unstable towards the formation of a Cooper pair even if there is only an infinitesimal attractive interaction between the electrons.
- To break a Cooper pair at the Fermi surface requires energy

$$2\Delta = 3.5 k_B T_c \tag{11.2}$$

to be supplied. This energy is known as the superconducting energy gap.
- In metallic (called "conventional") superconductors, the gap is small compared to the Fermi energy. The gap scale is a few meV; the Fermi energy several eV.
- Thermal energies break Cooper pairs and, because the gap is a collective effect, reduce the gap. It eventually reaches zero at T_c. The behavior (from both experiment and theory) is shown in Fig. 11.5 [216].
- Excitations above the gap are called *quasiparticles*. They have many of the properties of normal electrons, being spin-half Fermions, but are affected by the BCS coherence factors and have an energy-wave vector dispersion relation that differs significantly from that of ordinary electrons in the solid.

Ginzburg-Landau theory [212–214] is a phenomenological theory that is based on the Landau theory of second-order phase transitions. The free energy of a superconductor near T_c is written in terms of a complex order parameter, ψ, which is nonzero below T_c and zero above. The theory (when folded into BCS theory [217]) makes the following identifications,

$$\psi = \sqrt{n_c(T)} e^{i\phi} = \Delta(T), \tag{11.3}$$

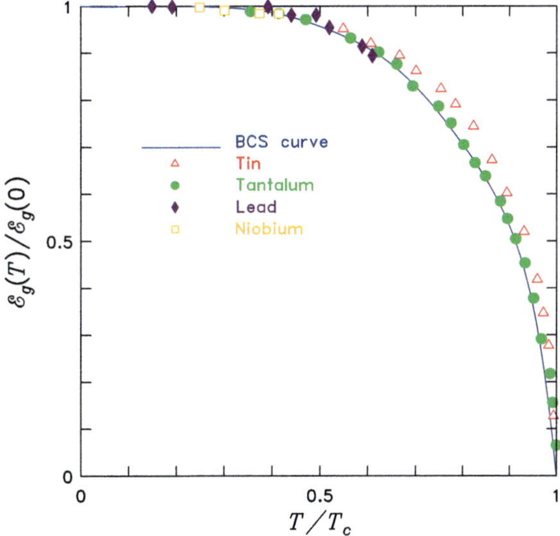

Fig. 11.5   Superconducting energy gap measured by tunneling as a function of temperature. (After Ref. 216.)

where ψ represents a coherent superposition of all Cooper pairs, the pair density is n_c, ϕ is the phase of the order parameter, and Δ is the BCS energy gap. The two-fluid model, in which there is a normal fluid (described perhaps by the Drude model) intermixed with the superfluid, is consistent with the Ginzburg-Landau theory. As n_c is reduced approaching T_c, (Fig. 11.5) n_n grows so that $n = 2n_c + n_n$.

11.3 The London Model

11.3.1 Heuristic Derivation

Dating from 1935, the London model [218] is a mix of electrodynamics and quantum mechanics. Often discussions just start by writing the London equation, but I can derive it in a relatively simple way. I'll start by reminding myself of the definition of the electrical current carried by a set of Cooper pairs. Classically it is

$$\mathbf{j} = n_c Q_c \mathbf{v} = \frac{n_c Q_c}{m_c}\mathbf{p}, \tag{11.4}$$

where $\mathbf{v}\,(\mathbf{p})$ is the average velocity (momentum) of the superconducting particles (Cooper pairs), n_c is their number density, Q_c is their charge, and m_c is their mass. Quantum mechanically, I would write something like

$$\mathbf{j}_s = \frac{Q_c}{2m_c}(\psi^*\mathbf{p}\,\psi - \psi\mathbf{p}\,\psi^*), \tag{11.5}$$

with $\mathbf{p}$ the momentum operator and ψ the wave function of the superconducting state, given by Eq. 11.3. In the presence of an electromagnetic field,

$$\mathbf{p} = -i\hbar\nabla - \frac{Q_c}{c}\mathbf{A}, \tag{11.6}$$

where $\mathbf{A}$ is the vector potential. The requirement of charge neutrality in a homogeneous superconductor causes n_c to be constant in space. Hence, from Eq. 11.3,

$$-i\hbar\nabla\psi = -i\hbar\sqrt{n_c}e^{i\phi}i\nabla\phi,$$

and Eq. 11.5 becomes

$$\mathbf{j}_s = \frac{n_c Q_c}{m_c}\left(\hbar\nabla\phi - \frac{Q_c}{c}\mathbf{A}\right). \tag{11.7}$$

Now, for Cooper pairs,

$$Q_c = -2e \qquad m_c = 2m \qquad \text{and} \qquad n_c = \frac{n_s}{2},$$

with n_s the superfluid density in a two-fluid model. At zero temperature in clean metals, $n_s = n$. Thus $n_c Q_c/m_c = -n_s e/2m$ and $n_c Q_c^2/m_c = n_s e^2/m$. With these replacements I arrive at

$$\mathbf{j}_s = -\frac{n_s e}{2m}\hbar\nabla\phi - \frac{n_s e^2}{mc}\mathbf{A}. \tag{11.8}$$

The two terms in Eq. 11.8 lead to different physics. The first term, proportional to the gradient of the phase, governs the Josephson effect, pair tunneling between two superconductors. The second term controls the electromagnetic properties of homogeneous superconductors in which the phase gradient is zero. Constant phase is expected for linear response; however, if there is a supercurrent $\mathbf{j}_s$ then there *is* a gradient of the phase. For the linear optical conductivity I can neglect as small the current from any phase gradient and write

$$\mathbf{j}_s = -\frac{n_s e^2}{mc}\mathbf{A}. \tag{11.9}$$

Eq. 11.9 is known as the London equation.

11.3.2 London Model Optical Conductivity

I'll subject the superconductor to an electric field $\mathbf{E} = \mathbf{E}_0 e^{-i\omega t}$, an AC field (or plane wave) of frequency ω. Recall that

$$\mathbf{E} = -\frac{1}{c}\frac{\partial\mathbf{A}}{\partial t} = \mathbf{E}_0 e^{-i\omega t},$$

meaning that

$$\mathbf{A} = -\frac{ic}{w}\mathbf{E}.$$

I put this vector potential into Eq. 11.9 to find

$$\mathbf{j}_s = i\frac{n_s e^2}{m\omega}\mathbf{E}.$$ (11.10)

As usual, Ohm's law is $\mathbf{j}_s = \sigma\mathbf{E}$. Therefore, the optical conductivity is

$$\sigma_s = i\frac{n_s e^2}{m\omega}.$$ (11.11)

I have seen this before! Look at Eq. 8.25 and 10.1. Equation 11.11 is formally the same as the finite-frequency conductivity of the perfect metal: the Drude metal, Eq. 4.8, in the limit where $\tau \to \infty$. Moreover, I can define a superconducting plasma frequency $\omega_{ps} = \sqrt{4\pi n_s e^2/m}$, write $\sigma_s = i\omega_{ps}^2/4\pi\omega$, and then (using $\epsilon_s = \epsilon_c + 4\pi i\sigma/w$) get the dielectric function:

$$\epsilon_s = \epsilon_c - \frac{\omega_{ps}^2}{\omega^2},$$ (11.12)

Eq. 11.12 is the dielectric response of perfectly free carriers with no scattering. It has a negative real part and zero imaginary part.

At this point I can list four questions that I must answer. They are

1. How do I get the infinite DC conductivity of the superconductor?
2. How does the superconductivity affect electromagnetic wave propagation and reflectance?
3. What about the Meissner effect?
4. This is cute, but why do I not take a bunch of Drude-like free carriers and just let their mean free time between collisions go to infinity?

11.3.3 Infinite DC Conductivity

I already know the answer to the first question; it was part of the introduction to the need for Kramers–Kronig relations. (See Section 10.1.2.) The real part of the conductivity is a delta function at $\omega = 0$. I'll write it as $\sigma_1 = \mathcal{A}\delta(\omega)$ and substitute that into the Kramers–Kronig relation, Eq. 10.20, obtaining

$$\epsilon_1(\omega) = 1 + 8\mathcal{P}\int_0^\infty d\omega'\,\frac{\mathcal{A}\delta(\omega')}{\omega'^2 - \omega^2}.$$ (11.13)

The integral is easy to do,* and I get

$$\epsilon_1(\omega) = 1 - \frac{4\mathcal{A}}{\omega^2},$$

or, equivalently, the imaginary part of the conductivity is

$$\sigma_2(\omega) = \frac{\mathcal{A}}{\pi\omega}.$$

---

* Because the lower limit of the integral is $\omega = 0$, which is where the delta function is located, the integral only gets half the area, giving $-\mathcal{A}/2\omega^2$ not $-\mathcal{A}/\omega^2$.

I compare this equation with Eq. 11.12 and see that $\mathcal{A} = \omega_{ps}^2/4 = \pi n_s e^2/m$. This value for the weight of the delta function satisfies the sum rule, Eq. 10.36

$$\int_0^\infty d\omega'\, \sigma_1(\omega') = \frac{\omega_{ps}^2}{8} = \frac{\pi n_s e^2}{2m}. \tag{11.14}$$

So, finally, the London optical conductivity is

$$\sigma(\omega) = \frac{\pi n_s e^2}{m}\delta(\omega) + i\frac{n_s e^2}{m\omega}. \tag{11.15}$$

If all the electrons are part of the superconducting condensate, $n_s = n$ and $\omega_{ps} = \omega_p$.

11.3.4　Electromagnetic Wave Propagation and Reflectance

The London dielectric function ($\omega > 0$) is purely real and negative. The refractive index in the London model is

$$N = n + i\kappa = \sqrt{\epsilon} = 0 + i\sqrt{\frac{\omega_{ps}^2}{\omega^2} - 1} \approx 0 + i\frac{\omega_{ps}}{\omega}. \tag{11.16}$$

So the real part of the refractive index* $n = 0$ and $\kappa \approx \omega_{ps}/\omega$. (The approximation in Eq. 11.16 is good. The range of frequencies where the London model describes the superconducting conductivity ends at the superconducting energy gap frequency, say 5 meV or 40 cm^{-1}, whereas the plasma frequency of a typical metal is, say, 5 eV or 40,000 cm^{-1}.)

The electric field in the superconductor decays as $\mathbf{E} = \mathbf{E}_0 e^{-\kappa\omega x/c} = \mathbf{E}_0 e^{-x/\lambda_L}$ with distance x away from the surface. The decay length is

$$\lambda_L \equiv \frac{c}{\kappa\omega} = \frac{c}{\omega_{ps}} \equiv \sqrt{\frac{mc^2}{4\pi n_s e^2}}, \tag{11.17}$$

where λ_L, independent of ω, is known as the London penetration depth.† The London penetration depth plays the role of the skin depth of the metal in the electrodynamics of the superconductor. Typically it is a lot shorter than the normal-state skin depth at frequencies below the superconducting gap, where the model applies.

Finally, the reflectance, given by Eq. 3.33, is unity. The London superconductor is a perfect reflector. Experimentally, superconductors are almost perfect reflectors at frequencies below their energy gap.

---

* I use n for both the carrier density and refractive index, here and elsewhere. Symbol duplication is an unavoidable consequence of the limited number of glyphs in English, the adherence to convention for symbol choice, and in the use of a limited number of fonts.

† The units dilemma rears its head here. ω_{ps} is an angular frequency in radians/s or s^{-1}. But, like other plasma frequencies, it is often reported in cm^{-1}, making the London length appear to be just the vacuum wavelength of light (λ_p) corresponding to this plasma frequency. But if I go through the conversions correctly (Appendix A) I get $\lambda_L = \lambda_p/2\pi$.

11.3.5 Meissner Effect

I now consider a superconductor with a static magnetic field $\mathbf{B}$ applied parallel to the surface so as to work out the Meissner effect. I will use the London gauge:

$$\nabla \cdot \mathbf{A} = 0.$$

Using the London gauge and the London equation (Eq. 11.9) I can write

$$\nabla \cdot \mathbf{A} \propto \nabla \cdot \mathbf{j} = -\frac{\partial \rho}{\partial t}$$

where I've used the continuity equation on the right-hand side. If I want to think about static fields, I want no time variation of the charge density. The London gauge enforces this requirement.

Before continuing, I must think a bit about Maxwell's equations. I am about to argue that the superconductor in the presence of an external static magnetic field $\mathbf{B}$ will have a supercurrent distribution $\mathbf{j}_s(\mathbf{r})$ which serves to make $\mathbf{B} = 0$ in the interior of the superconductor, producing the Meissner effect. Now, $\mathbf{j}_s$ is not a free current in the sense of the current in the macroscopic Ampère's law (Eq. 2.1d). It is a microscopic current that will generate a net magnetization of the superconductor. Therefore, I'll write Ampère's law in its microscopic form [2]:

$$\nabla \times \mathbf{B} = \frac{4\pi}{c}\mathbf{j}_s. \tag{11.18}$$

Because the Meissner effect is a static problem, the time derivative of the electric field is zero.

Now, I take the curl of Eq. 11.9,

$$\nabla \times \mathbf{j}_s = -\frac{n_s e^2}{mc}\nabla \times \mathbf{A} = -\frac{n_s e^2}{mc}\mathbf{B}, \tag{11.19}$$

and take the curl again,

$$\begin{aligned}
\nabla \times (\nabla \times \mathbf{j}_s) &= -\frac{n_s e^2}{mc}\nabla \times \mathbf{B} \\
&= -\frac{n_s e^2}{mc}\frac{4\pi}{c}\mathbf{j}_s \\
&= \nabla(\nabla \cdot \mathbf{j}_s) - \nabla^2 \mathbf{j}_s \\
&= -\nabla^2 \mathbf{j}_s,
\end{aligned} \tag{11.20}$$

where I used Ampère's law, Eq. 11.18, to go from the first to the second line, the properties of the double cross product to get the third line, and the London gauge condition to go from the third to the fourth line. Thus,

$$\nabla^2 \mathbf{j}_s = \frac{4\pi n_s e^2}{mc^2}\mathbf{j}_s = \frac{1}{\lambda_L^2}\mathbf{j}_s,$$

using Eq. 11.17. The physically reasonable solution is that $j_s = 0$ deep in the interior of the superconductor and that there are surface currents which decay away from the surface as

$$\mathbf{j}_s(x) = \mathbf{j}_0 e^{-x/\lambda_L}.$$

So the currents are on the surface and $\mathbf{j}_s(x) \to 0$ in the interior. What about the magnetic field $\mathbf{B}$? To answer, I return to Eq. 11.18 and take the curl of both sides

$$
\begin{aligned}
\nabla \times (\nabla \times \mathbf{B}) &= \frac{4\pi}{c} \nabla \times \mathbf{j}_s \\
&= -\frac{4\pi}{c} \frac{n_s e^2}{mc} \mathbf{B} \\
&= \nabla(\nabla \cdot \mathbf{B}) - \nabla^2 \mathbf{B} \\
&= -\nabla^2 \mathbf{B},
\end{aligned}
\tag{11.21}
$$

where I used Eq. 11.19 to go from the first to the second line, the properties of the double cross product to get the third line, and Eq. 2.4b to go from the third to the fourth line. Thus,

$$\nabla^2 \mathbf{B} = \frac{4\pi n_s e^2}{mc^2} \mathbf{B} = \frac{1}{\lambda_L^2} \mathbf{B}, \tag{11.22}$$

using Eq. 11.17 for λ_L. The physically reasonable solution is that $\mathbf{B} = 0$ deep in the interior of the superconductor and that it decays away from the surface as

$$\mathbf{B}(x) = \mathbf{B}_0 e^{-x/\lambda_L}, \tag{11.23}$$

as sketched in Fig. 11.6.

Both the field and the current decay in the same way as a function of depth into the superconductor. The current can be thought of as the current required to generate a field of $-\mathbf{B}_0$ in the interior, so that the sum of the applied field and the generated field is zero. Note that $\mathbf{B} \perp \mathbf{j}_s$. The current is into the page in Fig. 11.6.

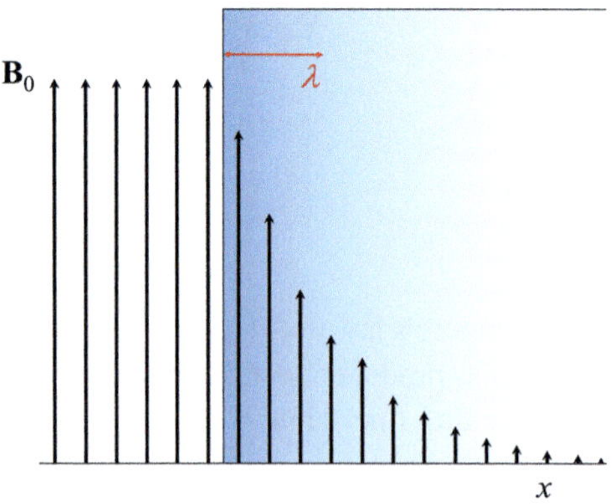

 The magnetic field falls from the external value to $1/e$ at a distance of λ_L below the surface of the superconductor.

11.3.6 Comparison to the Perfect Conductor

It is tempting (but not correct) to think of a superconductor as a metal in which (for some reason) scattering processes have stopped happening. The metal would have a Drude conductivity, Eq. 4.10

$$\sigma(\omega) = \frac{ne^2\tau/m}{1+\omega^2\tau^2} + i\frac{ne^2\omega\tau^2/m}{1+\omega^2\tau^2},$$

in which $\tau \to \infty$. The real part of the conductivity of this "perfect conductor" is the same as the London model conductivity: a delta function at the origin, infinite at DC and zero at all finite frequencies. At the same time, $\sigma_2 \to ne^2/m\omega$, either by taking the limit or by using the Kramers–Kronig relation for the conductivity.

To see what the magnetic field looks like in the interior of this perfect conductor, I will write Ampère's law in its microscopic form, Eq. 11.18, take the curl, and then the time derivative

$$\frac{d}{dt}\nabla \times (\nabla \times \mathbf{B}) = \frac{4\pi}{c}\nabla \times \frac{d\mathbf{j}}{dt}$$

where I have put the $\partial\mathbf{E}/\partial t$ term into the current. The current can be written $\mathbf{j} = -ne\mathbf{v}$. Writing out the double curl (as was done above) and using $\nabla \cdot \mathbf{B} = 0$, this equation becomes

$$-\frac{d}{dt}\nabla^2\mathbf{B} = -\frac{4\pi}{c}\nabla \times ne\frac{d\mathbf{v}}{dt}$$

But, $d\mathbf{v}/dt = \mathbf{a}$, the acceleration, and Newton's law for these carriers is $\mathbf{a} = -e\mathbf{E}/m$, so the right hand side becomes

$$-\frac{d}{dt}\nabla^2\mathbf{B} = \frac{4\pi ne^2}{mc}\nabla \times \mathbf{E}$$

Now, I can replace $\nabla \times \mathbf{E}$ using Maxwell's equation 2.4c, interchange the operations on the left hand side, and by analogy with Eq. 11.17 define the penetration length of the perfect conductor as $\lambda_P = \sqrt{mc^2/4\pi ne^2}$, arriving at

$$\nabla^2\frac{d\mathbf{B}}{dt} = \frac{1}{\lambda_P^2}\frac{d\mathbf{B}}{dt}. \tag{11.24}$$

The behavior of the time derivative of $\mathbf{B}$ for the perfect conductor is of the same form as Eq. 11.22 for the superconductor, except that it is for the derivative rather than for the field itself. So for the perfect conductor, a few penetration depths from the surface, the field can be present, but

$$\frac{d\mathbf{B}}{dt} = 0,$$

i.e., it cannot change. This behavior is the ultimate example of Lenz's law: a current is induced in the conductor by the changing magnetic field so directed as to oppose the change in flux. Of course, the decay length of the time-varying field is just the skin depth in the $\omega \gg 1/\tau$ limit.*

---

* The perfect conductor has $1/\tau = 0$ so any finite frequency is in the high-frequency limit.

Table 11.2 London penetration depth of superconductors.	
Material	λ_L Å
Al	160
Sn	340
Pb	370
Nb	390
$YBa_2Cu_3O_7$	1,440

11.4 Length Scales for Superconductors

11.4.1 London Penetration Depth

The penetration depth, $\lambda_L = \sqrt{mc^2/4\pi n_s e^2}$, is typically in the 100–2,000 Å range. Some values are shown in Table 11.2. Metals with high electron density (Al) have shorter penetration depths than those with lower electron density.

I know that as the temperature increases, the superfluid density decreases (to zero at T_c). (See Fig. 11.5.) The penetration depth follows $\lambda_L(T) \sim 1/\sqrt{n_s(T)}$, increasing towards T_c. If I use the Ginzburg-Landau temperature dependence (and the identification of $\sqrt{n_s} = \Delta$, I get $\lambda_L(T) = \sqrt{T_c/(T_c - T)}$.

The London penetration depth is also the "skin depth" of the superconductor. Light* falls off in the interior with characteristic decay length λ_L. It is interesting to compare the normal-state skin depth and the London length. The latter given in Eq. 11.17, repeated here:

$$\lambda_L = \frac{c}{\omega_{ps}} = \sqrt{\frac{mc^2}{4\pi n_s e^2}}.$$

Note that this penetration depth is frequency independent. Now, a Drude metal with $\sigma_{dc} = ne^2\tau/m$ has a skin depth given by Eqs. 4.38 and 4.39:

$$\delta = \begin{cases} \sqrt{\dfrac{c^2}{2\pi\omega\sigma_{dc}}} = \sqrt{\dfrac{mc^2}{2\pi ne^2\omega\tau}} & \omega \ll 1/\tau \\[2ex] \dfrac{c}{\omega_p} = \sqrt{\dfrac{mc^2}{4\pi ne^2}} & \omega \gg 1/\tau \end{cases}$$

As long as $n_s = n$ (which is the case for pure metals at low temperatures) the London penetration depth is the same as the high-frequency limit of the skin depth of the metal. The low-frequency skin depth of the metal is much longer that the London length and grows as the frequency decreases.

---

* The light must be in the far-infrared range, with photon energy smaller than the energy gap, for the superconductor to respond as a superconductor.

11.4.2 The Coherence Length

In addition to the penetration depth discussed already, a second length scale exists. It is the coherence length, first introduced by Pippard [219]. The coherence length also appears in Ginzburg-Landau theory [212, 213] and BCS theory [210, 215]. The three lengths are similar but not identical, either in concept or in value. I'll give a heuristic argument as follows. The Cooper-pair wave function extends over a finite distance; at longer distances the amplitude begins to decrease. The wave function ψ thus has an envelope

$$\psi \sim e^{-r/\xi_0}$$

where ξ_0 is the coherence length. A consequence of the finite length is that plane waves can no longer be correct for the wave function. Instead there must be a range of wave vectors, i.e., a q-dependent theory. The range is

$$\delta k_{sc} = k_F \pm q = k_F \pm \frac{1}{\xi_0}.$$

The range of wave vectors must be related to the superconducting binding energy Δ, leading to a spread of kinetic energies,

$$\frac{\hbar^2(\delta k_{sc})^2}{2m} = \frac{\hbar^2 k_F^2}{2m} \pm \frac{\hbar^2 k_F}{m\xi_0} = \mathcal{E}_F \pm 2\Delta,$$

where I have neglected the term containing $1/\xi_0^2$. Note that this argument basically invokes the uncertainty principle: limiting the space part to ξ_0 leads to a spread of momenta and corresponding spread of energies. Subtracting $\mathcal{E}_F$ from both sides gives $\hbar^2 k_F / m\xi_0 = 2\Delta$ or

$$\xi_0 = \frac{\hbar^2 k_F}{2m\Delta} = \frac{\hbar v_F}{2\Delta},$$

using $\hbar k_F = m v_F$.

The BCS theory gives

$$\xi_0 = \frac{2}{\pi} \frac{\hbar v_F}{2\Delta}, \tag{11.25}$$

close enough. Two materials parameters appear in Eq. 11.25. In the numerator is the Fermi velocity (or Fermi wave vector), which increases with carrier density; the denominator holds the gap (or, equivalently, the transition temperature: $2\Delta = 3.5kT_c$ in BCS theory). Thus high electron density, low T_c materials (think Al) will have long coherence lengths; low electron density, high T_c materials (think $YBa_2Cu_3O_7$) will have short coherence lengths. This expectation is in accord with observation. See Table 11.3.

11.5 Excitations in a Superconductor

The BCS theory [210, 211, 215] prescribes an energy gap Δ at the Fermi level; no electron may have an energy (measured relative to the Fermi energy – aka the chemical potential)

Table 11.3 Coherence length of superconducting materials.		
Materials	T_c (K)	ξ_0 (Å)
Aluminum	1.19	12,000
Indium	3.40	3,300
Tin	3.72	2,600
Gallium	5.90	1,600
Lead	7.22	800
Niobium	9.25	350
PbMo	15	25
Nb_3Sn	17	40
$C_{60}K_3$	19	30
$C_{60}Rb_3$	31	23
$YBa_2Cu_3O_7$	93	15

smaller than Δ. Similarly, no hole (the absence of an electron below the Fermi surface) may exist with an energy* smaller than Δ.

The gap is in the energy spectrum of single-particle excitations of the superconductor. It has an analog in the energy gap of a semiconductor. In the intrinsic (undoped) semiconductor, the Fermi level lies in the middle of the gap (with magnitude E_g) between valence and conduction bands. The minimum excitation energies are $E_g/2$. But superconductors are of course not semiconductors. The gap in the latter is due to the static potential of the crystal; it occurs for a crystal momentum $\mathbf{k}$ at the Brillouin zone boundary. The gap in the superconductor is tied to the Fermi surface; it occurs at a wave vector $\mathbf{k}$ whose magnitude equals the Fermi wave vector, k_F.

There is no gap in the values of the wave vectors $\{\mathbf{k}\}$! As in any solid, the wave vectors cover a continuous range, from zero to the Brillouin zone boundary or from zero to infinity depending on the picture one uses. I am writing this paragraph because many discussions of superconducting gaps and gap symmetry present a cartoon where the gap is shown as a shell – with nodes or without them depending on the material – around a Fermi sphere. The picture is wrong for three reasons: (1) The Fermi sphere is a k-space entity and there are no forbidden values of the wave vector $\mathbf{k}$. (2) The gap is most important for states near the Fermi surface, as I will discuss in a minute, but it affects (at least formally) all the quasiparticle states. (3) The gap applies equally to hole states below the Fermi level as to electron states above.

11.5.1 The Excitation Spectrum of a Normal Metal

Before discussing the excitation spectrum of a superconductor, let me think about the excitations of a normal metal. The ground state is the $T = 0$ Fermi sphere shown in

---

* Hole energy is measured *downwards* from the Fermi level.

Fig. 8.2. All single-particle states between zero and $\mathcal{E}_F$ are occupied; all above are empty. There are two possible kinds of excitations. Electron-like excitations occupy states above $\mathcal{E}_F$ and hole-like excitations correspond to vacant states below $\mathcal{E}_F$. If the excitations are made by (low-energy) optical absorption as intraband transitions or if they are due to finite temperatures (and the broadening of the Fermi–Dirac function), then they are made by promoting an electron from below to above $\mathcal{E}_F$.*

As I mentioned at the start of Chapter 8, I'll call the excitations, whether hole-like or electron-like, "quasiparticles." I can plot the quasiparticle excitation spectrum of the free-electron normal metal. The electron energies are purely kinetic

$$\mathcal{E}(k) = \frac{\hbar^2 k^2}{2m},$$

as in Eq. 8.4. Excitation energies are measured relative to the Fermi energy, so the energy of an electron-like or hole-like excitation

$$\mathcal{E}_n(k) = \left| \frac{\hbar^2 k^2}{2m} - \mathcal{E}_F \right| \equiv \left| \epsilon(k) \right|. \tag{11.26}$$

Here, $\epsilon(k) = \hbar^2 k^2/2m - \mathcal{E}_F$ is the kinetic energy of the electron, measured relative to the Fermi energy.

My plot of $\mathcal{E}_n(k)$ vs k will be almost a straight line, because I am interested in excitation energies on the scale of the superconducting energy gap. Superconducting gaps are typically a few meV, whereas the Fermi energy (Table 8.1) is typically more than 10 eV.[†] For meV variations around the Fermi energy, the curvature in Eq. 8.4 as a function of k is not noticeable. But to be sure that things are absolutely linear, I will plot the quasiparticle energy against the kinetic energy, measured relative to the Fermi energy, i.e., $\mathcal{E}_n(k)$ versus $\epsilon(k)$. Such a plot is shown in Fig. 11.7 as the red dashed line. The left side shows the hole-like quasiparticles and the right side the electron-like quasiparticles.

11.5.2 The Excitation Spectrum of a Superconductor

The excitation spectrum of a superconductor is modified from that of the normal metal by the presence of the energy gap. In BCS theory [210, 211, 215]

$$\mathcal{E}_s(k) = \sqrt{\epsilon(k)^2 + \Delta^2}. \tag{11.27}$$

The quantity $\mathcal{E}_s(k)$ is shown in Fig. 11.7 as the blue solid line. The minimum quasiparticle energy is the gap Δ. Because optical absorption breaks a Cooper pair and produces a pair of quasiparticles, the corresponding threshold for optical absorption is 2Δ. I only need a potential difference of Δ if I tunnel electrons in[‡] from a normal metal into the superconductor.

---

* There are other options. I may inject electrons by tunneling through a barrier. Unless I withdraw them somewhere else, the metal will become charged. In small structures, this process can be observed as single electron tunneling and leads to the Coulomb blockade [220].

† There are exceptions, including the cuprate superconductors and organic superconductors.

‡ Or out; tunneling electrons out corresponds to putting a hole in. This *single-particle* tunneling differs from the pair tunneling of the Josephson effect. In single-particle tunneling, I must have a potential of $\pm\Delta$ between

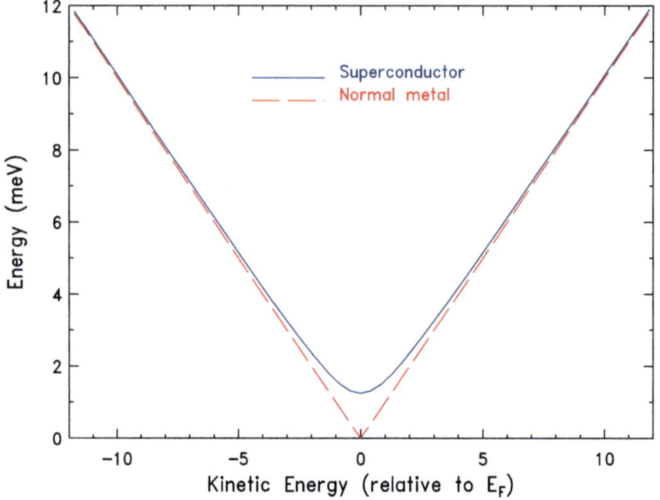

Fig. 11.7 The excitation spectrum of a normal metal (dashes) and a superconductor (solid). The superconducting gap was taken to be $\Delta = 1.2\,\text{meV}\ (10\,\text{cm}^{-1})$.

It is remarkable that the phenomenon of superconductivity, with such a dramatic effect on electrical transport and optical properties, affects the energy spectrum of the electrons in such a modest way. Once one moves several times the gap energy away from the Fermi surface, the energy spectrum is indistinguishable, and one might say that those deeply buried (and highly excited) states are indistinguishable from the normal metal. Yet, the London penetration depth, Eq. 11.17, involves all the conduction electrons.* The heat capacity also suggests that all the electrons participate in the superconducting condensation.

Of course, I can take the same point of view about the conduction by a normal metal, shown in Fig. 8.4. Instead of thinking that the entire Fermi sphere is displaced, I could say that a number of electrons are excited above the Fermi surface in the direction of the electron flow, and an equal number of holes exist on the opposite side, where occupied electron states once existed.[†]

For electrons far below $\mathcal{E}_F$, the states $|\mathbf{k},\ \uparrow\rangle$ and $|-\mathbf{k},\ \downarrow\rangle$ are both occupied with 100% probability. Are these Cooper pairs in the normal metal? Not really, because the BCS state is a *coherent* combination of Cooper pairs.

I can also represent the excitation energy in a semiconductor picture. In such a picture, the kinetic energy is just that of the electrons, so states below $\mathcal{E}_F$ have negative energies. I get the picture shown in Fig. 11.8. The electron-like excitations have positive energies and the hole-like ones have negative energies. That the superconducting gap is 2Δ is obvious

---

normal metal and superconductor, or I must have a potential of $\pm 2\Delta$ between two superconductors, in order to have a tunneling current. The DC Josephson effect is the tunneling of Cooper pairs between two superconductors at zero bias.

* In clean metals.

[†] This is in fact what the Boltzmann equation suggests.

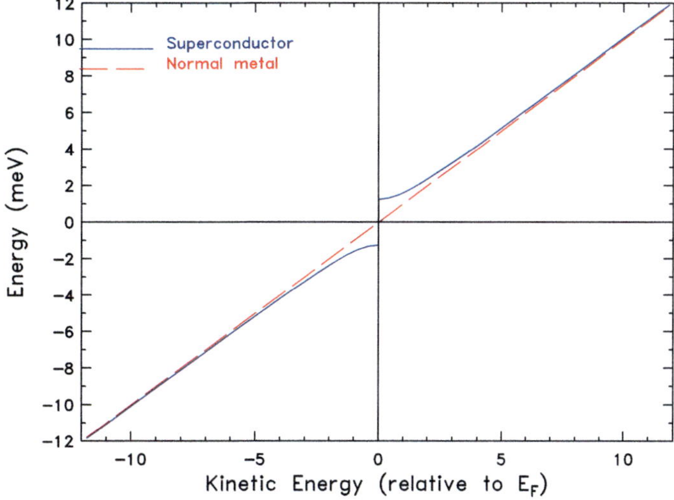

Fig. 11.8 The "semiconductor picture" of the excitation spectrum of a normal metal (dashes) and a superconductor (solid). The superconducting gap was taken to be $\Delta = 1.2$ meV (10 cm^{-1}). The Fermi energy is at $\mathcal{E} = 0$.

here, because I must promote an electron from the highest-energy state below $\mathcal{E} = 0$ (the Fermi energy) to the lowest energy state above $\mathcal{E} = 0$.

I like the excitation picture, Fig. 11.7, better. Many people do use the semiconductor picture, however.

11.5.3 The Density of States

There is a one-to-one correspondence between the states of the superconductor and those of the normal metal, indexed by the quantum numbers **k** and spin. Thus every superconducting state with energy $\mathcal{E}(k)$ is in correspondence with a normal-metal state with energy ϵ_k. The relation is of course given by Eq. 11.27. Thus, I have

$$\mathcal{N}_s(\mathcal{E})d\mathcal{E} = \mathcal{N}_n(\epsilon)d\epsilon$$

where $\mathcal{N}_s$ is the density of states of the superconductor and $\mathcal{N}_n$ the same for the normal metal. The free-electron density of states is $\mathcal{N}_n \sim \sqrt{\mathcal{E}_F + \epsilon}$, slowly varying for small ϵ around $\mathcal{E}_F$. The range of ϵ in Fig. 11.7 is 10 meV: the Fermi energy is 10 eV. I can without worrying about error take $\mathcal{N}_n$ as constant, equal to the value at the Fermi level, $\mathcal{N}_0$. so that,

$$\mathcal{N}_s(\mathcal{E}) = \mathcal{N}_0 \frac{d\epsilon}{d\mathcal{E}}.$$

From Eq. 11.27, $\epsilon(k) = \sqrt{\mathcal{E}(k)^2 - \Delta^2}$ and then

$$\frac{d\epsilon}{d\mathcal{E}} = \frac{\mathcal{E}}{\sqrt{\mathcal{E}^2 - \Delta^2}}.$$

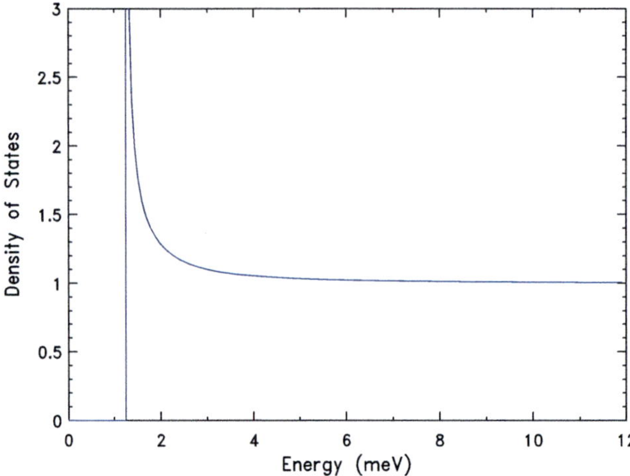

Fig. 11.9 The density of states $\mathcal{N}_s$ of a superconductor versus the energy $\mathcal{E}$. $\mathcal{N}_s(\mathcal{E})$ is normalized by the Fermi surface density of states of the normal metal, $\mathcal{N}_0$. The gap was taken as 1.2 meV.

The superconductor has no states below Δ and a square root singularity located at Δ:

$$\mathcal{N}_s(\mathcal{E}) = \begin{cases} 0 & \mathcal{E} < \Delta \\ \dfrac{\mathcal{E}\mathcal{N}_0}{\sqrt{\mathcal{E}^2 - \Delta^2}} & \mathcal{E} > \Delta. \end{cases} \tag{11.28}$$

This function is plotted in Fig. 11.9.

Generally, when there is a peak in the density of states for excitations, I would expect a peak in the optical conductivity at that energy. However, experiment [84, 85, 88, 221] shows that this peak is absent in a superconductor; indeed, the conductivity is zero exactly where the density of states diverges. See Fig. 11.12 for an example. It is worth asking why the conductivity does not follow the density of states. The answer is in the BCS coherence factors.

11.5.4 The Coherence Factors

The ground state of the superconductor is a linear combination of Cooper pair states where the pairs $|\mathbf{k} \uparrow, -\mathbf{k} \downarrow\rangle$ are either full [state $O(k)$] or empty [state $\bar{O}(k)$],

$$\psi_0 = v_k O(k) + u_k \bar{O}(k)$$

where v_k is the amplitude (square root of the probability) for the state $|\mathbf{k} \uparrow, -\mathbf{k} \downarrow\rangle$ to be full and u_k is the amplitude for the state $|\mathbf{k} \uparrow, -\mathbf{k} \downarrow\rangle$ to be empty. Normalization requires that $|u|^2 + |v|^2 = 1$. BCS [210, 215] find that

$$v_k = \sqrt{\frac{1}{2}\left(1 - \frac{\epsilon}{\mathcal{E}}\right)} \tag{11.29}$$

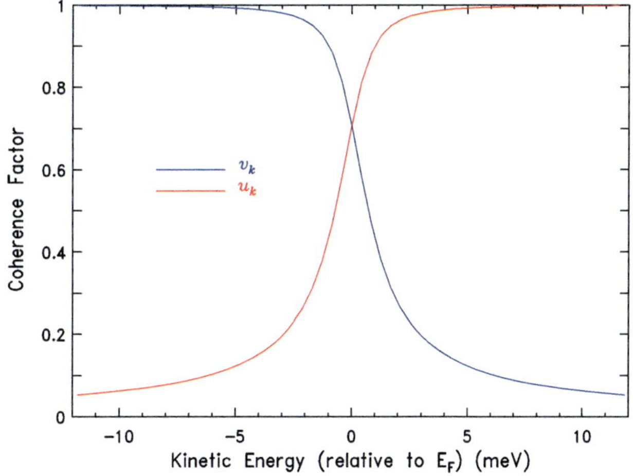

Fig. 11.10 The amplitude v_k for the state $|\mathbf{k}\uparrow, -\mathbf{k}\downarrow\rangle$ to be full and u_k for the state $|\mathbf{k}\uparrow, -\mathbf{k}\downarrow\rangle$ to be empty as a function of $\epsilon(k)$. The gap was taken as 1.2 meV.

and

$$u_k = \sqrt{\frac{1}{2}\left(1 + \frac{\epsilon}{\mathcal{E}}\right)} \tag{11.30}$$

with $\epsilon(k)$ defined in Eq. 11.26 and $\mathcal{E}(k)$ defined in Eq. 11.27. These functions are shown in Fig. 11.10.

The figures show that the superconductor (at $T = 0$) has a finite probability for Cooper-pair states with $k > k_F$ to be occupied and a finite probability for those with $k < k_F$ to be unoccupied.* This result should be contrasted with the Fermi–Dirac function of Eq. 8.11 which is perfectly sharp at $T = 0$. In fact, v_k qualitatively resembles $f(E)$ for $T \approx T_c$.

For states with $k = k_F$, the probability that the Cooper-pair state is empty and the probability that it is full are equal; I can think of this situation as implying that these Cooper pairs have 50% electron-like character and 50% hole-like character. The mixed character has important consequences for the optical properties, explaining why the conductivity at the gap is zero rather than maximum.

11.6 The Optical Conductivity of a Superconductor

11.6.1 Mattis-Bardeen Theory

The optical conductivity of a superconductor was first worked out by Mattis and Bardeen [222] and has been extended by Nam [223, 224] and others [225–227]. The Mattis-Bardeen calculation took the scattering of photoexcited quasiparticles in the superconductor and of

---

* $\epsilon(k) = 0$ for states at the Fermi energy.

normal carriers in the metal to be the same, calculating $\sigma_{1s}(\omega)/\sigma_n$ and $\sigma_{2s}(\omega)/\sigma_n$ in the dirty limit ($1/\tau \gg 2\Delta$) and the extreme anomalous limit ($\ell \gg \delta$). The result (at $T = 0$) is expressed in terms of complete elliptic integrals. The results at finite temperature may be written

$$\frac{\sigma_1(\omega)}{\sigma_n} = \theta(\hbar\omega - 2\Delta)\frac{1}{\hbar\omega}\int_{\Delta-\hbar\omega}^{-\Delta} d\mathcal{E}\frac{\left[\mathcal{E}(\mathcal{E}+\hbar\omega)+\Delta^2\right]}{\sqrt{\mathcal{E}^2-\Delta^2}\sqrt{(\mathcal{E}+\hbar\omega)^2-\Delta^2}} \times [1 - 2f(\mathcal{E}+\hbar\omega)]$$

$$+ \frac{2}{\hbar\omega}\int_{\Delta}^{\infty} d\mathcal{E}\frac{\mathcal{E}(\mathcal{E}+\hbar\omega)+\Delta^2}{(\sqrt{\mathcal{E}^2-\Delta^2}\sqrt{(\mathcal{E}+\hbar\omega)^2-\Delta^2}}[f(\mathcal{E}) - f(\mathcal{E}+\hbar\omega)] \qquad (11.31)$$

and

$$\frac{\sigma_2(\omega)}{\sigma_n} = \frac{1}{\hbar\omega}\int_{\max(-\Delta,\Delta-\hbar\omega)}^{\Delta} d\mathcal{E}\frac{\mathcal{E}(\mathcal{E}+\hbar\omega)+\Delta^2}{\sqrt{\Delta^2-\mathcal{E}^2}\sqrt{(\mathcal{E}+\hbar\omega)^2-\Delta^2}}[1 - 2f(\mathcal{E}+\hbar\omega)]$$

$$(11.32)$$

where $\Delta \equiv \Delta(T)$ is the temperature-dependent gap function, shown in Fig. 11.5, and $f(\mathcal{E})$ is the Fermi function at temperature T, enforcing the requirement that (in the semiconductor picture) the excitation is from occupied to unoccupied state. Note that I've written Eq. 11.31 as two integrals and in the order opposite to the way they are usually written [222]. The first integral only contributes for $\hbar\omega > 2\Delta$ and represents the contribution from photoexcited quasiparticles. The second integral represents the contribution from thermally excited quasiparticles. According to Eq. 11.27, the BCS excitation spectrum has a gap, $\mathcal{E}_s(k) = \sqrt{\epsilon(k)^2 + \Delta^2}$. The second integral therefore starts at Δ. The limits on the integral for $\sigma_2(\omega)$ seem perverse but it does all work out.

Figure 11.11 shows the Mattis-Bardeen conductivity for a "dirty-limit" superconductor. (I'll discuss what "dirty limit" means in the next section.) The real (σ_1) and imaginary (σ_2) parts of the conductivity are shown in the left panel. The real part is zero up to the

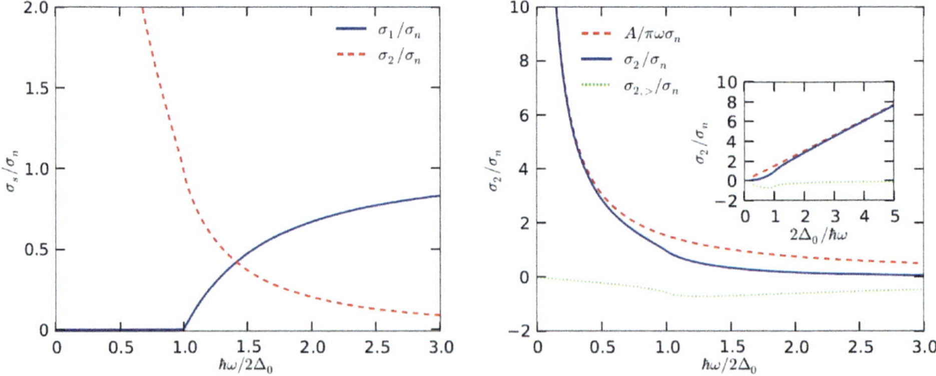

Fig. 11.11 Left: real (solid line) and imaginary (dashed line) parts of the optical conductivity at $T = 0$ K from Mattis-Bardeen theory. Right: the imaginary part of the optical conductivity (solid line), with one part related to the delta function in σ_1 (dashed line), and the other part related to the finite frequency part of σ_1 (dotted line). The inset shows the same quantities plotted on $1/\omega$ scale.

superconducting energy gap 2Δ and then rises smoothly to reach the normal-state value at a frequency that is several times the gap. The imaginary part looks very London-like at low frequency and falls as $1/\omega$ with increasing frequency. There is a little bit of structure near the gap but σ_2 continues to fall as frequency increases. The right panel compares the BCS (Mattis-Bardeen) curve for σ_2 with the London-model curve.

11.6.2 The Dirty Limit

Dirty metals, alloys, and thin films often have short mean free paths. The mean free path ℓ can be of order 1–2 nm, compared to perhaps micrometers in pure materials at low temperatures. When the mean free path is shorter than the coherence length, the coherence length itself is shortened, to [215, 228]

$$\frac{1}{\xi} = \frac{1}{\xi_0} + \frac{1}{\ell}$$

with limits of ξ_0 in the clean limit and ℓ in the dirty limit.

The short mean free path of the dirty metal means a broad spectrum for $\sigma_1(\omega)$. (See Eq. 4.10a and Fig. 4.1.) It also has an effect on the penetration depth and optical conductivity. To see how, let me start with a series of inequalities:

$$\ell < \xi_0$$
$$v_F \tau < \frac{\hbar v_F}{2\Delta} \tag{11.33}$$
$$2\Delta < \frac{\hbar}{\tau}$$

indicating that the Drude spectrum extends beyond (often well beyond) the gap in the conductivity spectrum.

Now, I invoke the conductivity sum rule, Eq. 9.29:

$$\int_0^{\omega_{\max}} d\omega\, \sigma_1(\omega) = \frac{1}{8}\omega_p^2$$

where I'll stop the integral at a frequency well above $1/\tau$ but below the onset of interband transitions, so that ω_p is the free-carrier plasma frequency of the metal. The sum rule tells me that the integral gives the same answer for the plasma frequency in normal and superconducting states. The normal state has the Drude free-carrier response. The superconducting state has an infinite DC conductivity [a delta function $\mathcal{A}\delta(\omega)$ at $\omega = 0$], a gap of 2Δ, and a conductivity that increases from zero at the gap to approach the normal state at high frequencies. An example of the conductivity of a dirty-limit superconductor at four temperatures is shown in Fig. 11.12 [221]. (The circles and triangles are data for σ_1 and σ_2 respectively, while the solid lines are a dirty-limit BCS theory calculation [222].) The conductivity is normalized by the normal-state conductivity, which is essentially equal to the DC conductivity, σ_{dc}, at all frequencies shown.

The superconducting state conductivity is zero up to the gap, 2Δ, after which it rises slowly to join the normal-state conductivity. The sum rule says that the "missing area" in

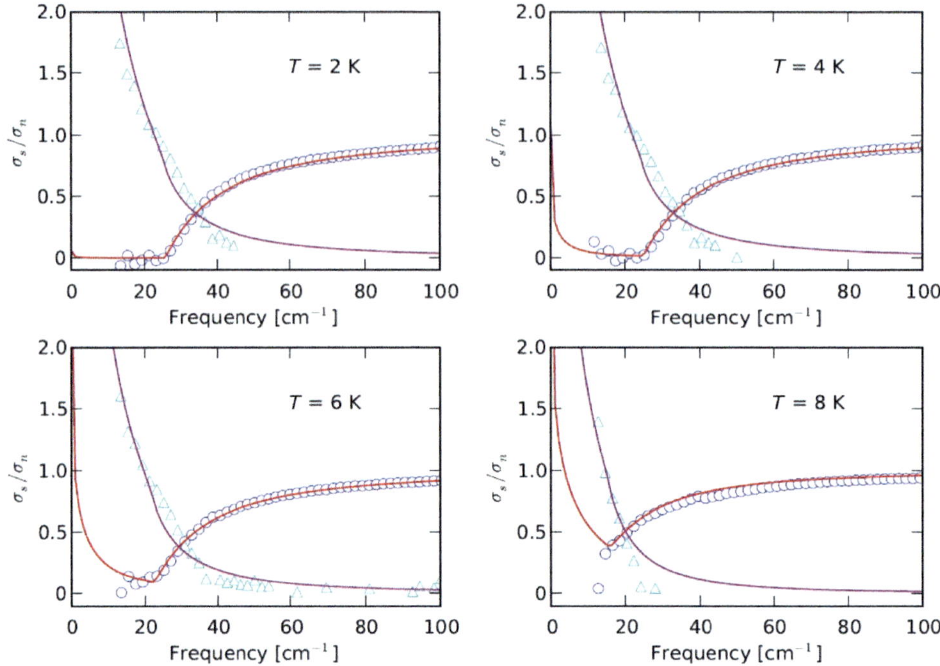

Fig. 11.12 Conductivity of a superconducting NbTiN film ($T_c \approx 10$ K) at four temperatures. The data are normalized to the (approximately constant) normal-state conductivity ($\sigma_N = 6{,}800\ \Omega^{-1}\text{cm}^{-1}$) [221].

σ_{1s} appears in the delta function DC conductivity.* I can estimate the strength of this delta function from the sum rule by calculating

$$\int_{0^+}^{\omega_{\max}} d\omega\, [\sigma_{1n}(\omega) - \sigma_{1s}(\omega)] = \frac{1}{2}\mathcal{A} = \frac{1}{8}\omega_{ps}^2$$

where I start the integral at 0^+ to miss the delta function and end it when the difference between normal and superconducting conductivities has become negligible. The sum rule tells me that the missing area in the finite frequency part of $\sigma_{1s}(\omega)$ is equal to $\mathcal{A}/2$, the integral of the delta function from 0 to 0^+. (The factor of 1/2 also can be seen earlier in this chapter, in Secton 11.3.3; it arises because the integral starting at $\omega = 0$ only picks up half of the area of the delta function.) The weight of the delta function gives $\sigma_2 = \omega_{ps}^2/4\pi\omega$.

The missing area in Fig. 11.12 is proportional to $\sigma_{\mathrm{dc}} = ne^2\tau/m$, to 2Δ, the width of the gap, and to a factor I will call Z that represents the way the conductivity rises to reach the normal-state value. If the conductivity stepped up to 1.0 at $\omega = 2\Delta$, I would have $Z = 1$. Because the conductivity rises slowly, Z is larger than one.

Thus,

$$\frac{1}{8}\omega_{ps}^2 = \frac{ne^2\tau}{m} \cdot 2\Delta \cdot Z$$

---

* The missing area in Fig. 11.12 is the area between a horizontal straight line at $\sigma_1/\sigma_N = 1.0$ and actual the σ_1/σ_N curve.

and hence,

$$\frac{1}{\lambda_L} = \frac{\omega_{ps}}{c} = \sqrt{\frac{8ne^2\tau}{m}}2\Delta Z.$$

The penetration depth increases as τ gets shorter, following $\lambda_L \sim 1/\sqrt{\tau}$.

11.6.3 The Clean Limit

It is easy to guess that the clean limit is the case opposite to that just discussed. The clean metal has a long mean free path, so that $\ell \gg \xi_0$. Then the coherence length is what it is, with examples in Table 11.3. The inequalities in Eq. 11.33 point in the opposite direction, and

$$2\Delta \gg \frac{1}{\tau}. \tag{11.34}$$

The full width at half-maximum of the Drude function, Eq. 4.10a is equal to $1/\tau$, so that all of the Drude spectral weight is below the gap and condenses to form the superconducting delta-function DC conductivity. The normal state conductivity evaluated at the gap frequency* $\omega = 2\Delta$ is

$$\sigma_D = \frac{ne^2/m}{(2\Delta)^2\tau} + i\frac{ne^2}{2m\Delta},$$

making σ_2 larger than σ_1 by a factor of $2\Delta\tau$.

If the normal metal is a really good conductor, then the anomalous skin effect applies in the normal metal (to be discussed in Chapter 14). Similar physics dominates the optical conductivity in the superconducting state. The conductivity is wave-vector dependent, with the q value in the range of $1/\lambda_L$. The anomalous limit has been described by Mattis and Bardeen [222] and Tinkham [215].

Other materials, especially those with lower electron density such as the cuprate superconductors and organic superconductors, have normal skin depth while being in the clean limit. Then, $\omega_{ps} = \omega_p$ and the conductivity will be

$$\sigma = \frac{\pi ne^2}{m}\delta(\omega) + i\frac{ne^2}{m\omega}. \tag{11.35}$$

11.6.4 A Hand-Waving Calculation

Here I will make a simple-minded calculation based on the ideas of the coherence factors and the mixing of electron-like and hole-like states near the gap energy. At high frequencies the mixing is minimal and the normal-state conductivity and the superconducting-state conductivity are essentially indistinguishable. This statement agrees with observation: there is no change in the color (as seen by humans) of the metal on making the transition from normal to superconducting.

---

* I will be cavalier in this chapter and use 2Δ either as an energy or as a frequency. I think it will be obvious from context. (I am no theorist, but setting $\hbar = 1$ is often convenient.)

For frequencies near the gap, I will use a Drude-like model for the quasiparticle scattering.* Note that the scattering is essential to give a dominant real part to the conductivity; if $\omega \gg 1/\tau$ the conductivity is mostly imaginary. So, I write

$$\sigma_{1s}(\omega) = \frac{nQ_k^2\tau/m}{1+\omega^2\tau^2}, \tag{11.36}$$

where $Q_k \equiv Q\left(\mathcal{E}(k)\right)$ is the effective charge of an excitation at energy $\mathcal{E}(k)$, given by Eq. 11.27. This equation should be compared to the ordinary Drude conductivity of Eq. 4.10a,

$$\sigma_{1n}(\omega) = \frac{ne^2\tau/m}{1+\omega^2\tau^2},$$

so that

$$\frac{\sigma_{1s}(\omega)}{\sigma_{1n}(\omega)} = \frac{Q_k^2}{e^2}.$$

What does $Q(\mathcal{E})$ look like? Well, remember the coherence factors, Eqs. 11.29 and 11.30 (Fig. 11.10), where v_k is the amplitude for a state to be full (electron-like) and u_k is the amplitude for a state to be empty (hole-like). Their squares are the probabilities, of course. Holes and electrons have opposite charges, so the effective charge is the difference between the probabilities. I'll write it as

$$Q_k = -e(v_k^2 - u_k^2).$$

Using Eqs. 11.29 and 11.30, I get

$$v_k^2 - u_k^2 = -\frac{\epsilon}{\mathcal{E}}$$

so that so long as the energies of both excitations[†] exceed the gap,

$$Q_k^2 = e^2\frac{\epsilon^2}{\mathcal{E}^2} = e^2\frac{\mathcal{E}^2 - \Delta^2}{\mathcal{E}^2}.$$

Now, the energy of the excitations is delivered by the photon energy $\hbar\omega$. I'll take the energy of each excitation as half of this energy.[‡] The conductivity is then

$$\sigma_{1s}(\omega) = \begin{cases} \mathcal{A}\delta(\omega) & \omega = 0 \\ 0 & 0 < \omega \le 2\Delta_\omega \\ \sigma_{1n}(\omega)\dfrac{\omega^2 - (2\Delta_\omega)^2}{\omega^2} & \omega \ge 2\Delta_\omega \end{cases} \tag{11.37}$$

where $2\Delta_\omega$ is the energy gap (for pairbreaking) in frequency units. This conductivity is shown in Fig. 11.13.

---

* As did Mattis and Bardeen.
† Remember that the excitations are created in pairs.
‡ This division of energy is where I am waving hands the most; in principle, I should integrate over a range of excitation energies, allowing one excitation to have an energy ranging from Δ to $\hbar\omega - \Delta$ while the other ranges from $\hbar\omega - \Delta$ to Δ, keeping the total energy absorbed equal to $\hbar\omega$.

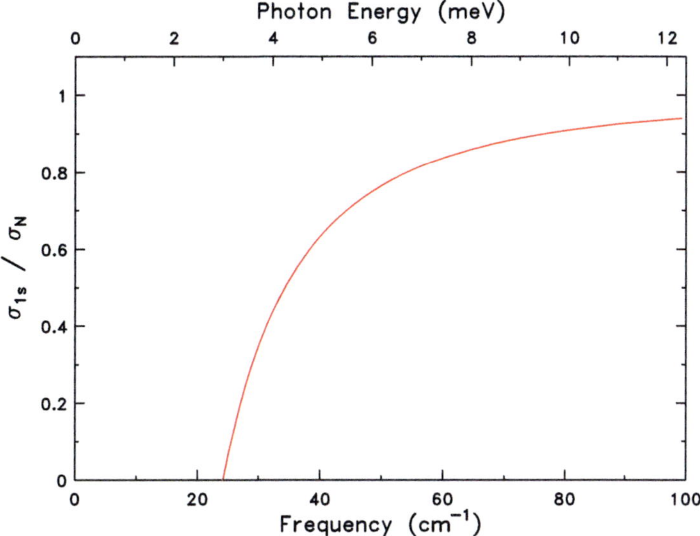

Fig. 11.13 The optical conductivity according to Eq. 11.37. The gap was taken to be $2\Delta = 3$ meV ($2\Delta_\omega = 24\,\text{cm}^{-1}$).

The conductivity is zero right at the gap (despite the divergence of the density of states there). The reason is that the effective charge of the gap-edge excitation is zero. The conductivity then rises smoothly until it becomes equal to the normal-state conductivity at $\omega \gg 2\Delta_\omega$.

By the sum rule, $\mathcal{A}$, the weight of the delta function, is twice the missing area in the superconducting-state conductivity. I can do the integral to find the missing area. In the dirty limit, where $2\Delta\tau \gg 1$, $\mathcal{A} = 4\sigma_{1n}(0)(2\Delta)$. I look back at section 11.6.2 and see that $Z = 2$.

11.7 Thin Film Superconductors

Many infrared studies of superconductors have used thin films [84, 85, 88, 221]. The transmittance spectrum of a thin-film superconductor is quite distinctive. The films are thin enough that the Glover-Tinkham formulas are appropriate. (See Chapter 7, Section 7.4.1.) The transmittance through the film is given by Eq. 7.30:

$$\mathcal{T}_f = \frac{4n}{(n+1+Z_0\sigma_1 d)^2 + (Z_0\sigma_2 d)^2},$$

where n is the refractive index of the substrate, $Z_0 = 377\,\Omega$ is the impedance of free space, and d the film thickness. I look at Figs. 11.11 and 11.13 and expect small transmittance at low frequencies because σ_2 is so large. (σ_1 is zero below the gap.) The transmittance increases with increasing frequency because σ_2 falls. This behavior continues until $\omega = 2\Delta$; here, σ_1 becomes finite and rises towards the normal state value and so the transmittance decreases. Eventually the frequency is high enough that the conductivity and

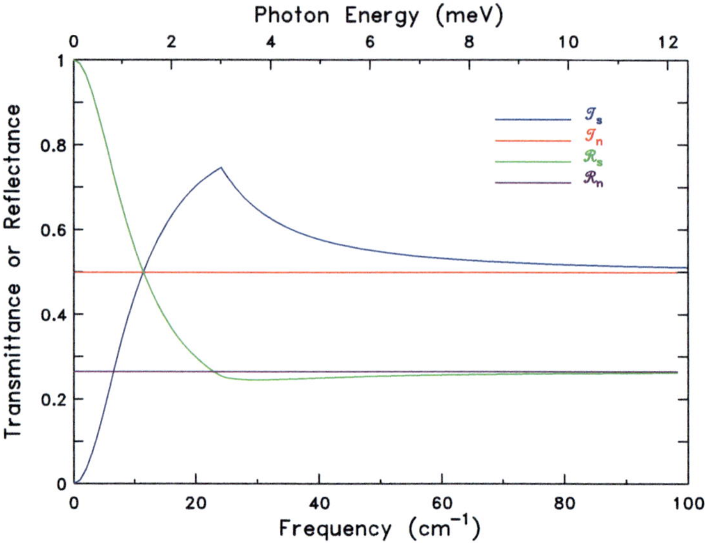

Fig. 11.14 The transmission through and reflectance from a thin superconducting and normal film on a transparent substrate. The gap has been taken to be $2\Delta = 3$ meV (24 cm^{-1}). The normal-state sheet resistance used in the calculation is $R_\square = 1/\sigma_{\mathrm{dc}}d = 377\ \Omega$. The substrate refractive index is $n = 2.1$.

transmittance are indistinguishable from the normal-state values. Calculated transmittance and reflectance curves are shown in Fig. 11.14 for both the superconducting and normal states. The normal-state curves are flat, because the calculations were done in the dirty limit, $\omega\tau \ll 1$. The superconducting state transmittance is larger than the normal state transmittance in the vicinity of the gap and $\mathcal{T}_s$ has a maximum at $\omega = 2\Delta$.

The reflectance is 100% at the lowest frequencies and falls as frequency increases towards the gap. $\mathcal{R}_s + \mathcal{T}_s = 1$ below the gap, whereas above the gap and in the normal state $\mathcal{R} + \mathcal{T} = 1 - \mathcal{A} < 1$.

Measurements of superconducting thin films often yield the ratio of superconducting to normal state transmissions rather than the absolute transmission. There are two motivations for this approach: (1) One can leave the sample in place and change temperature or magnetic field, reducing the effects of alignment, standing waves in the apparatus, and other effects which can bother sample-in/sample-out measurements. (2) The effects of the substrate (mostly) divide out. According to Eq. 7.30:

$$\frac{\mathcal{T}_s}{\mathcal{T}_n} = \frac{(n+1+Z_0\sigma_N d)^2}{(n+1+Z_0\sigma_1 d)^2 + (Z_0\sigma_2 d)^2}.$$

An example is shown in Fig. 11.15. This plot show the transmission ratio, $\mathcal{T}_s/\mathcal{T}_n$, for a NbTiN thin film at $T = 2$ K along with a fit to Mattis–Bardeen theory [229]. The film had a transition temperature of $T_c = 10$ K. The maximum transmittance occurs at the gap frequency, which for this sample is $2\Delta = 25$ cm^{-1}. The normal-state sheet resistance of the film is $R_\square = 1/\sigma_{\mathrm{dc}}d = 130\ \Omega$. The film is deposited on a crystal quartz substrate with refractive index $n = 2.12$. These parameters were used in the calculation.

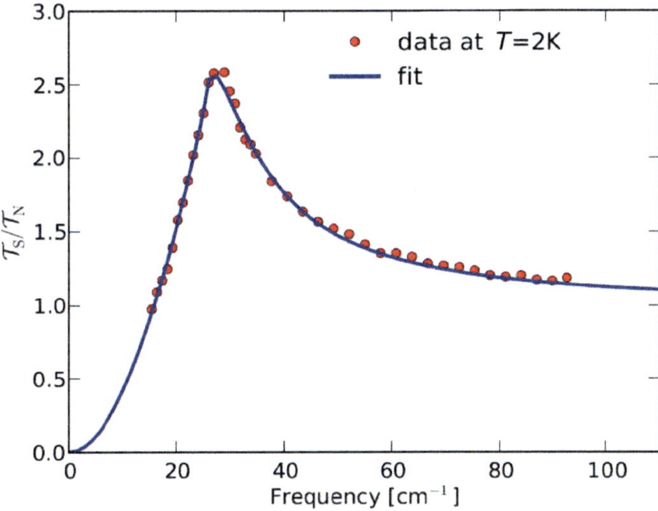

Fig. 11.15 The ratio of superconducting-state to normal state transmittance of a NbTiN thin film. From Ref. 229.

11.8 Unconventional Superconductors

At the center of superconducting research these days are the so-called "unconventional" superconductors [230–232]. The interest in these materials exploded at the discovery [190, 233] thirty years ago of superconductivity with $T_c > 90$ K in layered copper-oxide (cuprate) materials. See Fig. 11.2 and Table 11.1. Many experimenters and theorists caught the "high-T_c fever" and work continues almost unabated today. Examples of such materials include the heavy Fermions (UPt$_3$), organics [(TMTSF)$_2$PF$_6$], ruthenates (Sr$_2$RuO$_4$), charge-density-wave materials (2H-NbSe$_2$), cuprates (YBa$_2$Cu$_3$O$_{7-\delta}$), and ferropnictides (LaOFeAs). These materials differ in fundamental ways from the metallic (conventional) materials that I have addressed up to this point in this chapter.

Among the unconventional properties are anisotropic crystal and electronic structure, strong correlations among the charge carriers, the presence of magnetism, charge ordering, and observations of an unusual linear resistivity in some cases. Among the conventional properties still remaining are infinite DC conductivity, the Meissner effect,* flux quantization with the conventional flux quantum, Josephson tunneling, and Cooper pairs.

Many unconventional superconductors have anisotropic superconducting gaps. When I showed the excitation spectrum of a conventional superconductor in Fig. 11.7, I did not need to specify the direction of $\mathbf{k}$ because the value of Δ is the same everywhere on the Fermi surface. In contrast, the value of the excitation energy in unconventional superconductors often varies with $\mathbf{k}$; there are values where $\Delta = 0$ (nodal values) and others where it is a maximum (anti-nodal). The gap in the cuprates is believed to be of d-wave form, with four nodes equally spaced around the two-dimensional Fermi surface.

---

* Most unconventional superconductors are type-II materials.

The optical properties of unconventional superconductors have been widely studied. I'll give one example, the optical conductivity of $Bi_2Sr_2CaCu_2O_8$. There are a number of review articles that cover the optical properties of cuprates [234–237] and of other unconventional superconductors [238–243] for those who want more details.

The key structural element in $Bi_2Sr_2CaCu_2O_8$ and in the other cuprate superconductors are the two-dimensional CuO_2 (*ab*) planes. The layered structure (orthorhombic or occasionally tetragonal) leads to highly anisotropic properties, including electrical and optical conductivity. The mobile carriers (and the Cooper pairs) reside on the *ab* plane. Although the superconductivity is three dimensional, the superconducting properties (London penetration depth, coherence length, and critical current) are quite anisotropic. Anisotropy is large in the normal state conductivity too. The *ab* plane is metallic whereas the *c* axis appears almost insulating.

With this brief introduction complete, I'll discuss the optical conductivity of $Bi_2Sr_2CaCu_2O_8$, a cuprate superconductor with $T_c = 85$ K. The first single-crystal reflectance spectra of $Bi_2Sr_2CaCu_2O_8$ were reported by Reedyk *et al.*[244] Similar spectra were found by other workers [138, 177, 245–247]. The optical properties display marked anisotropy. The spectra for the *c* polarization (perpendicular to the CuO_2 planes) are typically those of an insulator. The *ab*-plane spectra appear metallic at first look: the reflectance is high in the far infrared and drops steadily throughout the infrared to a minimum around 10,000 cm^{-1} (1.3 eV). Very little temperature dependence is observed above 3,000 cm^{-1}. The reflectance is low in the visible; all the cuprates are black crystals.

Kramers–Kronig analysis of reflectance gives the optical conductivity (and other optical "constants") of $Bi_2Sr_2CaCu_2O_8$. The temperature evolution of the $\mathbf{E}\|a$ optical conductivity is shown in Fig. 11.16, after Ref. 138. There are several features to these

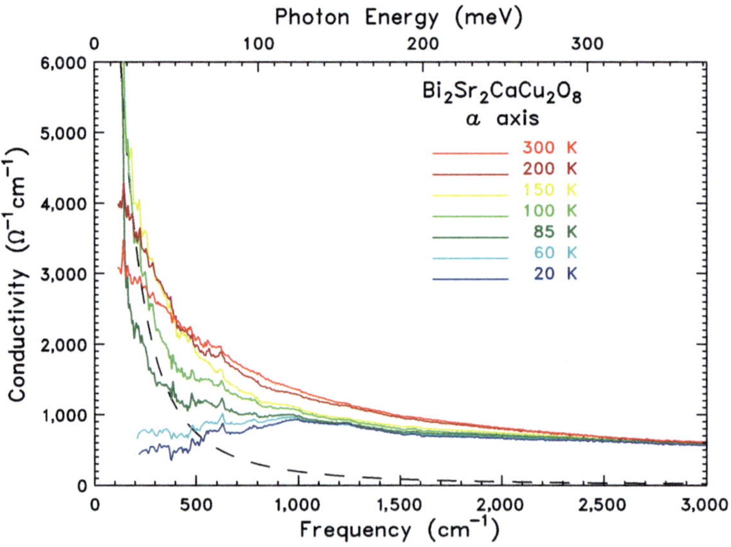

Fig. 11.16 $Bi_2Sr_2CaCu_2O_8$ optical conductivity for $\mathbf{E}\|a$ at temperatures between 20 and 300 K. The dashed line shows a Drude-model fit to the low-frequency, 100 K data. From Ref. 138.

spectra, features that occur in all the cuprate superconductors. First, in the normal state (100–300 K), the low-frequency optical conductivity extrapolates reasonably well to the DC conductivity. Second, with decreasing temperature, the low frequency conductivity increases in agreement with the T-linear resistivity. There is a characteristic narrowing of this far-infrared portion of the spectrum. This narrowing can be seen most easily by noting that the 300 K data is larger than the lower temperature data above about 500 cm^{-1} but then crosses through each of the lower temperature curves at progressively lower frequencies, becoming the smallest conductivity near 100 cm^{-1}. Third, at high frequencies, $\sigma_1(\omega)$ does not show much temperature variation; all the curves draw together around 3,000 cm^{-1}. Fourth, below T_c, the low-frequency conductivity is considerably reduced, so that the 20 K conductivity is smallest at all frequencies measured. The "missing area" in the far-infrared conductivity appears as the zero-frequency delta-function response of the superfluid.

Two approaches have been used to discuss the optical properties of the cuprate superconductors, called the "one-component" and the "two-component" pictures. In the first, there is only a single band or type of carrier, and the unusual frequency dependence in the midinfrared is attributed to interaction with a spectrum of (optically inactive) excitations. This interaction causes a frequency dependence to the carrier scattering rate and enhances the low-frequency effective mass. This approach, appropriate for strongly-interacting materials, is discussed in Chapter 13.

In the two-component approach, two different contributions are assumed: the DC conductivity comes from a set of free carriers, described by a Drude model; the midinfrared conductivity, nearly temperature independent, is due to one or more bands of "midinfrared" carriers. The properties of the free carriers determine the DC conductivity, including its temperature dependence, while the midinfrared carriers dominate the higher frequency conductivity. The superconductivity results from condensation of the free carriers below T_c.

The dashed line in Fig. 11.16 shows the Drude conductivity at 100 K, given by

$$\sigma_{1D}^{fit} = \frac{1}{4\pi} \frac{\omega_{pD}^2 \tau}{1 + \omega^2 \tau^2}, \tag{11.38}$$

with $\omega_{pD} = 9{,}040$ cm^{-1} (1.12 eV) and $1/\tau = 128$ cm^{-1} (16 meV). The Drude contributions at other temperature have essentially the same plasma frequency and scattering rates that are linear in temperature.

The two approaches are still discussed, though the majority opinion favors the one-component approach. Fortunately, the issue does not need to be settled in order to address the phenomenology of the superconducting-state. If the 100 K $1/\tau$ is used together with estimates [248] for the Fermi velocity v_f to compute the ab-plane mean free path ℓ, I get $\ell = v_f\tau \sim 100$ Å.* This number is considerably larger than the typical ab-plane coherence length [249], $\xi \sim 15$ Å. This comparison suggests that the cuprates are in the clean limit [250], i.e., $\ell > \xi$. In this limit, absorption associated with the superconducting gap is not observable because in the superconducting state most of the spectral weight moves to the zero frequency delta function. Thus the only signature of the condensate is

---

* The DC mean free path is essentially the same in the one- and two-component pictures.

its inductive response, seen in the imaginary part of the conductivity, $\sigma_2(\omega)$. The delta function conductivity gives, via the Kramers–Kronig relations,

$$\sigma_2 = \frac{\omega_{ps}{}^2}{4\pi\omega}, \tag{11.39}$$

where ω_{ps} is the plasma frequency of the superconducting condensate, $\omega_{ps} = 4\pi n_s e^2/m$, with n_s the density of superfluid carriers. From the low-frequency behavior at $T = 20$ K, (or, equivalently, from the missing area under $\sigma_1(\omega)$ in the superconducting state) $\omega_{ps} = 9,000$ cm^{-1} (1.11 eV), about the same as the plasma frequency of the free carriers in the two-component analysis. The same analysis gives the a-axis London penetration length $\lambda_L^a \sim 1,800$ Å, in agreement with measurements by other means. The London model, Eq. 11.15, is a good model for the low-frequency conductivity of superconducting cuprates, with nearly all of the Drude-like conductivity spectrum condensing.

12

Semiconductors and Insulators

I will address in this chapter interband transitions in semiconductors and insulators [54, 73, 98, 251, 252]. All solids show these interband optical transitions. In many ways they are analogs of the electric-dipole transitions between principal quantum numbers in atoms, such as $1s \rightarrow 2p$ in hydrogen. Although I made some progress towards the goal in Chapter 9, I mostly have treated these interband transitions in one of two ways. (1) Accumulate the low-energy contributions of these transitions into ϵ_c. (2) Include an electronic Lorentz oscillator as representing the transition between two levels separated by $\hbar\omega_e$. Neither approach is very satisfactory.*

12.1 Band Structure

Before considering the optical excitations, the so-called interband transitions, I first should say something about what these "bands" really are. Free electrons do not have bands. The energy is solely the kinetic energy $\hbar^2 k^2/2m$; I take the electrons and pour them into these free-electron states until I have run out of electrons.

In reality of course, no electron is actually free. All interact with the atomic nuclei and with other electrons. The innermost ones are deeply bound in atomic-like very low-energy states. The most energetic electrons may be nearly free, with the details determined by the chemical and crystallographic structure of the solid. Even states above the Fermi energy (which are unoccupied – or empty – at low temperatures) will be affected by the band structure. The electron energies turn out to be organized into a range of quasi-continuous energies, called "bands," separated by *forbidden energy gaps* where no states exist [19, 24, 98–100].

12.1.1 Some Qualitative Arguments

Let me start by considering an isolated atom. I know that the energies of electrons bound to this atom are quantized into discrete levels and that electromagnetic absorption (photons) will promote electrons from one level to another. Consider the ground state wave function of a simple atom, such as hydrogen. Two H atoms and their $1s$ orbitals are shown in the upper panel of Fig. 12.1. The wave function falls off quickly away from the central nucleus. When the two hydrogens are allowed to interact, the two wave functions can add

---

* Although if all I am interested in is the dielectric function at low frequencies – well below the energy of interband effects – they may prove adequate.

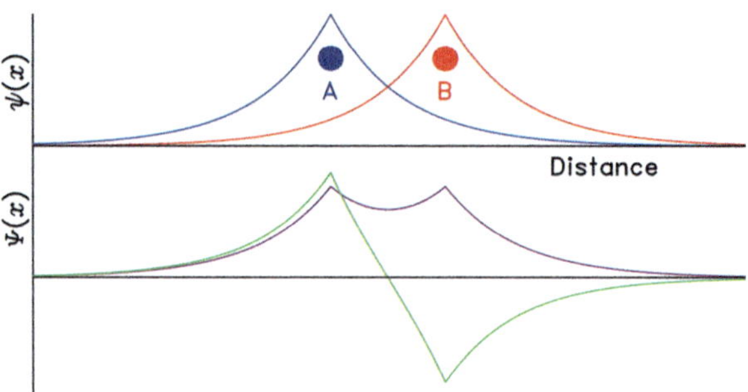

Fig. 12.1   Upper panel: Hydrogen ground-state wave function for two atoms: A and B. Lower panel: the two wave functions can either be added or subtracted. Addition (purple) leads to the bonding orbital; subtraction (green) to the antibonding orbital.

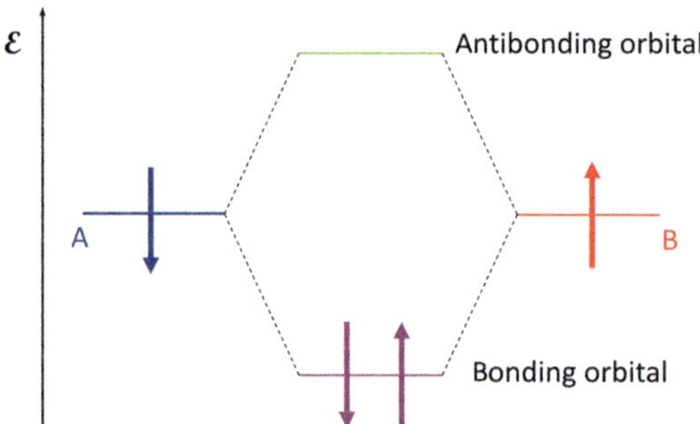

Fig. 12.2   Splitting of the bonding and antibonding levels in H_2.

either constructively or destructively. Constructive addition leads to the bonding orbital; destructive to the antibonding orbital.* It is evident that the bonding orbital has much more electronic amplitude in the space between the two protons. This counteracts the repulsion felt by the two protons and lowers the energy of the H_2 molecule. The antibonding orbital has small amplitude in the space between, which leads to a significant increase in the energy. These effects are sketched in the bottom panels of Fig. 12.1. The energy levels are shown in Fig. 12.2. The bonding level holds the two electrons and the antibonding level – which could hold two more – is empty.

Now consider a long chain of such atoms, containing N atoms, all separated by lattice constant a. There will be many ways to form linear combinations of the wave functions and so there will be many levels, ranging from one where all are added, so all in phase, to one where each is $180°$ out of phase with its two neighbors. Indeed there will be N such levels. If there is one electron/atom, half will be full and half empty. I claim that this linear chain would be a metal.[†]

---

* In this discussion I am leaving out the beautiful physics that comes from the spin states of the electrons and the symmetry that the two-electron wave function must have under exchange. The bonding orbital is a singlet, $S = 0$, and the antibonding orbital a triplet, $S = 1$.

† One-dimensional metallic chains are susceptible to a Peierls [253] or charge-density-wave transition, where the bond lengths become alternating, short, $a - u$, and long, $a + u$. This doubles the unit cell to $2a$, and gives two electrons per dimer, e.g., a chain of H_2 molecules. This dimerized state is an insulator.

If there were two electrons per atom, the uniform chain would have every level filled with two electrons, one spin up and one spin down. It seems natural to claim that this chain would be an insulator.

I have gone about as far as I can without doing some calculations. I could follow the ideas above and do a *tight-binding* model. But instead, let me return to my free electron model and add, as a perturbation, a periodic potential.

12.2 Nearly Free Electrons

The nuclei in the solid form a periodic lattice, so that the potential satisfies

$$V(\mathbf{r}) = V(\mathbf{r} + \mathbf{T}) \tag{12.1}$$

where $\mathbf{T}$ is the translation vector of the crystal. In the one-dimensional case I will consider, $\mathbf{T} = ja\hat{\mathbf{x}}$ where the lattice constant is a and j is an integer.[*]

Now if $V = 0$, $\mathcal{E} = \hbar^2 k^2 / 2m$ (free electrons) with $k = 2\pi j / L$ (periodic boundary conditions). What is the effect of the periodic potential? Well, the free electron states are plane waves, $\psi = e^{ikx}/\sqrt{L}$, and plane waves traveling in a periodic medium are diffracted. In one dimension, the incident wave and the diffracted wave must both travel along $\hat{\mathbf{x}}$. Diffraction effects thus produce waves traveling in both directions along the chain of atoms.[†]

In general these waves just continue along forever. (There is no absorption.) But for certain values of k, I can get strong interference effects.[‡] Consider a partial wave which travels forward for distance a, is reflected and travels backward for distance a, and is reflected again. If the total distance traveled is an integer times the wavelength, there will be constructive interference. So the condition for diffraction is $2a = \lambda$. But of course $k = 2\pi/\lambda$ so the condition is also

$$k = m\frac{\pi}{a}$$

where m, the diffraction order, is a positive or negative integer. Look at the case where $m = \pm 1$. (Other orders have higher energies.) Because of diffraction and constructive interference, the wave functions become standing waves:

$$\psi_1 = \frac{e^{i\pi x/a} + e^{-i\pi x/a}}{\sqrt{2L}} = \sqrt{\frac{2}{L}} \cos\left(\frac{\pi x}{a}\right)$$

and

$$\psi_2 = \frac{e^{i\pi x/a} - e^{-i\pi x/a}}{\sqrt{2L}} = i\sqrt{\frac{2}{L}} \sin\left(\frac{\pi x}{a}\right)$$

---

[*] I'll use j to index the sites just as I did in Chapter 8 (Section 8.2). The one-dimensional chain has length $L = Ja$. If I want to extend to a three-dimensional crystal, then the number of atoms is $N = J^3$.

[†] I can also consider this as specular reflection by each atom. Note that if there are very many atoms, the effect of very many scattering events, even very weak ones, is to have equal amplitudes flowing in both directions.

[‡] Interference is of course the basis of diffraction.

where I have done a side calculation for the normalization. The standing waves lead to a nonuniform density,

$$|\psi_1|^2 = \frac{2}{L}\cos^2\left(\frac{\pi x}{a}\right)$$

and

$$|\psi_2|^2 = \frac{2}{L}\sin^2\left(\frac{\pi x}{a}\right).$$

One probability density has crests at $x = 0,\ \pm a,\ \pm 2a,\ \ldots$ and the other has crests half a lattice constant away.

Because the potential $V(x)$ is periodic, it can be written as a Fourier series; because the period is a, the longest wavelength component is a, making the lowest Fourier component $2\pi/a$. Let me consider that component* alone,

$$V(x) = -V_0 \cos\left(2\pi\frac{x}{a}\right)$$

with a minimum at the lattice sites $0, \pm a$, etc. and a maximum between the atoms. I treat this as a perturbation,

$$\mathcal{E}_1^{(1)} = \langle\psi_1|V|\psi_1\rangle$$

$$= -\frac{2}{L}V_0\int_0^L dx\ \cos\left(2\pi\frac{x}{a}\right)\cos^2\left(\pi\frac{x}{a}\right)$$

but $\cos^2(\phi) = [1 + \cos(2\phi)]/2$ and so

$$= -\frac{1}{L}V_0\int_0^L dx[\cos\left(2\pi\frac{x}{a}\right) + \cos^2\left(2\pi\frac{x}{a}\right)$$

The first term in the integral is zero and the second is $L/2$, so that

$$\mathcal{E}_1^{(1)} = -\frac{V_0}{2}$$

and, for $\psi_2 \sim \sin(kx)$

$$\mathcal{E}_2^{(1)} = -\frac{2}{L}V_0\int_0^L dx\ \cos\left(2\pi\frac{x}{a}\right)\sin^2\left(\pi\frac{x}{a}\right)$$

$$= \frac{V_0}{2}.$$

For the states with $k = \pm\pi/a$, there is an energy difference between the two wave functions. One (the one that goes as $\sin(kx)$) is raised and the other (that goes as $\cos(kx)$) is lowered. That it goes like this is what one might have expected: the $\cos(kx)$ function has the electron most likely to be at the nucleii of the atoms, where the potential is a minimum while the $\sin(kx)$ function has the electron between the nucleii, where the potential is maximum.

The jump in energy of magnitude V_0 at $k = \pm\pi/a$ means that there is a range of energies for which there are no solutions to the Schrödinger equation [19, 24, 254]. Note

---

* There is a DC component of course, but I can suppress that by choosing the zero of energy equal to the DC component.

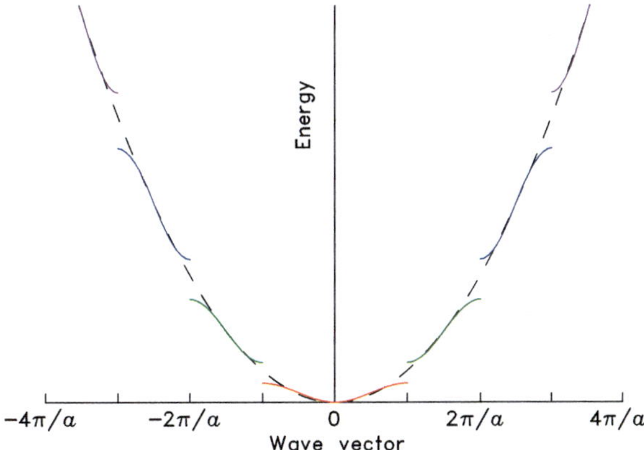

Fig. 12.3 Energy as a function of the wave vector for a solid with a periodic lattice potential. There are three bands shown completely, plus part of a fourth. There are gaps between the bands where no allowed energies exist. The free-electron parabola, $\mathcal{E} = \hbar^2 k^2 / 2m$, is shown as a dashed line.

that the number of k states is unchanged, for spin-1/2 electrons there are $2\frac{2\pi/a}{2\pi/L} = 2N$ states between $k = -\pi/a$ and $k = +\pi/a$.* Moreover, there are no disallowed values of k – only disallowed values of $\mathcal{E}(k)$. These disallowed energies are call *forbidden energy gaps*. Additional gaps appear at $k = \pm 2\pi/a, \pm 3\pi/a, \ldots$.

The interference/diffraction effect affects $\psi(k = \pm\pi/a)$, which has $\lambda = 2a$, the most but states with nearby wave vectors are also affected. A calculated dispersion relation is shown in Fig. 12.3. Three bands and part of a fourth are shown; there are gaps between the bands. The underlying parabolic free-electron curve is also shown; it is close to these curves in the middle of the bands, but passes below or above them near the band edges.

I can now understand the reason why some materials are metals and some insulators. Suppose the linear chain is made up of hydrogen, which has 1 electron per atom. Or suppose it is potassium or silver, with effectively one *valence* electron per atom. Then there are N electrons and $2N$ valid k values, so the Fermi wave vector is $k_F = \pm\pi/2a$; half the states in the lowest-energy band are occupied at $T = 0$ and half unoccupied.

If there are two electrons per atom (helium), this one-dimensional solid would have the first band full and a gap in energy between the highest occupied state (the most energetic state in the first band) and the lowest unoccupied state (the least energetic state in the second band). Then $k_F = \pm\pi/a$ and the electrons cannot be accelerated (with a corresponding increase in momentum) without acquiring additional energy. The solid would be an insulator.[†]

---

* Plus one. The range is more properly stated as $-\pi/a < k \le \pi/a$.

[†] On first hearing this, it is natural to ask why the divalent alkaline earth elements (beryllium, magnesium, calcium, …) are in fact metals. The reason is that the electrons have dispersion in the $\hat{x}$, $\hat{y}$, and $\hat{z}$ directions. Suppose I say that the symmetry of the periodic lattice is such that the diffraction occurs when $\{k_x, k_y, k_z\} = \pm\pi/a$, so that the surface in k-space where the gaps occur is a cube. For electrons with $\mathbf{k}\|\hat{x}$ there is a gap at $\{k_x = \pi/a, k_y = 0, k_z = 0\}$ at a free-electron energy of $\hbar^2\pi^2/2ma^2$. But if I think about electrons whose wave vector is such that $k_x = k_y = k_z$, I reach the cube corner where the free electron energy is $3\hbar^2\pi^2/2ma^2$, three times higher. In many materials, this energy exceeds $\hbar^2\pi^2/2ma^2 + V_0$, so that the Fermi surface, a

If there are 3 valence electrons per atom (aluminum), the one-dimensional Fermi wave vector is $k_F = \pm 3\pi/2a$. The first band is full but the second is half-full. Metal.

If 4 (silicon, germanium), then $k_F = \pm 2\pi/a$ and the first and second bands are both full. The material would be an insulator.*

12.3 Bloch's Theorem

Bloch's theorem is the basis for much of the picture solid-state physics has for electrons, phonons, and other quantum-mechanical objects in a crystal with translational invariance.[†] It states that the wave functions are of the form of a product of plane waves and a periodic function that respects the symmetry of the lattice. A statement of the theorem is

$$\psi_{\mathbf{k}}(\mathbf{r}) = e^{i\mathbf{k}\cdot\mathbf{r}} u_{\mathbf{k}}(\mathbf{r}), \tag{12.2}$$

where $u_{\mathbf{k}}(\mathbf{r})$ is periodic in $\mathbf{T}$, as in Eq. 12.1. So $u_{\mathbf{k}}(\mathbf{r})$ describes the variation of the wave function within the unit cell and the plane-wave exponential function gives the phase between unit cells. If $u_{\mathbf{k}} = $ constant, then Eq. 12.2 is the wave function for free electrons.

Equation 12.2 is stated for three-dimensional crystals, but for simplicity I'll discuss it only in one dimension. The Schrödinger equation is

$$-\frac{\hbar^2}{2m}\frac{\partial^2 \psi(x)}{\partial x^2} + V(x)\psi(x) = \mathcal{E}\psi(x), \tag{12.3}$$

where the potential energy function is periodic: $V(x + ja) = V(x)$ for j any integer and a the lattice constant. I use periodic boundary conditions, so that $\psi(x + L) = \psi(x)$ with $L = Ja$. Translational symmetry implies that the probability density satisfies

$$|\psi(x + ja)|^2 = |\psi(x)|^2. \tag{12.4}$$

Now let me translate from one unit cell to the next ($j = 1$). The points x and $x + a$ are equivalent, making the wave functions differ at most by a factor A, the eigenvalue of the translation operator:

$$\psi(x + a) = A\psi(x), \tag{12.5}$$

where Eq. 12.4 requires that $A^*A = 1$, so A has unit length and some phase. If I translate to the next neighbor along the chain,

$$\psi(x + 2a) = A\psi(x + a) = A^2\psi(x).$$

surface of constant energy, lies in the second band for $\mathbf{k}\|\hat{x}$ and in the first band for $\mathbf{k}$ along the diagonal. The material would then be a metal.

* As indeed are Si and Ge. They are called *semiconductors* for historical reasons. The band gap is not so large in these materials as in, for example, Al_2O_3 or diamond. So at room temperature the Fermi–Dirac distribution function is not zero in the third band, so there are thermally excited electrons that can conduct in a semi-efficient way. There are also vacancies in the second band, so that there are also *holes* to contribute to the conductivity.

[†] Here, translational invariance means that the structure repeats periodically and for ever and ever as I move through space.

Here comes the big step: translate by J lattice constants. Then

$$\psi(x + Ja) = A^J \psi(x) = \psi(x) \tag{12.6}$$

where the second equality comes from the periodic boundary conditions; the wave function must be single valued when I translate by $L = Ja$. This requirement means that $A^J = 1$ or $A = e^{2\pi i j/J}$ so that the Jth power of A is unity. I then rewrite Eq. 12.5 as

$$\psi(x + a) = e^{\frac{2\pi i j}{J}} \psi(x). \tag{12.7}$$

Let me try the following wave function

$$\psi(x) = e^{ikx} u_k(x). \tag{12.8}$$

I make no assumptions about the values of k or the nature of the function $u(x)$. A translation by one lattice constant gives

$$\psi(x + a) = e^{ik(x+a)} u(x + a) = e^{\frac{2\pi i j}{J}} e^{ikx} u(x)$$

where the first equality comes from Eq. 12.8 and the second from Eq. 12.7. I cancel e^{ikx} in the middle part and right side. Then, using quantization of k by the periodic boundary conditions, I find that

$$k = \frac{2\pi j}{L} = \frac{2\pi j}{Ja},$$

with $j = 0, \pm 1, \pm 2, \dots$ This value of k allows me to cancel the other complex exponential, leaving only

$$u(x + a) = u(x),$$

completing the proof of Bloch's theorem.

12.4 The Brillouin Zone

Bloch's theorem, the periodic lattice, and the boundary conditions all conspire to limit what values k can take. I began with it as a real number (a quantity along a continuous line) taking any value from negative to positive infinity. Then, periodic boundary conditions restricted it to a countable infinity, represented by integers. Now I will show that only a finite range of these integer-based k values has distinct physical significance.

I start by once again translating by a single lattice constant

$$\psi(x + a) = u_k(x + a) e^{ik(x+a)} = u_k(x) e^{ikx} e^{ika}$$
$$= e^{ika} \psi(x). \tag{12.9}$$

Now, define a wave vector k' by

$$k = k' + \frac{2\pi m}{a},$$

with m an integer. Substitute into Eq. 12.9 to find

$$e^{ika}\psi(x) = e^{ik'a}e^{2\pi im}\psi(x).$$

But $e^{2\pi im} = 1$ so

$$e^{ika}\psi(x) = e^{ik'a}\psi(x).$$

The state with wave vector k and wave vector $k + 2\pi m/a$, with m an integer, are the same. The wave function is the same as is the energy eigenvalue. k is periodic with period $2\pi/a$. Values of k in the range

$$-\frac{\pi}{a} < k \le \frac{\pi}{a} \tag{12.10}$$

are all I need.

I am free to modify Fig. 12.3 by adding or subtracting integer times $2\pi/a$ as many times as needed to bring the wave vector into the range specified by Eq. 12.10. This translation produces the dispersion relations shown in the left-hand panel of Fig. 12.4. The small price that I have to pay for this reduction is that there are many values of energy for any k in range, so that I have to supply a band index to specify about which I speak.

The range of k specified by Eq. 12.10 is said to be in the first Brillouin zone. The bands lie one above another. I'll have to supply a band index n to specify the wave function and the energy $\mathcal{E}_n$. Despite this drastic change, the basic discussion earlier in this chapter (Section 12.2) as to whether a given material is metal or insulator is unchanged. I may also apply translations by the range of the Brillouin zone, $2\pi m/a$, $m = 1, 2\ldots$, to extend each of the bands, as shown in the right panel of Fig. 12.4. The curves join smoothly, repeating once per Brillouin zone. All three representations (extended, reduced, and repeated) are equally valid.

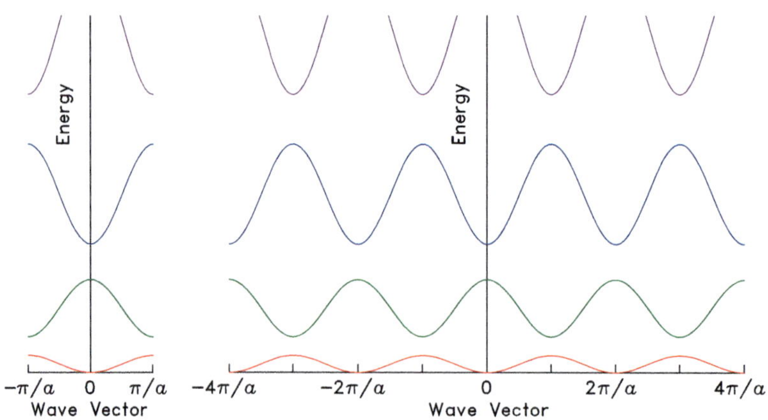

Fig. 12.4   Left: Energy as a function of the wave vector in the reduced zone scheme. Right: Energy as a function of the wave vector in the repeated zone scheme. Both diagrams show three bands completely, plus part of a fourth.

12.5 Band Gaps of Semiconductors

Table 12.1, based on data in Wikipedia [255], lists a large number of semiconductors, showing the group, material, chemical formula, band gap value, and type of gap. The data come from a variety of sources [24, 256–264]. In the table, the group represents the column in the periodic table. Group IV is the column headed by carbon. Group III–V are compounds of elements from the column headed by boron and the one headed by nitrogen,

Table 12.1 Semiconductor materials. (Continued on the next page.)

Group	Formula	Gap eV	Gap cm^{-1}	type
IV	C	5.47	44,100	Indirect
IV	Si	1.11	8,900	Indirect
IV	Ge	0.67	5,400	Indirect
IV	α-Sn	0.08	600	Indirect
IV	3C-SiC	2.3	18,500	Indirect
IV	4H-SiC	3.3	26,600	Indirect
IV	6H-SiC	3	24,100	Indirect
VI	S_8	2.6	20,900	
VI	Se	1.74	14,000	
VI	Te	0.33	2,600	
III–V	cubic BN	6.36	51,200	Indirect
III–V	hexagonal BN	5.96	48,000	Quasi-direct
III–V	BP	2	16,100	Indirect
III–V	BAs	1.5	12,000	Indirect
III–V	$B_{12}As_2$	3.47	27,900	Indirect
III–V	AlN	6.28	50,600	Direct
III–V	AlP	2.45	19,700	Indirect
III–V	AlAs	2.16	17,400	Indirect
III–V	AlSb	1.6	12,900	Indirect
III–V	AlSb	2.2	17,700	Direct
III–V	GaN	3.44	27,700	Direct
III–V	GaP	2.26	18,200	Indirect
III–V	GaAs	1.43	11,500	Direct
III–V	GaSb	0.726	5,800	Direct
III–V	InN	0.7	5,600	Direct
III–V	InP	1.35	10,800	Direct
III–V	InAs	0.36	2,900	Direct
III–V	InSb	0.17	1,300	Direct
II–VI	CdSe	1.74	14,000	Direct
II–VI	CdS	2.42	19,500	Direct
II–VI	CdTe	1.49	12,000	
II–VI	ZnO	3.37	27,100	Direct
II–VI	ZnSe	2.7	21,700	Direct

Table 12.1 Continued.

Group	Formula	Gap eV	Gap cm^{-1}	type
II–VI	ZnS	3.54	28.500	Direct
II–VI	ZnTe	2.25	18,100	Direct
IV–VI	PbSe	0.27	2,100	Direct
IV–VI	PbS	0.37	2,900	
IV–VI	PbTe	0.32	2,500	
IV–VI	SnS	1	8,000	Indirect
IV–VI	SnS$_2$	2.2	17,700	
I–VII	CuCl			Direct
I–VI	Cu$_2$S	1.2	9,600	Direct
II–V	Cd$_3$As$_2$	0.14	1,100	n-type
Oxide	TiO$_2$ anatase	3.2	25,800	Indirect
Oxide	TiO$_2$ rutile	3.02	24,300	Direct
Oxide	TiO$_2$ brookite	2.96	23,800	
Oxide	Cu$_2$O	2.17	17,500	
Oxide	CuO	1.2	9,600	p-type
Oxide	UO$_2$	1.3	10,400	High
Oxide	SnO$_2$	3.7	29,800	Oxygen-deficient
Oxide	BaTiO$_3$	3	24,100	Ferroelectric,
Oxide	SrTiO$_3$	3.3	26,600	Ferroelectric,
Oxide	LiNbO$_3$	4	32,200	Ferroelectric,
Oxide	La$_2$CuO$_4$	2	16,100	Superconductor
Layered	GaSe	2.1	16,900	Indirect
Other	CuInSe$_2$	1	8,000	Direct
Other	Ag$_2$S	0.9	7,200	
Other	FeS$_2$	0.95	7,600	
Other	Cu$_2$ZnSnS$_4$	1.49	12,000	Direct

and so forth. The gap type specifies generally whether the gap is direct or indirect. The band structures in Fig. 12.4 represent direct band gaps because the minimum of one band is right above the maximum of the one below. I'll address indirect band gaps later in this chapter (Section 12.9).

Table 12.2 displays properties of semiconducting alloys. These materials can be tuned to some extent by changing composition and so represent not a single material but a class of materials.

12.6 Effective Mass

The free electron kinetic energy depends on the electron mass,

$$\mathcal{E} = \frac{\hbar^2 k^2}{2m} = \frac{p^2}{2m}.$$

Table 12.2 Table of semiconductor alloy systems.			
Group	Formula	Low gap eV	High gap eV
IV	$Si_{1-x}Ge_x$	0.67	1.11
III–V	$Al_xGa_{1-x}As$	1.42	2.16
III–V	$In_xGa_{1-x}As$	0.36	1.43
III–V	$In_xGa_{1-x}P$	1.35	2.26
III–V	$Al_xIn_{1-x}As$	0.36	2.16
III–V	GaAsP	1.43	2.26
III–V	GaAsSb	0.7	1.42
III–V	AlGaN	3.44	6.28
III–V	AlGaP	2.26	2.45
III–V	InGaN	2	3.4
II–VI	CdZnTe	1.4	2.2
II–VI	HgCdTe	0	1.5
II–VI	HgZnTe	0	2.25
Other	$Cu(In,Ga)Se_2$	1	1.7

The kinetic energy is parabolic in the momentum, with the curvature of the function determined by m^{-1}, so

$$\frac{\partial^2 \mathcal{E}}{\partial k^2} = \frac{\hbar^2}{m},$$

or, observing (Fig. 12.4) that the second derivative may not be the same for all vales of k,

$$m_k^* = \frac{\hbar^2}{\frac{\partial^2 \mathcal{E}(k)}{\partial k^2}}.*$$

The free-electron parabola is concave upward, and the mass is everywhere equal to the free-electron mass. Looking at Fig. 12.4 (or Fig. 12.3), I see that the curvature at the bottom of the lowest band is nearly that of the free-electron band, but that near either $k = \pm\pi/a$ or $k = 0$ the curvature is considerably larger, making m_k^* smaller than the free-electron mass. Moreover, the top half of each band is *concave downward*, changing the sign of the second derivative, so that the effective mass of these electrons is negative!

Technologically important semiconductors (Si, Ge, GaAs, InSb, …) all have four electrons per atom (or three on one and five on the other) so the first two bands are full and the third empty. The second band is called the valence band and the third the conduction band. Now, in these semiconductors,* the gap between the top of the valence band and the bottom of the conduction band is small enough that at finite temperatures there are thermally excited carriers in the conduction band.[†] Because the total number of electrons is fixed, empty states are left behind when thermal energies promote carriers to the conduction band. These vacant states are called *holes*. The absence of a negatively charged electron is

---

* See also the footnote at the end of Section 12.2.

[†] In *doped* semiconductors, the majority carriers come from impurity levels located in the forbidden band gap, but here I consider intrinsic (undoped) semiconductors and insulators.

Table 12.3 Effective mass in selected semiconductors.			
Group	Material	Electron	Hole
IV	Si (4K)	1.06	0.59
	Si (300K)	1.09	1.15
	Ge	0.55	0.37
III–V	GaAs	0.067	0.45
	InSb	0.013	0.6
II–VI	ZnO	0.29	1.21
	ZnSe	0.17	1.44

a positive hole; the electron mass in the top half of the band is negative so the hole mass is positive. Hole energy is measured downwards from the top of the band. Holes contribute to the total electrical and optical properties. I write

$$\mathbf{j} = -ne\mathbf{v}_e + pe\mathbf{v}_h$$

where n is the electron number density, p is the hole number density, and $\mathbf{v}_e$ ($\mathbf{v}_h$) are the electron (hole) drift velocity.

12.7 Direct Interband Transitions

Now I turn my attention to calculating optical properties due to interband transitions. First, I'll do the simple case of a direct ($\Delta k = 0$) transition in a solid. To start, look at the left-hand panel of Fig. 12.4 and consider a solid with an average of 4 electrons per atom (such as GaAs or InSb) making the first two bands fully occupied and the third band empty. A sketch of the band structure of such a solid [265–267] is shown in the left panel of Fig. 12.5. The highest filled and the lowest empty states are both at the zone center. This energy difference will be the minimum photon energy for optical absorption. There is a much wider energy difference at X; the direct interband transitions span the minimum to maximum energies.

Another case is shown in the right panel of Fig. 12.5, representing the noble metals. The fully occupied d levels occur 2–4 eV below the Fermi level. Direct transitions shown as the vertical arrow, can occur from these d levels to conduction band energies above the Fermi level. Intraband transitions also occur, leading to the Drude conductivity of the conduction electrons.

The selection rules for these transitions are discussed in Chapter 9. The dielectric function is given in Eq. 9.21. I'll make a few changes in notation. The optical absorption promotes an electron from its ground state in the valence band to an excited state in the conduction band. So I'll call the ground state $|0\rangle$ a valence-band state $|v\rangle$. Similarly, the excited state $|n\rangle$ becomes a conduction-band state $|c\rangle$. In addition, I'll take $\gamma_{cv} \to 0$. Then,

$$\epsilon = 1 + \frac{8\pi e^2}{\hbar} \sum_{vc} \frac{\omega_{cv}}{\omega_{cv}^2 - \omega^2} |\langle c|x|v\rangle|^2. \tag{12.11}$$

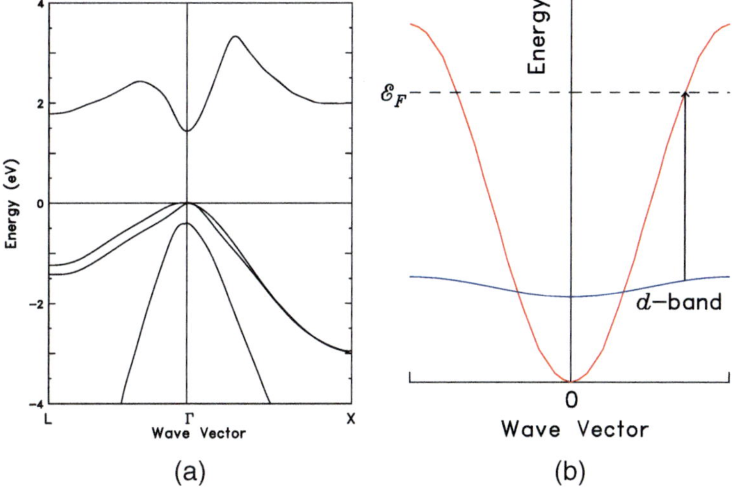

 (a) Band structure, $\mathcal{E}(\mathbf{k})$, of GaAs for energies around the band gap [265–267]. The horizontal axis is marked with special values of $\mathbf{k}$ in two directions. Γ is the zone center, $\mathbf{k} = 0$ while X is the zone boundary in the k_x direction (or, on account of the cubic symmetry, the k_y or k_z direction). L is the zone boundary in the body diagonal of the cubic lattice. The three, concave down, lower bands are the (filled) valence bands and the upper band is the (empty) conduction band. The forbidden energy gap is the range from the top of the valence band (at Γ) to the bottom of the conduction band (also at Γ). (b) Schematic band structure, $\mathcal{E}(\mathbf{k})$, of a noble metal (Ag, Cu, or Au) for conduction-band energies. The horizontal axis is the wave vector. The Fermi energy is shown. (Note that the Fermi surface in these metals is not a sphere, and the Fermi wave vector in some directions – the so-called "necks" – is at the zone boundary.) The d band is the narrow band well below the Fermi level. Optical transitions from the d band to the conduction band occur for conduction band states above $\mathcal{E}_F$.

As discussed in Chapter 9 (Section 9.6), this dielectric function has an imaginary part that is a series of delta functions,

$$\epsilon_2 = \frac{4\pi^2 e^2}{\hbar\omega} \sum_n \delta(\omega_{cv} - \omega)\, \omega_{cv} |\,\langle c|x|v\rangle\,|^2. \tag{12.12}$$

I now convert this into an equation that uses the momentum matrix element:

$$\langle c|x|v\rangle = \frac{1}{i\omega_{cv}m}\,\langle c|p_x|v\rangle$$

so

$$\omega_{cv}|\,\langle c|x|v\rangle\,|^2 = \frac{1}{m^2}\frac{|\,\langle c|p_x|v\rangle\,|^2}{\omega_{cv}}$$

where I use p_x because the electric field is in the $\hat{x}$ direction.

12.7.1　The Matrix Element

This matrix element is an integral,

$$\langle c|p_x|v\rangle = \int_V d^3r\,\psi_c^*\,p_x\,\psi_v. \tag{12.13}$$

I write these wave functions in Bloch form

$$\psi_{ck''} = \frac{1}{\sqrt{\Omega}}\, u_{ck''}(\mathbf{r})\, e^{i\mathbf{k}''\cdot\mathbf{r}}$$

and

$$\psi_{vk'} = \frac{1}{\sqrt{\Omega}}\, u_{vk'}(\mathbf{r})\, e^{i\mathbf{k}'\cdot\mathbf{r}}$$

with Ω the volume and $u(\mathbf{r}) = u(\mathbf{r}+\mathbf{T})$ (periodic). Here, $\mathbf{k}$ specifies the place in momentum space where the wave function exists. The prime and double prime have no deep meaning; they are different because the k-space coordinate of ground and excited state could be different.[*]

Now I begin to compute the matrix element. To consider a specific model, I will do this for the semiconductor whose band structure is like that shown in Fig. 12.5. Moreover, in the crystal I may arrive at a particular location $\mathbf{r}$ by first translating to a specific unit cell (i) by the translation vector of the lattice $\mathbf{T}_i$ and then by moving within that specific cell by a vector $\mathbf{r}_i$. This process requires

$$\mathbf{r} = \mathbf{T}_i + \mathbf{r}_i$$

and the integral in Eq. 12.13 can thus be written

$$\langle c|p_x|v\rangle = \sum_{\text{cells}} \int_{\text{cell}} d^3 r_i\, \psi_c^*\, p_x\, \psi_v.$$

Next, I install the explicit wave functions for conduction and valence bands:

$$\langle c|p_x|v\rangle = \frac{1}{\Omega} \sum_{\text{cells}} \int_{\text{cell}} d^3 r_i\, u_{ck''}^*(\mathbf{T}_i + \mathbf{r}_i)\, e^{-i\mathbf{k}''\cdot(\mathbf{T}_i+\mathbf{r}_i)}\, p_x\, u_{vk'}(\mathbf{T}_i + \mathbf{r}_i)\, e^{i\mathbf{k}'\cdot(\mathbf{T}_i+\mathbf{r}_i)}.$$

I collect terms, bring those that are do not depend on $\mathbf{r}_i$ outside the integral, and make use of the periodicity of the Bloch states to find

$$\langle c|p_x|v\rangle = \frac{1}{\Omega} \sum_{\text{cells}} e^{i(\mathbf{k}'-\mathbf{k}'')\cdot\mathbf{T}_i} \int_{\text{cell}} d^3 r_i\, e^{-i\mathbf{k}''\cdot\mathbf{r}_i}\, u_{ck''}^*(\mathbf{r}_i)\, p_x\, u_{vk'}(\mathbf{r}_i)\, e^{i\mathbf{k}'\cdot\mathbf{r}_i}.$$

I'll look at the pieces of this equation. First:

$$\sum_{\text{cells}} e^{i(\mathbf{k}'-\mathbf{k}'')\cdot\mathbf{T}_i} = \begin{cases} N_{\text{cells}} & \text{if } \mathbf{k}' - \mathbf{k}'' = 0 \\ 0 & \text{otherwise,} \end{cases}$$

so I can replace $\mathbf{k}''$ with $\mathbf{k}'$ everywhere the former appears. Moreover $N_{\text{cells}}/\Omega = 1/V_{\text{cell}}$. Second, $p_x = -i\hbar\nabla_x$ so

$$p_x u_{vk'}(\mathbf{r}_i)\, e^{i\mathbf{k}'\cdot\mathbf{r}_i} = -i\hbar\, e^{i\mathbf{k}'\cdot\mathbf{r}_i}[ik_x u_{vk'}(\mathbf{r}_i) + \nabla_x u_{vk'}(\mathbf{r}_i)].$$

Third,

$$e^{-i\mathbf{k}''\cdot\mathbf{r}_i}\, e^{i\mathbf{k}'\cdot\mathbf{r}_i} = 1$$

because $\mathbf{k}' = \mathbf{k}''$.

[*] I allow for them to be different; the selection rule ($\mathbf{k}'' = \mathbf{k}'$) I have argued for up to now will come out of this calculation. The choice of prime and double prime rather than, for example, unprimed is designed to make the final answer come out with the notation I want.

Fourth, because of orthogonality ($\langle c|v \rangle = 0$)

$$\int_{\text{cell}} d^3r_i \, u^*_{ck'}(\mathbf{r}_i) \, k_x \, u_{vk'}(\mathbf{r}_i) = 0,$$

allowing the matrix element to be written

$$\langle c|p_x|v \rangle = \frac{1}{V_{\text{cell}}} \int_{\text{cell}} d^3r_i \, u^*_{ck'}(\mathbf{r}_i) \, p_x \, u_{vk'}(\mathbf{r}_i) \equiv p_{vc}(\mathbf{k}').$$

12.7.2 The Imaginary Part of the Dielectric Function

Now I return to Eq. 12.12, converting the sum to an integral:

$$\sum_n \rightarrow 2 \int \frac{d^3k}{(2\pi)^3}.$$

The imaginary part of the dielectric function becomes

$$\epsilon_2 = \frac{e^2}{\pi m^2 \hbar \omega^2} \int d^3k \, \delta(\omega_{cv}(\mathbf{k}) - \omega)|p_{vc}(\mathbf{k})|^2, \qquad (12.14)$$

with

$$\omega_{cv}(\mathbf{k}) = \frac{\mathcal{E}_c(\mathbf{k}) - \mathcal{E}_v(\mathbf{k})}{\hbar}.$$

12.7.3 The Absorption Edge

I will make a crude estimate of the absorption edge near the minimum energy. To do so, I will need to have the band structure (to know how $\omega_{cv}(\mathbf{k})$ varies with $\mathbf{k}$) and the matrix element p_{vc} (to know how it varies with $\mathbf{k}$).

For the first, I consider parabolic bands separated by $\mathcal{E}_g = \hbar\omega_g$, the minimum forbidden energy gap. See Fig. 12.6 for a sketch of the band structure. The conduction band energy is

$$\mathcal{E}_c = \hbar\omega_g + \frac{\hbar^2 k^2}{2m_c}$$

where m_c is the effective mass in the conduction band. The minimum energy is at $k = 0$, where $\mathcal{E}_c = \hbar\omega_g$ The valence band energy is

$$\mathcal{E}_v = -\frac{\hbar^2 k^2}{2|m_v|}$$

where m_v is the effective mass in the valence band. I take the absolute value because the effective mass is in fact negative ($\mathcal{E}_v$ is concave downwards) and I have put the minus sign in explicitly. The zero of energy is the $k = 0$ energy of the valence band maximum, shown as the horizontal line in Fig. 12.6.

For the second, I take p_{vc} as independent of $\mathbf{k}$. It then may be taken outside the integral in Eq. 12.14, which becomes

$$\epsilon_2 = \frac{e^2}{\pi m^2 \hbar \omega^2} |p_{vc}|^2 \int d^3k \, \delta(\omega_{cv}(\mathbf{k}) - \omega). \qquad (12.15)$$

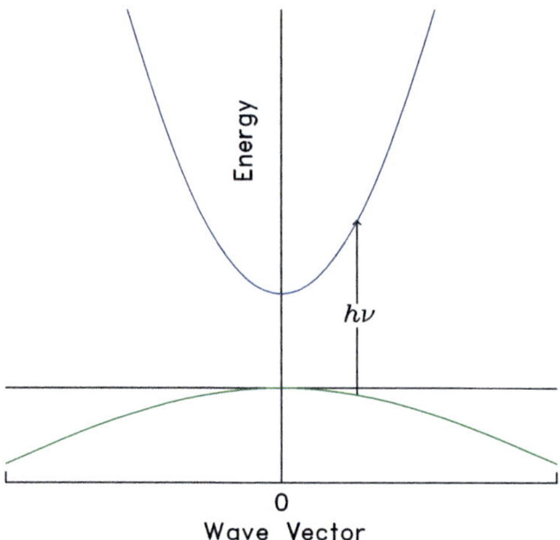

Fig. 12.6 Energy band structure, $\mathcal{E}(\mathbf{k})$. A vertical optical transition from the valence band to the conduction band is shown. I put the zero of energy at the top of the valence band.

Now, the energy difference between valence and conduction bands depends on k according to:

$$\hbar\omega_{cv}(\mathbf{k}) = \hbar\omega_g + \left(\frac{\hbar^2 k^2}{2m_c} + \frac{\hbar^2 k^2}{2|m_v|}\right).$$

I will define an average effective, or reduced, mass as

$$\frac{1}{m_r} = \frac{1}{m_c} + \frac{1}{|m_v|} \tag{12.16}$$

so that the frequency difference becomes

$$\omega_{cv}(\mathbf{k}) = \omega_g + \frac{\hbar k^2}{2m_r}. \tag{12.17}$$

Turning Eq. 12.17 equation around, I find

$$k = \frac{1}{2}\sqrt{\frac{2m_r}{\hbar}}\sqrt{\omega_{cv}(\mathbf{k}) - \omega_g}.$$

Taking the differential gives

$$dk = \sqrt{\frac{2m_r}{\hbar}}\,\frac{d\omega_{cv}(\mathbf{k})}{\sqrt{\omega_{cv}(\mathbf{k}) - \omega_g}}.$$

The integral in Eq. 12.15 is a three-dimensional integral in $d^3k = k^2\,dk\,d\Omega$ where $d\Omega$ is the angular integral over the 4π solid angle. But nothing depends on the angular co-ordinates, so I may immediately write

$$d^3k = 4\pi k^2 dk = 2\pi \left(\frac{2m_r}{\hbar}\right)^{3/2}\sqrt{\omega_{cv}(\mathbf{k}) - \omega_g}\,d\omega_{cv}(\mathbf{k}).$$

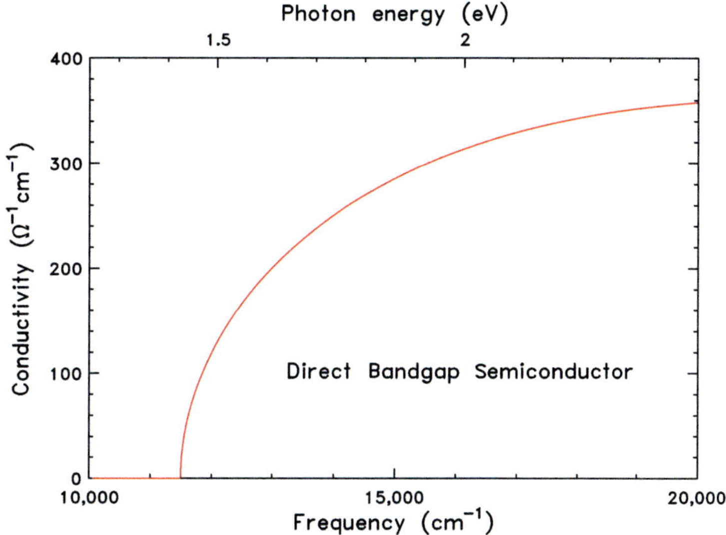

Fig. 12.7 Calculated conductivity edge of a direct band gap semiconductor, showing the $\sqrt{\omega - \omega_g}$ behavior. The parameters are representative of GaAs. Note the suppressed zero of the frequency scale.

The lower limit of the integral at $k = 0$ is equivalent to $\omega_{cv}(\mathbf{k}) = \omega_g$. Thus, the imaginary part of the dielectric function becomes

$$\epsilon_2 = \frac{e^2}{\pi m^2 \hbar \omega^2} |p_{vc}|^2 2\pi \left(\frac{2m_r}{\hbar} \right)^{3/2} \int_{\omega_g}^{\infty} d\omega_{cv}(\mathbf{k}) \, \delta(\omega_{cv}(\mathbf{k}) - \omega) \sqrt{\omega_{cv}(\mathbf{k}) - \omega_g}.$$

The integral is easy to do (thanks to the delta function). Doing it, and collecting terms, I get*

$$\epsilon_2 = \begin{cases} 0 & \omega \leq \omega_g \\ \dfrac{2e^2}{m^2 \hbar} \left(\dfrac{2m_r}{\hbar} \right)^{3/2} |p_{vc}|^2 \dfrac{\sqrt{\omega - \omega_g}}{\omega^2} & \omega \geq \omega_g. \end{cases} \tag{12.18}$$

I can simplify this a bit by using the oscillator strength f, where

$$f = \frac{2\langle |p_{vc}|^2 \rangle}{m\hbar\omega} = \frac{2m\omega \langle |x_{vc}|^2 \rangle}{\hbar} \tag{12.19}$$

according to Eq. 9.22. I also write $\sigma_1 = \omega\epsilon_2/4\pi$ and then

$$\sigma_1 = \begin{cases} 0 & \omega \leq \omega_g \\ \dfrac{e^2}{2\pi m} \left(\dfrac{2m_r}{\hbar} \right)^{3/2} f \sqrt{\omega - \omega_g} & \omega \geq \omega_g. \end{cases} \tag{12.20}$$

Fig. 12.7 shows a plot of the absorption edge using parameters appropriate to GaAs.

---

* A similar equation is oftern written for α. It needs the refractive index n also, because $\alpha = \omega\epsilon_2/nc$.

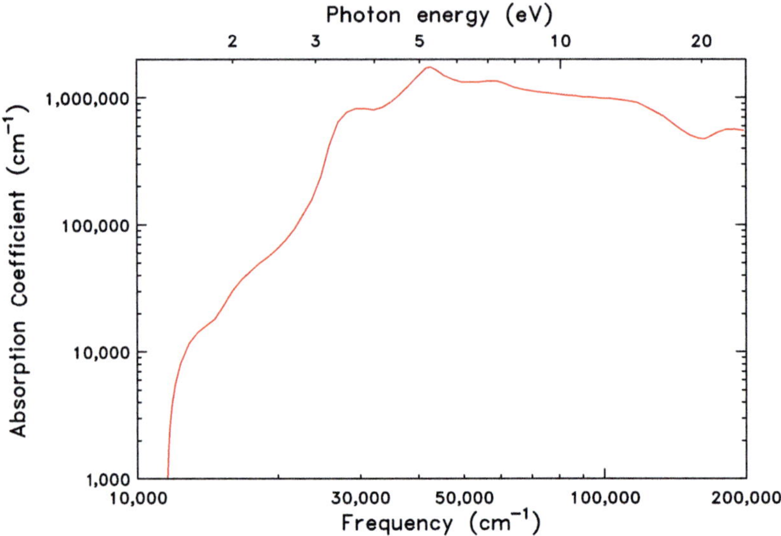

Fig. 12.8   Room temperature absorption coeffficient of GaAs. Note the suppressed zero of the frequency scale.

The experimental absorption coefficient* of GaAs [68, 268–271] is shown in Fig. 12.8. The sharp rise at 1.42 eV is due to direct interband transitions. (Exciton effects, evident in the low-temperature spectra of GaAs, also play a role. See Section 12.10, later in this chapter.) Other transitions from valence to conduction band occur at higher frequencies. The peaks at 3.5 and 5 eV come from energies where there are maxima in the joint density of states, as discussed briefly in the next section.

12.8 The Joint Density of States and Critical Points

Once the energy of the interband transition increases significantly above the minimum band gap, the parabolic dispersions of valence and conduction bands end. I still can calculate the conductivity, but to do so requires the details of the band structure. It is however possible to write the conductivity in a compact form that connects closely to the band structure. I start with Eq. 12.14, which relates ϵ_2 to the integral of the square of a matrix element and an energy-conserving delta function. The integral is over d^3k and the range is the entire Brillouin zone. This range in Fig. 12.4, which shows the k_x direction (or k_y or k_z), is $-\pi/a \to \pi/a$. For general $\mathbf{k}$ direction the range depends in detail on the crystal structure.

I'll assert now that the matrix element $|p_{vc}(\mathbf{k})|$ is independent of $\mathbf{k}$ and take it outside the integral as $|p_{vc}|$. I convert to conductivity and use $\mathcal{E}_{cv}(\mathbf{k}) = \mathcal{E}_c(\mathbf{k}) - \mathcal{E}_v(\mathbf{k}) = \hbar\omega_{cv}(\mathbf{k})$:

$$\sigma_1(\omega) = \frac{e^2}{4\pi^2 m^2 \omega} |p_{vc}|^2 \int d^3k \; \delta(\mathcal{E}_{cv}(\mathbf{k}) - \hbar\omega). \tag{12.21}$$

---

* See Appendix A; the conductivity is related to the absorption coefficient α by $\sigma = n\alpha/377\ \Omega$, with $n \approx 3\text{–}4$ being the refractive index.

The argument of the delta function in Eq. 12.21 is a function $g(\mathbf{k}) = \mathcal{E}_{cv}(\mathbf{k}) - \hbar\omega$. The delta function picks out the places where $g(\mathbf{k}) = 0$. This zero is equivalent to where energy is conserved in making a vertical transition from valence to conduction band: $\hbar\omega = \mathcal{E}_{cv}(\mathbf{k}) = \mathcal{E}_c(\mathbf{k}) - \mathcal{E}_v(\mathbf{k})$. The delta function has the property that [54]

$$\int d^3x f(\mathbf{x})\delta(g(\mathbf{x})) = \int_{g(\mathbf{x})=0} dS \, \frac{f(\mathbf{x})}{|\nabla g(\mathbf{x})|}$$

where the integral on the right is over the two-dimensional surface defined by $g(\mathbf{x}) = 0$. The application of this property to Eq. 12.21 (with $f(\mathbf{x}) = 1$) then reads

$$\sigma_1(\omega) = \frac{e^2}{4\pi^2 m^2 \omega} |P_{vc}|^2 \int_{\mathcal{E}_{cv}=\hbar\omega} dS \, \frac{1}{|\nabla_k \mathcal{E}_{cv}(\mathbf{k})|}. \tag{12.22}$$

When multiplied by $1/4\pi^3$, the integral in Eq. 12.22 is called the *joint density of states*, defined as

$$J_{cv} = \frac{1}{4\pi^3} \int_{\mathcal{E}_{cv}=\hbar\omega} dS \, \frac{1}{|\nabla_k \mathcal{E}_{cv}(\mathbf{k})|}. \tag{12.23}$$

The surface over which the integral is taken is analogous to the Fermi surface. The Fermi surface is a constant-energy surface in $\mathbf{k}$ space defined by the dispersion relation of the conduction electrons where the energy at every value of $\mathbf{k}$ on the surface equals the Fermi energy. The surface in Eq. 12.23 is a constant-energy-difference surface in $\mathbf{k}$ space where the difference between the conduction band and valence band energies at every value of $\mathbf{k}$ on the surface is equal the photon energy.

The conductivity in terms of the joint density of states is

$$\sigma_1(\omega) = \frac{\pi e^2}{m^2 \omega} |P_{vc}|^2 J_{cv}(\omega).$$

The conductivity will have structure where the joint density of states has structure. These are called critical points or van Hove singularities. These clearly occur when both valence band and conduction band dispersions have zero slope, so that $\nabla_k \mathcal{E}_c(\mathbf{k}) = \nabla_k \mathcal{E}_v(\mathbf{k}) = 0$. If I look at Fig. 12.4, I see critical points at $k = 0$ and at $k = \pm\pi/a$. There are also critical points when both valence and conduction bands have identical (but nonzero) slope. These parallel-band effects occur in many solids [54, 252, 272].

12.9　Indirect Band Gap Absorption

Indirect band gap semiconductors, as the name suggests, are materials in which the k-space minimum of the (empty) conduction band is *not* directly above the k-space maximum of the (filled) valence band. The smallest energy difference $\mathcal{E}_g$ between valence and conduction bands is still the forbidden energy gap,* but now there is an offset – often substantial – in k values between the max and the min. An example of an electronic band structure with an

---

* It is forbidden because there are no eigenstates in the crystal with $\mathcal{E}$ in this gap.

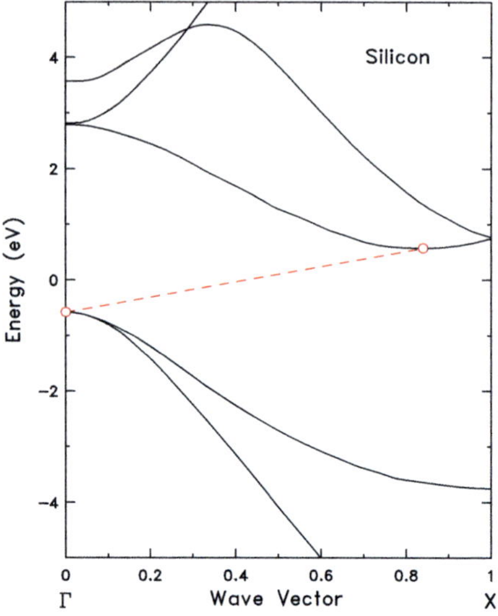

Fig. 12.9   Band structure of an indirect band gap semiconductor. The Fermi level is at $E = 0$. Bands with positive energies are empty; these are the conduction bands. Bands with negative energies are full; the valence bands. The energy difference between the circles is the energy gap, $\mathcal{E}_g$.

indirect gap is shown in Fig. 12.9. This diagram is drawn to represent the band structure of silicon [73, 76, 273–275], but would serve for germanium [73, 273, 274] and diamond [276] with modest changes.

Because optical transitions have $\Delta k = 0$ as one of their selection rules, light cannot promote by itself an electron from valence maximum to conduction minimum. Direct vertical transitions are allowed of course, but I will pay a larger photon energy price to make them.

The momentum $\hbar k$ necessary for the process of excitation from valence band maximum to conduction band minimum can come from the phonons. A single phonon has energy considerably less than the energy gap of the semiconductor, so the energy for the process has to be supplied by the photon. But the momentum can come from the emission or absorption of a phonon. At $T = 0$ only emission is possible but at finite temperatures thermal phonons can be absorbed.

So now I will consider a process where I absorb a photon with $\hbar\omega \approx \mathcal{E}_g$ and $q = 0$, emit or absorb a phonon with energy $\hbar\Omega \ll \mathcal{E}_g$ and with $\mathbf{k}_\Omega = \mathbf{k}_v - \mathbf{k}_c$, and wind up with a conduction-band electron with wave vector $\mathbf{k}_c$ as well as with a valence-band hole with wave vector $\mathbf{k}_v$. (If the band structure is that of Fig. 12.9 then the valence-band wave vector is zero, but this is not the only possibility.) The process contains two steps and I will have to go to second order in perturbation theory.

The full calculation is rather complicated; I will of necessity be rather sketchy here. Chapter 9 laid out the quantum mechanical basis for optical absorption. As there, I start with a two-term Hamiltonian, $\mathcal{H} = \mathcal{H}_0 + \mathcal{H}_1$, where $\mathcal{H}_0 \psi_n(\mathbf{k}) = \mathcal{E}_n \psi_n(\mathbf{k})$ describes the (solved) electronic structure of the solid and $\mathcal{H}_1$ is the electromagnetic perturbation. Then, following the steps that led to Eq. 9.12, I can write

$$i\hbar \dot{a}_n^{(1)} = \mathcal{H}_{1n0}\left(e^{i(\omega_{n0}-\omega)t} + e^{i(\omega_{n0}+\omega)t}\right), \tag{12.24}$$

for direct processes in the semiconductor. The superscript (1) is to remind me that this equation is first-order in the perturbation expansion. On the right hand side, I did not write out the perturbation in terms of the field as I did in Eq. 9.9. Equation 12.24 is a perfectly good equation and it leads to a threshold for optical conductivity at frequencies where direct transitions, whose wave vectors satisfy Eq. 9.2, occur. But if $\hbar\omega < \mathcal{E}_{n0}$ with $\mathbf{k}_n = \mathbf{k}_0$, then $\dot{a}_n^{(1)} = 0$.

To address indirect transitions, I'll need to go to second order in perturbation theory, so that the time derivative of a_n depends on a sum over all excited states m. The ground state, $m = 0$, is omitted because it does not give a finite transition rate in first order.* I start with Eq. 9.7; second-order perturbation theory consists of using the results of first-order perturbation theory for the time-dependent coefficients $a_m^{(1)}(t)$ on the right hand side.

$$i\hbar \dot{a}_n^{(2)} = \sum_{m>0} a_m^{(1)} \mathcal{H}_{1nm}\left(e^{i(\omega_{nm}-\omega)t} + e^{i(\omega_{nm}+\omega)t}\right). \tag{12.25}$$

I'll need to include in my perturbation interactions of the electrons with both photons (as $V^{e\gamma}$) and phonons (as V^{ep}). The perturbation Hamiltonian then looks like

$$\mathcal{H}_1 = \frac{1}{2}V^{e\gamma}\left(e^{-i\omega t} + e^{i\omega t}\right) + \frac{1}{2}V^{ep}\left(e^{-i\Omega_k t} + e^{i\Omega_k t}\right), \tag{12.26}$$

where I have pulled out the time dependent parts of the perturbation to show that one term depends on the photon frequency and one on the phonon frequency. I have hidden everything else (the dipole moment, the Coulomb interaction between electron and ion cores, interatomic springs, etc.) in $V^{e\gamma}$ and V^{ep}. The frequencies in Eq. 12.26 are ω the frequency of the photon and Ω_k, the frequency of a phonon with wave vector k.

So now I have to calculate the first-order coefficients $a_m^{(1)}$ that go into Eq. 12.26. The matrix elements that I need are $V_{m0}^{e\gamma} = \langle m|V^{e\gamma}|0\rangle$ and $V_{m0}^{ep} = \langle m|V^{ep}|0\rangle$. Then Eq. 12.24 for the first order time derivative becomes

$$i\hbar \dot{a}_m^{(1)} = V_{m0}^{e\gamma}\left(e^{i(\omega_{m0}-\omega)t} + e^{i(\omega_{m0}+\omega)t}\right) + V_{m0}^{ep}\left(e^{i(\omega_{m0}-\Omega_k)t} + e^{i(\omega_{m0}+\Omega_k)t}\right). \tag{12.27}$$

The four terms represent photon absorption ($\omega_{m0}-\omega$), photon emission ($\omega_{m0}+\omega$), phonon absorption ($\omega_{m0} - \Omega_k$), and phonon emission ($\omega_{m0} + \Omega_k$). Photon emission is needed to study photoluminescence, electroluminescence, LEDs, lasers, and other process where the semiconductor emits light. But for absorption, I can omit it. In contrast, I need to allow for phonon emission *and* phonon absorption to understand the temperature dependence of the absorption. So the time derivative integrates† to become

$$a_m^{(1)} = \frac{V_{m0}^{e\gamma}}{\hbar}\left(\frac{1 - e^{i(\omega_{m0}-\omega)t}}{\omega_{m0} - \omega}\right) + \frac{V_{m0}^{ep}}{\hbar}\left(\frac{1 - e^{i(\omega_{m0}-\Omega_k)t}}{\omega_{m0} - \Omega_k} + \frac{1 - e^{i(\omega_{m0}+\Omega_k)t}}{\omega_{m0} + \Omega_k}\right).$$

$$\tag{12.28}$$

---

* If the $m = 0$ term did give a finite transition rate, I could and would have stopped there.

† Change t to t' in Eq. 12.27 and calculate $\int_0^t dt' \dot{a}_m^{(1)}(t')$.

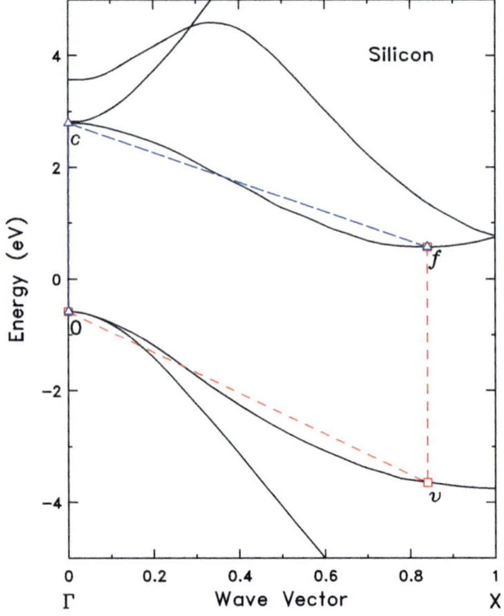

Fig. 12.10 Transitions from valence band maximum (marked with 0) to conduction band minimum (marked with f) via two intermediate states v (in the valence band) and c (in the conduction band). In one case the photon goes first, a direct transition from valence band to conduction band, and then the phonon takes the excited electron to the conduction band minimum. In the second case, the phonon goes first and the photon second.

Now I could substitute Eq. 12.28 for $a_m^{(1)}$ into Eq. 12.25 right now. Equation 12.25 has a sum over m, an infinite number of ways to go from initial to final state. It behooves me to think about how many of these actually will contribute. A little thought (and Fig. 12.10) tells me that there are exactly two ways in second order to go from valence-band maximum to conduction-band minimum. Any other path is at least third order. I can make a vertical transition to the conduction-band state at $k = 0$ and then use a phonon to get to the final state, or I can have a phonon take me to a valence-band state under the conduction-band minimum and then make a vertical transition.*

With this restriction, Eq. 12.25 has only two terms, with $m = v, c$ and $n = f$. I have to understand that the matrix elements there are those of the perturbation and the frequencies ω are the frequencies of the time dependent perturbation. Thus: $\mathcal{H}_{1fv} = V_{fv}^{e\gamma}$ with ω; $\mathcal{H}_{1fc} = V_{fc}^{ep}$ with $\omega = \Omega_k$. Then, continuing to omit the photon emission term, I have

$$i\hbar\dot{a}_f^{(2)} = a_v^{(1)} V_{fv}^{e\gamma} e^{i(\omega_{fv}-\omega)t} + a_c^{(1)} V_{fc}^{ep}\left(e^{i(\omega_{fc}-\Omega_k)t} + e^{i(\omega_{fc}+\Omega_k)t}\right). \qquad (12.29)$$

I am ready to substitute Eq. 12.28 into Eq. 12.29, but I should do this with some forethought [54]. Both Eq. 12.29 and Eq. 12.28 contain basically two terms, one with

---

* If I get asked about the Pauli principal and the phonon momentum-changing transition using an already occupied state, I will run the process backwards: take a hole from the conduction band minimum – where there are lots of holes – to the valence band right "above" it in hole speak, and then use a phonon to move the hole to valence-band maximum.

the electron–photon matrix element and one with the electron–phonon matrix element. When everything is multiplied out, there will be two terms containing both matrix elements (which I want) and two containing one or the other squared (which I do not want). These will not contribute to the optical absorption, because in one case (two phonons) I do not have enough energy and in the other (two photons) not enough momentum. So I'll keep only the terms that do contribute, and find

$$i\hbar \dot{a}_f^{(2)} = \frac{V_{v0}^{ep}}{\hbar} \left(\frac{1 - e^{i(\omega_{v0} - \Omega_k)t}}{\omega_{v0} - \Omega_k} + \frac{1 - e^{i(\omega_{v0} + \Omega_k)t}}{\omega_{v0} + \Omega_k} \right) V_{fv}^{e\gamma} e^{i(\omega_{fv} - \omega)t}$$

$$+ \frac{V_{c0}^{e\gamma}}{\hbar} \left(\frac{1 - e^{i(\omega_{c0} - \omega)t}}{\omega_{c0} - \omega} \right) V_{fc}^{ep} \left(e^{i(\omega_{fc} - \Omega_k)t} + e^{i(\omega_{fc} + \Omega_k)t} \right). \qquad (12.30)$$

I'll collect terms and simplify as much as I can. One (certainly incorrect) simplification is to assume that the two electron–photon and the two electron–phonon matrix elements are the same. For them to be the same, the electric dipole moment $\langle c|x|v \rangle$ for valence to conduction band optical transitions is independent of k. An assumption of constant matrix elements is often made to take them outside of integrals and I will do it here. I'll also assume that the electron–phonon interaction is the same for valence-band and conduction-band carriers despite their different effective masses. Also, I'll make use of knowing the form of the frequencies $\omega_{nm} = (\mathcal{E}_n - \mathcal{E}_m)/\hbar$ so I can write $\omega_{v0} + \omega_{fv} = (\mathcal{E}_v - \mathcal{E}_0 + \mathcal{E}_f - \mathcal{E}_v)/\hbar = (\mathcal{E}_f - \mathcal{E}_0)/\hbar = \omega_g$, where $\omega_g = \mathcal{E}_g/\hbar$ is the energy gap frequency of the semiconductor. As well, $\omega_{c0} + \omega_{fc} = (\mathcal{E}_c - \mathcal{E}_0 + \mathcal{E}_f - \mathcal{E}_c)/\hbar = (\mathcal{E}_f - \mathcal{E}_0)/\hbar = \omega_g$. With these simplifications, Eq. 12.30 becomes

$$i\hbar \dot{a}_f^{(2)} = \frac{V^{ep} V^{e\gamma}}{\hbar} \left[\frac{e^{i(\omega_{fv} - \omega)t} - e^{i(\omega_g - \Omega_k - \omega)t}}{\omega_{v0} - \Omega_k} + \frac{e^{i(\omega_{fv} - \omega)t} - e^{i(\omega_g + \Omega_k - \omega)t}}{\omega_{v0} + \Omega_k} \right.$$

$$\left. + \frac{e^{i(\omega_{fc} - \Omega_k)t} + e^{i(\omega_{fc} + \Omega_k)t} - e^{i(\omega_g - \Omega_k - \omega)t} - e^{i(\omega_g + \Omega_k - \omega)t}}{\omega_{c0} - \omega} \right]. \qquad (12.31)$$

Now look at some of the terms in Eq. 12.31. There are three of them that do not contain ω_g: $e^{i(\omega_{fv} - \omega)t}$ and $e^{i(\omega_{fc} \pm \Omega_k)t}$. These do not contribute to the optical conductivity. In the first, $\omega_{fv} > \omega_g \approx \omega$ and, hence, the term oscillates rapidly; it will produce on integration something like $\sin^2[\frac{1}{2}(\omega_{fv} - \omega)t]/(\omega_{fv} - \omega)^2$. As time goes on this function becomes a very narrow function of frequency centered at $\omega = \omega_{fv}$. But I am calculating the onset of absorption at $\omega = \omega_g < \omega_{fv}$. I can neglect this term. A similar argument applies to $e^{i(\omega_{fc} \pm \Omega_k)t}$, where $\Omega_k \ll \omega_{fc}$. So I will just take these away before I do the integration. Finally, energy conservation gives $\omega = \omega_g \pm \Omega_k = \omega_{fv} + \omega_{v0} \pm \Omega_k$, where I use the plus sign for phonon emission and the minus sign for phonon absorption. I can write two of the denominators in Eq. 12.31 as $\omega_{v0} \pm \Omega_k = \omega - \omega_{fv}$. Then after some algebra

$$i\hbar \dot{a}_f^{(2)} = -\frac{V^{ep} V^{e\gamma}}{\hbar} \left(\frac{1}{\omega - \omega_{fv}} + \frac{1}{\omega_{c0} - \omega} \right) \left[e^{i(\omega_g - \Omega_k - \omega)t} + e^{i(\omega_g + \Omega_k - \omega)t} \right]. \qquad (12.32)$$

I'm finally ready to integrate Eq. 12.32 to find

$$a_f^{(2)} = \frac{V^{ep} V^{e\gamma}}{\hbar} \left(\frac{1}{\omega - \omega_{fv}} + \frac{1}{\omega_{c0} - \omega} \right) \times \left[\frac{e^{i(\omega_g - \Omega_k - \omega)t} - 1}{\omega_g - \Omega_k - \omega} + \frac{e^{i(\omega_g + \Omega_k - \omega)t} - 1}{\omega_g + \Omega_k - \omega} \right].$$

(12.33)

This equation contains two terms of the form

$$\frac{e^{i\beta t} - 1}{\beta} = \frac{e^{i\beta t/2} - e^{-i\beta t/2}}{\beta} e^{i\beta t/2} = 2i \frac{\sin \frac{\beta t}{2}}{\beta} e^{i\beta t/2}.$$

When I calculate $|a_f^{(2)}|^2$, the probability for an electron to appear in the final state, the i and the complex exponential go away. I am left with

$$\frac{4 \sin^2 \frac{\beta t}{2}}{\beta^2} = 2\pi t \delta(\beta),$$

where I've used $\mathrm{sinc}^2(x) = \sin^2(x)/x^2$ as a representation of the delta function. The delta function enforces energy conservation in the optical process. In addition, some cross terms will drop out when the amplitude is squared, as they do not grow linearly in time.

So I finally write the transition rate, $W_{vc} = d|a_f^{(2)}|^2/dt$ as

$$W_{vc} = \frac{2\pi |V^{ep}|^2 |V^{e\gamma}|^2}{\hbar^4} \left[\frac{1}{(\omega_{fv} - \omega)^2} + \frac{1}{(\omega_{c0} - \omega)^2} \right] \times$$
$$\times \left[\delta(\omega_g - \Omega_k - \omega) + \delta(\omega_g + \Omega_k - \omega) \right].$$

(12.34)

This equation describes the rate at which light, with the help of the phonons, can promote electrons from the valence band maximum to the conduction band minimum. The strength of the absorption depends on the interaction of the electrons with the phonons (via V^{ep}) and on the interaction of the electrons with the light (via $V^{e\gamma}$). When I compare the first half of Eq. 12.27 with Eq. 9.12, I see that $V^{e\gamma} = (e/2) \langle c|x|v \rangle E_0$, where $|v\rangle$ and $|c\rangle$ are the relevant valence- and conduction-band wave functions and E_0 is the electric field amplitude. The matrix element comes in squared, so the transition rate is proportional to the intensity, just as in Eq. 9.16. This proportionality is the behavior I expect: the higher the intensity, the more photons per second arrive at the sample and so the higher the rate of absorption.

As in Eq. 9.19 I can convert to optical conductivity using the absorption coefficient as an intermediate step. The result is

$$\sigma_1 = \frac{2\hbar\omega W_{vc}}{|E_0|^2}.$$

(12.35)

There are still a number of things left to do, however. I need to discuss the other matrix element in Eq. 12.34, V^{ep}, the coupling of the electrons to the phonons. The two delta functions enforce energy conservation among the electrons, the phonons (emitted or absorbed), and the light. The condition is

$$\omega = \begin{cases} \omega_g - \Omega_k & \text{phonon absorption} \\ \omega_g + \Omega_k & \text{phonon emission.} \end{cases} \tag{12.36}$$

I also have momentum (or wave vector) conservation. (See Fig. 12.10.)

$$\mathbf{k}_f - \mathbf{k}_0 = \mathbf{k}_\Omega \tag{12.37}$$

where $\mathbf{k}_0$ and $\mathbf{k}_f$ are the initial and final electron wave vector and $\mathbf{k}_\Omega$ is the phonon wave vector. The band structure of the semiconductor determines the left side of Eq. 12.37 and then the phonon dispersion determines the phonon frequency. The absorption process occurs a bit below the semiconducting gap, because it uses both the photon energy and the phonon energy to promote the electron. The emission process occurs a bit above the semiconducting gap, because the photon energy must be used both to create the phonon and to promote the electron.

Two simple possibilities for the phonon dispersion Ω_k are*

$$\Omega_k = \begin{cases} c_s k & \text{Debye,} \\ \Omega_0 & \text{Einstein} \end{cases}$$

where c_s is the speed of sound. (See Fig. 5.4 for the results of a slightly more realistic model.) All phonon branches have k-values from 0 to the zone boundary, so wave vector conservation is always possible. When k is large, the frequency is also relatively large, at least on the scale of the phonon frequencies. It is almost always small on the scale of ω_g.

Phonons obey Bose–Einstein statistics; in thermal equilibrium the number of phonons in a given state with a given frequency is given by the Bose–Einstein distribution [19, 24]. The probability of phonon absorption is proportional to

$$P_A = n(\Omega_k, T) = \frac{1}{e^{\left(\frac{\hbar\Omega_k}{k_B T}\right)} - 1}, \tag{12.38}$$

where the phonon modes are indexed by wave vector k. (A polarization and branch index should also be attached here.) The probability of emission is

$$P_E = 1 + n(\Omega_k, T) = \frac{e^{\left(\frac{\hbar\Omega_k}{k_B T}\right)}}{e^{\left(\frac{\hbar\Omega_k}{k_B T}\right)} - 1}.$$

I'll need to include these probabilities in the electron–phonon matrix element for each process. At finite temperatures I can have either absorption or emission of the phonon. Equation 12.36 tells me that the threshold for absorption is lower than that for emission (by $2\Omega_k$). But the probability of emission dominates for $T < \hbar\Omega_k/k_B$. The scale of temperatures is the Debye temperature, which is 370 K for Ge and 645 K for Si. At low temperature,

$$P_A = e^{-\left(\frac{\hbar\Omega_k}{k_B T}\right)}$$

$$P_E = 1 + e^{-\left(\frac{\hbar\Omega_k}{k_B T}\right)}.$$

---

* Another possibility of course is to use the actual phonon dispersion curves as measured by neutron scattering or as calculated by some theoretical method.

In the following, I use these low-temperature probabilities; it is straightforward to insert the more general probabilities for higher temperature cases.

I'll treat various absorption processes as independent; the optical conductivity will be a sum of all possible processes at the given frequency. As frequency increases from zero, phonon absorption will turn on first, then phonon emission, and then direct interband transitions.

So now I have all the components needed to assemble the optical conductivity from indirect transitions from the valence-band maximum to the conduction-band minimum. But this is only one of many possible indirect transitions. Suppose I increase the photon energy a bit. Looking at Fig. 12.10, it is obvious that I could go from the valence band maximum to states in the conduction band a bit above the minimum. Of course, I could alternatively start a bit below the valence-band maximum and make a transition to the conduction-band minimum. Considering all directions in **k**-space, there are many states with energy differences of $\hbar\omega$ with ω a bit above ω_g. The calculation at this point is very similar to the derivation in Section 12.7, so I will be even more sketchy here than I was earlier in the section. I'll abandon all the various constants (e, $\hbar$, etc.) but keep other terms.

First, I'll deal with the pesky $1/(\omega_{fv} - \omega)^2 + 1/(\omega_{c0} - \omega)^2$ term in Eq. 12.34. This term looks like a resonant denominator but it is not. The frequency ω is near the minimum value, $\omega \approx \omega_g$. The two frequencies ω_{fv} and ω_{c0} are respectively the frequencies of direct interband transitions at the conduction band minimum and at the valence band maximum. These frequencies differ a bit but are much larger than the gap frequency. So I'll set $\omega_{fv} \approx \omega_{dg} \approx \omega_{c0}$ with $\hbar\omega_{dg}$ the direct band gap energy. The other things to do are

- Sum over states in both conduction band (cb) and valence band (vb): $\Sigma_{(k \text{ in cb})} \cdot \Sigma_{(k \text{ in vb})} \cdot$
- Convert to integrals: $\Sigma_k \rightarrow 2 \int d^3k/(2\pi)^3$.
- Convert to energy integrals assuming parabolic behavior around the band extremes. If Fig. 12.10 describes the band structure, the valence band follows $\mathcal{E} = \mathcal{E}_0 - \hbar^2 k^2/2m_h^*$ and the conduction band follows $\mathcal{E} = \mathcal{E}_f + \hbar^2(k - k_f)^2/2m_h^*$.
- Put in the density of energy states: $\mathscr{D}(\mathcal{E}) = \sqrt{\mathcal{E}}$ away from the band extremes.
- Replace one factor of ω with ω_g because I really am only interested in evaluating things at $\omega = \omega_g$.
- Do one energy integral, consuming the delta functions: $\mathcal{E} = \hbar(\omega \pm \Omega_k - \omega_g)$.
- Use $V^{e\gamma} = (e/2)|x_{vc}|E_0$.

After these steps, the conductivity is proportional to

$$\sigma_1 \propto |V^{ep}|^2 |x_{vc}|^2 \frac{\omega_g}{(\omega_d - \omega_g)^2} \omega P_{\{A, E\}} \sqrt{\hbar\omega \pm \hbar\Omega_k - \mathcal{E}_g} \int_0^{\hbar\omega \pm \hbar\Omega_k - \mathcal{E}_g} \sqrt{\mathcal{E}} d\mathcal{E}$$

where the $+$ sign is for the case of phonon absorption and the $-$ for phonon emission. I choose the appropriate probability when I pick which sign I want. The integral is straightforward and the result gives the following regimes. First, transparent:

$$\sigma_1 = 0, \tag{12.39}$$

for $\omega \leq \omega_g - \Omega_k$. Second, phonon absorption:

$$\sigma_1 \propto |V^{ep}|^2 |x_{vc}|^2 \frac{\omega_g}{(\omega_d - \omega_g)^2} e^{-\left(\frac{\hbar\Omega_k}{k_B T}\right)} (\omega + \Omega_k - \omega_g)^2 \equiv \sigma_{1A}, \tag{12.40}$$

for $\omega_g - \Omega_k \leq \omega \leq \omega_g + \Omega_k$. Third, phonon emission *and* phonon absorption:

$$\sigma_1 \propto |V^{ep}|^2 |x_{vc}|^2 \frac{\omega_g}{(\omega_d - \omega_g)^2} \left[1 + e^{-\left(\frac{\hbar\Omega_k}{k_B T}\right)} \right] (\omega - \Omega_k - \omega_g)^2 + \sigma_{1A} \equiv \sigma_{1E} + \sigma_{1A} \tag{12.41}$$

for $\omega_g + \Omega_k \leq \omega \leq \omega_{dg}$, where ω_{dg} is the direct band gap. Finally, (Eq. 12.20)

$$\sigma_1 \propto f \sqrt{\omega - \omega_{dg}} + \sigma_{1E} + \sigma_{1A} \tag{12.42}$$

for $\omega \geq \omega_{dg}$.

Once the absorption turns on, there are two regions where the absorption increases quadratically in the frequency (or energy) difference above their threshold. The first, phonon absorption, is exponentially small at low temperatures. The second, phonon emission, is weakly temperature-dependent at low temperatures. At high temperatures, both terms are strong. Both are negligible once the direct absorption turns on at ω_{dg}.

12.10 Excitons

The picture of interband transitions that began with Section 12.7 puts the electron–hole pair produced by optical absorption into delocalized states, extending in principal throughout the crystal. This point of view, energy band theory, is justified in ignoring the Coulomb interaction between electron and hole, inasmuch as this may well have been left out or approximated in deriving the band structure ($\mathcal{H}_0$) in the first place. If, in contrast one begins with a localized tight-binding theory using linear combinations of atomic orbitals, the electron and hole might be located on a single atomic site. In this second case, it is likely that the Coulomb energy between electron (negative) and hole (positive) must be included.

Reality is as ever somewhere between these two extremes. Effects of the Coulomb interaction are observed in many solids and go by the name of *excitons*. The exciton is a bound electron–hole pair. A photon may create an exciton if the photon energy equals the semiconductor's energy gap minus the exciton binding energy $\mathcal{E}_{ex}$: $\hbar\omega = \mathcal{E}_g - \mathcal{E}_{ex}$. Because the exciton is neutral (one electron plus one hole) it does not carry electrical current. The exciton can migrate through the crystal, hopping from site to site, and carrying therefore an energy current.

Many authors [54, 105, 252] distinguish two exciton types: Frenkel excitons are small (one atomic site) and tightly bound. Wannier excitons are large (with a radius considerably larger than the lattice spacing) and have a corresponding smaller binding energy. A third type, similar to Frenkel excitons, is the charge-transfer exciton where the electron and hole occupy neighboring molecules. Unlike Frenkel and Wannier excitons, charge-transfer excitons have a net dipole moment.

Excitons are observed in many solids. The excitons in KCl (and other ionic crystals), anthracene (and other organic materials), and other large band gap solids are interpreted as Frenkel excitons. Charge-transfer excitons occur in low-dimensional organic conductors, La_2CuO_4, and other cuprate materials. Wannier excitons occur in many semiconductors including Ge, GaAs, CuO_2, Xe, and ZnTe.

I'll work out an extremely simple model for Wannier excitons, based on the Coulomb interaction in the crystal. To start, I'll consider that the electron and hole are in band states with little relative difference in their group velocities. In vacuum their potential energy would be $V(r) = -ee/r$. Using either the Bohr model or the Schrödinger equation they would be bound with energy $\mathcal{E}_n = -R/n^2$ with $n = 1, 2, 3 \ldots$ the principal quantum numbers and $R = \mu e^2/2\hbar^2$ the Rydberg of positronium, about 6.8 eV. ($\mu = m_e/2$ is the reduced mass.) The radius of the $n = 1$ wave function (the Bohr radius) is $a = \hbar^2/\mu e^2 \approx 1$ Å.

Two changes need to be made on account of the crystal. First, the dielectric medium screens the electric field, so I need to make the substitution $e^2 \rightarrow e^2/\epsilon_1(0)$, with $\epsilon_1(0)$ the static dielectric constant.* Second, the effective mass of the electron and hole are not the free electron mass but instead the band mass at the band edges, m_e^* and m_h^* respectively.

The Hamiltonian for the excited electron hole pair is

$$\mathcal{H} = -\frac{\hbar^2}{2m_e^*}\nabla_e^2 - \frac{\hbar^2}{2m_h^*}\nabla_h^2 - \frac{e^2}{e_1(0)|\mathbf{r}_e - \mathbf{r}_h|}. \tag{12.43}$$

The solution is easy.†

$$\mathcal{E} = \mathcal{E}_g + \frac{\hbar K^2}{2(m_e^* + m_h^*)} - \frac{R}{n^2}\frac{\mu^*}{m[\epsilon_1(0)]^2}, \tag{12.44}$$

where μ^* is the reduced mass, $1/\mu^* = 1/m_e^* + 1/m_h^*$, and $\mathbf{K} = \mathbf{k}_e + \mathbf{k}_h$ is the total wave vector of the two particles. The second term is the kinetic energy of the center of mass of the electron–hole pair and the third is their binding energy.

The exciton "Rydberg" is

$$R_{ex} = R\frac{\mu^*}{m[\epsilon_1(0)]^2}.$$

Because the dielectric constant is big (of order 10) and the reduced mass is small (perhaps $0.05m$) the exciton Rydberg is $R_{ex} \approx 7$ meV ~ 50 cm^{-1}. At the same time, the exciton radius for $n = 1$ is $a_{ex} = a_0 m\epsilon_1(0)/\mu^* \approx 100$ Å. This size is much larger than the 2–3 Å interatomic spacing of most semiconductors. The large size justifies the use of the Wannier picture. Some values [277, 278] are shown in Table 12.4.

There are some evident trends. As the band gaps increase, so do the exciton binding energies; the exciton dimensions decrease with increasing gap.

It is not uncommon to see plotted on a diagram like Fig. 12.4 or Fig. 12.5 exciton dispersion relations that follow Eq. 12.44. The dispersion curves appear just below the conduction band, offset by R_{ex}/n^2. Such a diagram can be helpful in visualizing the situation but is in fact incorrect. The band diagrams show the energies of single electrons

---

* I use the static dielectric constant because I am seeking a stationary state for the exciton wave function.
† Because I know how to solve the hydrogen atom!

Table 12.4 Exciton binding energy and size for certain semiconductors.

Material	$\mathcal{E}_g$ eV	R_{ex} meV	a_{ex} Å
InSb	0.237	0.5	860
$Cd_{0.3}Hg_{0.7}Te$	0.26	0.7	640
Ge	0.89	1.4	360
InP	1.42	5.0	140
GaAs	1.52	4.1	150
CdTe	1.61	10.6	80
Cu_2O	2.17	97.2	38
CdS	2.6	28	27
ZnSe	2.82	20.4	60
GaN	3.51	22.7	40

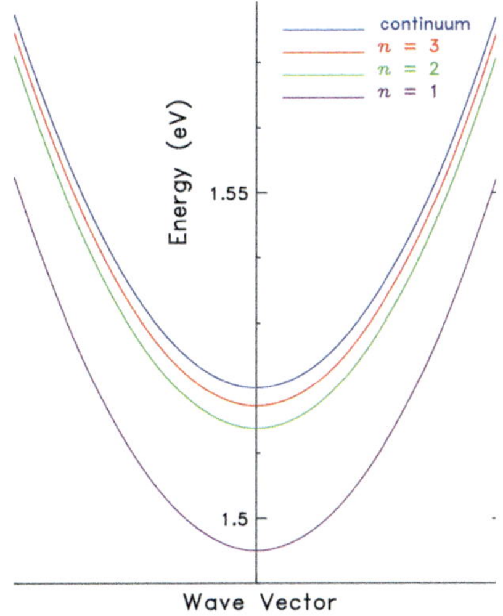

Fig. 12.11 Excitation energy in a semiconductor. Three exciton levels are shown as well as the continuum of excitations above these.

in valence and conduction bands. I can put an electron in the conduction band by marking the energy $\mathcal{E}_e$ and wave vector $\mathbf{k}_e$. I can indicate a hole by marking the energy $\mathcal{E}_v$ and wave vector $\mathbf{k}_v$ of the valence band state from whence it came, understanding that the hole energy $\mathcal{E}_h = -\mathcal{E}_v$ and the hole wave vector $\mathbf{k}_v = -\mathbf{k}_v$. I cannot put the energy of an electron–hole pair on such a diagram.

A better way to approach such a diagram is to make an excitation spectrum, just as I did for a superconductor, in Chapter 11 (Section 11.5). Such a diagram is shown in Fig. 12.11 The ground state, $\mathcal{E} = 0$, is the semiconductor with full valence band and empty conduction band. A photon of frequency ω and essentially zero wave vector is absorbed. If

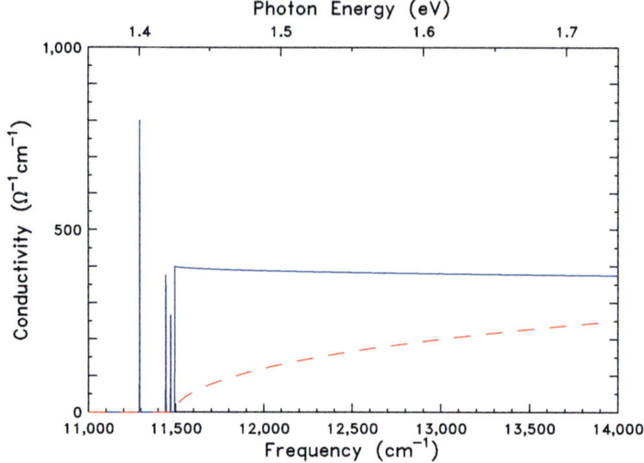

Fig. 12.12   Sketch of the optical conductivity of a semiconductor with several exciton lines below the onset of interband transitions.

there is no Coulomb interaction, assuming parabolic dispersion, and if $\hbar\omega \geq \mathcal{E}_g$, then an electron–hole pair is created with energy

$$\mathcal{E}_g + \frac{\hbar^2 k_e^2}{2m_e} + \frac{\hbar^2 k_h^2}{2m_h} = \hbar\omega. \tag{12.45}$$

The wave vector selection rule is satisfied by requiring

$$\mathbf{k}_e + \mathbf{k}_h = 0$$

which will be true if the valence band state and conduction band state have the same wave vector.

If now I turn on the Coulomb interaction, there is an attraction of electrons and holes. I create excitons, electron–hole pairs with energies given by Eq. 12.44. Three such levels are shown in Fig. 12.11.

Excitons contribute to the optical conductivity when photons create excitons with one or another value of n. Now the photon has $q = 0$ so cannot give the exciton center of mass any momentum. Consequently there is a peak in $\sigma_1(\omega)$ when $\omega = \omega_g - R_{ex}/\hbar n^2$ for a number of values of n. Just above the exciton lines, which can carry considerable oscillator strength, the continuum of electron–hole excitations (interband transitions) begins. A cartoon of the spectrum is in Fig. 12.12 and an example of exciton peaks in Cu_2O is in Fig. 12.13.

12.11 Impurity-Induced Absorption

Silicon, GaAs, Ge, and other semiconductors are a dominant force in the world economy for one reason: their electrical properties may be determined by incorporating controlled amounts of other atoms. These atoms are called "impurities" and the process is called

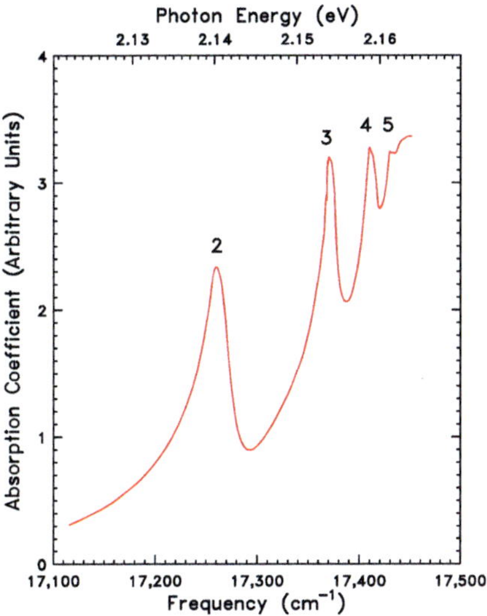

Fig. 12.13 Absorption coefficient [$-\ln(\mathscr{T})$ actually] of Cu_2O showing a series of exciton lines. From Ref. 279.

"doping." Neither term is particularly well chosen; in particular the dopants are not undesirable impurities but instead are carefully controlled as to type, concentration, and location.

Si is typically doped with atoms from periodic table columns on either side of Si itself. Si (group IV A) has four valence electrons ($3s^2\,3p^2$) which bond covalently to four neighboring Si atoms, forming a tetrahedrally coordinated cubic crystal structure. Atoms from the neighboring columns have one more or one less valence electron. B, to the left (group III A), has three and P, to the right (group V A), has five. When a B atom substitutes for Si, it accepts an electron from the valence band to complete the bonding, leaving a hole behind. When a P atom is substituted, it uses four of its electrons for the covalent bonds and donates the fifth to the conduction band. Thus doping controls the free-carrier density, and hence the electrical conductivity. It also controls the energy location of the chemical potential. And finally it controls the sign of the charge carriers. Hole doped Si (e.g., with B) has positive carriers and is called p-type silicon. Ga, In, N, and Al also can be used to produce p-type Si for special applications. Electron doped Si (e.g., with P) is called n-type silicon. As and Sb are also used to make n-type Si.

If the above were the whole story, I would be almost done with this section. I would just say that the free carriers give a Drude-like conductivity with carrier density that depends on doping and with a scattering rate determined by a number of processes, some of which could be doping dependent.

But this is not the whole story. It is correct at very high doping, where the doped Si is actually a metal. (The Fermi level is in either the conduction or valence band.) It is approximately true at very high temperatures, although in that case there are also free carriers thermally populated in both conduction and valence bands. But the dopant atoms have a net charge. P and other donors have given up one electron and so have a charge $+e$. B, and other acceptors, have accepted one, and have a charge $-e$. There is a Coulomb interaction between the dopant ions and an electron or hole in the solid. The result is another hydrogen atom problem.

I'll start by looking at the quantum problem at $T = 0$. The Hamiltonian is

$$\mathcal{H} = -\frac{\hbar^2}{2m^*}\nabla^2 - \frac{e^2}{\epsilon_1(0)|\mathbf{r}|}, \tag{12.46}$$

where m^* is the effective mass, either hole or electron depending on which case is in play, $\mathbf{r}$ is the separation of the electron or hole from the impurity, and $\epsilon_1(0)$ is the static dielectric constant. Like the exciton, Eq. 12.46 is a scaled version of the Hamiltonian for the hydrogen atom, which I know how to solve. In fact, shallow impurities are an interesting analog to the hydrogen atom. The energy levels for electrons in the conduction band at the level of the Bohr model are

$$\mathcal{E}_e = -\frac{R}{n^2}\frac{m_e^*}{m[\epsilon_1(0)]^2} = -\frac{R_e}{n^2}. \tag{12.47}$$

Here $R = me^4/4\pi\hbar^3 c = 13.6$ eV is the Rydberg energy, $n = 1, 2, 3 \ldots$ is the principal quantum number, and $m_e^*(= 0.26m$ for Si) is the effective mass. This mass is an average of transverse and longitudinal effective masses for the Si conduction band minima. If I use $\epsilon_1(0) = 11.7$ as the Si dielectric constant, then $R_e = 25.8$ meV. The energy is measured from the bottom of the conduction band, so the donor level is about 26 meV below the conduction band edge. At the same time, the radius for $n = 1$ is $a_e = a_0 m \epsilon_1(0)/m_e^* \approx 22$ Å. The electron wave function extends over many lattice constants.

A full quantum-mechanical picture of the energy levels of the donor atom includes many effects [252, 280]. The bound electron wave function is expanded in terms of Bloch functions for the crystal and the anisotropy of the conduction band and valley-orbit coupling are included. The results of a calculation along with experimental values for the levels are shown in Fig. 12.14. The values of the ground state energy splitting from the conduction-band minimum are given in Table 12.5.

For holes in the valence band, Eq. 12.46 gives

$$\mathcal{E}_v = -\frac{R}{n^2}\frac{m_h^*}{m[\epsilon_1(0)]^2}. \tag{12.48}$$

Here $m_h^*(= 0.39m$ for Si) is an effective average of light-hole and heavy-hole masses (a detail of the band structure of Si). Using these numbers then $R_h = 38.7$ meV. The energy of the hole states are measured downward from the valence-band maximum, so this

Table 12.5 Ionization energy ($\mathcal{E}_c - \mathcal{E}_d$ or $\mathcal{E}_a - \mathcal{E}_v$, in meV) of donors and acceptors in silicon [252, 280–282].

	Donors			Acceptors		
Dopant	P	As	Sb	B	Al	In
Ionization energy	46	54	43	45	57	160

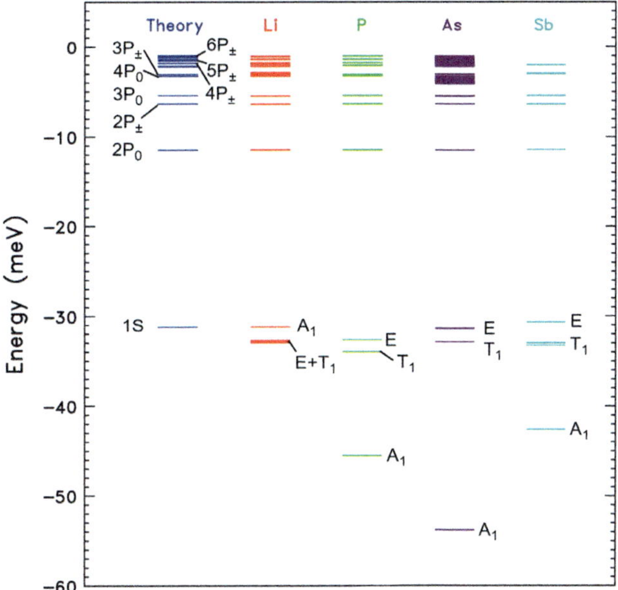

Fig. 12.14　Calculated and measured donor levels in Si [280].

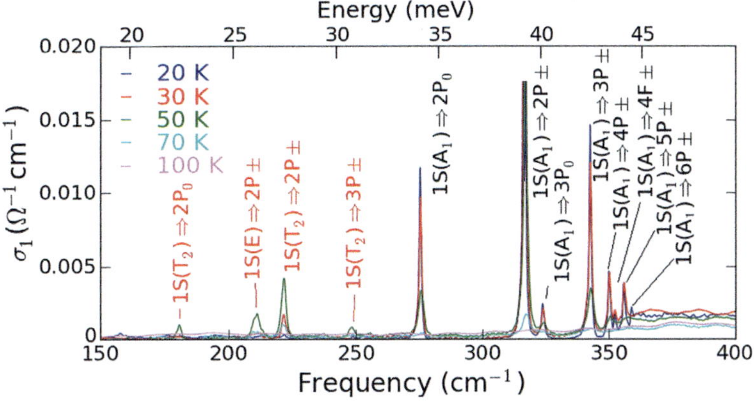

Fig. 12.15　Optical conductivity of lightly doped Si at five temperatures.

negative energy level is about 39 meV *above* the top of the valence band. Measurements put it a bit further away [252, 281, 282].

The low-temperature spectra of lightly doped Si (and other semiconductors) resembles that of hydrogen, after accounting for the scaling by dielectric constant and effective mass discussed above. An example is shown in Fig. 12.15. This diagram shows the optical conductivity measured at a number of temperatures on a Si sample with about 6×10^{13} P atoms/cm^3 (a dopant concentration of 1.2 parts per billion). The lines tagged in black correspond to transitions from the ground state 1S(A$_1$) to higher nP states. Transitions up to $n = 6$ are identified.

As temperature is increased, the intensity of these narrow conductivity peaks decreases, because carriers are being excited thermally to the conduction band. (And in fact at higher temperatures a Drude-like free-carrier peak is seen below 50 cm^{-1} or 6 meV.) Another effect that is seen is the appearance of new lines (tagged in red) at intermediate temperatures. These lines are strongest around 50 K. They are due to carriers thermally promoted above the ground state being excited to higher levels. The analogy to the behavior of hydrogen is good. The features seen at the lowest temperature are comparable to the Lyman series, where the starting point is the ground state. The features that appear as temperature is increased are like the Balmer series, where the starting point is the first level above the ground state.

Much of the richness of solid-state physics arises from the interactions among electrons, phonons, magnons, and other entities in the solid. These interactions are responsible for superconductivity, magnetism, metal–insulator transitions, and many other effects. They are also responsible for temperature-dependent transport properties and for frequency-dependent optical properties.

I have worked out the dielectric function and optical conductivity of a Drude metal already. I first did it classically (Chapter 4), then quantum mechanically, and then quantum mechanically again using the Boltzmann equation (Chapter 8). All three times I got the same results for the conductivity (Eqs. 4.8, 8.21, and 8.39) and also for the dielectric function (Eqs. 4.11 and 8.22). The scattering rate $1/\tau$ appears as a parameter in these equations. The charge carriers are assumed to scatter off of *something* at a rate given by $1/\tau$. There are many such somethings; two that are important for simple metals are impurities and phonons, both of which break the translational symmetry of the crystal. Many other possibilities exist. The assumption implicit (if unstated at the time) is that $1/\tau$ is just a constant and the same value should be used at zero frequency and at petahertz frequencies.

Now, I know that $1/\tau$ is a function of temperature. I showed as an example the resistivity of the good metal silver in Fig. 8.6. I write here the Drude formula for the DC resistivity: $\rho_{dc} = 1/\sigma_{\rm dc} = (m/ne^2)(1/\tau)$, where the mass m, density n, and charge e of the carriers are essentially constant in temperature.* The strongly temperature-dependent ρ is then caused by a strongly temperature-dependent $1/\tau$. In pure ordinary metals, such as Ag, the temperature-dependent transport is attributed to the temperature dependence of the phonons.

It is worth asking this: Can the scattering rate be temperature dependent without also being frequency dependent? I think that the answer is no, although I cannot provide a proof. I can give an example of a case where I do expect both temperature and frequency dependences. The example is the interaction of electrons and phonons. The phonons have a range of energies or frequencies, from zero to the Debye frequency in the simplest case; more complicated phonon spectra exist in multi-atom solids. The occupied energies, however, are determined by the temperature and the distribution function. Roughly speaking, the range is from zero to $\hbar\omega \approx k_B T$ where k_B is Boltzmann's constant and T the temperature. The zero-frequency (DC) transport involves scattering by these

---

* This statement is not true for a semiconductor, where the carrier density n is thermally activated and thus varies exponentially with $1/T$. It is also not true for quasiparticles in a superconductor, whose charge q consists of a mixture of electron-like and hole-like states.

thermal excitations and varies with temperature because the number of thermally-generated phonons varies with temperature. There is another allowed phonon–electron interaction at finite photon frequencies. The electrons, driven at the frequency of the electric field, may emit (rather than absorb) a phonon.* The emission process may occur at zero temperature, but clearly the energy of the emitted excitation must come from the electromagnetic field – the photon. Phonon emission is therefore a finite frequency effect and should be largest at frequencies at and above the maximum frequency of the phonons. Emission leads to scattering – and current relaxation – because the emitted phonon has a momentum, requiring that the electron change its momentum. Phonon energies are much less than Fermi energies, so that the change of energy is typically small but the change of momentum can be large. A consequence of the emission process is a frequency dependence to the scattering rate and conductivity that is not captured by the Drude model with scattering rate $1/\tau(T)$ as obtained from DC resistance.

I will cover briefly in this chapter this and another of other optical effects in interacting electron systems. The interactions may be between (or among) the charge carriers themselves or may be with the phonons or other bosonic spectra. This topic is the realm of quantum many-body theory and of numerical approaches. This fact leads to a somewhat unsatisfactory approach: I'll present results from workers in the field, cite their papers, try to give a discussion of the underlying physics, work out in a couple of cases simple toy models, and then say what optical effects are obtained from these concepts. But derivations from first principles will be lacking.

13.1 Notation: The Generalized Drude Model

I need to define the notation used before I can address the ways in which interactions can affect optical properties. It is not as straightforward as it might seem. I cannot simply replace $1/\tau$ with $1/\tau(\omega)$ and put this function into Eqs. 4.11 and 4.8. The real and imaginary parts of the dielectric function and conductivity satisfy Kramers–Kronig relations and the scattering rate appears in different ways in the real and the imaginary parts. (See Eqs. 4.10a and 4.10b or Eqs. 4.14a and 4.14b.) This substitution would violate causality and I don't want to do that.

There are in reality two (equivalent) way to resolve this violation but there are a bewildering number of variations on these two. The two are: (1) I make another quantity (the plasma frequency or the effective mass) also frequency dependent [283], choosing the frequency dependence so that Kramers–Kronig is satisfied. (2) I make the scattering rate a complex quantity [284], and relate the real and imaginary parts of this quantity so that the conductivity or dielectric function obey Kramers–Kronig.† A related variation is to define an optical self energy, the usual self energy averaged over the Fermi surface [285].

---

* Emission of other other kinds of excitations is also possible.

† Perhaps not surprisingly, this obedience is achieved by making the complex scattering rate itself obey a Kramers–Kronig relation.

13.1.1 Frequency-Dependent Mass (Or Plasma Frequency)

To make the plasma frequency or the effective mass also frequency dependent, I'll rewrite Eq. 4.13 as

$$\epsilon(\omega) = \epsilon_{ni}(\omega) - \frac{[\omega_{eff}(\omega)]^2}{\omega^2 + i\omega/\tau^*(\omega)} \tag{13.1}$$

where $1/\tau^*$ is the frequency dependent scattering rate (written with an asterisk because the effective mass will be starred too), $\omega_{eff}(\omega)$ is the effective plasma frequency, and I have gathered all the other contributors to the dielectric function (core electrons, interband transitions, optical phonons – the noninteracting parts) into ϵ_{ni}.

The plasma frequency of the metal in the noninteracting case is $\omega_{pb} = \sqrt{4\pi n e^2/m_b}$ (Eq. 4.12). I've added the subscript b to the plasma frequency and to the mass to stand for "bare" or "band," the mass in the absence of interactions but allowing for band structure effects.* It thus makes good sense to define $\omega_{eff}(\omega)$ as containing a frequency-dependent effective mass $m^*(\omega)$ via

$$\omega_{eff}(\omega) = \sqrt{\frac{4\pi n e^2}{m^*(\omega)}} \tag{13.2}$$

and writing the equivalent of Eq. 4.11 for the interacting case as

$$\epsilon(\omega) = \epsilon_{ni}(\omega) - \frac{4\pi n e^2/m^*(\omega)}{\omega^2 + i\omega/\tau^*(\omega)} \tag{13.3}$$

where n and e, the carrier density and electronic charge respectively, are not functions of frequency.[†]

Another variation is to define a dimensionless mass enhancement factor $\lambda(\omega)$ and write

$$m^*(\omega) = m_b[1 + \lambda(\omega)]. \tag{13.4}$$

Of the three conceptually identical equations, I like Eq. 13.3 the best and will adopt it as a standard. I like Eq. 13.1 the least and won't use it any more.

Papers discussing optical effects in correlated or interacting electrons often show or calculate the real part of the optical conductivity. One advantage of using the conductivity is that various processes generally add in parallel; I can write $\sigma = \sigma_{ni} + \sigma_i$ where the subscript i means "interacting" and then only write the interacting term. I may even drop the subscript. A disadvantage is that I have to get the units right. Note that the contributions of the noninteracting electrons (e.g., interband transitions) may occur at high frequencies and thus may be ignored for the real part of the conductivity. However, the low-frequency limiting behavior generates a linear-in-frequency addition to the imaginary part: $\sigma_{ni2} = -\omega(\epsilon_{ni1} - 1)/4\pi$. (See Eq. 3.8.) *This additional term rarely can be neglected.* To do so is equivalent to neglecting ϵ_c in Eq. 4.11. The result is the loss of the plasma minimum in

---

* In the simple case it of course is just the free electron mass.

[†] The density and charge may also depend on the frequency. Indeed, I discussed in Section 11.6.4 the excitation spectrum of a superconductor in terms of an effective charge for the quasiparticles, interpreted as mixing of electrons and holes. I kept the mass constant, equal to m_b.

reflectance that I plotted in Fig. 4.7; the reflectance simply decreases monotonically starting at the frequency where $\epsilon_1(\omega)$ becomes positive.

With frequency-dependent effective mass and scattering rate, the Drude conductivity becomes

$$\sigma_i = i\,\frac{ne^2/m^*(\omega)}{\omega + i/\tau^*(\omega)}, \tag{13.5}$$

where I've relocated τ^* and i from where they are in the standard expression, Eq. 4.8, so that τ^* only appears once, and as a rate.

If I have the complex dielectric function or complex conductivity for some material, *and* I have properly subtracted ϵ_{ni} or σ_{ni}, then I can see what the complex scattering rate and effective mass (or $\lambda(\omega)$ or ω_{eff}) look like by inverting the real and imaginary part of Eqs. 13.5 or 13.3. The result is

$$1/\tau^* = \omega\,\frac{\sigma_{i1}}{\sigma_{i2}} \tag{13.6}$$

and

$$\frac{m^*}{m_b} = \frac{\omega_{pb}^2}{4\pi\omega}\left(\frac{\sigma_{i2}}{\sigma_{i1}^2 + \sigma_{i2}^2}\right). \tag{13.7}$$

I need to know ω_{pb} in order to evaluate Eq. 13.7. I use the sum rule on the conductivity, Eq. 9.30,

$$\int_0^\infty d\omega\, \sigma_{i1}(\omega) = \frac{1}{8}\omega_{pb}^2 = \frac{\pi}{2}\frac{ne^2}{m_b}.$$

The idea of the sum rule is that interactions (including phonon scattering or superconductivity or any other) can change the functional form of the optical conductivity and can move spectral weight from one frequency location to another, but that the overall area under the curve is unchanged. Its result can be interpreted as giving the plasma frequency of the bare or band carriers, before turning on the interactions.

13.1.2 Complex Scattering Rate

An alternate approach is to make the scattering rate a complex number and leave the plasma frequency as the bare plasma frequency [284]. The dielectric function then looks like

$$\epsilon(\omega) = \epsilon_{ni}(\omega) - \frac{\omega_{pb}^2}{\omega^2 + \omega M(\omega)}. \tag{13.8}$$

The complex function $M(\omega)$ is known as the "memory function" and contains the information about the interactions. If there is only scattering from impurities, one finds $M_1 = 0$ and $M_2 = 1/\tau_i^*$.

I actually see a variation on this notation more than the original. The variation is to define $M = i\Gamma$ and write the dielectric function as

$$\epsilon(\omega) = \epsilon_{ni}(\omega) - \frac{\omega_{pb}^2}{\omega^2 + i\omega\Gamma(\omega)}, \tag{13.9}$$

and the conductivity as

$$\sigma_i = i\,\frac{ne^2/m_b}{\omega + i\Gamma(\omega)}. \tag{13.10}$$

The quantity Γ is also known as the memory function.

Many authors [237, 246, 286, 287] use a different notation, writing

$$\sigma_i = i\,\frac{ne^2/m_b}{\omega(1 + \lambda^{\mathrm{op}}) + i/\tau^{\mathrm{op}}} \tag{13.11}$$

where $1/\tau^{\mathrm{op}} = \Gamma_1$ is the optical (or "unrenormalized") scattering rate and $\lambda^{\mathrm{op}} = -\Gamma_2/\omega$ is the mass enhancement factor. Note that although λ^{op} is the same quantity that appears in Eq. 13.4, $1/\tau^{\mathrm{op}}$ differs from $1/\tau^*$. I'll discuss the differences below.

I started this section by asserting that the Kramers–Kronig relations for the conductivity or the dielectric function required a complex scattering rate.* Let me write Ohm's law in a way closer to the way it is done in circuits. Instead of $\mathbf{j}(\omega) = \sigma(\omega)\mathbf{E}(\omega)$ I'll use $\mathbf{E}(\omega) = \rho(\omega)\mathbf{j}(\omega)$. The resistivity ρ is a response function relating the field to the current and must satisfy Kramers–Kronig. But according to Eq. 13.10,

$$\rho_1 + i\rho_2 = \frac{m_b}{ne^2}[\Gamma_1 + i(\Gamma_2 - \omega)] \tag{13.12}$$

so Γ_1 and $\Gamma_2 - \omega$ satisfy the same Kramers–Kronig relations as ρ_1 and ρ_2.

The real and imaginary parts of Γ are easy to obtain from Eq. 13.12

$$\Gamma_1 = \frac{ne^2}{m_b}\rho_1 = \frac{ne^2}{m_b}\frac{\sigma_{i1}}{\sigma_{i1}^2 + \sigma_{i2}^2}$$

and

$$\Gamma_2 = \omega + \frac{ne^2}{m_b}\rho_2 = \omega - \frac{ne^2}{m_b}\frac{\sigma_{i2}}{\sigma_{i1}^2 + \sigma_{i2}^2}.$$

To extract the complex scattering rate (or memory function) from optical conductivity data, I need to know the ratio of carrier density n to bare mass m_b. This ratio appears in the saturation value of the oscillator strength sum rule on the conductivity, and many workers use the value it reaches at the onset of interband transitions to estimate the ratio. Of course, considerations of crystal chemistry and electronic structure calculations also can give insight into these quantities.

13.1.3 Optical Self Energy

Littlewood and Varma [285] have written the dielectric function in the following form

$$\epsilon(\omega) \approx \epsilon_{ni} - \frac{\omega_{pb}^2}{[\omega^2 - 2\omega\Sigma_{op}(\omega/2)]}, \tag{13.13}$$

where Σ_{op} is the optical self energy governing quasiparticle excitations. The factors of two appear because quasiparticle excitations (as in superconductors or electron–hole pairs in

---

* Or a frequency-dependent plasma frequency.

semiconductors) are made in pairs. The division of the photon frequency by two in the argument of Σ_{op} assumes equal energy supplied to each of the excitations and is often not explicitly included. Sometimes it is explicitly excluded. A discussion of the relation between the self energy and the memory function, including how to average over the Fermi surface, has been given by Allen [288]. I compare Eq. 13.13 with Eqs. 13.8 and 13.10 and find that $M(\omega) = i\Gamma(\omega) = -2\Sigma_{op}(\omega/2)$.

13.1.4 Relations among These Approaches

The first comment I can make is that I cannot see a fundamental reason to use one or another of these approaches. All accomplish the goal of considering frequency-dependent or energy-dependent effects in the optical conductivity of correlated electron systems. Moreover, it is easy to translate among them. Thus, with a bit of algebra I get

$$\Gamma_1 = \frac{m^*}{m_b}\frac{1}{\tau^*},$$

$$\Gamma_2 = \omega\left(1 - \frac{m^*}{m_b}\right),$$

and

$$\frac{1}{\tau^*} = \frac{\omega\Gamma_1}{\omega - \Gamma_2},$$

$$\frac{m^*}{m_b} = \frac{\omega - \Gamma_2}{\omega}. \tag{13.14}$$

The terms "renormalized" and "unrenormalized" modify "scattering rates" in many papers. Let me see how the two rates appear. I start with Eq. 13.4 where I've written the effective mass of the charge carriers as $m^* = m_b(1 + \lambda)$. Here, m^* is a mass that has been renormalized from the bare/band value by the interactions of the carriers with other excitations in the solid. I think the term comes from Landau's Fermi liquid theory, where he showed that the states in strongly-interacting systems are one-to-one with those in the noninteracting Fermi gas as long as one renormalizes the parameters.

Now, I substitute m^* into Eq. 13.5, and multiply top and bottom by $1 + \lambda$ to get

$$\sigma_i = i\frac{ne^2/m_b}{\omega(1 + \lambda(\omega)) + i(1 + \lambda(\omega))/\tau^*(\omega)}. \tag{13.15}$$

I'll now write Eq. 13.10 again, but expanding $\Gamma = \Gamma_1 + i\Gamma_2$:

$$\sigma_i = i\frac{ne^2/m_b}{\omega - \Gamma_2(\omega) + i\Gamma_1(\omega)}. \tag{13.16}$$

When I compare the real parts of the denominators of Eqs. 13.15 and 13.16 I learn nothing that is not in Eq. 13.14. But the imaginary parts show that

$$\frac{1}{\tau^*(\omega)} = \frac{\Gamma_1(\omega)}{1 + \lambda(\omega)}. \tag{13.17}$$

The quantity $1/\tau^*$ is called the renormalized scattering rate and Γ_1 the unrenormalized rate.

Note that the distinction between renormalized and unrenormalized rates evaporates at zero frequency. The DC limit of Eq. 13.5 is

$$\sigma_{\rm dc} = \frac{ne^2\tau^*}{m^*} = \frac{ne^2}{\Gamma_1 m_b}.$$

The enhanced mass is canceled by the smaller scattering rate.

Finally, let me comment that I have seen Γ_1 and gG_2 described as causal and m^* and $1/\tau^*$ described as non causal. Do not be deceived by this. The equations using m^* and $1/\tau^*$ do not violate causality. What is meant is that the two functions are not Kramers–Kronig pairs.

13.2 Electron–Electron Interactions

Electrons are charged particles and interact via the Coulomb interaction. An electron that is separated by the Bohr radius from another electron sees an electrostatic potential of 27 eV. Of course the average separation in a solid of the conduction electrons one from another is a bit more than this, but still I might expect to see eV scale Coulomb energies. These energies are not too different from the Fermi energy.

Given this large energy, it seems surprising that the free-electron gas, neglecting completely interactions among the electrons, describes well the electrical, thermal, magnetic, and optical properties of metals. The answer lies in Fermi liquid theory. In this picture, interactions between electrons are turned on adiabatically in a way that there is a one-to-one mapping between electron states in the noninteracting picture and electron-states in the interacting picture. The free-electron states are mapped into Fermionic quasiparticle states in the interacting system. The quantum numbers of the quasiparticle (charge, momentum, and spin) remain the same as they were before interactions were turned on. The interacting system also has a Fermi surface with the same volume (number of states) but the energy, the effective mass, and the scattering rate differ from those of the free particles. Despite the different parameters, Fermi liquid theory allows the strongly interacting system to be described in a noninteracting picture.

13.2.1 The Fermi Liquid

The transport processes in the Fermi liquid should be affected by electron–electron scattering. Here, I'll try to estimate the rate of such scattering.*

The qualitative argument is that at temperature T two electrons can interact effectively only if their energies are within the $k_B T$ energy range around the Fermi energy $\mathcal{E}_F$ where the Fermi–Dirac distribution function is neither one nor zero. The probability that one electron has an such an energy is of order $k_B T/\mathcal{E}_F$. The probability that the other also

---

* I'll do this in a free-electron picture. I actually do not need any of the Fermi-liquid machinery to do the calculation at the level at which I'll do it.

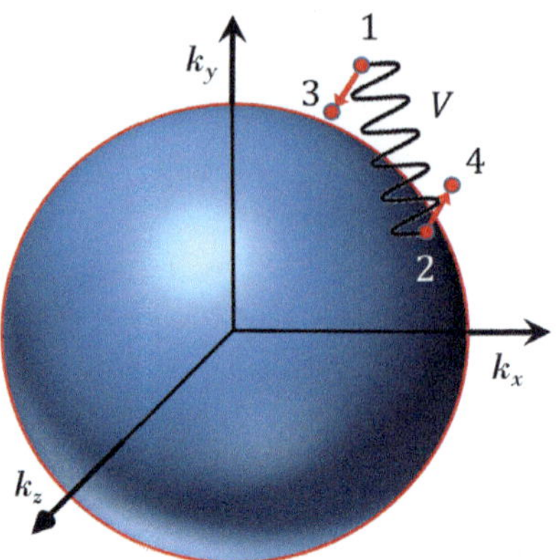

Fig. 13.1 Electron 1 (above the Fermi level) scatters with electron 2 (below it) and relaxes towards the Fermi level. A hole is left in the Fermi sphere and there are two electrons above the Fermi surface. Momentum and energy are conserved in the process.

is within this range is the same. Because the particles are independent, the scattering probability is proportional to $(k_B T / \mathcal{E}_F)^2$.

I can flesh this idea out a bit more by considering the zero temperature Fermi sphere with an electron in a state close to but just above the Fermi surface. This electron has energy $\mathcal{E}_1 > \mathcal{E}_F$. (I'll suppose in a moment that the electron acquired this extra energy by absorbing a photon with energy $\hbar\omega = \mathcal{E}_1 - \mathcal{E}_F$.) As indicated in Fig. 13.1, this electron can scatter from a second electron with energy $\mathcal{E}_2 < \mathcal{E}_F$ (just below the Fermi energy).

The discussion is simplified if I consider the energies of the initial and final states relative to the Fermi level: $\epsilon_i = \mathcal{E}_i - \mathcal{E}_F$. Electron 1 has relative energy $\epsilon_1 > 0$ and electron 2 has relative energy $\epsilon_2 < 0$. The scattered electrons have relative energies ϵ_3 and ϵ_4; the Pauli principle requires both energies to be just above the Fermi energy. Moreover, energy conservation requires the distance electron 2 is below the Fermi surface to be less than the distance electron 1 was above it: $|\epsilon_2| < \epsilon_1$.

Therefore, the fraction f_i of electrons in the Fermi sphere that actually can interact with electron 1 is the ratio of the volume a shell of thickness ϵ_1 to the volume of the whole Fermi sphere:

$$f_i \approx \frac{V(\mathcal{E}_F) - V(\mathcal{E}_F - \epsilon_1)}{V(\mathcal{E}_F)} \approx 1 - \left(\frac{\mathcal{E}_F - \epsilon_1}{\mathcal{E}_F}\right)^{3/2} \approx \frac{\epsilon_1}{\mathcal{E}_F}, \qquad (12.5)$$

where I have used $4\pi k_F^3 / 3$ with $k_F = \sqrt{2m\mathcal{E}_F}/\hbar$ for the volume of the sphere. The fraction f_i is small.

Now I look at the final states after the interaction process. Energy conservation requires that the sum of final relative energies satisfies $0 \le \epsilon_3 + \epsilon_4 \le \epsilon_1$. I'll assume upper limits of ϵ_1, and see that the fraction f_f of states above the Fermi level that are accessible to the scattered electrons is of order

$$f_f \approx \frac{\epsilon_1}{\mathcal{E}_F}. \qquad (12.6)$$

Consequently the phase space f for an electron–electron scattering event becomes

$$f \approx \left(\frac{\epsilon_1}{\mathcal{E}_F}\right)^2.$$ (13.18)

If I include finite temperatures, there is a width of energies of order $k_B T$ around the Fermi energy to include in this phase-space argument. I also set $\epsilon_1 \approx \hbar\omega$, the energy available from a photon, to get

$$f \approx \left(\frac{\hbar\omega}{\mathcal{E}_F}\right)^2 + \left(\frac{k_B T}{\mathcal{E}_F}\right)^2.$$ (13.19)

This is a very small number, say 10^{-8} at 10 K or 10 cm^{-1}. The scattering rate will be proportional to f, so that it follows that $1/\tau \propto T^2$ at low frequencies and $1/\tau \propto \omega^2$ at low temperatures.

There is however, an annoying flaw in this argument: The scattering process conserves momentum, which means that $\mathbf{k}_1 + \mathbf{k}_2 = \mathbf{k}_3 + \mathbf{k}_4$. I suppose that before the scattering event the current was given by Eq. 8.20, $\mathbf{j} = -(ne\hbar/m)\delta\mathbf{k}_i$, with n the electronic density, e the electronic charge, m the effective mass of the charge carriers, and $\delta\mathbf{k}_i$ the displacement of the Fermi sphere from its equilibrium position. Then, after the scattering has taken place I have a new $\delta\mathbf{k}_f = \delta\mathbf{k}_i + (\mathbf{k}_3 + \mathbf{k}_4 - \mathbf{k}_1 - \mathbf{k}_2)/N = \delta\mathbf{k}_i$ where N is the total number of electrons. So the current is unchanged because the scattering is within the electron system alone and the total momentum is unchanged.

For electron–electron scattering to affect the conductivity, it must involve "umklapp scattering," a process where at least one of the two scattering wave vectors lies outside the Brillouin zone. By the arguments surrounding Fig. 12.4, I can reduce this vector to the first zone by the addition of whatever reciprocal lattice vector $\mathbf{G}$ that accomplishes this task. This umklapp scattering process does change the total momentum of the electron system and contribute to the optical conductivity.

As an illustration, consider a one-dimensional example so that k is a scalar. As in the left panel of Fig. 12.4, k has a range of $-\pi/a < k \leq \pi/a$. Suppose $k_3 = (1+\delta)\pi/a$, placing it a small amount δ past the zone boundary, which is at $+\pi/a$. I can equally well represent it as $k_3' = (1 + \delta)\pi/a - 2\pi/a$, using $G = 2\pi/a$. Then $k_3' = -(1 - \delta)\pi/a$, close to the zone boundary on the opposite side of the Brillouin zone. Umklapp scattering has significantly affected the electron's momentum. Electron–electron interactions plus umklapp scattering will lead to ω^2 and T^2 terms in the scattering rate.

Many-body calculations of the scattering rate due to electron–electron scattering have been done by a number of workers [289–295]. The scattering rate is

$$\frac{1}{\tau} = \frac{2}{3\pi\hbar\mathcal{E}_F}\left[(\hbar\omega)^2 + (2\pi k_B T)^2\right],$$ (13.20)

or in more generic form

$$\frac{1}{\tau} = a\left[(\hbar\omega)^2 + b(\pi k_B T)^2\right].$$ (13.21)

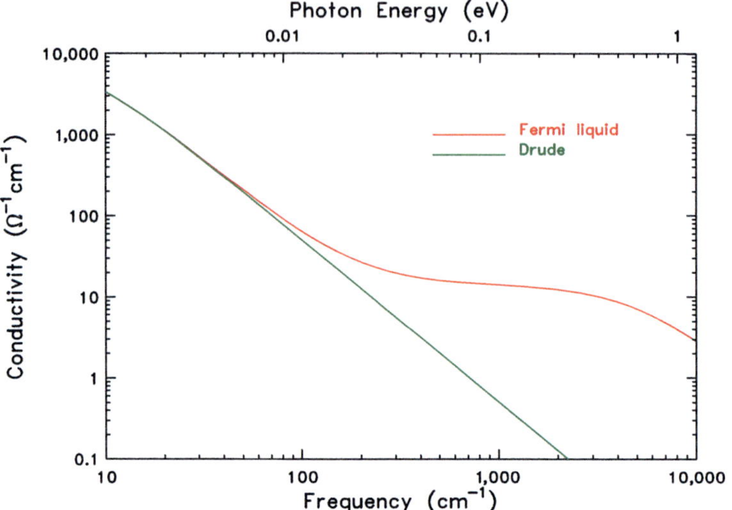

Fig. 13.2 Optical conductivity of a Fermi-liquid metal. A Drude spectrum is shown for comparison.

The factor b should be $b = 4$ but experiments find it varying from below 1 to almost 6 [295]. These discrepancies are an active area of research. In principal, the ω^2 behavior continues up to photon energies of order the Fermi energy, above which $1/\tau$ should become constant.*

The real part of the Fermi-liquid conductivity [289, 295] is then

$$\sigma_1(\omega) = \sigma_0 \frac{\omega^2 + (2\pi k_B T/\hbar)^2}{\omega^2}.$$

When $\hbar\omega \gg 2\pi k_B T$ the conductivity is constant, called the "Fermi foot." In this it differs from the high-frequency part of the Drude formula, where $\sigma_{1D} \sim 1/\omega^2$. A plot of the conductivity is in Fig. 13.2. The conductivity spectrum has a low frequency part which is equal to the DC conductivity, a Drude rolloff, and then a long foot beginning when $\hbar\omega \approx 2\pi k_B T$ (about 190 cm^{-1} in this calculation, which is for $T = 50$ K). Once $\hbar\omega > \mathcal{E}_F$ (about 5,000 cm^{-1}), the Drude spectrum is regained but with a much larger scattering rate. A Drude spectrum, having the same DC conductivity and low-frequency scattering rate is also shown in Fig. 13.2.

13.2.2 The Marginal Fermi Liquid

Varma *et al.* [296] proposed a phenomenological model for the oxide superconductors that they call a "marginal Fermi liquid." This model assumes that excited charge carriers interact with a spectrum of (optically inactive) excitations which is flat over $T < \omega < \omega_c$, where ω_c is a high energy scale (the bandwidth, for example) that cuts off the spectrum. Evidence advanced in support of this hypothesis describing the cuprates includes the Raman spectrum [297], where a broad continuum of electronic excitations extending to

---

* At such frequencies, usually, many other optical processes have also turned on and the detailed Fermi-liquid behavior would be masked anyway.

1 eV (8,000 cm^{-1}) is observed, and the angle-resolved photoemission [298, 299], where a broad, asymmetric peak at the Fermi energy is seen.

According to Varma *et al.* [296] and Littlewood and Varma [285] the quasiparticle self energy Σ of the marginal Fermi liquid has an imaginary part which qualitatively goes as

$$-\,\mathrm{Im}\,\Sigma(\omega) \sim \begin{cases} \pi^2 \lambda T, & \omega < T \\ \pi \lambda \omega, & \omega > T \end{cases} \tag{13.22}$$

where λ is the strength of the interaction, an integral of square of a matrix element and a density of states. There is a corresponding logarithmic (in ω/ω_c or T/ω_c, whichever is greater) behavior in the effective mass, with the mass enhancement proportional to λ.

The inverse lifetime (or scattering rate) of the charge carriers is given by the imaginary part of the self energy, which at zero frequency is linear in T. This result is in accord with the measured resistivity of optimally-doped cuprate superconductors [300, 301]. At high frequencies, there is no temperature dependence in Eq. 13.22 and the scattering rate is proportional to ω; the lifetime to $1/\omega$. Thus, at high frequencies the conductivity becomes $\sigma_1(\omega) = \omega_p^2 \tau/4\pi\,(1+\omega^2\tau^2) \sim \omega_p^2/4\pi\omega$. It goes like $1/\omega$ in contrast to the $1/\omega^2$ dependence of an ordinary metal with frequency-independent scattering. This behavior is reasonably in accord with the conductivity spectrum of Fig. 11.16.

13.2.3 The Hubbard Model

In both the Fermi liquid and the marginal Fermi liquid, the transport and optics are basically metallic in the qualitative sense. The DC conductivity decreases as temperature is raised from the lowest T. Analysis discloses that the carrier density is roughly constant over a wide temperature range. The optical conductivity has a maximum at $\omega = 0$ and falls with increasing frequency. (The frequency dependence differs from the Drude metal, but has a superficial resemblance and often may be fit by a Drude–Lorentz model.)

In contrast, the Hubbard model [302] allows for strong Coulomb repulsion between electrons. If this repulsion is sufficiently strong, and if the electron occupancy is right (about one per site), the Hubbard model gives an insulating ground state in a system that otherwise would be a metal. This insulating state is a specific example of a Mott transition – an insulating state driven by correlations of the electrons [303].

Thus there are two different mechanisms that can affect solids which otherwise would be free-electron metals and make them insulators. In *band insulators,* the electron interacts with the periodic potential of the ions, giving a gap in the single-particle spectrum when the Fermi surface coincides with the Brillouin-zone boundary. Mott Insulators are insulators due to the electron–electron interaction. Many-body physics leads in Mott systems to a gap in the charge excitation spectrum. The system has to pay too high a price to have electrons be mobile and thereby approach each other too closely.

In it's simplest form, the Hubbard Hamiltonian is written in second-quantized notation as

$$\mathcal{H} = t \sum_{\langle ij \rangle} \sum_{\sigma} c_{i\sigma}^{\dagger} c_{j\sigma} + U \sum_{i} n_{i\downarrow} n_{i\uparrow} \tag{13.23}$$

where t is a transfer or hopping integral, $c_{i\sigma}^{\dagger}$ creates an electron on site i with spin σ, $c_{j\sigma}$ destroys an electron on site j with spin σ (so that in concert they transfer an electron from site j to site i). The first sum is a restricted sum over sites and their neighbors, organized so that the electron can transfer between j and any neighbor. The second sum (on σ) is a sum on spin. In the second term, U is the strength of the Coulomb repulsion between two electrons on the same site and $n_{i\sigma} = c_{i\sigma}^{\dagger} c_{i\sigma}$ counts the number of electrons on site i with spin σ. The Pauli principal ensures that the count is either 0 or 1. The second term is summed over sites, adding U to the energy eigenvalue every time two electrons occupy the same site.

If $U = 0$, then Eq. 13.23 is the Hamiltonian of a tight-binding metal. It is easy to solve by writing the Fourier transform of the creation/annihilation operators

$$c_{i\sigma}^{\dagger} = \frac{1}{\sqrt{N}} \sum_{\mathbf{k}} e^{-i\mathbf{k}\cdot\mathbf{a}}$$

where N is the number of sites, $\mathbf{k}$ is the wave vector, proportional to (integer)$\cdot 2\pi/L$, and the sum would be over these integers. Using this operator and its Hermitian conjugate, it is easy to show that the energy is

$$\mathcal{E}_k = -2t\cos(ka). \tag{13.24}$$

The larger is the transfer integral, the greater the width of the band. This relationship makes sense: strong overlap between sites leads to large bandwidth. Here, the width is $W = 4t$.

The nonlinear behavior of the Hubbard model make it intractable in general. An exact solution has been made for the case of half filling (exactly one electron per site) in one space dimension [304]. The result is that the one dimensional chain is an insulator for all finite U. (However the charge gap is quite small for $U < 2t$.) Other results come from numerical solutions of clusters and from various approximations. There are lots of these, done in all space dimensions (from one to infinity) and sometimes for more than a single band. Almost all results do show a Mott–Hubbard metal–insulator transition as U is increased from zero. Typically, $U \gtrsim 4t$ is needed in order to render the solid an insulator.*

The picture that emerges is sketched in Fig. 13.3. On the left is the metal ($U = 0$), with bandwidth W. The metallic band is half filled. Because a single band can generally hold two electrons per site and still make the Pauli principal satisfied, the half-filled metal has one electron per site.

On the right is the insulator, with U large enough to make it so. There are now two bands: the "lower Hubbard band," which is completely filled and the "upper Hubbard band," completely empty. Both bands are narrower than the single band of the metal. There is an energy gap between the two bands; the system is an insulator.

It is relatively simple to imagine the optical properties of such a system. The metal would have a Drude spectrum although probably with a frequency-dependant scattering rate and effective mass. The insulator would have the spectrum of a semiconductor or insulator with a gap $\mathcal{E}_g \sim U$.

---

* A band filling near 1/2 is also needed. The Hubbard model can be extended with next-nearest-neighbor Coulomb repulsion V If U and V are both large enough, the 1/4 filled chain is insulating. Extension to even longer-ranged Coulomb terms are possible too. A number of interesting cases are discussed is in Ref. 305.

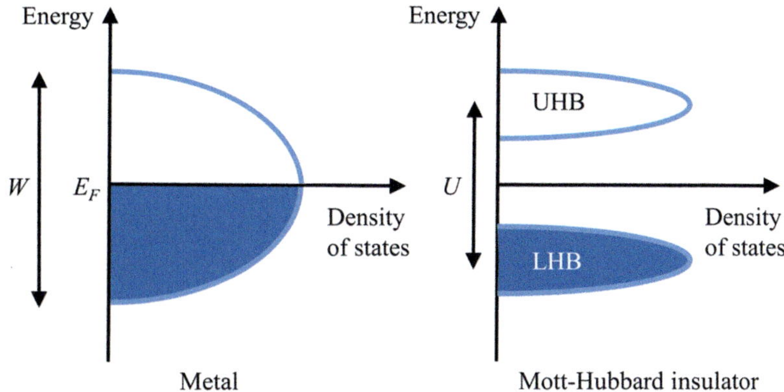

Fig. 13.3   Cartoon of energy band densities of states in (left) half-filled-band metal and (right) highly correlated "metal" with large Coulomb energy U.)

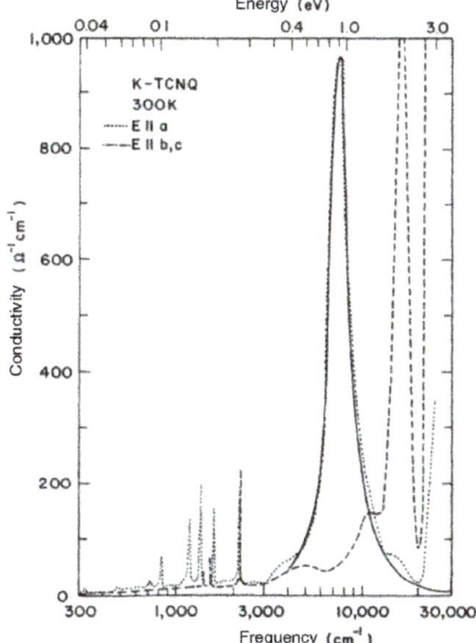

Fig. 13.4   Optical conductivity of K-TCNQ at 300 K. The dotted line shows the a-axis (stacking axis) spectrum. The spectrum perpendicular to the stacking axis (b or c) is dashed and a fit to Lorentzian oscillators is the solid line.

An example is shown in Fig. 13.4. This figure shows the optical conductivity of potassium-tetracyanoquinodimethane (K-TCNQ or K-$(NC)_2CC_6H_4C(CN)_2$), a quasi one dimensional organic solid [306]. The material is highly anisotropic, with a nearly tetragonal crystal structure. (See Chapter 15.) This material is a charge-transfer salt, in which an electron is removed from the K atom, rendering it a closed-shell K$^+$ ion, and transferred to the TCNQ molecule, making it a singly-charged ion. TCNQ is a planar molecule and stacks face-to-face, with the stacking axis (also called the chain axis) along the a crystal axis. With one electron per molecule, K-TCNQ would be expected to be an organic metal. Instead. it is an insulator, and the insulating behavior is generally interpreted as due to a strong Coulomb interaction, exactly as in the Hubbard model.

The dotted line in Fig. 13.4 shows the a-axis (stacking axis) spectrum. The most prominent feature is a strong peak at about 7,000 cm^{-1} (0.9 eV). This peak is attributed to transitions across a Mott–Hubbard gap, with $U \approx 0.9$ eV. There are a number of sharp peaks at lower frequencies, where the vibrational modes of the molecule occur. These are due to the Rice effect; see Section 13.3.2.

To see how the Hubbard model works, I will work through the simple problem of two electrons on two sites. This is the "half-filled" case and actually contains much of the physics [307–309]. My system consists of two sites, each with a single molecular orbital. Because the electrons have spin, the two sites may be occupied with 0–4 electrons, subject to obeying the Pauli principal. The only interesting case is the one with two electrons.

Applied to two sites, numbered 1 and 2, Eq. 13.23 is

$$\mathcal{H} = t \sum_{\sigma} \left(c_{1\sigma}^{\dagger} c_{2\sigma} + c_{2\sigma}^{\dagger} c_{1\sigma} \right) + U \left(n_{1\uparrow} n_{2\downarrow} + n_{2\uparrow} n_{1\downarrow} \right). \tag{13.25}$$

The first term allows an electron to hop from site 1 to 2 and from 2 to 1. The second term represents the cost of having the two electrons on the same site. There can also be a site energy $\mathcal{E}_n$ which I'll take to be zero. I have two electrons and I can place them on either of the two sites and with either spin up or spin down. (I of course cannot place them both on the same site with the same spin.) An example basis state is $|\uparrow\downarrow, 00\rangle$, where both electrons are on site 1, none on site 2. Another example, $|\uparrow 0, 0\downarrow\rangle$, has an up-spin electron on site 1 and a down-spin electron on site 2. There are a total of six possibilities: (a) two for double occupancy of one or the other site; (b) two with spins antiparallel and with one on each site; and (c) two for spins parallel and with one on each site. I could start with these, but it helps to form linear combinations that respect the symmetry of the system [307, 308]. So I write the following six basis states:

$$|a_+\rangle = \frac{1}{\sqrt{2}} \left(|\uparrow\downarrow, 00\rangle + |00, \uparrow\downarrow\rangle \right)$$

$$|a_-\rangle = \frac{1}{\sqrt{2}} \left(|\uparrow\downarrow, 00\rangle - |00, \uparrow\downarrow\rangle \right)$$

$$|b_+\rangle = \frac{1}{\sqrt{2}} \left(|\uparrow 0, 0\downarrow\rangle + |0\downarrow, \uparrow 0\rangle \right) \tag{13.26}$$

$$|b_-\rangle = \frac{1}{\sqrt{2}} \left(|\uparrow 0, 0\downarrow\rangle - |0\downarrow, \uparrow 0\rangle \right)$$

$$|c_u\rangle = |\uparrow 0, \uparrow 0\rangle$$

$$|c_d\rangle = |0\downarrow, 0\downarrow\rangle$$

The states are normalized and orthogonal.

I operate with the Hamiltonian of Eq. 13.25 on $|a_+\rangle$ $|a_-\rangle$, and $|b_+\rangle$:

$$\mathcal{H} |a_+\rangle = \frac{1}{\sqrt{2}} t \left(|\uparrow 0, 0\downarrow\rangle + |\uparrow 0, 0\downarrow\rangle + |0\downarrow, \uparrow 0\rangle + |0\downarrow, \uparrow 0\rangle \right)$$

$$+ \frac{1}{\sqrt{2}} U \left(|\uparrow\downarrow, 00\rangle + |00, \uparrow\downarrow\rangle \right)$$

$$= 2t |b_+\rangle + U |a_+\rangle.$$

$$\mathcal{H}\,|a_-\rangle = \frac{1}{\sqrt{2}}t\,\left(|\!\uparrow 0,0\downarrow\rangle - |\!\uparrow 0,0\downarrow\rangle + |0\downarrow,\uparrow 0\rangle - |0\downarrow,\uparrow 0\rangle\right)$$

$$+ \frac{1}{\sqrt{2}}U\,\left(|\!\uparrow\downarrow,00\rangle - |00,\uparrow\downarrow\rangle\right)$$

$$= U\,|a_-\rangle.$$

$$\mathcal{H}\,|b_+\rangle = +\frac{1}{\sqrt{|b_+\rangle 2}}t\,\left(|\!\uparrow\downarrow,00\rangle + |00,\uparrow\downarrow\rangle\, |\!\uparrow\downarrow,00\rangle + |00,\uparrow\downarrow\rangle\right)$$

$$= 2t\,|a_+\rangle.$$

The remaining wave functions in Eq. 13.26 are already eigenfuctions, with energy $\mathcal{E} = 0$. (If I had included the site energy, every diagonal term would have an additional $2\mathcal{E}_n$.) The Hamiltonian matrix is

$$\mathcal{H} = \begin{array}{c} \\ a_+ \\ b_+ \\ a_- \\ b_- \\ c_u \\ c_d \end{array}
\begin{array}{c} \begin{array}{cccccc} a_+ & b_+ & a_- & b_- & c_u & c_d \end{array} \\
\left(\begin{array}{cccccc}
U & 2t & 0 & 0 & 0 & 0 \\
2t & 0 & 0 & 0 & 0 & 0 \\
0 & 0 & U & 0 & 0 & 0 \\
0 & 0 & 0 & 0 & 0 & 0 \\
0 & 0 & 0 & 0 & 0 & 0 \\
0 & 0 & 0 & 0 & 0 & 0
\end{array}\right) \end{array} \tag{13.27}$$

Thanks to the fact that I wrote linear combinations of the basic wave functions in Eq. 13.26, it is trivial to diagonalize Eq. 13.27. I have only the 2×2 part at the upper right to address, a result of mixing of the $|a_+\rangle$ and $|b_+\rangle$ states. The characteristic equation is

$$\begin{vmatrix} U - \mathcal{E} & 2t \\ 2t & -\mathcal{E} \end{vmatrix} = \mathcal{E}^2 - U\mathcal{E} - 4t^2 = 0$$

and the eigenvalues are

$$\mathcal{E} = \frac{U}{2} \pm \frac{1}{2}\sqrt{U^2 + 16t^2}. \tag{13.28}$$

The negative sign corresponds to the energy of the ground state and the positive sign the energy of one (out of five) of the excited states.

It is straightforward to find the wave functions corresponding to these energies. The ground-state wave function, with energy $\mathcal{E}_g = U/2 - \sqrt{U^2/4 + 4t^2}$, is

$$|g\rangle = \frac{1}{\sqrt{4t^2 + \mathcal{E}_g^2}}\left(2t\,|b_+\rangle + \mathcal{E}_g\,|a_+\rangle\right). \tag{13.29}$$

For the upper level, with high energy $\mathcal{E}_h = U/2 + \sqrt{U^2/4 + 4t^2}$, the wave function is

$$|h\rangle = \frac{1}{\sqrt{4t^2 + \mathcal{E}_h^2}}\left(\mathcal{E}_h\,|a_+\rangle + 2t\,|b_+\rangle\right). \tag{13.30}$$

I leave it to you to show that $\langle g|h\rangle = 0$.

Both solutions contain an admixture of the two initial basis states. Their relative weights depend on the ratio of t to U. For $U \gg t$, the ground-state energy is

$$\mathcal{E}_g = -4t^2/U,$$

slightly negative. The corresponding wave function is

$$|g\rangle = C_g \left(|b_+\rangle - \frac{2t}{U} |a_+\rangle \right),$$

where C_g is for normalization.* The amplitude of the double-occupied term $|a_+\rangle$ is small. The highest-energy excited-state in the large U limit has

$$\mathcal{E}_h = U + \frac{4t^2}{U}$$

while

$$|h\rangle = C_h \left(|a_+\rangle + \frac{2t}{U} |b_+\rangle \right).$$

with C_h another normalization factor. There is also a high-energy state with energy U and wave function $|a_-\rangle$. I'll return to this state in a moment.

In the other limit, $t \gg U$, $\mathcal{E}_g = -2t + U/2$ and $\mathcal{E}_h = 2t + U/2$. The wave functions become – in the case where $U = 0$ – just the bonding and antibonding states of the two-site molecule.

The states $|b_-\rangle$, $|c_u\rangle$, and $|c_d\rangle$ all have total spin $S = 1$. These are the triplets, with energies $\mathcal{E}_T = 0$; their energy is between the singlet $S = 0$ ground state and the excited state energies $\mathcal{E}_h$ and U of the two other singlet states.

I have not yet said much about the state $|a_-\rangle$. This state is the one that is connected to the ground state by the electric-dipole matrix element, and hence the state that governs the dielectric function. If the vector joining site 1 to site 2 is $\mathbf{a}$, then I write the dipole moment as

$$\mathbf{p} = -\frac{e\mathbf{a}}{2}(n_1 - n_2) \equiv -\frac{e\mathbf{a}}{2}\delta n$$

where $n_i = n_{i\uparrow} + n_{i\downarrow}$ is the number operator for site i. Now,

$$\delta n \, |b_+\rangle = 0,$$

because the two electrons are with equal amplitude on each site. However

$$\delta n \, |a_+\rangle = 2 \, |a_-\rangle,$$

because $|a_+\rangle$ is made up of states where there is double occupancy. Thus the dipole matrix element couples the ground state to state $|a_-\rangle$. It is a charge-transfer excitation where an electron is transferred from one site to another; the energy of the excited state is U and the photon energy required to make the transition is

$$\hbar\omega_{ct} = U - \mathcal{E}_g = U/2 + \sqrt{U^2/4 + 4t^2}. \tag{13.31}$$

---

* I get C_g from requiring that $\langle g|g\rangle = 1$.

If $U \gg t$, the energy difference is $\hbar\omega_{ct} \approx U$, which is what I would expect in the strongly interacting case. For large U the dipole matrix element is

$$\langle a_- | p | g \rangle \approx ea\frac{2t}{U},$$

so that charge-transfer transitions are suppressed by the energy cost of double occupancy.

The optical conductivity (or $\epsilon - \epsilon_c$, with ϵ_c the high-frequency "core" part of the dielectric function) is proportional to the square of this matrix element. Following Rice [307], I write it as

$$\epsilon = \epsilon_c + \frac{4\pi n e^2 a^2}{\hbar} \frac{4t^2}{U} \frac{1}{\omega_{ct}^2 - \omega^2 + i\omega\gamma}, \tag{13.32}$$

where n is the number density of molecules (*not* dimers), I took $\omega_{ct} = U/\hbar$, and I added by hand a finite line width γ to the charge transfer excitation.

13.3 Electron–Phonon Interactions

13.3.1 The Holstein Effect

The temperature-dependence of the resistivity of simple metals is well understood as arising from electron scattering by thermally-generated phonons. In pure metals, the mean free path and scattering time τ can decrease by several orders of magnitude on warming from near zero Kelvin to room temperature. At first, it is easy to imagine that the same decrease would be observed in the optical regime, at least for photon energies below the onset of interband transitions, say at midinfrared frequencies of a thousand cm^{-1} or 0.1 eV. This strong temperature dependence is not observed; instead, the optical measure of electron scattering finds a time τ considerably shorter than what one infers from DC resistance [310, 311].

The cause of the difference between DC and optical measures is explained by the Holstein model [92, 312–316]. In fact, both the frequency and temperature dependence of the optical conductivity can be described in this model. The interaction of the electrons with the phonons can be shown to depend on $\alpha^2 F(\Omega)$, the electron–phonon spectral density, a function of Ω, the phonon frequency.* The overall electron–phonon coupling constant λ is given by

$$\lambda = 2 \int_0^\infty \frac{\alpha^2 F(\Omega)}{\Omega} \, d\Omega. \tag{13.33}$$

Now, at DC (and very low frequencies) the only scattering processes available are the absorption of or scattering by a phonon, with a change of electron momentum and energy removed from the electric current density. The temperature dependence of the

---

* This spectral density is in fact the product of two things: a coupling constant α (which enters through its square) and a density of states for the phonons $F(\Omega)$.

resistance, therefore, is set by the number of thermally generated phonons (a Bose–Einstein distribution) combined with their momentum. Phonons with small momentum do not scatter electrons by very large angles. Above the Debye temperature, the entire acoustic branch is excited and [19, 98, 101] $\hbar/\tau = 2\pi\lambda k_B T$. At temperatures well below the Debye temperature, only low energy (and small momentum) phonons exist, and the resistivity is expected to follow T^5.

If, however, I turn the frequency up, another process can come into play: the electrons may also *emit* phonons. In this Holstein emission process, a charge carrier absorbs a photon of energy ω, emits an excitation of energy Ω and momentum $\hbar\mathbf{k}_\Omega$ and scatters, giving rise to absorption. (The excitation is generally a phonon but in fact could be any boson that couples linearly to the charge carriers.) Momentum is conserved in the electron–phonon system; energy largely in the photon–phonon system. Only phonons with energy less than the photon energy play a role. These considerations lead to the following expression for the $T = 0$ electron–phonon scattering rate due to Holstein phonon emission [312],

$$1/\tau_h(\omega) = \frac{2\pi}{\omega}\int_0^\omega \alpha^2 F(\Omega)(\omega - \Omega)d\Omega. \tag{13.34}$$

A Holstein sideband due to electron–phonon coupling has been observed in the infrared spectra of conventional strong-coupling metals, such as lead [313–316]. Figure 13.5 shows an example of the effect in Pb. The figure shows the difference between the surface resistance in the superconducting state and the surface resistance in the normal state, $R_s - R_n$. There is a three times smaller scale below 26 cm^{-1}. The measurements are on an absolute scale and agree very well with the calculations by Shaw and Swihart [317], which are shown in the figure, and also with calculations by Scher [318] and Allen [312].

When the metal becomes superconducting and an energy gap appears in the excitation spectrum (shown in Fig. 11.7), the process also includes breaking a Cooper pair into its

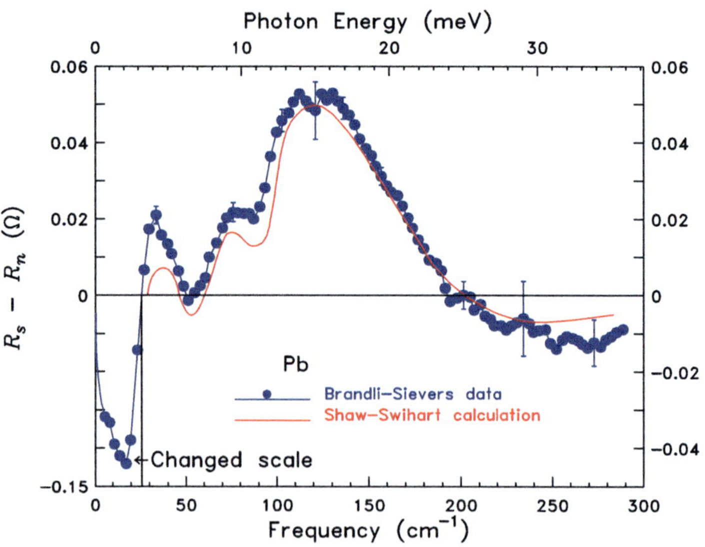

 Holstein sidebands in Pb. After Ref. 314.

component electrons. Thus energy balance requires $\omega \geq \Omega + 2\Delta/\hbar$. All the phonon-related features are shifted up by the gap frequency. Such a shift is a pretty unambiguous proof that the Holstein effect is observed. The features in Fig. 13.5 are in accord with this picture. The step upward in $R_s - R_n$ at 21 cm^{-1} occurs at the superconducting gap 2Δ. The smaller upward steps at about 55 and 90 cm^{-1} are respectively from the Holstein emission at the transverse and longitudinal phonon density of states maxima, shifted by 2Δ. The downward steps at 35 and 65 cm^{-1} are the Holstein emission in the normal state, which of course are unshifted.

I'll use a simple toy model to illustrate the phenomenon. Suppose my phonon spectrum consists of a single Einstein mode: $F(\Omega) = D\delta(\Omega - \Omega_0)$. I evaluate Eq. 13.33 first to find that $\lambda = 2D\alpha^2/\Omega_0$. Then Eq. 13.34 gives

$$1/\tau_h(\omega) = \begin{cases} 0 & \omega \leq \Omega_0 \\ \pi\lambda\Omega_0\dfrac{\omega - \Omega_0}{\omega} & \omega \geq \Omega_0 \end{cases} \tag{13.35}$$

The scattering rate is strongly frequency dependent. I show a plot in Fig. 13.6. I used $\Omega_0 = 200$ cm^{-1} and $\lambda = 0.5$ or 2 for the calculations. The scattering rate turns on at the phonon frequency, rises smoothly and then flattens out. The limiting high frequency value is $1/\tau_h = \pi\lambda\Omega_0$, comparable to the DC scattering rate at $T = \hbar\Omega_0/2k_B$. In accord with the discussion in Section 13.1 of this chapter, the frequency-dependent scattering rate is accompanied by a frequency-dependent mass enhancement factor $\lambda(\omega)$; both must be included in calculating via Eq. 13.11 the conductivity [286].

Figure 13.7 shows the optical conductivity at low temperatures of a metal due to the Holstein emission process. I included a modest impurity scattering here, $1/\tau_i = 10$ cm^{-1}, and employed Matthiessen's law: $1/\tau = 1/\tau_H(\omega) + 1/\tau_i$. The inset shows the conductivity on a logarithmic scale. Below Ω_0, the conductivity has Drude form, falling to half its DC value at 10 cm^{-1} and falling like $1/\omega^2$ above this frequency. Then, at 200 cm^{-1}, the phonon

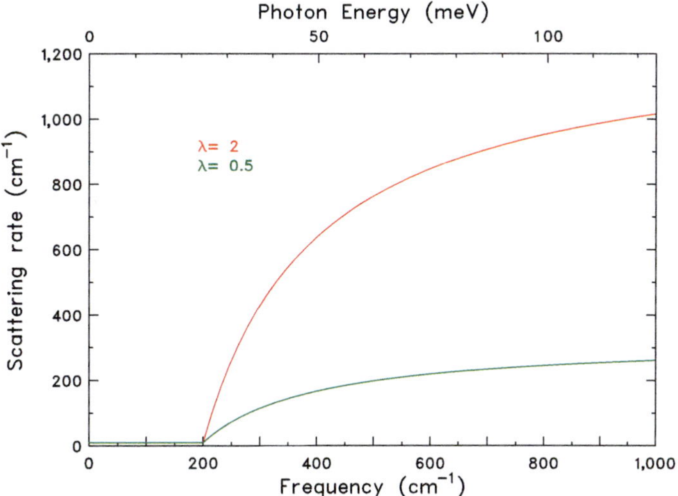

Fig. 13.6   Phonon emission scattering rate due to coupling to an Einstein phonon at 200 cm^{-1}.

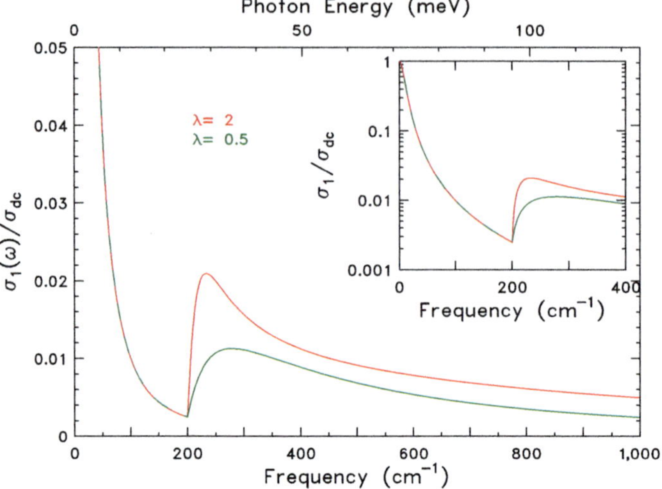

Fig. 13.7   Real part of the optical conductivity (normalized by the DC value) of a metal due to the Holstein emission process.

emission processes begin. The scattering rate increases, but so does the conductivity. The reason is that the metal is in the high frequency limit, where $\sigma_1 \approx ne^2/mw^2\tau$, so an increase in $1/\tau$, or a decrease in τ, makes the conductivity bigger. As the frequency continues to increase and the scattering rate approaches its limiting high-frequency value, the conductivity begins to fall as $1/\omega^2$ again, but now the width of the Drude-like spectrum is 314 cm^{-1} rather than 10 cm^{-1}.

There are differences in details if I were to use a more realistic phonon spectrum. Typical metals have $F(\Omega) \sim \omega^2$ at low frequencies. There are two maxima in the density of states, one at the transverse zone-boundary frequency and one at the longitudinal zone-boundary frequency. The width of the $F(\Omega)$ softens the onset of the increase in $1/\tau(\omega)$ but the basic behavior is the same as in Fig. 13.7.

13.3.2 The Rice Effect

Many materials, including linear-chain organic and inorganic conductors, electro-active polymers, fullerenes, and cuprates exhibit charge ordering, charge-density waves, or dimerized ground states.* The charge density wave is generally understood as being a Peierls transition [253]. The Peierls transition occurs on account of the electron–phonon interaction. A variation of charge occurs along the chain of molecules with a wave vector of $2k_F$. This charge density wave causes (or is caused by) a lattice distortion with the same wave vector. In consequence, the electron states get a zone boundary at $\pm k_F$, causing occupied states just below the Fermi wave vector to have their energies lowered (and empty states just above to have their energy raised). The empty states do not contribute to the total

---

* By "dimerized" I mean a system where two identical molecular units are in some way bonded to each other more strongly than to other units in the solid. Other systems occur as trimers (three molecules) or tetramers (four); the Rice effect can occur in all these materials.

energy but the filled states do; consequently the overall system energy is lowered. A gap occurs at the Fermi surface, rendering the material an insulator.

If the molecules making up the solid have internal degrees of freedom or if there is a basis of atoms for the un-dimerized solid, then the infrared spectrum can display unusual phonon features that I like to call the "Rice effect" [307, 319–322].

Other names used for this effect are phase phonons, charged phonons, electron-molecular vibration (EMV) interactions, and infrared active vibrations (IRAV). The Rice effect is observed in organic linear-chain conductors [306, 323–326], polyacetylene [126, 327–330], doped C_{60} [331, 332], and possibly cuprate superconductors [333, 334].

I'll discuss this in the context of the organic linear-chain conductors, where linear coupling of charge carriers to the totally symmetric (A_g) phonons of the molecules leads to structure in the conductivity spectrum at the phonon frequencies. This mechanism occurs because the electron energies depend on the bond lengths while at the same time the bond lengths depend on the local charge density. Infrared radiation at the A_g phonon frequencies can pump charge over long distances, giving rise to absorption that can approach electronic oscillator strength and that is polarized in the direction of easy electronic motion. This direction is in the quasi-one-dimensional stacking axis in the organics, which nearly is perpendicular to the planes of the molecules. (The A_g modes are normally infrared inactive and their normal modes are in the plane of the molecule. Thus, the Rice effect causes infrared features from the "wrong" modes, with the "wrong" polarization, and with the "wrong" oscillator strength.) The Rice effect is responsible for the narrow peaks in the spectrum of Fig. 13.4. When the electronic continuum overlies the phonon frequencies, characteristic Fano lineshape antiresonances occur in the continuum spectrum.

As an example, Fig. 13.8 shows the optical conductivity [326] of the quasi-one-dimensional organic semiconductor N-dimethyl thiomorpholinium(TCNQ)$_2$. This spectrum was measured with the electric field vector polarized along the c crystallographic axis, which is the stacking axis or the chain axis of the TCNQ molecule and also the most highly conducting direction. The spectrum has an electronic excitation at $\omega_e \approx 3,200$ cm^{-1} and a second (not shown at $\approx 10,500$ cm^{-1}. The mid-infrared band contains a majority of

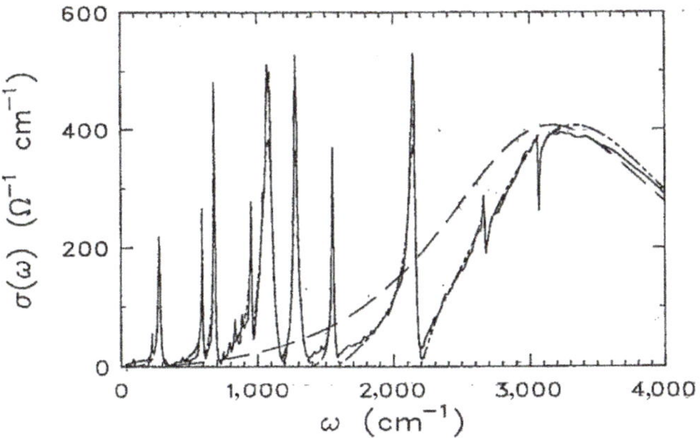

Optical conductivity of semiconducting N-dimethyl thiomorpholinium(TCNQ)$_2$.

the oscillator strength. These charge-transfer absorptions are attributed to the transfer of charge between TCNQ molecules in the stack [335–337] and can be understood in terms of strong Coulomb interactions using an extended Hubbard model with both a large on-site interaction U and a only somewhat weaker next-nearest-neighbor interaction V.

At low frequency, the spectrum displays many sharp vibrational features. Ten of these modes are A_g vibrations of the TCNQ molecule, activated by coupling to the charge transfer excitations [319,320,326,335]. These phonon features in the conductivity generally take on the appearance of a derivative of the ordinary Lorentz oscillator phonon absorption, known as a Fano lineshape.

The Rice effect leads to the following dielectric function:

$$\epsilon(\omega) = \epsilon_c + \cdots + \frac{4\pi \chi_e(\omega)}{1 - D_0(\omega)\chi_e(\omega)/\chi_0}, \tag{13.36}$$

where ϵ_c is the high frequency dielectric constant, $\cdots$ represents other contributions (interband transitions, ordinary optical phonons, and non-interacting free carriers, for example), and the last term is the effect of the Rice effect phonons. In it, χ_e is the electronic susceptibility of the carriers in the absence of coupling, χ_0 is the zero-frequency susceptibility, and D_0 is the phonon propagator:

$$D_0(\omega) = \sum_\alpha \frac{\lambda_\alpha \omega_\alpha^2}{\omega_\alpha^2 - \omega^2 - i\omega\gamma_\alpha}, \tag{13.37}$$

with λ_α the dimensionless electron–phonon coupling constant, ω_α the unperturbed frequency, and γ_α the width of the αth phonon mode. The total dimensionless electron–phonon coupling constant is $\lambda = \sum_\alpha \lambda_\alpha$ and the usual electron–phonon coupling function $\alpha^2 F(\omega)$ is related to D_0 by

$$\alpha^2 F(\omega) = \pi^{-1} \operatorname{Im} D_0(\omega).$$

Ohm's law is *local:* the current **j** at location **r** depends on the electric field at location **r** and not on the field anywhere else. In pure metals at low temperatures the response to the field becomes nonlocal. The current at location **r** depends on fields at other locations **r'**. This nonlocal response is not unusual. Anyone who has played shuffleboard (or hockey) knows that the puck will slide considerable distance away from the location where the force was applied to it.

The principal optical consequence of nonlocality is the anomalous skin effect. I'll first review the normal skin effect, already introduced in Chapter 4 (Section 4.5.2). Then I'll discuss the conditions necessary for the appearance of nonlocal effects and make a simple argument of how to calculate the skin depth in the nonlocal regime.

14.1 The Normal Skin Effect

Recall Eqs. 4.38 and 4.39, which specify the penetration depth of a metal. The amplitude of the field decays with distance along the path of the light as $e^{-x/\delta}$, with $\kappa\omega/c = 1/\delta$. At low frequencies, $\omega\tau \ll 1$, I found

$$\delta = \frac{c}{\omega\kappa} \approx \sqrt{\frac{c^2}{2\pi\omega\sigma_{\mathrm{dc}}}} = \frac{c}{\omega_p}\sqrt{\frac{2}{\omega\tau}},$$

where $\sigma_{\mathrm{dc}} = \omega_p^2\tau/4\pi$. In the other limit, high frequencies, $\omega\tau \gg 1$, and the penetration depth becomes

$$\delta \approx \frac{c}{\omega_p}.$$

This length, c/ω_p, is as short as the penetration depth gets.

The skin effect is not only an optical effect. It also affects electrical currents in wires carrying high-frequency* signals. Figure 14.1 shows the field or current distribution flowing in a wire with circular cross section. The current density is high near the surface and falls exponentially to near zero in the center. Figure 14.2 shows the room temperature skin depth for several metals at frequencies from 1 Hz to 10 THz. Lines are shown for metals with a range of conductivities: good metals (Ag, Cu, Au), fair metals (Pd, Pt, Ta),

---

* Or any frequency. Even at 60 Hz, good conductors have a finite penetration of fields and currents. The skin depth in copper is about 1 cm at 60 Hz and 300 K.

The current in a conductor carrying an alternating current. The current flows only near the surface, dying out exponentially towards the center.

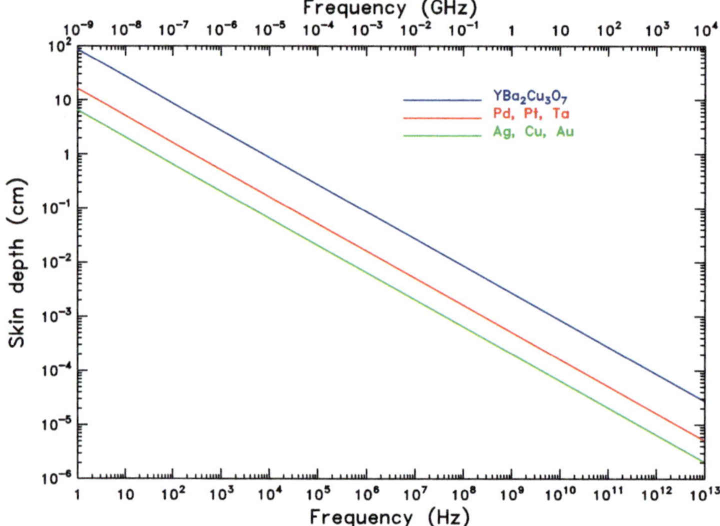

Room-temperature classical skin depth as a function of frequency for metals with a range of conductivities or resistivities, varying from 300 $\mu\Omega$-cm (YBa$_2$Cu$_3$O$_7$) to around 2 $\mu\Omega$-cm (Ag, Cu, Au) with Pd, Pt, and Ta ($\rho \approx 10\ \mu\Omega$-cm) in the middle. The frequency range is 1 Hz–10 THz. Note that 10 THz $==$ 333 cm^{-1}.

and a "bad metal" (YBa$_2$Cu$_3$O$_7$). The skin depth decreases (very approximately) from 10 cm to 100 nm over these frequencies. Note that 1 THz is 33 cm^{-1} or about 4 meV. On this many-decades plot, the differences among metals whose conductivity varies by less than $\pm50\%$ is not large.

The classical skin depth formula, Eq. 4.38, is valid at low frequencies ($\omega \ll 1/\tau$). As frequency increases, one must use the exact expression, $\delta = c/\omega\kappa$, which is easy enough if one knows the complex dielectric function. Figure 14.3 shows the skin depth for silver in the radio-frequency to ultraviolet range, 3 MHz to 1.2 PHz, 10^{-4} cm^{-1} to 40,000 cm^{-1}, or just over 10^{-8} eV to 5 eV.* This range includes $\omega\tau = 1$, so the skin depth initially

---

* The calculation does not make an approximation for δ. The dielectric function model consists of a Drude free-carrier part, plus Lorentz oscillators above 4 eV to account for the contribution of the d electrons. I then calculated ϵ, followed by $N = n + i\kappa = \sqrt{\epsilon}$ and $\delta = c/\omega\kappa$.

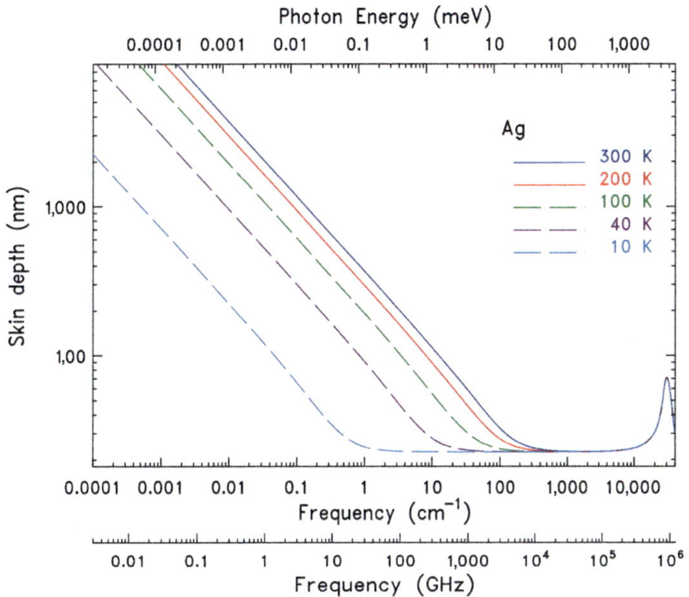

Fig. 14.3 Classical skin depth for silver from RF to uv frequencies at 5 temperatures. Three frequency/energy scales are given. The curves shown with dashed lines are not realistic; silver would be in the anomalous skin effect regime and not the classical regime over much of the frequency range.

falls as $\sqrt{1/\omega}$ before flattening out in the high frequency limit, at $\delta \approx c/\omega_p$ according to Eq. 4.39. There is a significant increase in the skin depth near the plasma minimum. This rise is due to the sign change of ϵ at this frequency and a corresponding decrease in κ. Then, for silver, the interband transition from the d levels turn on, and the classical penetration depth decreases again.

At room temperature, the high frequency skin depth is comparable to the 40 nm mean free path. It is much larger that the mean free path at frequencies below $1/\tau$. However, as temperature is reduced, the conductivity increases (because the mean free path is getting longer and longer), making the skin depth (Eq. 4.38) shorter and shorter. I have included in Fig. 14.3 the classical skin depth for silver assuming the conductivity varies as in Fig. 8.6. It does not take long in the case of silver and other good conductors to get into a situation where the mean free path is much larger than the skin depth.

The mean free path is the length scale (just as the relaxation time is the time scale) for relaxation of the current. But if the mean free path is larger than the skin depth, then those electrons moving at angles to the surface see an electric field which varies enormously over the trajectory. Electrons moving parallel to the surface see a constant field. This situation is illustrated in Fig. 14.4. The current is a complicated and nonlocal response to the field. The dashed curves for temperatures in the 10–100 K range in Fig. 14.3 do not apply to this physical situation.

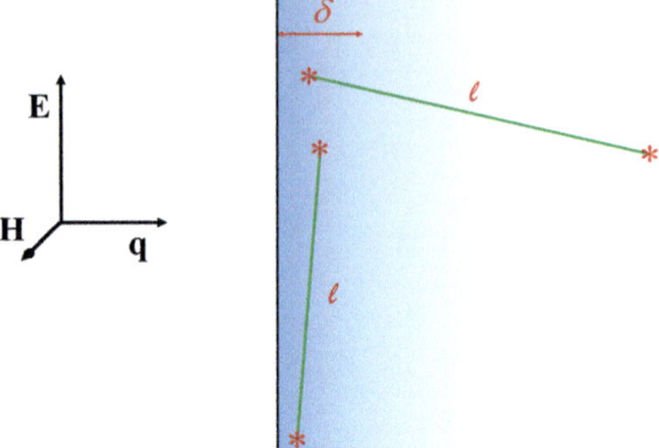

Fig. 14.4 The electric field of light which is incident on the surface of a clean metal accelerates the electrons that are within the skin depth. Two electron trajectories are shown. One has the electron staying within the skin and seeing the external field over its entire mean free path. The second shows the electron going into the depths of the metal. Each mean free path starts and ends with a momentum-changing collision.

14.2 The Anomalous Skin Effect

This anomalous skin effect was studied in the 1950s by Pippard [338], Chambers [339], and many others. As I will discuss, the physical situation is:

1. $\delta \neq \delta_{\mathrm{cl}}$.
2. $\delta_{\mathrm{ASE}} > \delta_{\mathrm{cl}}$.
3. The current is nonlocal, so Ohm's law becomes a convolution of the form

$$\mathbf{j}(\mathbf{r}) = \int_V \Sigma(\mathbf{r} - \mathbf{r}')\mathbf{E}(\mathbf{r}')d^3\mathbf{r}'$$

where $\Sigma(\mathbf{r} - \mathbf{r}')$ is the Fourier transform of $\sigma(q,\omega)$ with $\hbar q$ a momentum.
4. The fields are q dependent with $\langle q \rangle \approx 1/\delta_{\mathrm{ASE}}$.

In the anomalous skin effect regime, the effect of an external field on a specific electron state $\mathbf{k}$ depends on how $\mathbf{k}$ is oriented relative to the surface. Two distinct electron trajectories are shown in Fig. 14.4. Each electron starts off immediately after a collision and travels for the mean free path before another collision occurs. The electrons are accelerated by any electric field as they move along their trajectories. The electron whose path is entirely within the skin feels this force over its entire trajectory. (The field does work during the entire time that the electron is flying free between collisions; the work gives energy to the electron and causes it to contribute to the electric current. At the collision, the electron motion is randomized and no longer contributes to the current.) In contrast, the electron that departs deep into the metal only sees the field for part of its total free path. The field therefore does less work on this second electron.

14.3 The Extreme Anomalous Limit

The concepts in the previous paragraph form the basis of Pippard's effectiveness concept for the skin effect. The electric field amplitude decays from the surface with a length δ. The decay occurs because energy is absorbed by the electrons. To be most effective, the entire trajectory must be where the field is finite, i.e., within the skin. The path length is the mean free path ℓ. When $\ell \gg \delta$, this requirement restricts the angle of the $\mathbf{k}$ to a small value relative to the surface. The "effective" number of electrons is smaller than the total number:

$$n_{\text{eff}} = \frac{\delta}{\ell} n. \tag{14.1}$$

The metal then acts as if the current is only carried by the effective electrons*

$$\sigma_{\text{eff}} = \frac{n_{\text{eff}} e^2 \tau}{m} = \frac{n_{\text{eff}} e^2 \ell}{m v_{\text{F}}}$$

where I have used $\tau = \ell / v_{\text{F}}$. Now replace n_{eff} using Eq. 14.1:

$$\sigma_{\text{eff}} = \frac{n e^2 \delta}{m v_{\text{F}}}. \tag{14.2}$$

I now calculate the skin depth self consistently:

$$\delta_{\text{ASE}} = \frac{c}{\sqrt{2\pi \sigma_{\text{eff}} \omega}} = \frac{c}{\sqrt{2\pi \left(\frac{n e^2 \delta_{\text{ASE}}}{m v_{\text{F}}} \right) \omega}}$$

where I have substituted using Eq. 14.2. Square both sides, multiply by δ_{ASE}, and take cube root to find

$$\delta_{\text{ASE}} = \left(\frac{c^2 m v_{\text{F}}}{2\pi n e^2 \omega} \right)^{1/3}. \tag{14.3}$$

The skin depth in the extreme anomalous regime goes as $\omega^{-1/3}$ compared to $\omega^{-1/2}$ in the normal skin regime. If I use the definition of the plasma frequency, Eq. 8.23, I can recast this for comparison with Eqs. 4.38 and 4.39:

$$\delta_{\text{ASE}} = \frac{c}{\omega_p} \left(\frac{2 v_{\text{F}}}{c} \right)^{1/3} \left(\frac{\omega_p}{\omega} \right)^{1/3}. \tag{14.4}$$

The plasma frequency, ω_p, appears *twice* in Eq. 14.4; I put the c/ω_p factor in front to allow comparison to the normal skin effect formulae. Two cube roots follow this scale factor. The first cube root is smaller than unity[†] but the second is *much* larger than unity. The plasma frequency is 70,000 cm^{-1} or 9 eV whereas the frequency is smaller than $1/\tau = 0.1$ cm^{-1} or 10 μeV in pure metals[‡] at low temperatures.

---

* I write the conductivity as a DC conductivity. Hence, this picture is a low frequency, $\omega \ll 1/\tau$, picture. The high frequency penetration length c/ω_p is a kinetic inductance or inertial effect and does not depend on the mean free path.

[†] Recall: $v_{\text{F}} = c/200$.

[‡] The anomalous skin effect *only* occurs in pure metals at low temperatures.

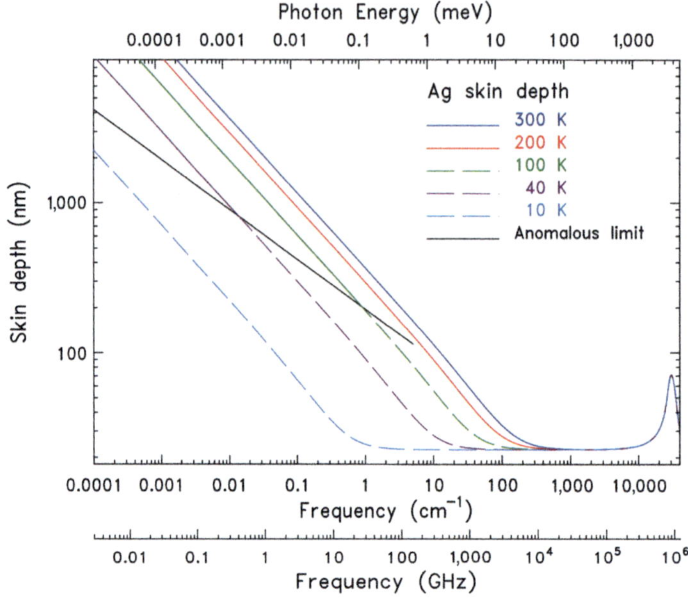

Fig. 14.5 Skin depth for silver from RF to uv frequencies at 5 temperatures. Three frequency/energy scales are given. The extreme anomalous limit is shown as a solid line. The skin depth calculated from the classical equation is shown with dashed lines where it is incorrect. Below the frequency where the classical and anomalous skin effect curves cross, silver is in the classical regime.

None of the terms in Eq. 14.3 have any significant temperature dependence; consequently, neither does δ_{ASE}. The temperature-dependent family of curves calculated for the normal skin effect will all merge into a single, frequency-dependent curve at low temperatures. This behavior is illustrated in Fig. 14.5, where I show the classical skin depth for silver in the optical range, 3 MHz to 1.2 PHz, 10^{-4} cm^{-1} to 40,000 cm^{-1}, or just over 10^{-8} eV to 5 eV. The anomalous limit (which is temperature independent) is shown as a black solid line. With increasing frequency the skin depth passes from classical to anomalous and then back to a high-frequency relaxation regime as $\omega \sim 1/\tau$. At the lowest temperatures, the crossover is at 70 kHz.

14.4 ω–τ Plot

The anomalous regime has both a high-frequency and a low-frequency boundary. The low-frequency boundary occurs where the anomalous skin depth equals the normal skin depth and also equals the mean free path: $\delta_{\mathrm{ASE}} = \delta_{\mathrm{cl}} = \ell$. If $\delta = \ell$, then all electrons are effective.

I can calculate the frequency where the transition occurs by setting $\ell = \delta_{\mathrm{cl}}$ to find

$$\ell = v_{\mathrm{F}}\tau = \left(\frac{c^2}{\omega_p^2} \frac{2}{\omega\tau} \right)^{1/2}$$

or

$$\omega = \frac{2c^2}{\tau^3 v_{\mathrm{F}}^2 \omega_p^2} = \frac{2c^2 v_{\mathrm{F}}}{\ell^3 \omega_p^2}. \tag{14.5}$$

This is also the frequency where $\delta_{\mathrm{cl}} = \delta_{\mathrm{ASE}}$.

Equation 14.5 represents a boundary between the normal skin effect (low ω or short τ) and the anomalous skin effect (higher ω or long τ). Both of these boundaries are computed in the low-frequency limit, $\omega \ll 1/\tau$. Once the frequency increases enough that $\omega \gg 1/\tau$, both normal and anomalous skin effects are modified. In this high frequency region, the field oscillates many times, accelerating and decelerating the electron each cycle until it has a collision. This relaxation regime in the local limit is where $\omega \gg 1/\tau$, $\sigma_1 \ll \sigma_2$, and $\delta_{\mathrm{cl}} = c/\omega_p$, (Eq. 4.39). As the mean free path becomes longer than the skin depth, many electrons will enter the skin, be accelerated and decelerated each cycle of the field, collide once with an impurity or the surface, and exit the skin. The dissipation in this "anomalous relaxation" regime is less than when the mean free path is short, but the differences in the values of the skin depth in the local relaxation and anomalous relaxation regimes are small.

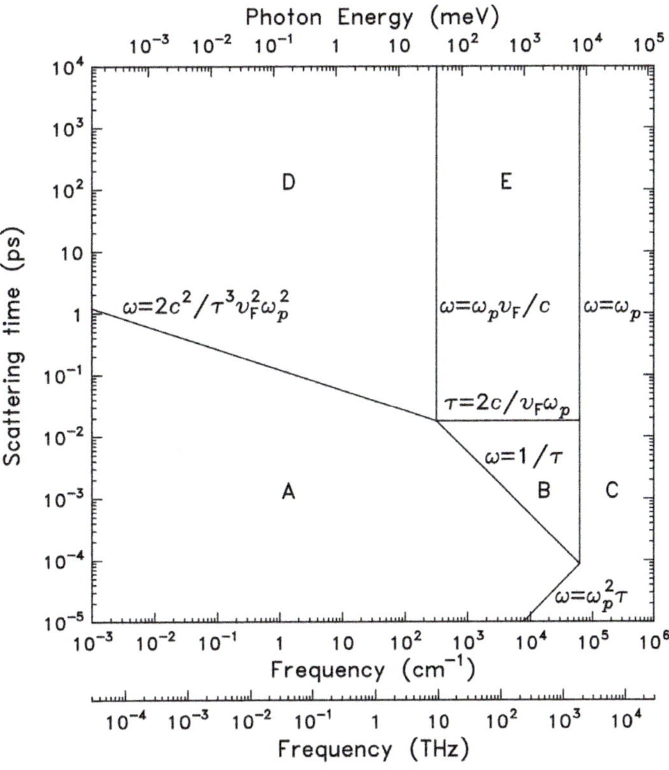

Fig. 14.6 Roadmap [340] of the skin effects in the ω–τ plane. The materials parameters used are those of silver or copper. The roadmap frequency scales are THz and cm^{-1} (bottom) and meV (top); the interesting range is 1–1,000 cm^{-1}, .03–30 THz, or 0.1–100 meV. Note that the angular frequency ω is found by multiplying the frequency in Hz by 2π or the frequency in cm^{-1} by $2\pi c$.

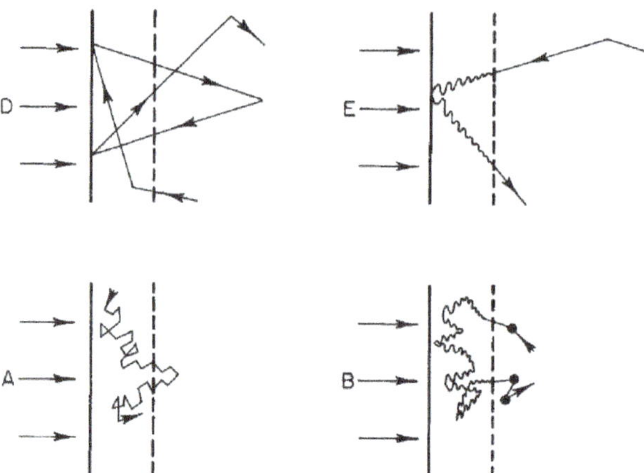

Fig. 14.7   Sketch [340] of electron trajectories in region A (normal skin effect), region B (relaxation regime), region D (anomalous skin effect), and region E (anomalous relaxation).

Figure 14.6 shows a roadmap [340] of the skin effects in the ω–τ plane. The horizontal axis is the frequency ω while the vertical axis is the scattering time τ. The materials parameters used are those of silver or copper. For low frequencies and short mean free paths (high temperatures), one is in the normal skin regime (Region A). If the mean free path (and relaxation time) get longer and longer, one eventually crosses the boundary given by Eq. 14.5 and is in the anomalous skin regime (Region D). If instead, I keep τ short but increase the frequency, I enter the relaxation regime, (Region B) unless τ is really short, in which case the metal becomes partially transparent (Region C), because ϵ_1 is positive.* At high frequencies and in clean metals the relaxation regime and the anomalous skin effect regimes cross the boundary into the anomalous relaxation regime (Region E). Finally, once $\omega > \omega_p/\epsilon_c$, the metal is partially transparent. Figure 14.7 shows [340] electron trajectories in four of the five regimes. (In Region C the skin depth is very large, so that the trajectories are as in A.)

14.5 The Surface Impedance

I want to find a material-related function to describe the optical properties of a nonlocal system. Of course, it should be also defined and reasonable for the local limit where Ohm's law is valid. Moreover, it should be related to observables such as the absorption or reflectance. The skin depth is such a quantity but it specifies the behavior of the fields inside the metal. I would like something that can be related to measured quantities outside

---

* Write $\epsilon_1 = \epsilon_c - \omega_p^2/(\omega^2 + 1/\tau^2) = 0$. If $1/\tau$ is a large frequency, the zero of ϵ_1 is well below the usual $\omega = \omega_p/\epsilon_c$. In fact, it may not exist at all; for a really dirty metal, $\epsilon_1 = \epsilon_c - \omega_p^2\tau^2$ can be everywhere positive.

the metal. The surface impedance $Z = R + iX$ is one such measure. As I will discuss, its real part, the surface resistance R, is related to loss in the metal and the surface reactance, X, to dispersion.

I'll start by reminding myself of the resistance and reactance of an electrical circuit. At zero frequency (DC currents), the resistance is given by freshman physics (Ohm's law No. 2): $R = V/I$ where V is the voltage drop of the circuit and I is the current passing through it.* At finite frequencies, I must replace the real R with a complex Z. Now, if the current is organized into a surface sheet, with surface current K, then the same law is [2] $Z = E/K$. The skin is such a sheet, except that the current is not in an infinitesimally thick layer but in a layer a few times the skin depth thick. Then[†]

$$K = \int_0^\infty j(x)dx$$

and

$$Z = \frac{E(0)}{\int_0^\infty j(x)dx}, \tag{14.6}$$

where $E(0)$ is the field at the surface.

Wherever the normal skin effect is in effect, I can use Ohm's law $j = \sigma E$. I also know that the field goes as $E = E(0)e^{-x/\delta}$. Then

$$Z = \frac{E(0)}{\sigma E(0) \int_0^\infty e^{-x/\delta}dx}. \tag{14.7}$$

The integral is easy to do; it integrates to $-(\delta)e^{-x/\delta}$. This contributes zero at the upper limit and δ at the lower limit. Hence

$$Z = \frac{1}{\sigma\delta}. \tag{14.8}$$

At low frequencies, $\sigma = \sigma_1 = \sigma_{dc}$ and the surface impedance is real: $R = 1/\sigma_{dc}\delta$ is the resistance of a slab of metal with thickness δ and equal length and width. This is often called the square resistance or the sheet resistance $R_\square$. At higher frequencies, both σ and δ are complex, and thus so is Z.

Using Eq. 4.38 I find (low frequencies!)

$$Z = \sqrt{\frac{2\pi\omega}{\sigma_{dc}c^2}} \quad \text{when} \quad \omega \ll 1/\tau. \tag{14.9}$$

The surface impedance grows as $\omega^{1/2}$ and decreases if the conductivity increases.[‡]

---

* #1: $V = IR$; #3: $I = V/R$. ☺
[†] Fields and currents are vectors, but if I take the ratio of two of these, I must take the magnitudes. For isotropic and homogeneous materials, the current and electric field are in the same direction so I just take ratios of the amplitudes.
[‡] But only as the square root of the conductivity. At DC the resistance decreases linearly in the conductivity.

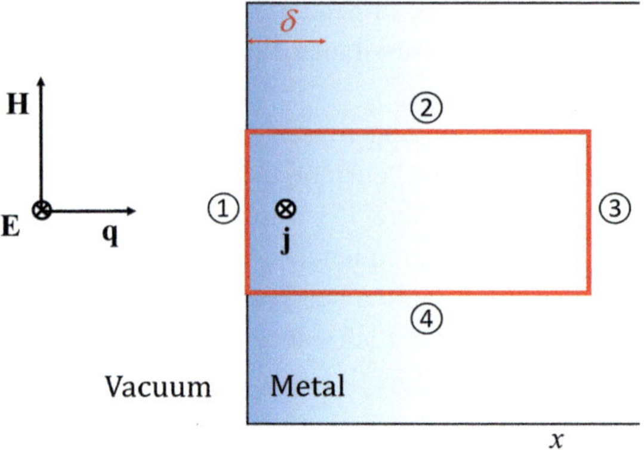

Fig. 14.8   Geometry for calculating the surface impedance. The four numbered line segments will be where I convert a surface integral to a line integral. Segment ③ is deep enough that the field is zero at its location. The $\hat{y}$ axis, **E**, and **j** are into the plane of the figure.

14.5.1 The General Case (At Low Frequencies)

At frequencies low enough that I can neglect the inductive response (because $\sigma_1 \gg \sigma_2$) and the displacement current, Maxwell's fourth equation (Eq. 2.4d), is

$$\nabla \times \mathbf{H} = \frac{4\pi}{c}\mathbf{j}. \tag{14.10}$$

Now look at the geometry of Fig. 14.8. The electromagnetic wave of frequency ω is incident from the left, with **q** normal to the surface, **H** in the plane of the paper, and **E** into the paper. As usual, I will take $\hat{q}$, $\hat{e}$, and $\hat{h}$ respectively along the x, y, and z Cartesian axes. After the wave crosses the surface, it will decay with a skin depth δ (equation for and magnitude of to be specified). The current **j** (which also decays with decay length δ) is into the plane of the paper, parallel to **E**. A rectangular surface is shown outlined in red. It is bounded by 4 line segments: ①, ②, ③, and ④. ③ is very many skin depths below the surface, where fields are zero.

So I will integrate Eq. 14.10 over the area of this surface using $d\mathbf{S} = \hat{y}\,dz\,dx$

$$\int_S d\mathbf{S} \cdot \nabla \times \mathbf{H} = \frac{4\pi}{c}\int_S d\mathbf{S} \cdot \mathbf{j}.$$
$$= \frac{4\pi}{c}\int dz \int dx\,\hat{y} \cdot \mathbf{j}.$$

Using the definition of Z, Eq. 14.7, I can evaluate easily the right-hand side:

$$\int_S d\mathbf{S} \cdot \nabla \times \mathbf{H} = \frac{4\pi}{c}\int dz\,\frac{E(0)}{Z}$$
$$= \frac{4\pi}{c}\frac{E(0)}{Z}L_①,$$

where $L_①$ is the length of line segment $①$. But also I can transform the surface integral on the left side to a line integral around the perimeter of the surface $\mathbf{S}$ using Stokes theorem! It becomes

$$\int_S d\mathbf{S} \cdot \nabla \times \mathbf{H} = \oint \mathbf{H} \cdot d\mathbf{l}$$

$$= \int_① dz\, H(0) + \int_② dx\, \mathbf{H} \cdot \hat{\mathbf{x}}$$

$$- \int_③ dz\, \mathbf{H}(x \to \infty) - \int_④ dx\, \mathbf{H} \cdot \hat{\mathbf{x}}$$

$$= H(0) L_①.$$

Here, the differential vector $d\mathbf{l}$ runs clockwise around the perimeter, following $①$ … $④$. The dot product of $\mathbf{H}$ and $d\mathbf{l}$ makes the integrals $②$ and $④$ zero; the magnetic field $\mathbf{H}$ is zero along $③$. Hence, the surface impedance is

$$Z = \frac{4\pi}{c} \frac{E(0)}{H(0)}, \tag{14.11}$$

the ratio of the electric field at the surface to the magnetic field at the surface.

14.5.2 Impedance of Free Space

Suppose I replace the metal with vacuum. The fields are still defined in any y–z plane. In vacuum $E = H$ so the impedance of free space is*

$$Z_0 = \frac{4\pi}{c} = 377 \ \Omega. \tag{14.12}$$

14.5.3 Fields at the Surface

I'll start by looking back at the discussion of normal-incidence reflection in Chapter 3, Section 3.6. I know that the electric field is reversed on reflection (because the metal is the more dense medium). Thus, with the incident field along $\hat{\mathbf{y}}$, the reflected field is along $-\hat{\mathbf{y}}$. The reflectance is nearly 100% in good metals, so the resultant field is $1 - (-r)$ with the amplitude reflectivity r given in Eq. 3.32.[†] Because the $\mathbf{E}$ fields are antiparallel, the $\mathbf{H}$ fields are parallel, adding to $1 + (-r)$, larger than then incident $\mathbf{H}$ field by almost a factor of two. Thus the surface impedance is much smaller than 377 Ω. Indeed, from above I get

$$Z = \frac{4\pi}{c} \frac{1+r}{1-r}$$

with $r = (1 - N)/(1 + N)$. ($r < 0$.) Substituting, I find

$$Z = \frac{4\pi}{c} \frac{1}{N}. \tag{14.13}$$

---

* See Appendix A.

[†] r is negative (the field is reversed) so I write it as $1 - (-r)$ to indicate a difference between unity and a number just above -1, e.g., -0.99. The difference is 0.01.

But I knew this! The magnitude of the magnetic field inside the metal (or any local, nonmagnetic, linear, homogeneous, and isotropic material) is $H = NE$. The tangential components are continuous at the interface, so the fields at the surface are the same as the fields just inside the metal.

14.5.4 Back to the Dielectric Function

Because $N = \sqrt{\epsilon}$, I can write $Z = (4\pi/c)\sqrt{1/\epsilon}$. Moreover, I can write $\epsilon = \epsilon_c + 4\pi i\sigma/\omega$ and at low frequencies (compared to ω_p) neglect ϵ_c and obtain

$$Z = \frac{4\pi}{c}\sqrt{\frac{\omega}{4\pi i\sigma}}. \tag{14.14}$$

Now $\sqrt{1/i} = (1-i)/\sqrt{2}$, so I can also write

$$Z = R + iX = (1-i)\sqrt{\frac{2\pi\omega}{c^2\sigma}}. \tag{14.15}$$

At low frequencies $\sigma \approx \sigma_{\mathrm{dc}}$ and $\delta = \delta_{\mathrm{cl}}$ and*

$$R = -X = \frac{1}{\sigma_{\mathrm{dc}}\delta_{\mathrm{cl}}}.$$

14.5.5 The Surface Impedance in the Anomalous Regime

The same approach works for the anomalous skin effect:

$$R = \frac{1}{\sigma_{\mathrm{eff}}\delta_{\mathrm{ASE}}},$$

with $\sigma_{\mathrm{eff}} = ne^2\delta_{\mathrm{ASE}}/mv_{\mathrm{F}}$ and $\delta_{\mathrm{ASE}} = (c^2 mv_{\mathrm{F}}/2\pi ne^2\omega)^{1/3}$. Note that δ_{ASE} comes in twice, once directly and once replacing the mean free path in σ_{eff}. Then

$$R = \frac{mv_{\mathrm{F}}}{ne^2}\left(\frac{2\pi ne^2\omega}{c^2 mv_{\mathrm{F}}}\right)^{2/3},$$

or, after some algebra

$$R = \left(\frac{(2\pi)^2 mv_{\mathrm{F}}\omega^2}{ne^2 c^4}\right)^{1/3}. \tag{14.16}$$

So the anomalous-limit surface impedance goes as $\omega^{2/3}$. It can be rewritten as:

$$R = Z_0\left(\frac{v_{\mathrm{F}}}{4c}\right)^{1/3}\left(\frac{\omega}{\omega_p}\right)^{2/3}, \tag{14.17}$$

with $Z_0 = 4\pi/c = 377\ \Omega$, according to whether you want cgs or SI units.

---

* Consider a thin film with thickness $d \ll \delta$. Then the currents are confined to a distance d not a distance δ, and

$$Z = \frac{1}{\sigma d} = \frac{\rho}{d} \equiv Z_\square.$$

14.5.6 Reflectance in Terms of the Surface Impedance

According to Eq. 3.33 the reflectance is

$$\mathscr{R} = \left| \frac{1 - N}{1 + N} \right|^2 .$$

Now $N = Z_0/Z$ so

$$\mathscr{R} = \left| \frac{Z - Z_0}{Z + Z_0} \right|^2 . \tag{14.18}$$

Eq. 14.18 is very familiar to the discussion of energy flow in waveguides and coaxial transmission lines. It says that a discontinuity in the line impedance causes a reflection in the line. The most extreme discontinuity is a short circuit ($Z = 0$) or an open circuit $Z = \infty$); both give 100% reflection.

14.5.7 Numerical Values

I use as an example the classical surface impedance (Eq. 14.15) with $\sigma_{dc} = 6 \times 10^5 \ \Omega^{-1}\mathrm{cm}^{-1}$ at 300 K. If the frequency is $\omega = 10^{10} \ \mathrm{s}^{-1}$ (0.05 cm^{-1}), then $R = 4$ m Ω. Compare this very low value with the impedance of free space, $Z_0 = 377 \ \Omega$. The metal looks like a short circuit. Just as a shorted transmission line reflects the signals sent down it, the metallic surface is an excellent reflector.* Indeed, I can rewrite Eq. 14.18 as $\mathscr{R} = 1 - 4Z/Z_0 \sim 1 - 10^{-5}$.

---

* I find this transmission line analogy very useful in understanding why a high absorber is also a high reflector. The short circuit ought to draw very large currents and dissipate lots of power. It does not. Instead, it returns the electromagnetic energy back toward the source.

15

Anisotropic Crystals

15.1 Optics of Crystals

The study of anisotropic materials, the subject of this chapter, is often called "crystal optics," because many crystalline solids exhibit anisotropic properties. In fact, only one of the seven systems known to crystallography, the cubic crystal system, has isotropic optical properties. I'll define an anisotropic material as one having properties that depend on direction. "Direction of what?" is a fair question when faced with the definition I just gave. The answer: it is essentially always the direction of the electric field which controls wave propagation, charge transport, and optical absorption.

Despite the large numbers of anisotropic crystals, anisotropic materials are not widely experienced in everyday life. Many persons have seen or used "polaroid," plastic polarizing films [341]. These polarizing films are used as sunglasses and in photography; these devices are *not* single crystals but they are anisotropic. Polaroid sheet contains linear chains of absorbing molecules in a transparent polymer host. It absorbs light with linear polarization parallel to the chains and has maximum transmission for light with polarization rotated by $90°$.

Research life differs from everyday life. Many of the solids that have attracted the most attention in condensed-matter physics over the past several decades are anisotropic. The list includes cuprate superconductors, Fe-based superconductors, certain topological insulators, organic linear-chain conductors, electroactive polymers, graphene, graphite, carbon nanotubes, and others.

15.2 Polarized Light

Looking around in textbooks or on the Internet, I sometimes encounter statements like the following: "A polarizer is an optical device that passes light of a specific polarization and blocks waves of other polarizations." This statement is (at best) incomplete: the polarizer will block (by absorption, by refraction out of the beam, or by reflection) light with its electric field directed along one specific axis.* It will transmit more or less of the light in all other polarizations, with maximum transmission for light polarized perpendicular to the

---

* In real polarizers the blockage is rarely 100%; however, the leakage can in the best cases be only a few parts per million.

direction of blockage. Because the superposition principle applies to these fields, the light incident on the device can be treated as a superposition of two orthogonal field directions: one is parallel to the blocking direction and the other is perpendicular to it. The transmitted beam is this second component.

15.2.1 Linear Polarization

It should come as no surprise that I now assert that the electromagnetic waves I introduced in Chapter 2 are linearly polarized waves. The fields point in specific directions, unchanged as the waves propagate. I'll take the electric field direction as representing the plane of polarization. The plane-wave solutions are given by Eqs. 2.5 and 2.6; the direction of the electric field for waves in vacuum was defined right after Eq. 2.9 as $\hat{e}$ for the electric field and $\hat{h}$ for the magnetic field, with $\hat{e} \cdot \hat{h} = 0$. A cartoon of the fields may be seen in Fig. 2.1. The fields for waves in local, nonmagnetic, linear, isotropic, and homogeneous media are the solutions to Eqs. 3.3a–3.3d, written as a polarized plane wave in Eq. 3.11.

To be specific, let me orient the propagation vector $\mathbf{q}$ along the x axis and the electric field along the y axis. My linearly-polarized plane wave is then

$$\mathbf{E} = E_0 \hat{y} e^{i(qx-\omega t)}, \tag{15.1}$$

where E_0 is the (complex) amplitude of the field.

15.2.2 Unpolarized Light

It is easy to write a formula for linearly polarized light. It is not so easy to write one for *unpolarized* light. Suppose I try to add a second field, at right angles to the field in Eq. 15.1, thinking that I will then have the field along both y and z, which would seem to fill the bill of having the field along no single axis. The second field is

$$\mathbf{E}_2 = E_0 \hat{z} e^{i(qx-\omega t)}. \tag{15.2}$$

If I add the fields in Eqs. 15.1 and 15.2 I get the sum to be

$$\mathbf{E}_s = E_0(\hat{y} + \hat{z}) e^{i(qx-\omega t)} = \sqrt{2} E_0 \hat{e} e^{i(qx-\omega t)}, \tag{15.3}$$

with $\hat{e} = (\hat{y} + \hat{z})/\sqrt{2}$. The sum is still linearly polarized; the direction of polarization ($\hat{e}$) is at $45°$ to the y and z axes. The superposition of the two linearly polarized beams has produced another linearly polarized beam.

So what is "unpolarized light" and how do I write the fields for it? Let me start the answer by saying that the linearly polarized fields in Eqs. 15.1–15.3 are perfectly coherent. I can delay one or the other by seconds or years or even the age of the Universe and still find coherent superposition because the frequency and wavelength are unvarying.* Unpolarized light consists of the superposition of uncorrelated fields oriented in random directions. (The

---

* No actual source exists with coherence times of years, much less the age of the Universe. The initial LIGO laser interferometer had coherence times of 2×10^5 s, about two days [342]. More typical values for very stable lasers are a few milliseconds.

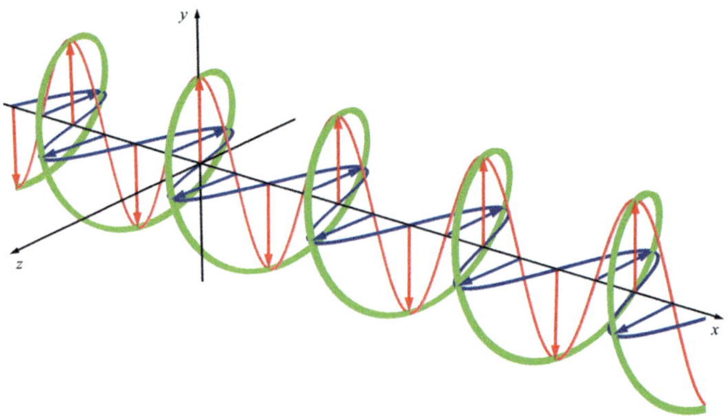

Fig. 15.1   Right circular polarized waves. The red cosine wave is the $\hat{y}$ component while the blue sine wave is the $\hat{z}$ component. The resultant is a green helix that rotates once per wavelength.

fields are perpendicular to the propagation direction, of course.) There is enough variation in frequency that there is no coherent superposition of the constituent fields.

The fields may also be partially polarized, in which case a polarizer finds power variations as it is rotated through $180°$ but not to zero at any orientation. I can represent partially polarized light as a superposition of unpolarized light and perfectly polarized light.

The method of writing unpolarized and partially polarized light uses Stokes vectors and Mueller matrices [2, 343, 344], 4×4 matrices acting on 4-element vectors. This approach does the job, but is more than needed here. From the point of view of understanding, measuring, and interpreting the optical properties of solids, I can employ the perfectly polarized light introduced above.

15.2.3 Circular Polarization

Photons have helicity; the two possible states are $\pm\hbar$ and correspond to circularly polarized light. Circular polarization is the sum of two linearly polarized fields shifted in phase by $90°$. See Fig. 15.1. I write

$$\mathbf{E}_\pm = E_0(\hat{y} \pm i\hat{z})e^{i(qx-\omega t)}, \tag{15.4}$$

where the minus sign is for right circular polarization and the plus is for left circular polarization.*

I've said that the observable is the real part of the complex field. It is worth writing the real part of Eq. 15.4:

$$\mathbf{E}_\pm = E_0[\hat{y}\cos(qx - \omega t) \mp \hat{z}\sin(qx - \omega t)].$$

---

*   There are two conventions in use. These are generally described by saying that you put your right or left thumb along the direction of the wave the fingers then point in the direction of rotation of the corresponding circular polarization (right or left). The problem is that sometimes the thumb points towards the source of the radiation and sometimes away. I have used the second convention.

The electric field vector of the circular polarized wave at fixed time follows a helical path along the direction of propagation. circular polarization

Plane polarized light can be written as a sum of equal amplitudes of right and left circularly polarized light at various relative phases. If the relative phase is zero, I add them as

$$1/2(\hat{\boldsymbol{y}} + i\hat{\boldsymbol{z}}) + 1/2(\hat{\boldsymbol{y}} - i\hat{\boldsymbol{z}}) = \hat{\boldsymbol{y}},$$

to get polarization along the x axis with unit amplitude. If the relative phase is π, $180°$, then I add them as

$$1/2(\hat{\boldsymbol{y}} + i\hat{\boldsymbol{z}}) + 1/2(\hat{\boldsymbol{y}} - i\hat{\boldsymbol{z}})e^{i\pi} = 1/2(\hat{\boldsymbol{y}} + i\hat{\boldsymbol{z}}) - 1/2(\hat{\boldsymbol{y}} - i\hat{\boldsymbol{z}}) = i\hat{\boldsymbol{z}}.$$

(To avoid shifting the phase of the plane polarized light, I would need to retard the phase of the left circular polarization by $90°$ and advance the right by the same amount.) In general, a phase difference of ϕ rotates the plane of polarization by $\theta = \phi/2$.

15.2.4 Elliptical Polarization

Looking up or down the beam, the E field of plane polarized light is a line along the polarization direction. The E field of circular polarized light is a circle. Elliptical polarized light is a linear combination of plane and circular polarizations. Looking along the beam, the E field of elliptical polarized light traces an ellipse. The ratio of major to minor axis of the ellipse is set by the field strengths of the plane and circular constituents.

15.3 Crystal Symmetry

Symmetry is a fundamental concept in physics. The basic idea is that any physical system which is unchanged – the correct word is "invariant" – under some operation will have observables that must respect the symmetry. Mathematically, the symmetry operations define a mathematical group; group theory [28, 29] is the mathematical theory of these groups and their manipulation. Most books on solid-state physics cover crystal symmetry. I particularly like the discussion by Burns [23] but Kittel [24] and Ashcroft and Mermin [19] are also good.

15.3.1 Translational Symmetry

Translational symmetry is the basic symmetry that defines a crystal. Each crystal structure can be constructed by starting at some point in space and moving by the translation vector (actually a set of vectors) that defines the crystal to generate all the points of the crystal lattice. For example, I can define the NaCl lattice by starting at the center of an Na atom and then moving by a translation vector to the center of a neighboring Na atom, then by a translation vector to the center of another Na atom, and so on. Now, if I move to the nearest Cl atom, the same set of translation vectors will trace out all the Cl atoms in the crystal.

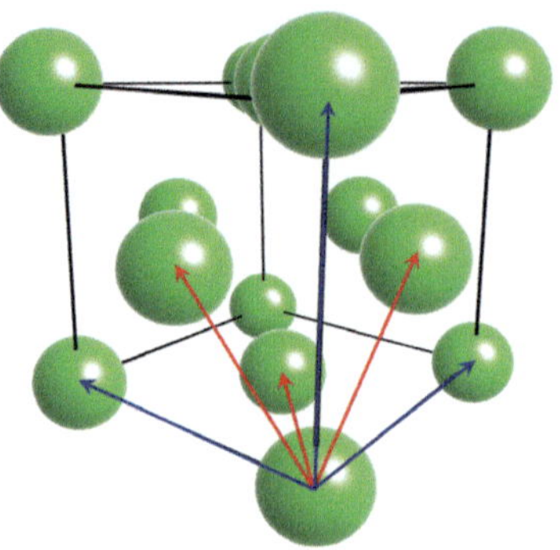

Fig. 15.2   A sketch of a face centered cubic lattice (for a monatomic crystal), showing the primitive translation vectors (red) and the conventional translation vectors (blue).

The translation vector is usually written as

$$\mathbf{T} = n\mathbf{a} + m\mathbf{b} + p\mathbf{c}$$

where n, m, and p are (positive and negative) integers and the vectors $\mathbf{a}$, $\mathbf{b}$, and $\mathbf{c}$ define the unit cell of the crystal. The set of points defined by $\mathbf{T}$ as the integers run over the range from minus to plus infinity is known as a Bravais lattice. The name is that of French crystallographer Auguste Bravais (1811–1863), who proved that in three-dimensional space only fourteen distinct lattices may be constructed. To each point of the Bravais lattice one attaches a basis of atoms. (The attachments is done in an identical fashion for each point in order to preserve translational symmetry.) The basis in the example above is one Na and one Cl atom.

Note that there are usually two ways to set up the vectors $\mathbf{a}$, $\mathbf{b}$, and $\mathbf{c}$. In one, they are the "primitive vectors," the smallest vectors possible; therefore, they reach every point in the Bravais lattice. The second way uses the conventional vectors. These define a unit cell which is larger than the primitive cell; however, the conventional cell generally clearly exhibits the symmetry of the lattice. The difference is perhaps best illustrated with an example. Figure 15.2 shows a face-centered cubic (fcc) lattice. The primitive vectors go from the origin at the front bottom corner of the cube across the three adjacent faces to the atoms – nearest neighbors all – in the centers of the faces. The conventional vectors go from the origin along the cube edges to the second-neighbor atoms at the adjacent corners.

If the primitive vectors are used, all the points of the Bravais lattice will be reached. If the conventional vectors are used, only a fraction are reached by movements by $\mathbf{T}$; the other points are assigned to the basis of the structure. There are four atoms in the basis in the conventional fcc cell. Although the need to define a basis even in monatomic crystals seems like an added complication, the conventional cell is much easier to visualize and to use to illustrate the crystal symmetry; consequently, I'll use the conventional cell.

The vectors $\mathbf{a}$, $\mathbf{b}$, and $\mathbf{c}$ define a rectangular prism with edges of length a, b, and c; there are angles α, β, and γ defined respectively in the following way. α is the angle between $\mathbf{b}$ and $\mathbf{c}$. β is the angle between $\mathbf{c}$ and $\mathbf{a}$. γ is the angle between $\mathbf{a}$ and $\mathbf{b}$.

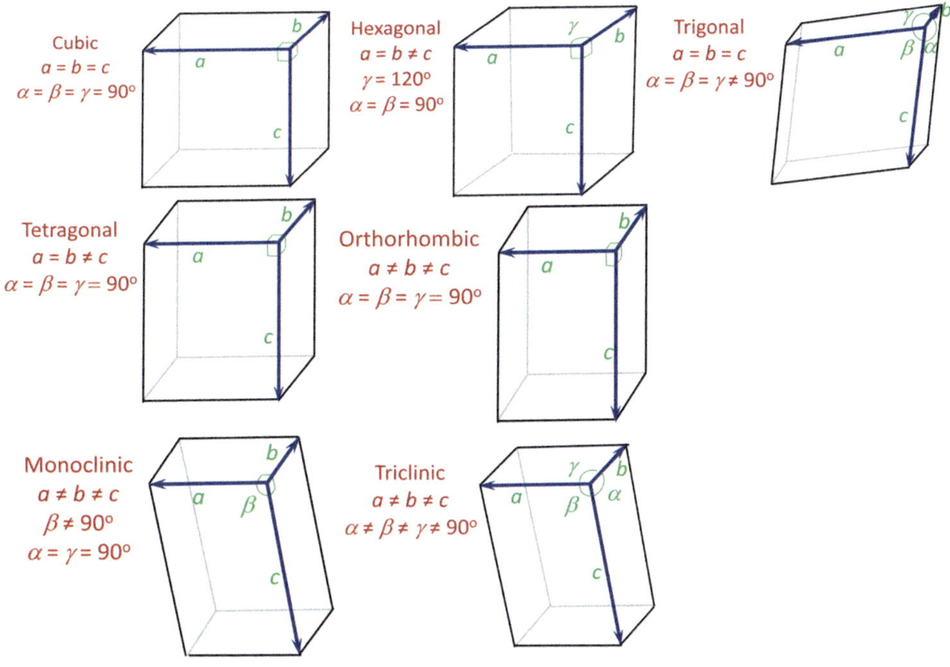

Fig. 15.3   The seven crystal systems.

15.3.2 Seven Crystal Systems

Every crystal belongs to one of seven crystal systems. Each system is defined in terms of the relative lengths of the unit vectors a, b, and c of the lattice unit cell, as well as the angles between these vectors. The seven crystal systems are cubic, tetragonal, orthorhombic, monoclinic, triclinic, hexagonal, and trigonal. They are illustrated in Fig. 15.3.

Cubic. I start with the system with the highest symmetry: cubic. The cubic crystal has all axes equal ($a = b = c$) and all angles $90°$. The rectangular prism of the cubic system is, not surprisingly, a cube.

Tetragonal. If I take the cubic crystal and expand (or compress) it along one axis, I get a tetragonal crystal. Here, $a = b \neq c$; all angles are $90°$.

Orthorhombic. If I take the tetragonal crystal and expand (or compress) it along a second axis, I get an orthorhombic crystal. Here, $a \neq b \neq c$; all angles are $90°$.

Monoclinic. If now I take the orthorhombic crystal and tip one axis away from the normal to the plane defined by the other two in a way that preserves two right angles, I get a monoclinic crystal. For the monoclinic crystal, $a \neq b \neq c$; the angle $\beta \neq 90°$ while the other two are $90°$.

Triclinic. To make triclinic, I take monoclinic and tip the other axes so that none are perpendicular to any other, making $\alpha \neq \beta \neq \gamma \neq 90°$. It remains true that $a \neq b \neq c$ also. I've now made the lattice as low in symmetry as I can.

Hexagonal. I return to the tetragonal system and open the angle γ to $120°$. The formerly square *ab* plane is now a hexagonal figure. The hexagonal crystal has $a = b \neq c$ and $\alpha = \beta = 90°$; $\gamma = 120°$.

Trigonal. Finally, I return to the cube and pull it along one body diagonal, rendering all angles different from $90°$ but all still equal. The axes also remain equal, $a = b = c$. Trigonal sometimes called rhombohedral.

The seven crystal systems contain the fourteen Bravais lattices.* The criterion for defining a Bravais lattice is that the surroundings of the lattice points be identical. The Bravais lattices are shown in many books on solid-state physics and crystallography [19, 23, 24]. All of the Bravais lattices have lattice points located at the corners of the prism and some have other lattice points located at symmetric locations, in the middle of a face or at the body center. For example, the cubic system has simple cubic (sc), body-centered cubic (bcc), and face-centered cubic (fcc) Bravais lattices. It is worth noting that the hexagonal Bravais lattice (the only member of that crystal system) requires that there be a lattice point at the center of the hexagon, so that the lattice points on the *ab* plane form a triangular mesh.

15.4 The Dielectric Tensor

Crystal symmetry has a dominant effect on the dielectric function. In a general crystal, the relation between **D** and **E** is a 3×3 tensor[†] as I wrote already in Eq. 2.11:

$$\mathbf{D} = \overset{\leftrightarrow}{\epsilon} \cdot \mathbf{E}$$

where (still in general)

$$\overset{\leftrightarrow}{\epsilon} = \begin{pmatrix} \epsilon_{xx} & \epsilon_{xy} & \epsilon_{xz} \\ \epsilon_{yx} & \epsilon_{yy} & \epsilon_{yz} \\ \epsilon_{zx} & \epsilon_{zy} & \epsilon_{zz} \end{pmatrix}. \tag{15.5}$$

The off-diagonal terms allow an electric field along $\hat{x}$ to produce a response (polarization or current) along $\hat{y}$ or $\hat{z}$ as well as $\hat{x}$.

15.4.1 The Principal Axes

This dielectric tensor is a symmetric tensor;[‡] hence, it is possible to find a Cartesian coordinate system in which it can be rendered diagonal. In this coordinate system, I have

---

* When I take into account reflection, rotation, inversion, and "improper rotation" symmetry operations, I will find that there are thirty-two different crystal classes. Furthermore, the combination of all possible symmetry operations results in a total of 230 different space groups. Fortunately, for considering optical anisotropy, I only need to think about the seven crystal systems. If I want to consider selection rules, such as which vibrational modes appear in an infrared spectrum, then I would need to dig deeper.

[†] It is a 3×3 tensor because there are three directions in space.

[‡] The symmetry may be proved in a number of ways: from thermodynamics, using Onsager's reciprocity relation, by energy conservation [23].

$$\overleftrightarrow{\epsilon} = \begin{pmatrix} \epsilon_x & 0 & 0 \\ 0 & \epsilon_y & 0 \\ 0 & 0 & \epsilon_z \end{pmatrix}. \tag{15.6}$$

The axes of the coordinate system in which $\overleftrightarrow{\epsilon}$ is diagonal are known as the *principal axes*. The diagonal components are the principal components. I will write diagonal component with a single subscript, as ϵ_x rather than ϵ_{xx}. An alert reader might ask at this point: Does the fact that the dielectric function is diagonal in its principal axes mean that I have removed the situation where polarization or current are not in the same direction as the electric field? The answer is "yes" in the special case where the field is along one of the principal axes; in this case the polarization and current are also along the principal axis. However, if the field is in an arbitrary direction, then the answer is definitely "no!" Polarization and current will be in directions different from the electric field.

As an example, let me take $\mathbf{E} = (E_0/\sqrt{2})(\hat{\mathbf{y}} + \hat{\mathbf{z}})$, a field of amplitude E_0 at $45°$ to the y and z principal axes. The displacement vector is the dot product of $\overleftrightarrow{\epsilon}$ and $\mathbf{E}$, i.e.,*

$$\mathbf{D} = \overleftrightarrow{\epsilon} \cdot \mathbf{E} = \frac{E_0}{\sqrt{2}}(\epsilon_y \hat{\mathbf{y}} + \epsilon_z \hat{\mathbf{z}}),$$

making $\mathbf{D}$ not parallel to $\mathbf{E}$ if $\epsilon_y \neq \epsilon_z$, i.e., if the y–z plane is not isotropic.

15.4.2 Complex Dielectric Tensor

That the dielectric function is a tensor does not stop it from being a complex function, and I'll write it the same way I typically did when it was a scalar:

$$\overleftrightarrow{\epsilon} = \overleftrightarrow{\epsilon}_1 + \frac{4\pi i}{\omega} \overleftrightarrow{\sigma}_1, \tag{15.7}$$

where both $\overleftrightarrow{\epsilon}_1$ and $\overleftrightarrow{\sigma}_1$ are 3×3 diagonalizable tensors with real elements. For example the conductivity (in its principal axes) is

$$\overleftrightarrow{\sigma}_1 = \begin{pmatrix} \sigma_{1x} & 0 & 0 \\ 0 & \sigma_{1y} & 0 \\ 0 & 0 & \sigma_{1z} \end{pmatrix}. \tag{15.8}$$

15.4.3 Crystal Symmetry Restricts the Dielectric Tensor

The dielectric tensor must respect the symmetry of the crystal. If the crystal is indistinguishable along $\hat{\mathbf{x}}$ and $\hat{\mathbf{y}}$, then a field in those two directions will induce identical polarizations and currents; the corresponding elements in $\overleftrightarrow{\epsilon}$ must be identical.

---

* The process for multiplying the dielectric tensor $\overleftrightarrow{\epsilon}$ and the (column vector) electric field $\mathbf{E}$ is as follows:

$$\overleftrightarrow{\epsilon} \cdot \mathbf{E} = \begin{pmatrix} \epsilon_{xx} & \epsilon_{xy} & \epsilon_{xz} \\ \epsilon_{yx} & \epsilon_{yy} & \epsilon_{yz} \\ \epsilon_{zx} & \epsilon_{zy} & \epsilon_{zz} \end{pmatrix} \cdot \begin{pmatrix} E_x \\ E_y \\ E_z \end{pmatrix}$$

$$= \hat{\mathbf{x}}(\epsilon_{xx}E_x + \epsilon_{xy}E_y + \epsilon_{xz}E_z) + \hat{\mathbf{y}}(\epsilon_{yx}E_x + \epsilon_{yy}E_y + \epsilon_{yz}E_z)$$
$$+ \hat{\mathbf{z}}(\epsilon_{zx}E_x + \epsilon_{zy}E_y + \epsilon_{zz}E_z).$$

There are a number of ways I could understand why for example the dielectric tensor of an orthorhombic crystal has three different elements and where the principal axes point. I can use group theory. I can consider the properties of tensors. What I will do here is make a (perhaps somewhat hand-waving) argument based on things I can do to the crystal that leave it unchanged. The operations I need are rotation and reflection.

I'll state a sort of Murphy's law for the optics of crystals: if the dielectric tensor can possibly display anisotropy, it will do so. Murphy affects the principal-axis values: if they can be different, they will be different.* Moreover, if the symmetry does not fix the directions of the principal axes, the axes will change direction as the wavelength varies.

Cubic. I start again with the cubic crystal system. The cubic system has three orthogonal four-fold (90°) rotation axes along the a, b, and c crystal axes. (There are also three orthogonal mirror planes, all normal to the crystal axes, four 120° axes along the body diagonals, a center of inversion, and others, making a total of 36 cubic space groups [23]. The amount of detail can become quite large.) The three four-fold axes mean that the crystal properties are unchanged for 90° rotations; the diagonal components of the dielectric tensor must be the same. The tensor can then be written as

$$\overset{\leftrightarrow}{\epsilon} = \epsilon \begin{pmatrix} 1 & 0 & 0 \\ 0 & 1 & 0 \\ 0 & 0 & 1 \end{pmatrix} = \epsilon \overset{\leftrightarrow}{1}, \qquad Cubic. \tag{15.9}$$

where ϵ is a complex scalar and $\overset{\leftrightarrow}{1}$ is the identity tensor. But the dot product of the identity tensor and a vector is just the vector, so I have $\mathbf{D} = \overset{\leftrightarrow}{\epsilon} \cdot \mathbf{E} = \epsilon \overset{\leftrightarrow}{1} \cdot \mathbf{E} = \epsilon \mathbf{E}$. The dielectric function of a cubic crystal is a scalar; the material is isotropic.

Tetragonal. The tetragonal lattice has mirror planes in the a–b, a–c, and b–c planes. A reflection leaves unchanged a vector lying in the plane of the mirror but reverses one perpendicular to the mirror. To be unchanged under reflection, the axes must lie in the mirror or be perpendicular to it. Hence the principal axes are along the crystallographic axes. The structure has one fourfold rotation axis along the c axis. The components of the dielectric tensor for a and b are therefore identical; the c axis component is different. I then have

$$\overset{\leftrightarrow}{\epsilon} = \begin{pmatrix} \epsilon_a & 0 & 0 \\ 0 & \epsilon_a & 0 \\ 0 & 0 & \epsilon_c \end{pmatrix}. \qquad Tetragonal. \tag{15.10}$$

The dielectric tensor is said to be *uniaxial*: there is an optic axis along c and if the electric field is along c, then $\mathbf{D} = \hat{c}\epsilon_c E$. For electric field direction $\hat{e}$ in the a–b plane, $D = \hat{e}\epsilon_a E$.

Orthorhombic. The orthorhombic lattice has mirror planes in the a–b, a–c, and b–c planes. Hence the principal axes are along the crystallographic axes. There are only twofold axes, which take an axis into itself. So the three components are different,

---

* There are of course counterexamples; Murphy's laws are not theorems. For example, the components of the dielectric function vary with frequency and it is possible that if I graph them over a wide frequency range they will cross one or more times. At the frequencies where they cross, the anisotropy does not exist. However, as the mathematicians say, this is a set of measure zero.

$$\overleftrightarrow{\epsilon} = \begin{pmatrix} \epsilon_a & 0 & 0 \\ 0 & \epsilon_b & 0 \\ 0 & 0 & \epsilon_c \end{pmatrix}. \qquad \textit{Orthorhombic.} \qquad (15.11)$$

The orthorhombic system is *biaxial*.

Monoclinic. Monoclinic crystals are also biaxial. The b axis is at right angles to both a and c, making one of the principal axes parallel to b. The other two axes are not fixed relative to the crystal axes and will change directions as the frequency of the electromagnetic field changes. I must re-diagonalize the tensor for *each* frequency of the light that encounters the crystal.

$$\overleftrightarrow{\epsilon} = \begin{pmatrix} \epsilon_x & 0 & 0 \\ 0 & \epsilon_y & 0 \\ 0 & 0 & \epsilon_z \end{pmatrix}, \qquad \textit{Monoclinic.} \qquad (15.12)$$

with $\epsilon_y = \epsilon_b$.

Triclinic. The only symmetry operations are identity and inversion. The crystal is biaxial with axes all changing direction with frequency.

$$\overleftrightarrow{\epsilon} = \begin{pmatrix} \epsilon_x & 0 & 0 \\ 0 & \epsilon_y & 0 \\ 0 & 0 & \epsilon_z \end{pmatrix}. \qquad \textit{Triclinic.} \qquad (15.13)$$

Hexagonal. The hexagonal class has a six-fold axis perpendicular to the hexagonal or basal plane. This is as good as a four-fold axis for forcing the response of the plane to be independent of the direction of the in-plane field. Hexagonal crystals are uniaxial with the optic axis perpendicular to the basal plane

$$\overleftrightarrow{\epsilon} = \begin{pmatrix} \epsilon_a & 0 & 0 \\ 0 & \epsilon_a & 0 \\ 0 & 0 & \epsilon_c \end{pmatrix}. \qquad \textit{Hexagonal.} \qquad (15.14)$$

Trigonal. The optic axis is fixed to the direction the cube was pulled or pushed. Trigonal (rhombohedral) crystals are uniaxial.

$$\overleftrightarrow{\epsilon} = \begin{pmatrix} \epsilon_x & 0 & 0 \\ 0 & \epsilon_x & 0 \\ 0 & 0 & \epsilon_z \end{pmatrix}. \qquad \textit{Trigonal.} \qquad (15.15)$$

15.5 Plane-Wave Propagation in Anisotropic Materials

15.5.1 Coordinates

I need to define a coordinate system in which to describe electromagnetic wave propagation in a crystal. It may seem that I have a choice in how to align the axes of my coordinate system. On the one hand, I could, as I did in Section 2.8 for electromagnetic waves in vacuum, align the wave vector $\mathbf{q}$ and the fields $\mathbf{D}$ and $\mathbf{H}$ along the Cartesian axes and let the dielectric tensor be in nondiagonal form. On the other hand, I can choose as my Cartesian system the principal axes and let the fields fall on the crystal as they may, with the wave vector and field vectors of the electromagnetic wave possibly having components along two or three axes. In practice, the choice is not there. I will have to use the principal axes as my coordinate axes.

15.5.2 Maxwell's Equations Revisited

I will continue to write the electric field $\mathbf{E}$ and the magnetic field $\mathbf{H}$ of the light as complex plane waves, as in Eqs. 2.5 and 2.6. I will also write

$$\mathbf{D}(\mathbf{r}, t) = \mathbf{D}_0 e^{i(\mathbf{q}\cdot\mathbf{r} - \omega t)}. \tag{15.16}$$

I can start with Maxwell's equations given in Eqs. 2.4a–2.4d and then make use of the simplifications of local, nonmagnetic, linear, and homogeneous materials, leaving out only the assumption of isotropy. I will plan to use $\mathbf{D} = \overset{\leftrightarrow}{\epsilon} \cdot \mathbf{E}$ with $\overset{\leftrightarrow}{\epsilon} = \overset{\leftrightarrow}{\epsilon}_1 + (4\pi i/\omega)\overset{\leftrightarrow}{\sigma}_1$, so that, after substituting the plane-wave fields into Maxwell's equations, I have

$$\mathbf{q} \cdot \mathbf{D} = 0 \tag{15.17a}$$

$$\mathbf{q} \cdot \mathbf{H} = 0 \tag{15.17b}$$

$$\mathbf{q} \times \mathbf{E} = \frac{\omega}{c}\mathbf{H} \tag{15.17c}$$

$$\mathbf{q} \times \mathbf{H} = -\frac{\omega}{c}\mathbf{D}. \tag{15.17d}$$

These Maxwell's equations for anisotropic materials tell me that it is $\mathbf{q}$, $\mathbf{D}$, and $\mathbf{H}$ that determine my right-handed coordinate system. The electric field $\mathbf{E}$ is not in general parallel to $\mathbf{D}$, so the electric field will not in general be perpendicular to $\mathbf{q}$. Equation 15.17c tells me that the field $\mathbf{E}$ *is* perpendicular to $\mathbf{H}$.

15.5.3 Field Directions

From Maxwell's equations, Eqs. 15.17a–15.17d, I can construct the picture of the field directions shown in Fig. 15.4. The coordinates here are those of the fields, $\mathbf{q}$, $\mathbf{D}$, and $\mathbf{H}$, so that in general the dielectric tensor is not diagonal. The $\mathbf{E}$ field is inclined to these axes and has a component along either $\pm\mathbf{q}$ as shown in the left or right sides of the figure.

The direction of energy flow is determined by the Poynting vector $\mathbf{S}$, given (as I have said before) by $\mathbf{S} = (c/8\pi)\,\mathrm{Re}\,\mathbf{E} \times \mathbf{H}^*$. In vacuum and in isotropic materials, the wave vector and Poynting vector point in the same direction. This is not the case in anisotropic crystals. The wave vector $\mathbf{q}$ is related to the complex refractive index in the usual way:

$$\mathbf{q} = N\frac{\omega}{c}\hat{q}. \tag{15.18}$$

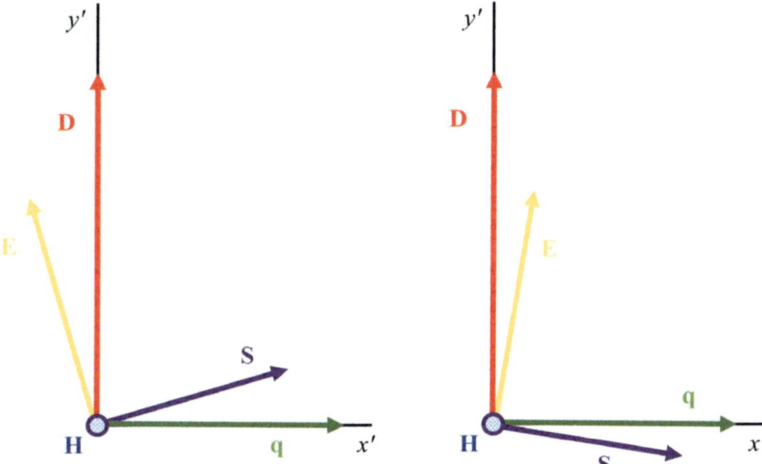

Fig. 15.4 Fields for a wave propagating along x' in an anisotropic medium with principal axes at an arbitrary angle with respect to the fields. The vectors **q**, **D**, and **H** form a right handed set and determine the primed coordinate system.

I combine Eqs. 15.17c and 15.18 to write $\mathbf{H} = N\hat{q} \times \mathbf{E}$. With this, and using the identity for the triple cross product, the Poynting vector is

$$\mathbf{S} = (\mathrm{Re}\,N)\frac{c}{8\pi}|E|^2[\hat{q} - \hat{e}(\hat{q}\cdot\hat{e})]. \tag{15.19}$$

Thus **S**, **E**, and **H** form a right-handed set and **S** points away from **q** by the same angle that **E** points away from **D**. Energy flow and the ray direction directions differ, as shown in Fig. 15.4.

15.5.4 The Refractive Index

Now, I find a formal solution for the refractive index (and, hence, for the wave vector) of plane-polarized light in anisotropic crystals. The easiest way to continue is to take the cross product of **q** with Eq. 15.17c and use the vector identity for the triple cross product on the left side and Eq. 15.17d on the right side:*

$$\mathbf{q} \times (\mathbf{q} \times \mathbf{E}) = \frac{\omega}{c}\mathbf{q} \times \mathbf{H}$$

$$\mathbf{q}(\mathbf{q} \cdot \mathbf{E}) - q^2\mathbf{E} = -\frac{\omega^2}{c^2}\mathbf{D}.$$

I then obtain (after using Eq. 15.18, canceling ω^2/c^2, using $\mathbf{D} = \overleftrightarrow{\epsilon} \cdot \mathbf{E}$, changing signs, and swapping left and right sides):

$$\overleftrightarrow{\epsilon} \cdot \mathbf{E} = N^2\left[\mathbf{E} - \hat{q}(\hat{q}\cdot\mathbf{E})\right]. \tag{15.20}$$

---

* An equally valid approach would be to take the curl of Eqs. 2.4c and 2.4d, use some vector identities, and obtain a wave equation for **E**. The use of plane waves form for the fields leads to the same equation for the refractive index that I get here.

As a comment, the amplitude and the $e^{i(\mathbf{q}\cdot r - \omega t)}$ terms in the electric field $\mathbf{E}$ will divide out; I could replace $\mathbf{E}$ with $\hat{e}$ in Eq. 15.20 if I wanted to do so. The refractive index N is determined by the dielectric tensor, by the wave propagation direction $\hat{q}$, and by the electric field direction $\hat{e}$. Before outlining the solution for the most general anisotropic crystal, I'll look at the important case of the uniaxial crystal.

15.6 The Uniaxial Crystal

The tetragonal, hexagonal, or trigonal systems are uniaxial: there is a single "optic axis," typically taken as being in the z direction.* The dielectric tensor is given by one of Eqs. 15.10, 15.14, or 15.15. I'll use the notation of the first two of these. Two of the diagonal components are the same (ϵ_a) and one is different (ϵ_c).

My goal is to obtain an equation for N in terms of the components of $\overleftrightarrow{\epsilon}$ for a given ray direction $\hat{q}$. My second goal is to find the electric field (to within an arbitrary constant). My third goal is to calculate the Poynting vector and to determine its direction relative to $\hat{q}$. The equation for N will be the equivalent for anisotropic materials of Eq. 3.10, which reads $N = \sqrt{\epsilon}$. It is equivalent but in no way as simple.[†]

The a–b plane of the uniaxial crystal is isotropic; the properties of the crystal do not change as it is rotated around the optic (z) axis. So I can set $\hat{q}$ in the x–z plane and at an angle θ to the optic axis, so that $\hat{q} \cdot \hat{y} = 0$ while $\hat{q} \cdot \hat{z} = \cos\theta$ and $\hat{q} \cdot \hat{x} = \sin\theta$, making

$$\hat{q} = \sin\theta\hat{x} + \cos\theta\hat{z} \qquad (15.21)$$

If I think now about the electric field direction, there seem to be two possibly distinct cases: (1) the field is perpendicular to the x–z direction (along $\hat{y}$) and (2) the field lies in the x–z plane and hence has no $\hat{y}$ component. In case (1) the waves are transverse whereas in case (2) they may have an $\mathbf{E}$-field component parallel to $\hat{q}$.

15.6.1 The Refractive Index

15.6.1.1 Case 1: E Perpendicular to the Optic Axis

If $\mathbf{E}$ is perpendicular to the optic axis (along $\hat{y}$) then it is perpendicular to $\mathbf{q}$ also. Equation 15.20 becomes

$$\epsilon_a E_y = N^2 E_y,$$

so that $N = \sqrt{\epsilon_a}$. The magnetic field is given by Eq. 15.17c; $\hat{q}$, $\hat{e}$, and $\hat{h}$ form a right-handed set just as in the isotropic case, Eq. 3.3. Moreover $\mathbf{D}(= \epsilon_a \mathbf{E})$ points in the same

---

* It is not the "optical axis." The optical axis is the axis about which an optical system, such as a lens, has a certain degree of symmetry.

[†] It is complicated enough for uniaxial materials; for biaxial crystals it is more so.

direction as **E**. This polarization sees no hint of the anisotropy. It is known as the "ordinary" ray or "ordinary" wave propagation:

$$N_o = \sqrt{\epsilon_a}. \tag{15.22}$$

15.6.1.2 Case 2: E in the Plane Containing the Optic Axis and q

For the "extraordinary" wave, **q** and **E** both have $\hat{x}$ and $\hat{z}$ components; neither has $\hat{y}$ components. Equation 15.20 thus becomes two equations:

$$\epsilon_a E_x = N^2 E_x - N^2 \sin\theta (\hat{q} \cdot \mathbf{E})$$
$$\epsilon_c E_z = N^2 E_z - N^2 \cos\theta (\hat{q} \cdot \mathbf{E}).$$

I can rewrite these in the following way (where I've multiplied the first by $\sin\theta$ and the second by $\cos\theta$):

$$E_x \sin\theta = \frac{N^2 \sin^2\theta}{N^2 - \epsilon_a} (\hat{q} \cdot \mathbf{E})$$
$$E_z \cos\theta = \frac{N^2 \cos^2\theta}{N^2 - \epsilon_c} (\hat{q} \cdot \mathbf{E})$$

I can add these, and then $\hat{q} \cdot \mathbf{E} = E_x \sin\theta + E_z \cos\theta$ cancels out, leaving

$$\frac{1}{N^2} = \frac{\sin^2\theta}{N^2 - \epsilon_a} + \frac{\cos^2\theta}{N^2 - \epsilon_c}.$$

This equation appears to be quadratic in N^2, but, when fractions are cleared, the N^4 terms cancel, leaving only the quadratic term. After a minute amount of algebra I get

$$N(\theta) = \frac{\sqrt{\epsilon_a \epsilon_c}}{\sqrt{\epsilon_c \cos^2\theta + \epsilon_a \sin^2\theta}}, \tag{15.23}$$

where I have written explicitly that the refractive index is a function of the angle the ray makes to the optic axis.

The refractive index moves smoothly between two extremes,

$$N(\theta) = \begin{cases} \sqrt{\epsilon_a} \equiv N_o & \theta = 0° \\ \sqrt{\epsilon_c} \equiv N_e & \theta = 90°, \end{cases}$$

where N_e is known as the "extraordinary" refractive index. Note that in fact the extraordinary wave has a refractive index $N(\theta)$ and a propagation velocity that depends on the direction of **q**. $N(\theta)$ equals what one sees in the literature and in tables of materials properties for N_e when $\theta = 90°$. This is the propagation direction for which $N(\theta)$ differs the most from N_o.

15.6.2 "Ray Surfaces"

As I have just shown, the refractive index of a uniaxial crystal has two values; the electric field directions (or polarizations) that correspond to these two values are orthogonal. One polarization, the extraordinary ray, has a refractive index that varies with the direction of $\mathbf{q}$. Consequently, light traveling in the crystal moves with a speed that depends on polarization and on direction of travel.

Imagine that there is a point source of light at the center of the crystal capable of emitting in all directions a short flash of unpolarized light. The flash expands out from the point with a speed set by c/n where n is the real part of the refractive index N. In an isotropic medium, the light will be located in an expanding spherical shell. In an anisotropic material, there are two shells of light. One, generated by ordinary rays, travels in a spherical shell at c/n_o and with electric field direction perpendicular to the optic axis. The other set of rays, the extraordinary ones, give an ellipsoid of revolution about the optic axis. The distance from the origin of this ellipsoid is $c/n(\theta)$, where θ is the angle of the ray to the optic axis. The sphere and ellipsoid touch along the optic axis ($\theta = 0$) where $n(\theta) = n_o = \sqrt{\epsilon_a}$. They have a maximum separation for propagation in the ab plane, where n_o still equals $\sqrt{\epsilon_a}$ and $n(\theta) = n_e = \sqrt{\epsilon_c}$.

There are two possibilities for the ellipsoidal shell. If $n_e > n_o$ (a *positive* uniaxial crystal), the extraordinary waves travel more slowly than the ordinary, and the extraordinary surface is a prolate ellipsoid inside the ordinary sphere. If, in contrast, $n_e < n_o$ (a *negative* uniaxial crystal), the extraordinary waves travel more quickly than the ordinary, and the extraordinary surface is an oblate ellipsoid surrounding the ordinary sphere. These surfaces are illustrated in Fig. 15.5 for both negative and positive uniaxial materials. The electric field for the ordinary waves are everywhere perpendicular to the optic axis. The electric field for the extraordinary wave is parallel to the optic axis when this wave is

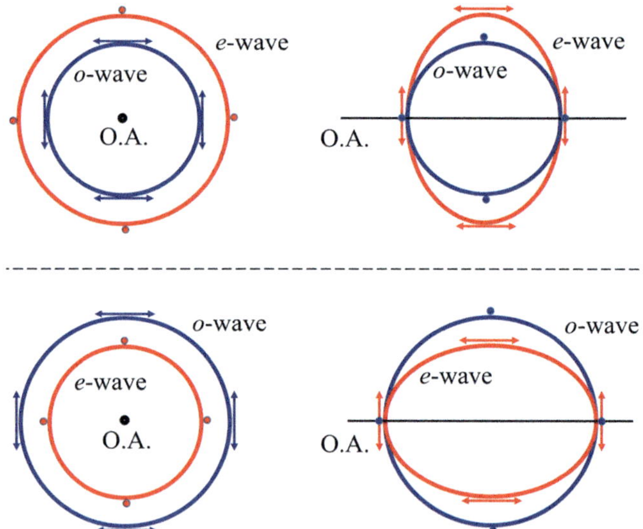

 Ray surfaces for negative (top) and positive (bottom) uniaxial materials. The polarization vectors are indicated on the surfaces. The surfaces are shown looking down the optic axis (O.A.) (left side) and with the optic axis lying in the plane of the figure (right side). Field directions are also shown; dots indicate fields perpendicular to the drawing; arrows show fields in the plane of the drawing.

propagating ab plane but rotates to remain tangent to the ellipse as the propagation direction moves out of this plane.

15.6.3 The Fields

Now I'll address the electric field, magnetic field, and Poynting vector in the uniaxial crystal. I used Faraday's law, Eq. 15.17c, and Ampère's law, Eq. 15.17d, to obtain the equation for $N(\theta)$ and will use one of these again to find the relationship between magnetic and electric fields. What I need now is Gauss' law, Eq. 15.17a. I will write

$$\nabla \cdot \mathbf{D} = 0 = \mathbf{q} \cdot \mathbf{D} \quad \text{or} \quad \hat{\mathbf{q}} \cdot \overset{\leftrightarrow}{\epsilon} \cdot \mathbf{E} = 0$$

where I have used that $\mathbf{D} = \overset{\leftrightarrow}{\epsilon} \cdot \mathbf{E}$ and divided out either $N_o \omega / c$ or $N(\theta) \omega / c$ depending on whether the $\mathbf{E}$ direction is for the ordinary or extraordinary direction. The ray direction is $\hat{\mathbf{q}} = \sin\theta\hat{\mathbf{x}} + \cos\theta\hat{\mathbf{z}}$.

15.6.3.1 Case 1: E Perpendicular to the Optic Axis: Electric and Magnetic Fields

The $\mathbf{E}$ field is the $\hat{\mathbf{y}}$ direction by construction. The refractive index is $N_o = \sqrt{\epsilon_a}$ as given in Eq. 15.22. Then, I can find the $\mathbf{H}$ field using Eq. 15.17c, which I write as $\mathbf{H} = N_o \hat{\mathbf{q}} \times \mathbf{E}$. The cross product is

$$\mathbf{H} = N_o \hat{\mathbf{q}} \times \mathbf{E} = N_o \begin{vmatrix} \hat{\mathbf{x}} & \hat{\mathbf{y}} & \hat{\mathbf{z}} \\ \sin\theta & 0 & \cos\theta \\ 0 & E_y & 0 \end{vmatrix} = -N_o E_y \cos\theta\hat{\mathbf{x}} + N_o E_y \sin\theta\hat{\mathbf{z}}.$$

The magnitude of this field is

$$H = N_o E_y \sqrt{\cos^2\theta + \sin^2\theta} = N_o E_y.$$

So the fields are in every way ordinary. The wave is a transverse electromagnetic wave with the same relation between $\mathbf{q}$, $\mathbf{E}$, and $\mathbf{H}$ as in an isotropic materials. The refractive index is $N = N_o$. The Poynting vector is parallel to the wave vector, so the waves and energy flow in the same direction.

15.6.3.2 Case 2: E in the Plane Containing the Optic Axis and q: Electric Field

Light with this polarization is called the "extraordinary" wave. I'll start by writing a general electric field vector in the x–z plane:

$$\mathbf{E} = E_x\hat{\mathbf{x}} + E_z\hat{\mathbf{z}}. \tag{15.24}$$

What I want to know is the direction of $\mathbf{E}$ as a function of θ; to know this field requires me to find E_x and E_z in terms of the overall magnitude of the electric field* E. Now the displacement vector $\mathbf{D}$ is

---

* There is no constraint on the magnitude of the field. I can always make the overall electric field larger or smaller by turning up or down my lamp, or my laser, or my RF generator.

$$\mathbf{D} = \overleftrightarrow{\epsilon} \cdot \mathbf{E} = \epsilon_a E_x \hat{\mathbf{x}} + \epsilon_c E_z \hat{\mathbf{z}},$$

and then Gauss' law is

$$\hat{\mathbf{q}} \cdot \mathbf{D} = 0 = \epsilon_a E_x \sin\theta + \epsilon_c E_z \cos\theta.$$

After a couple of lines of algebra I get

$$E_z = -E_x \frac{\epsilon_a}{\epsilon_c} \tan\theta. \tag{15.25}$$

Although Eq. 15.25 describes the ratio of the two components of the electric field, it is not very illuminating. So I'll write

$$\mathbf{E} = E_x \hat{\mathbf{x}} - E_x \frac{\epsilon_a}{\epsilon_c} \tan\theta \hat{\mathbf{z}}$$

and write the magnitude of the electric field $E = \sqrt{\mathbf{E} \cdot \mathbf{E}}$. Several lines of algebra later I arrive at

$$E_x = \frac{\epsilon_c \cos\theta}{\sqrt{\epsilon_a^2 \sin^2\theta + \epsilon_c^2 \cos^2\theta}} E. \tag{15.26}$$

I now put this into Eq. 15.25 and find

$$E_z = -\frac{\epsilon_a \sin\theta}{\sqrt{\epsilon_a^2 \sin^2\theta + \epsilon_c^2 \cos^2\theta}} E. \tag{15.27}$$

I said that the electric field is not perpendicular to the wave vector. Let me check that by calculating the dot product of $\hat{\mathbf{q}}$ and $\mathbf{E}$, which I'll call $E_\parallel$:

$$E_\parallel = \hat{\mathbf{q}} \cdot \mathbf{E} = E_x \sin\theta + E_z \cos\theta.$$

I substitute Eqs. 15.26 and 15.27 to find

$$E_\parallel = \frac{(\epsilon_c - \epsilon_a) \sin\theta \cos\theta}{\sqrt{\epsilon_a^2 \sin^2\theta + \epsilon_c^2 \cos^2\theta}} E. \tag{15.28}$$

I like Eq. 15.28 very much indeed. It shows that the $\hat{\mathbf{q}}$-parallel (extraordinary) part of the electric field in uniaxial crystals is zero if (1) $\epsilon_c = \epsilon_a$ (the material is not anisotropic) or (2) $\theta = 0$ (the wave is propagating along the optic axis) or (3) $\theta = 90°$ (the wave is propagating perpendicular to the optic axis). In all other cases, there is a component parallel to $\hat{\mathbf{q}}$.

15.6.3.3 Case 2: E in the Plane Containing the Optic Axis and q: Magnetic Field

Now that I have an equation for $\mathbf{E}$, I can calculate $\mathbf{H}$ from Ampère's law: $\mathbf{H} = N(\theta)\hat{\mathbf{q}} \times \mathbf{E}$. I do this calculation, using Eqs. 15.21 (for $\hat{\mathbf{q}}$), and (to start) 15.24 (for $\mathbf{E}$):

$$\mathbf{H} = N(\theta)(E_x \cos\theta - E_z \sin\theta)\hat{\mathbf{y}}.$$

Now, E_z is negative for θ in the range 0–90°. (See Eq. 15.27.) Thus, $\mathbf{H}$ is definitely in the plus $\hat{\mathbf{y}}$ direction. I substitute Eqs. 15.26, 15.27 (for fields), and 15.23 (for $N(\theta)$) to obtain

$$\mathbf{H} = \sqrt{\frac{\epsilon_a \epsilon_c (\epsilon_a \sin^2 \theta + \epsilon_c \cos^2 \theta)}{\epsilon_a^2 \sin^2 \theta + \epsilon_c^2 \cos^2 \theta}} \, E \hat{\mathbf{y}}. \tag{15.29}$$

I do not like Eq. 15.29 very much but see no way to clean it up. It has the advantage of being dimensionally correct and of giving

$$\mathbf{H} = \begin{cases} N_o E \hat{\mathbf{y}} & \theta = 0 \\ N_e E \hat{\mathbf{y}} & \theta = 90° \end{cases} \tag{15.30}$$

where $N_o = \sqrt{\epsilon_a}$ is the ordinary refractive index and $N_e = \sqrt{\epsilon_c}$ is the extraordinary refractive index. As ever, it is the $\mathbf{E}$-field direction that determines which refractive index is employed.

15.6.3.4 Case 2: E in the Plane Containing the Optic Axis and q: Poynting Vector

The next (and last) step in the treatment of wave propagation in uniaxial materials is to calculate the time-average Poynting vector, $\mathbf{S} = (c/8\pi)\mathbf{E} \times \mathbf{H}^*$. I'll use $\mathbf{E} = E_x \hat{\mathbf{x}} + E_z \hat{\mathbf{z}}$ (Eq. 15.24) and $\mathbf{H} = H \hat{\mathbf{y}}$. Then

$$\mathbf{S} = \frac{c}{8\pi}(-E_z H^* \hat{\mathbf{x}} + E_x H^* \hat{\mathbf{z}})$$

and, using Eqs. 15.26, 15.27,

$$\mathbf{S} = \frac{c}{8\pi} \frac{\epsilon_a \sin \theta \, \hat{\mathbf{x}} + \epsilon_c \cos \theta \, \hat{\mathbf{z}}}{\sqrt{\epsilon_a^2 \sin^2 \theta + \epsilon_c^2 \cos^2 \theta}} \, E H^*. \tag{15.31}$$

The magnitude of the Poynting vector is $S = (c/8\pi)E H^*$ (which I get either by calculating $\sqrt{\mathbf{S} \cdot \mathbf{S}}$ or by noting that the angle between the electric and magnetic fields is 90°).

The direction of the Poynting vector is more interesting than the magnitude; the unit vector in the direction of $\mathbf{S}$ is

$$\hat{s} = \frac{\epsilon_a \sin \theta \, \hat{\mathbf{x}} + \epsilon_c \cos \theta \, \hat{\mathbf{z}}}{\sqrt{\epsilon_a^2 \sin^2 \theta + \epsilon_c^2 \cos^2 \theta}} \tag{15.32}$$

and

$$\hat{q} \cdot \hat{s} \equiv \cos \phi = \frac{\epsilon_a \sin^2 \theta + \epsilon_c \cos^2 \theta}{\sqrt{\epsilon_a^2 \sin^2 \theta + \epsilon_c^2 \cos^2 \theta}}$$

where ϕ is the angle between $\hat{q}$ and $\hat{s}$. The angle is zero for $\theta = 0$ and 90° and not zero for other values.

As sketched in Fig. 15.4, the Poynting vector can lie either between $\mathbf{q}$ and $\mathbf{D}$ or on the other side of $\mathbf{q}$ from $\mathbf{D}$, depending on whether $E_\parallel = \hat{q} \cdot \mathbf{E}$ is negative or positive.

(See Eq. 15.28.) So I'll also specify the angle α between $\mathbf{S}$ and the optic axis, given by $\cos\alpha = \hat{z} \cdot \hat{s}$. The result is

$$\cos\alpha = \frac{\cos\theta}{\sqrt{\left(\frac{\epsilon_a}{\epsilon_c}\right)^2 \sin^2\theta + \cos^2\theta}}.$$

15.6.4 Some Uniaxial Materials

Many materials (quartz, sapphire, ice) are uniaxial crystals. The refractive indices of a number of these materials are in Table 15.1, from Ref. 345. The refractive index is for a frequency in the yellow (590 nm, 16,950 cm^{-1}, 2.10 eV), close to the wavelength of low-pressure sodium lamps, like those used in some street lamps. Most of these materials are reasonably transparent at this wavelength. The table gives the name, chemical formula, crystal system, n_o, n_e, and $\delta n = n_e - n_o$. All three of the uniaxial crystal systems are represented. Calcite has a reasonably large δn and good optical transparency, a reason why calcite is used in many polarizers supplied by commercial vendors.

Table 15.1 Uniaxial crystals, at 590 nm.

Material	Formula	Crystal system	n_o	n_e	Δn
Barium borate	BaB_2O_4	Tri	1.678	1.553	-0.124
Beryl	$Be_3Al_2(SiO_3)_6$	Hex	1.602	1.557	-0.045
Calcite	$CaCO_3$	Tri	1.658	1.486	-0.172
Calomel	Hg_2Cl_2	Tet	1.973	2.656	$+0.683$
Cinnabar	HgS	Tri	2.905	3.256	$+0.351$
Hematite	Fe_2O_3	Tri	2.940	3.220	$+0.287$
Ice	H_2O	Hex	1.309	1.313	$+0.004$
Lithium niobate	$LiNbO_3$	Tri	2.272	2.187	-0.085
Magnesium fluoride	MgF_2	Tet	1.380	1.385	$+0.006$
Quartz	SiO_2	Tri	1.544	1.553	$+0.009$
Rutile	TiO_2	Tet	2.616	2.903	$+0.287$
Sapphire	Al_2O_3	Tri	1.768	1.760	-0.008
Sodium nitrate	$NaNO_3$	Tri	1.587	1.336	-0.251
Silicon carbide	SiC	Hex	2.647	2.693	$+0.046$
Tourmaline	(a silicate)	Tri	1.669	1.638	-0.031
Zircon, high	$ZrSiO_4$	Tet	1.960	2.015	$+0.055$
Zircon, low	$ZrSiO_4$	Tet	1.920	1.967	$+0.047$

Tri = trigonal; Hex = hexagonal; Tet = tetragonal.

15.7　The Biaxial Crystal

My goal now is to obtain a general equation for N in terms of the components of $\overleftrightarrow{\epsilon}$ without restricting the symmetry. The algebra is similar to what I did for the uniaxial crystal. First, I write Eq. 15.20 in components, using the form in Eq. 15.6 for the dielectric tensor and dotting from the left with the appropriate unit vector:

$$\epsilon_x E_x = N^2 E_x - N^2(\hat{x} \cdot \hat{q})(\hat{q} \cdot \mathbf{E})$$
$$\epsilon_y E_y = N^2 E_y - N^2(\hat{y} \cdot \hat{q})(\hat{q} \cdot \mathbf{E}) \tag{15.33}$$
$$\epsilon_z E_z = N^2 E_z - N^2(\hat{z} \cdot \hat{q})(\hat{q} \cdot \mathbf{E})$$

where $(\hat{x} \cdot \hat{q})$ is a clumsy way to write "the x component of the unit vector $\hat{q}$." (q_x is already taken as the x component of the wave vector $\mathbf{q}$.) In order to be slightly less clumsy,[*] I will define $\hat{x} \cdot \hat{q} \equiv \alpha$, $\hat{y} \cdot \hat{q} \equiv \beta$, and $\hat{z} \cdot \hat{q} \equiv \gamma$, noting that the dot product of two unit vectors is just the cosine of the angle between them.[†]

These three equations can be reorganized (solve for E_i and then multiply by α, β, and γ respectively) to give

$$\alpha E_x = \frac{N^2 \alpha^2}{N^2 - \epsilon_x}(\hat{q} \cdot \mathbf{E})$$
$$\beta E_y = \frac{N^2 \beta^2}{N^2 - \epsilon_y}(\hat{q} \cdot \mathbf{E})$$
$$\gamma E_z = \frac{N^2 \gamma^2}{N^2 - \epsilon_z}(\hat{q} \cdot \mathbf{E}).$$

Now, I sum these three equations. I recognize the sum of the terms on the left as $\hat{q} \cdot \mathbf{E} = \alpha E_x + \beta E_y + \gamma E_z$, so that the sum is

$$\hat{q} \cdot \mathbf{E} = N^2 \left[\frac{\alpha^2}{N^2 - \epsilon_x} + \frac{\beta^2}{N^2 - \epsilon_y} + \frac{\gamma^2}{N^2 - \epsilon_z} \right] \hat{q} \cdot \mathbf{E}.$$

If I were working with isotropic systems, I would be in trouble because $\hat{q} \cdot \mathbf{E}$ would be zero and N^2 would equal ϵ. But this trouble does not arise for anisotropic materials. I can cancel $\hat{q} \cdot \mathbf{E}$ from both sides and obtain the desired equation for N^2 in terms of $\overleftrightarrow{\epsilon}$:

$$\frac{1}{N^2} = \frac{\alpha^2}{N^2 - \epsilon_x} + \frac{\beta^2}{N^2 - \epsilon_y} + \frac{\gamma^2}{N^2 - \epsilon_z}. \tag{15.34}$$

Eq. 15.34 appears to be a cubic[‡] equation for N^2. Fortunately the N^6 term cancels on both sides because $\alpha^2 + \beta^2 + \gamma^2 = 1$. The quadratic equation for N^2 is

---

[*]　But maybe a bit more obscure.

[†]　I'm using the conventional notation for direction cosines and hoping that the use of α, β, and γ earlier for the angles in the unit cell prism does not lead to too much confusion. In my discussion of uniaxial crystals, $\cos\theta = \gamma$.

[‡]　Or a sixth-order equation for N. But it is N^2 I want.

$$N^4\left[\epsilon_x\alpha^2 + \epsilon_y\beta^2 + \epsilon_z\gamma^2\right] - N^2\left[\epsilon_x\epsilon_y\{\alpha^2 + \beta^2\}\right.$$
$$\left. + \epsilon_y\epsilon_z\{\beta^2 + \gamma^2\} + \epsilon_z\epsilon_x\{\gamma^2 + \alpha^2\}\right] + \epsilon_x\epsilon_y\epsilon_z = 0. \tag{15.35}$$

Equation 15.35 is of the form $a(N^2)^2 + b(N^2) + c = 0$. I know the solution because I learned the quadratic equation some years ago.*

The three terms that go into the quadratic formula are

$$a = \epsilon_x\alpha^2 + \epsilon_y\beta^2 + \epsilon_z\gamma^2$$
$$b = -\epsilon_x\epsilon_y\{\alpha^2 + \beta^2\} - \epsilon_y\epsilon_z\{\beta^2 + \gamma^2\} - \epsilon_z\epsilon_x\{\gamma^2 + \alpha^2\} \tag{15.36}$$
$$c = \epsilon_x\epsilon_y\epsilon_z.$$

If ever I need it, I can write several pages of algebra and have a result. There will in fact be two results and, when they are plugged back into Eq. 15.20, I will find that the two results correspond to orthogonal directions of **E**.

The solutions depend also on direction of propagation, so that there are two ellipsoidal ray surfaces. These surfaces intersect at four points in pairs along a line joining the intersection points and the origin. At these points the two refractive indices are equal. The two lines are the two optic axes of the biaxial crystal.

15.7.1 Rays and Fields along Principal Axes

Optical effects in the biaxial crystal are considerably simplified if **q** and **E** are along principal axes. Let me set $\mathbf{E}\|\hat{x}$ but allow (for the moment) **q** to be arbitrary. With this direction for the electric field, the dot product $\hat{q} \cdot \mathbf{E} = \alpha E_x$ where α is the cosine of the angle between **q** and $\hat{x}$. I can then rewrite Eq. 15.33 as

$$\epsilon_x E_x = N^2 E_x - N^2\alpha^2 E_x$$
$$0 = N^2\beta\alpha E_x$$
$$0 = N^2\gamma\alpha E_x.$$

If I suppose that $\alpha \neq 0$, then the second and third of these require $\beta = \gamma = 0$. But then $\alpha = 1$ and the first equation leads to $E_x = 0$, contradicting my starting with $\mathbf{E} = E_x\hat{x}$. Thus, I have to take $\alpha = 0$, putting **q** in the y–z plane, perpendicular to **E**. Then, Eq. 15.33 become

$$N_a^2 = \epsilon_x. \tag{15.37}$$

Similar arguments give $N_b^2 = \epsilon_y$ when the field is in the $\hat{y}$ direction and $N_c^2 = \epsilon_z$ for field in the $\hat{z}$ direction. The anisotropy is evident in the different values of N affecting wave propagation for the three field directions but the wave propagation is purely ordinary. Moreover, the arguments that led to Eq. 15.30 tell me that the magnitudes of the corresponding magnetic field are in each case $H = N_i E$, just as in the ordinary case.

---

* There is a nice discussion at the Wolfran website on the method to preserve numerical accuracy when $b^2 \gg 4ac$ so one of the roots involves the difference between two almost identical numbers. See http://mathworld .wolfram.com/QuadraticEquation.html.

15.8 Anisotropic Material with Boundaries: $\mathscr{R}$ and $\mathscr{T}$

Now let me turn to the case of light incident on the surface of an anisotropic crystal. The general case is quite complicated. The surface may be cut at an arbitrary angle to the principal axes; the angle of incidence may vary, as may the polarization. I'll discuss some important cases that allow more or less simplification. A complete discussion of the general case can be found in Born and Wolf [8].

The transmission and reflection of light at an interface are determined by the boundary conditions.* As I've already shown, the electric field in the anisotropic medium generally consists of two orthogonal contributions: the ordinary wave and the extraordinary wave. Each has an associated magnetic field and $\mathbf{q}$ vector. So I will need to decompose the incident fields along these two directions and apply the boundary conditions to both.

15.8.1 Normal Incidence with Field along a Principal Axis

I will start with a simple case: The field is normally incident on a crystal that was grown (or cut and polished) with two principal axes lying in the surface and the third parallel to the normal. For definiteness, let me consider an orthorhombic crystal (biaxial) with the a axis along the normal, parallel to $\hat{x}$, and the b and c axes in the surface, parallel to $\hat{y}$ and $\hat{z}$, respectively. I'll then have $\mathbf{q}$ parallel to $\hat{x}$ and can polarize the incoming light in the y–z plane. Suppose that the incident field is polarized $\mathbf{E} \| \hat{y}$. Continuity of the electric field means that the field inside is also along $\hat{y}$ and that it is traveling in a medium characterized by N_b. (See Eq. 15.37 and what immediately follows that.) The field inside is smaller on account of the fact that there is also a reflected wave moving out along the normal.

The situation is in fact exactly the same as the one I worked out back in Chapter 3 (Section 3.3). See also Fig. 3.2. Using the notation of the current chapter[†] and letting the incident field be in vacuum, I can just write

$$r_b = \frac{1 - N_b}{1 + N_b}.$$

$$(15.38)$$

What I measure is the reflectance,

$$\begin{aligned}
\mathscr{R}_b &= \left| \frac{1 - N_b}{1 + N_b} \right|^2 \\
&= \frac{(n_b - 1)^2 + \kappa_b^2}{(n_b + 1)^2 + \kappa_b^2},
\end{aligned}$$

$$(15.39)$$

with n_b and κ_b the real and imaginary parts of $N_b = \sqrt{\epsilon_b}$.

Now I can rotate the polarizer 90° and measure $\mathscr{R}_c$. I can find a surface (or cut and polish one) containing the a axis, polarize the light appropriately, and measure $\mathscr{R}_a$[‡]

---

* In the absence of surface charges or currents, the boundary conditions are the continuity of the tangential component of $\mathbf{E}$, the normal component of $\mathbf{D}$, the normal component of $\mathbf{B}$, and the tangential component of $\mathbf{H}$.

[†] I used a and b for the two media in Chapter 3; here a and b represent crystal axes.

[‡] I can turn the polarizer again and measure one of the other two. In principle, neglecting edge states, this fourth measurement is redundant.

These three measurements are done for the cases where the wave propagation is the simplest. But they are *exactly* what I want if I am interested in the dielectric tensor of the anisotropic material. Causality still applies, and so I can use the Kramers–Kronig method to estimate the principal components of the dielectric tensor in the same way as for isotropic solids.

Transmittance is equally straightforward. The transmittance discussion in Chapter 7 applies to the polarized transmittance. The light polarized along the a axis travels at speeds and with absorption governed by N_a.

I say that it is simple or straightforward but there are subtleties that should be mentioned. For instance, reflection measurements are typically made at near normal incidence and not at exactly normal incidence.* Normally the reflectance from the Fresnel equations does not vary much with angle out to $10°$ or so. (See Appendix E.) In some cases, however, the reflectance for anisotropic media can show anomalies, especially with p-polarized light (light where the electric field is in the plane of incidence and thus light at nonnormal incidence has a vertical component). If the dielectric function for the principal axis parallel to the surface normal direction has a zero at some frequency, anomalies may appear in the p-polarized spectrum. A good example has been given by Duarte, Sanjurjo, and Katiyar [346]. The fix is to use s-polarized light as much as possible.

15.8.2 Normal Incidence with Field in a Plane Containing an Optic Axis

I've just seen that reflectance and transmittance is pretty simple if the field is along a principal axis. What about the case if the field direction is between these two directions? Let me suppose that surface contains both the b and c axes, as above. The incident field has $\mathbf{E}_i$ at an angle ϕ to the c axis, so that $\mathbf{E}_i = E_i \sin(\phi)\hat{\mathbf{y}} + E_i \cos(\phi)\hat{\mathbf{z}}$ with E_i the incident electric field strength. The reflected field has $\mathbf{E}_r = E_i r_b \sin(\phi)\hat{\mathbf{y}} + E_i r_c \cos(\phi)\hat{\mathbf{z}}$, making the reflected intensity be

$$\mathcal{R}(\phi) = \mathcal{R}_b \sin^2(\phi) + \mathcal{R}_c \cos^2(\phi).$$

One of $\mathcal{R}_b$ and $\mathcal{R}_c$ is larger than the other. The reflectance interpolates smoothly between these two values as ϕ is varied. The principal axes correspond to the direction where the reflectance is a maximum and the (orthogonal) direction where it is a minimum.

Note that the outgoing waves will not be polarized at ϕ. (Consider the case where $\mathcal{R}_b = 0$. The reflected field will be polarized always along $\hat{\mathbf{z}}$, increasing in intensity as $\cos^2(\phi)$ as the incident polarization direction is varied.) I could place an analyzing polarizer in the optical train behind the sample also at angle ϕ. I would then need to take the dot product $\hat{\mathbf{e}} \cdot \mathbf{E}_r$, where $e = \sin(\phi)\hat{\mathbf{y}} + \cos(\phi)\hat{\mathbf{z}}$ before squaring to get the intensity.

---

* Angles of incidence of $5°$–$10°$ are common in order that the reflected light can take a path distinct from the incident light. Even if a beamsplitter is used to pick off the reflected light, the light is typically focused on the sample; only the central ray in the pencil of incident light is at normal incidence.

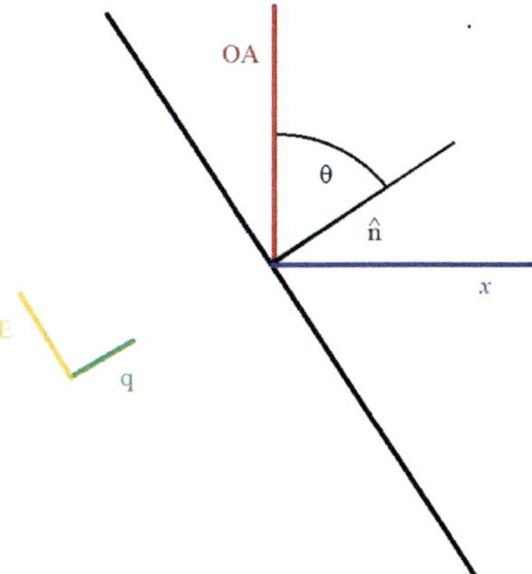

Fig. 15.6   Light is normally incident on a uniaxial crystal with optic axis at an angle θ to the surface normal. The electric field of the incident light is oriented so as to excite the extraordinary wave in the crystal.

15.8.3 Normal Incidence Reflectance with an Inclined Optic Axis

I now turn to the case of a uniaxial crystal cut so that the optic axis dips below the surface. I've simplified to a uniaxial crystal (tetragonal, hexagonal, or trigonal) and will make use of the fields that I calculated in Section 15.6.3 of this chapter. I'll continue to use a coordinate system in which the optic axis is along $\hat{z}$; consequently, the incoming wave vector and fields will be inclined relative to the coordinates. A diagram of the geometry is in Fig. 15.6.

The optic axis is along the $\hat{z}$ direction and the normal to the surface lies in the x–z plane, at an angle θ to the optic axis. The normal vector is then $\hat{n} = \hat{z}\cos(\theta) + \hat{x}\sin(\theta)$. The wave vector direction $\hat{q}$ will equal $\hat{n}$ for normal incidence.

15.8.3.1 The Ordinary Wave

If I orient my electric field along $\hat{y}$, not as shown in Fig. 15.6 but instead perpendicular to the plane of the diagram, I will have ordinary waves inside the crystal, with wave propagation and reflectance governed by $N_o = \sqrt{\epsilon_a}$ according to Eq. 15.22. The reflected field and reflectance follow as usual.

15.8.3.2 The Extraordinary Wave

If I orient my electric field in the x–z plane, as shown in Fig. 15.6, I will launch the extraordinary wave in the crystal. For simplicity (and without loss of generality in this linear medium) let me set the amplitude of the incident field to unity. Then I have

$$\mathbf{E}_i = \hat{z}\sin(\theta) - \hat{x}\cos(\theta).$$

The reflected field is written

$$\mathbf{E}_r = \hat{z}r_e\sin(\theta) - \hat{x}r_e\cos(\theta),$$

where r_e is the complex amplitude reflectivity.

For the field inside, I can write

$$\mathbf{E}_t = \hat{z}\,E_{tz} + \hat{x}\,E_{tx},$$

where E_{tz} is the strength of the field component along $\hat{z}$ and E_{tx} is the strength along $\hat{x}$. *I know these fields.* I worked out the components of the extraordinary wave earlier in this chapter; they are given in Eqs. 15.26 and 15.27. I also know the magnetic field (along $\hat{y}$ in this geometry); it is in Eq. 15.29.

The electric field and magnetic fields are continuous across the boundary, so of course all I now need to do is to equate the fields on the two sides of the boundary. After a moderate amount of algebra I obtain

$$r_e = \frac{1 - N(\theta)}{1 + N(\theta)} \tag{15.40}$$

where $N(\theta) = \sqrt{\epsilon_a \epsilon_c}/\sqrt{\epsilon_c \cos^2\theta + \epsilon_a \sin^2\theta}$, as given in Eq. 15.23. I think that Eq. 15.40 is both surprising and not surprising. It is surprising that things worked out so nicely to give a solution formally like the isotropic case. It is not surprising in that I'd expect a wave in the medium characterized by a particular value $N(\theta)$ of the refractive index to have a reflectance

$$\mathcal{R}_e = \left| \frac{1 - N(\theta)}{1 + N(\theta)} \right|^2$$
$$= \frac{[n(\theta) - 1]^2 + \kappa(\theta)^2}{[n(\theta) + 1]^2 + \kappa(\theta)^2}, \tag{15.41}$$

with $n(\theta)$ and $\kappa(\theta)$ the real and imaginary parts of $N(\theta)$.

15.9 Polarizers and Waveplates

Anisotropic materials are needed to produce or manipulate polarized light. There are a variety of devices, but the basics are linear polarizers, quarter-wave plates, and half-wave plates. I'll discuss each of these briefly.

15.9.1 Polarizers

A linear polarizer is a device which is opaque (or reflecting) for one specific polarization direction and transparent for the orthogonal polarization. (See Section 15.2 earlier in this chapter.) The simplest way to achieve this property is to have a slab geometry made from a uniaxial material with the optic axis ($\hat{z}$) in the plane of the slab. Suppose the corresponding component of the dielectric tensor, ϵ_c, has a relatively large imaginary part while the other two diagonal components have zero imaginary parts. Then a light field polarized along $\hat{z}$ sees an absorption coefficient $\alpha_c = \omega\epsilon_{c2}/nc$ and as long as the slab's thickness d is large enough, the transmittance, which – if reflection can be ignored – will be $\mathcal{T} = e^{-\alpha_c d} \to 0$. The transmittance for light polarized in the nonabsorbing direction will – neglecting reflection and residual absorption – be nearly 100%. Suppose the incident field is of the form $\mathbf{E} = E_0(\cos\theta\,\hat{\mathbf{y}} + \sin\theta\,\hat{\mathbf{z}})$ with $\mathbf{q}\|\hat{\mathbf{x}}$, having been linearly polarized by an upstream polarizer. Then the component parallel to y is transmitted and that parallel to z is not. The transmitted intensity is then

$$I_t = I_o \cos^2\theta, \tag{15.42}$$

where I_0 is the incident intensity. This $\cos^2$ behavior is known as the law of Malus.

The most common dichroic material of the type described above is known as "polaroid," a polymeric material containing aligned chains of absorbing molecules. The chains are absorbing for parallel polarization. Other polarizers are made of a parallel grid of wires; these reflect light parallel to the wire and transmit the component perpendicular to the wire.

The highest-quality polarizers operate on other principles. The refractive indices for ordinary and extraordinary rays differ, and so light incident on a crystal at an oblique angle is refracted according to Snell's law, with n_o or n_e coming into play depending on polarization. (See Appendix E.) So a prism with the optic axis parallel to the base of the prism will separate polarizations into two beams that will separate after traveling some distance behind the prism.

The critical angle for total internal reflection also depends on polarization. (See also Appendix E for a discussion of total internal reflection.) There are a number of polarizer designs that use total internal reflection to reflect one polarization while allowing the other polarization to be transmitted. These devices achieve a very high degree of polarization.

15.9.2 Quarter-Wave Plate

A quarter-wave plate is a device that converts linearly polarized light to circularly polarized light. It typically comes as a thin transparent slab made of a birefringent crystal. The optic axis lies in the plane of the slab. The user transmits light through the slab at normal incidence. The transmitted light gains phase according to the refractive index and distance traveled in the wave plate.

I set the $\hat{\mathbf{x}}$ direction as the normal into the slab and orient $\mathbf{q}$ along this normal. The optic axis, with refractive index n_e, is along $\hat{\mathbf{z}}$. Fields in the $\hat{\mathbf{y}}$ direction see refractive index n_o. I'll choose a material for which $n_e > n_o$.[*] The incoming field is plane polarized light with the plane of polarization at an angle θ to the optic axis:

$$\mathbf{E}_{\text{in}} = E_0\left[\hat{\mathbf{y}}\sin(\theta) + \hat{\mathbf{z}}\cos(\theta)\right] e^{i(qx-\omega t)},$$

where E_0 is the complex amplitude and I'll ignore reflection at the slab surfaces (or ask that the wave plate be antireflection coated). The two components of the electric field propagate with wavevectors $q_e = n_e\omega/c = 2\pi n_e/\lambda$ and $q_o = n_o\omega/c = 2\pi n_o/\lambda$ for the components respectively parallel and perpendicular to the optic axis.[†] The light leaving the wave plate is

$$\mathbf{E}_{\text{out}} = E_0\left[\hat{\mathbf{y}}\sin(\theta)e^{i2\pi n_o d/\lambda} + \hat{\mathbf{z}}\cos(\theta)e^{i2\pi n_e d/\lambda}\right] e^{i(qx-\omega t)},$$

---

[*] This requirement would allow many possible choices of materials; see Table 15.1.
[†] Here, λ is the vacuum wavelength of the light.

where d is the thickness of the plate. Because $n_o \neq n_e$ the two components have gained different phases on going through the wave plate. The phase difference is what matters; I'll define it as

$$\Delta\phi = \frac{2\pi (n_e - n_o)d}{\lambda} \tag{15.43}$$

so that the transmitted light is

$$\mathbf{E}_{\text{out}} = E_0 e^{i\phi_o} \left[\hat{\mathbf{y}} \sin(\theta) + \hat{\mathbf{z}} \cos(\theta) e^{i\Delta\phi} \right] e^{i(qx-\omega t)}, \tag{15.44}$$

where $\phi_o = 2\pi n_o d/\lambda$.

Now let me orient the polarization of the incoming plane polarized light so that $\theta = 45°$ and at the same time, make the thickness of the wave plate such that

$$(n_e - n_o)d = \frac{\lambda}{4}, \tag{15.45}$$

making $\Delta\phi = \pi/2$ and $e^{i\Delta\phi} = i$. In this case the transmitted field is left circularly polarized light:

$$\mathbf{E}_{\text{out}} = \frac{E_0 e^{i\phi_o}}{\sqrt{2}} \left[\hat{\mathbf{y}} + i\hat{\mathbf{z}}\right] e^{i(qx-\omega t)}. \tag{15.46}$$

This result is worth a few remarks. First, Eq. 15.45 makes clear the basis of the name "quarter-wave plate." If I want right circular polarization, I rotate the plate until the optic axis is along $\hat{\mathbf{y}}$. I can also use materials with $n_e < n_0$, in which case $\Delta\phi = -\pi/2$ and the analysis following Eq. 15.44 leads to right circular polarization. If I rotate things such that $\theta \neq 45°$, I get elliptically polarized light with an ellipticity that goes to zero when $\theta = 0$ or $90°$. Finally, if I make the quarter-wave plate from sapphire ($n_e - n_o = 0.008$ at $\lambda = 5,900$ Å), the thickness of a *zero-order* quarter-wave plate is rather thin, of order 18 μm. Multiple-order wave plates increase the thickness by as many factors of 2π in $\Delta\phi$ as needed to make a more robust device.

15.9.3 Half-Wave Plate

A half-wave plate is a device that rotates the polarization plane of linearly-polarized light. It differs from the quarter-wave plate only in the thickness d. The name even suggests the the difference; the half-wave plate is twice the thickness of the quarter-wave plate. The thickness satisfies

$$(n_e - n_o)d = \frac{\lambda}{2}, \tag{15.47}$$

making $\Delta\phi = \pi$ and $e^{i\Delta\phi} = -1$. Equation 15.44 then becomes

$$\mathbf{E}_{\text{out}} = E_0 e^{i\phi_o} \left[\hat{\mathbf{y}} \sin(\theta) - \hat{\mathbf{z}} \cos(\theta)\right] e^{i(qx-\omega t)}. \tag{15.48}$$

The light leaving the half-wave plate is clearly not parallel to the light entering it. I find the rotation angle Ψ by computing the dot product of the unit vectors of incident and transmitted light:

$$\cos(\Psi) = [\hat{y}\sin(\theta) + \hat{z}\cos(\theta)] \cdot [\hat{y}\sin(\theta) - \hat{z}\cos(\theta)]$$
$$= \sin^2(\theta) - \cos^2(\theta)$$
$$= \cos(2\theta).$$

If the incident polarization is at $45°$ to the optic axis, the plane of polarization is $90°$ away from the incident direction. Any other rotation angle may be obtained by adjusting the angle of the optic axis (or the incoming polarization) between 0 and $45°$.

15.10 Things Not in This Chapter

I'll now conclude the discussion of crystal optics by mentioning a couple of things that I have not covered. I only sketched the solution for biaxial crystals and omitted the extra richness associated with the ray surfaces of these materials. Also, I only briefly discussed the difference between optical waves in crystals and optical rays in crystals. Figure 15.7 illustrates the difference: the Poynting vector $\mathbf{S}$, which points in the direction of energy flow, is not parallel to the wave vector $\mathbf{q}$. Therefore, even for normal incidence, where Snell's laws tells me that the angle of refraction is zero (regardless of index of refraction), the energy of the extraordinary ray can propagate at an angle to the normal. This effect is usually demonstrated by looking at printed text through a birefringent crystal and observing two images of the text; the images have orthogonal polarizations.

I have not mentioned liquid crystals, where an applied electric field can induce anisotropy, or the Pockels electro-optic effect, where an electric field modifies the refractive index. These are both nonlinear processes of course, but once the field is set, the material acts as a linear anisotropic solid.

16

Magneto-Optics

This chapter explores the effect of a strong static magnetic field on the optical properties of metals and of insulators. The effects are substantial. One reason is that the static fields are really large, compared to the fields from ordinary incoherent light.* If for instance I have a 100 W light bulb, emitting in all directions, I can estimate that the magnetic field of the light at a distance of 1 m is about 2.6 mG or 2.6×10^{-7} T. This field is much smaller than the 0.5 G field at the Earth's surface.[†] Earth's field in turn is basically negligible compared to the 8–12 T superconducting solenoids existing at many laboratories or the 45 T hybrid magnet at the National High Magnetic Field Laboratory.[‡]

16.1 Charged Particle in a Magnetic Field

The field interacts with the charge motion in the solid through the Lorentz force,

$$\mathbf{F} = \frac{q}{c}\mathbf{v} \times \mathbf{B}_0 \tag{16.1}$$

for a particle with charge q. The charge for electrons is negative, $q = -e$. $\mathbf{B}_0$ is the strength of the (static) magnetic field. Note that when I need to be specific about directions, $\mathbf{B}_0$ will be in the $\hat{z}$ direction, so that I will write $\mathbf{B}_0 = B_0\hat{z}$ whenever I need to specify a direction. The velocity $\mathbf{v}$ is substantial; remember that the Fermi velocity is of order $c/200$. Figure 16.1 shows the three vectors when $\mathbf{v}$ is in the x–y plane.

Among the effects that I can address starting with Eq. 16.1 are the Hall effect, magnetoresistance, Cyclotron resonance, helicons (whistlers), and the Faraday effect. I'll start by considering free carriers (metals or doped semiconductors). Then, at the end of the chapter, I will work out Faraday rotation in a transparent insulator.

---

* Lasers, especially short-pulsed lasers are a different story.
[†] Wikipedia says 25–65 µT.
[‡] Fifty years or so ago, people reported the fields produced by magnets in their laboratories in G (Gauss) or kG (thousand Gauss). But at some time in the last few decades, reports of field in T (Tesla) became more common. This change may have been due to pressure to use SI units everywhere but may instead be due to the appearance of too many zeros in the cgs numbers; 12 T is easier to say than 120,000 G. In this chapter, I'll often specify fields in T, but I (and you) must remember that I am working in cgs-Gaussian units, so whenever I need to compute some quantity containing the field, I will need to enter field values in Gauss.

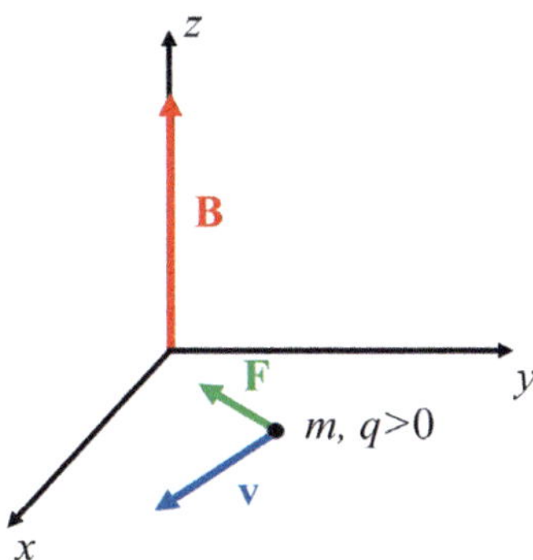

Fig. 16.1 Force **F** on a (positive) charge moving in a magnetic field **B** with velocity **v**.

16.1.1 Cyclotron Orbits

I'll start with the motion of a free electron in a metal when a magnetic field is applied. As I discussed in Chapter 8, the electrons are specified by a wave vector $\mathbf{k}$, related to the velocity by $\mathbf{v}_k = \hbar\mathbf{k}/m$ with m the mass. Thus the Lorentz force is

$$\mathbf{F}_k = -\frac{\hbar e}{mc}\mathbf{k} \times \mathbf{B}_0. \tag{16.1}$$

The force is orthogonal to both $\mathbf{k}$ and $\mathbf{B}_0$. Now the force leads to an acceleration, which in this case changes the direction of the velocity (or $\mathbf{k}$) vector without changing its magnitude. (Magnetic fields do no work; the kinetic energy is unchanged.) I use Eq. 8.13 and include the damping force from Eq. 8.14,

$$\hbar\frac{d\mathbf{k}}{dt} = \mathbf{F}_k - \hbar\frac{\delta\mathbf{k}}{\tau}$$

$$= -\frac{\hbar e}{mc}\mathbf{k} \times \mathbf{B}_0 - \hbar\frac{\delta\mathbf{k}}{\tau} \tag{16.2}$$

where τ is the mean free time between collisions.

Now, I take (for the moment) τ to be long enough that I can ignore it. The only force is the Lorentz force,

$$\frac{d\mathbf{k}}{dt} = -\frac{eB_0}{mc}\mathbf{k} \times \hat{\mathbf{z}}. \tag{16.3}$$

I'll write the components of $\mathbf{k}$, defining as I do ω_c, the *cyclotron frequency*,*

$$\omega_c \equiv \frac{eB_0}{mc} \tag{16.4}$$

---

* I'm making ω_c a positive quantity. I could equally well have made it negative (for electrons) by bringing the minus sign along. I'll have to remember that I did not do so when the question of the sense of the motion comes up.

so that

$$\frac{dk_z}{dt} = 0 \tag{16.5a}$$

$$\frac{dk_x}{dt} = -\omega_c k_y \tag{16.5b}$$

$$\frac{dk_y}{dt} = \omega_c k_x. \tag{16.5c}$$

The equations tell the story and some of it is pretty simple. The $\hat{z}$ component is constant because components of the velocity parallel to the magnetic field do not contribute to the Lorentz force. The other two are coupled; I solve them by taking another derivative of Eq. 16.5b to find that $\ddot{k}_x = -\omega_c \dot{k}_y = -\omega_c^2 k_x$. The solution is either a sine or cosine of $\omega_c t$. I look at the signs in Eqs. 16.5b and 16.5c and decide that

$$k_x = k_\perp \cos(\omega_c t)$$
$$k_y = k_\perp \sin(\omega_c t), \tag{16.6}$$

where $k_\perp$ is the magnitude of the component of k perpendicular to $\hat{z}$, the field direction; $k_\perp = \sqrt{k^2 - k_z^2}$. The in-plane part of the electron wave vector, $\mathbf{k}_\perp$ (and, equivalently, the in-plane velocity vector) remains tangent to the electron's circular orbit as it moves in a counterclockwise sense around the magnetic field.* The angular frequency of the orbital motion is given by Eq. 16.4. For electrons with a mass equal to m_e, the free-electron mass, the frequency $f_c = \omega_c/2\pi$ is

$$f_c = 2.80 \text{ MHz}\frac{B_0}{1 \text{ G}} = 28.0 \text{ GHz}\frac{B_0}{10 \text{ kG/1 T}} = 0.933 \text{ cm}^{-1}\frac{B_0}{10 \text{ kG/1 T}}.$$

If the effective mass of the charge carriers is higher or lower than the free-electron mass, the frequency will be correspondingly lower or higher. But the effect clearly occurs in the radio-frequency to far-infrared spectral region. Fields of 100 kG (10 T) provide cyclotron frequencies of just under 10 cm^{-1} (300 GHz/1.2 meV) for free electrons. If the effective mass is $m = 0.013m_e$, the electron effective mass in InSb, the frequency increases to $f_c = 720$ cm^{-1} (89 meV).

I convert from wave vector to velocity to find that $v_x = v_\perp \cos(\omega_c t)$ and $v_y = v_\perp \sin(\omega_c t)$ with $v_\perp = \hbar k_\perp/m$. Integrating once gives $x = x_0 + (v_\perp/\omega_c) \sin(\omega_c t)$ and $y = y_0 - (v_\perp/\omega_c) \cos(\omega_c t)$. The radius of the orbit is $R_c = (v_\perp/\omega_c)$. The velocity varies from zero to the Fermi velocity. So the maximum orbital radius is for electrons moving in the plane perpendicular to the field with the Fermi velocity; this radius is

$$R_c = \frac{v_F}{\omega_c} = \frac{m v_F c}{e B_0}.$$

At a field of 100 kG, the radius for an orbit for a Fermi surface electron moving in the x–y plane of a metal like silver is $R_c = 7{,}900$ Å. This is quite a bit longer than the mean free path at room temperature. (I usually use $\ell = 300$ Å for the mean free path at 300 K for a good metal like silver or copper. Thus $\ell/R_c = \omega_c \tau \approx 0.04$ at 300 K.) However at low

---

* If my charge carrier are holes, with $q = +e$, the sense is clockwise. All else is the same.

temperatures, the mean free path of pure metals can be 100–1,000 times longer, allowing many orbits before a scattering event upsets the electron. As I work through the various optical effects in a magnetic field, I'll see that it is the factor $\omega_c \tau$ that determines how heavy the hand of the magnetic field is on the properties of the solid. Large $\omega_c \tau$, big effect; small $\omega_c \tau$, modest effect.

16.1.2 Quantum Mechanical Picture

After the magnetic field is applied, the quantum numbers of the free electron metal, which in zero field were $\{k_x, k_y, k_z, \sigma\}$ where σ is the spin, are no longer good. I can see that they no longer represent stationary states of the Schrödinger equation because Eqs. 16.5a and 16.5b show that k_x and k_y are time dependent.

Calculating the energy levels and wave functions of the electrons in a magnetic field [347] has been of great interest since the excitement surrounding the quantum Hall effect. Most of the interest centers on the two-dimensional, low-density electron gas, where the quantum Hall effect is observed. But of course the physics is the same in metals, with their much larger Fermi energies.*

I'll consider completely free electrons, as in Chapter 8. But the momentum operator differs from that in Eq. 8.1, becoming $\mathbf{p} = -i\hbar\nabla - (e/c)\mathbf{A}$, where $\mathbf{A}$ is the magnetic vector potential. The Hamiltonin is $\mathcal{H} = p^2/2m$ so that the Schrödinger equation is

$$\frac{1}{2m}\left(-i\hbar\nabla - \frac{e}{c}\mathbf{A}\right)^2 \psi = E\psi. \tag{16.7}$$

Once I have a vector potential, I have the enjoyment of specifying a gauge. This problem is most simply approached in the Landau gauge, where the vector potential is oriented along $\hat{\mathbf{y}}$: $\mathbf{A} = B_0 x\,\hat{\mathbf{y}}$. If I take the curl, I find $\nabla \times \mathbf{A} = B_0\hat{\mathbf{z}}$, which is what I want.

In Cartesian coordinates, Eq. 16.7 becomes

$$-\frac{\hbar^2}{2m}\frac{\partial^2\psi}{\partial x^2} - \frac{\hbar^2}{2m}\left(\frac{\partial}{\partial y} - i\frac{eB_0}{\hbar c}x\right)^2\psi - \frac{\hbar^2}{2m}\frac{\partial^2\psi}{\partial z^2} = E\psi. \tag{16.8}$$

I separate variables. Try

$$\psi = u(x)e^{i\beta y}e^{ik_z z}$$

where the z dependence gives plane-wave solutions along z, with contribution to energy $\hbar^2 k_z^2/2m$. This result is no surprise; the magnetic field exerts no force on particles moving in the field direction

I substitute, cancel the complex exponentials, do a little arrangement of constants in the middle term, and arrive at

$$-\frac{\hbar^2}{2m}\frac{\partial^2 u(x)}{\partial x^2} + \frac{m}{2}\left(\frac{eB_0}{mc}x - \frac{\hbar\beta}{m}\right)^2 u(x) = E'u(x), \tag{16.9}$$

---

* The energies in question are the Fermi energy, of order eV in metals but meV in doped semiconductors, and $\hbar\omega_c$, of order meV to perhaps 100 meV, depending on the effective mass and the magnetic field.

with $E' = E - \hbar^2 k_z^2/2m$. I'm glad to see $\omega_c = eB_0/mc$ appearing in Eq. 16.9. This equation is that of a simple harmonic oscillator. I can write it in the conventional form using ω_c and by defining $x_0 = \hbar\beta/m\omega_c$ as the equilibrium point of the oscillator. Then

$$-\frac{\hbar^2}{2m}\frac{\partial^2 u(x)}{\partial x^2} + \frac{m\omega_c^2}{2}(x - x_0)^2 u(x) = E' u(x).$$

The function $u(x)$ is a product of a Gaussian centered at x_0 and Hermite polynomials. The energy eigenvalues are

$$E = \hbar\omega_c\left(n + \frac{1}{2}\right) + \frac{\hbar^2 k_z^2}{2m}, \tag{16.10}$$

with $n = 1, 2, 3, \ldots$ the Landau level index.

I can view the radius of a Landau orbit as being the classical turning point $\ell_n = x_{max} - x_0$ of the harmonic oscillator. The turning points are found by equating the total energy to the potential energy function.* I have $\hbar\omega_c(n + 1/2) = (m\omega_c^2/2)\ell_n^2$ or

$$\ell_n = \sqrt{\frac{2\hbar c}{eB_0}\left(n + \frac{1}{2}\right)} = \sqrt{\frac{2\Phi_0}{\pi B_0}\left(n + \frac{1}{2}\right)},$$

with $\Phi_0 = hc/2e$ the flux quantum. I see that the flux through the lowest Landau level $(n = 0)$, $\Phi = B_0(\pi\ell_0^2)$ is $\Phi = \Phi_0$. In fact there is a single flux quantum through the annulus between level $n - 1$ and level n for all n. Each Landau orbit n collects all the allowed $\{k_x, k_y\}$ points in the zero-field case that exist between it and the one below it in energy. (See Chapter 8.) Because each annulus has the same area, each has the same occupancy. The highest occupied level is that for which $\hbar\omega_c(n + 1/2) = E_F$.

There are orbits in k-space as well; see Eq. 16.6. These are oriented in a plane perpendicular to the field, i.e., in the x–y plane. The radius of the orbit with perpendicular wave vector $k_{\perp n}$ is given by $\hbar^2 k_{\perp n}^2/2m = \hbar\omega_c(n + 1/2)$. The orbital radius in k-space is then

$$k_{\perp n} = \sqrt{\frac{2eB_0}{\hbar c}\left(n + \frac{1}{2}\right)} = \sqrt{\frac{2\pi B_0}{\Phi_0}\left(n + \frac{1}{2}\right)}$$

with Φ_0 again the flux quantum. Note that $k_{\perp n}\ell_n = 2(n + \frac{1}{2})$.

The Fermi sphere reconstructs into a series of cylindrical shells, separated in energy by $\hbar\omega_c$. The energy separation increases as the magnetic field becomes larger; the number of Landau levels with energy below the Fermi energy becomes correspondingly smaller. As field is increased, a given Landau level's energy increases from below E_F to above E_F. The passing of Landau levels through the Fermi energy is responsible for oscillations in susceptibility (de Haas-van Alphen effect) and transport properties (Shubnikov-de Haas effect) as a function of field.[†]

Spin adds another magnetic energy, a Zeeman term, $E_\sigma = g\mu_B B_0 m_s$ where g is the Landé g-factor of the electron (just over 2 for free electrons but quite different from 2 in

---

* There are better ways to do this based on correspondence principle or on finding the maxima of the quantum mechanical probability.

[†] Actually the oscillations are periodic in $1/B_0$.

multi-electron atoms), $\mu_B = e\hbar/2mc$ is the Bohr magneton, and m_s the z-component of the electron spin. For free electrons $m_s = \pm 1/2$ and I take $g \approx 2$, making the Zeeman energy become

$$E_\sigma = \pm \hbar \frac{e B_0}{2mc} = \pm \frac{\omega_c}{2}. \tag{16.11}$$

So the spin down electrons are lowered in energy by an amount equal to the zero-point energy of the Landau levels while the spin up electrons are raised by the same amount. (In solids where there are many electrons in each atom, in cases when spin-orbit interactions are large, and in situations where there are strong crystal field effects, g differs very much from the free-electron value.)

16.2 Magneto-Conductivity Tensor of a Metal

Now I want to consider the optical properties of a free-electron metal in a magnetic field. I'll use the concepts of the familiar Drude model as developed in Chapter 8. The metal is subject to an electric field, so that I must modify Eq. 16.2 by including $\mathbf{E}$. The three forces to which an electron is subject are the accelerating electric field. the deflecting magnetic field, and the stopping damping force. And this time I cannot neglect damping, characterized by a relaxation time τ. Newton's law is

$$\hbar \frac{d\mathbf{k}}{dt} = -e\mathbf{E} - \frac{\hbar e}{mc}\mathbf{k} \times \mathbf{B}_0 - \hbar \frac{\delta \mathbf{k}}{\tau}. \tag{16.12}$$

I'll now make a few comments and state a few assumptions:

1. I can't do what I did in Eq. 8.12 and have the electric field along $\hat{\mathbf{y}}$ and the wave traveling along $\hat{\mathbf{x}}$. The reason is that the magnetic field is fixed to $\hat{\mathbf{z}}$, and I want to allow for cases where the electric field and propagation directions each have components parallel or perpendicular to the magnetic field. Hence, I'll write

$$\mathbf{E} = \mathbf{E}_0 e^{i(\mathbf{q}\cdot\mathbf{r} - \omega t)} \tag{16.13}$$

   and specify the directions of the electric-field vector $\mathbf{E}_0$ and wave vector $\mathbf{q}$ later.
2. I am operating in the linear* regime. As in the case without a magnetic field, this assumption means that after any start-up transient decays (requiring time τ), the time dependence of the electron's wave vector $\mathbf{k}$ follows $\mathbf{E}$, so that

$$\frac{d\mathbf{k}}{dt} = -i\omega\mathbf{k}.$$

3. The field is along $\hat{\mathbf{z}}$, so that $\mathbf{k} \times \mathbf{B}_0 = B_0 \mathbf{k} \times \hat{\mathbf{z}}$.
4. I can average over the $\mathbf{k}$ values of the occupied states, so that

$$\langle \mathbf{k} \rangle = \langle \delta \mathbf{k} \rangle = \frac{m}{\hbar} v_{\text{drift}}.$$

---

* And isotropic and homogeneous!

Moreover, $\delta\mathbf{k} = \mathbf{k} - \mathbf{k}_0$ with $\mathbf{k}_0$ the wave vector of the electron in equilibrium and $\langle\mathbf{k}_0\rangle = 0$. See the discussion around Eq. 8.14.

5. The current is just the charge density times the average velocity (or the number density n times the electronic charge $-e$ times the drift velocity),

$$\mathbf{j} = -ne\mathbf{v}_{\text{drift}} = -ne\frac{\hbar}{m}\langle\delta\mathbf{k}\rangle.$$

6. As ever, the cyclotron frequency is $\omega_c = eB_0/mc$.

Equation 16.12 becomes

$$-i\omega\langle\mathbf{k}\rangle = -\frac{e}{\hbar}\mathbf{E} - \omega_c\langle\mathbf{k}\rangle\times\hat{z} - \frac{\langle\delta\mathbf{k}\rangle}{\tau}.$$

I can work this problem at least two ways. One is to solve for $\delta\mathbf{k}$, taking into account assumption 4 and then use assumption 5 to get the current. The other is to use assumption 5 to turn the equation of motion into an equation for the current. I'll do the latter, substituting $\langle\delta\mathbf{k}\rangle = \langle\mathbf{k}\rangle = -(m/ne\hbar)\mathbf{j}$. I'll multiply through by $ne\hbar\tau/m$, leave the term containing $\mathbf{E}$ on the right, and move all terms containing $\mathbf{j}$ to the left, to find that my equation of motion has become

$$(1 - i\omega\tau)\mathbf{j} + \omega_c\tau\,\mathbf{j}\times\hat{z} = \frac{ne^2\tau}{m}\mathbf{E}. \tag{16.14}$$

Were it not for the magnetic field, this equation would become the Drude formula, Eq. 8.21 or Eq. 4.8.

But I do have a magnetic field. And I see that Eq. 16.14 tells me that whatever the direction of $\mathbf{j}$ (other than along $\hat{z}$), there will be a component of $\mathbf{E}$ that is parallel to $\mathbf{j}$ and one that is perpendicular to $\mathbf{j}$. Moreover, the factor $\omega_c\tau$ determines (at low frequencies at least) whether the parallel or perpendicular component is the larger.

To work out the relation between $\mathbf{j}$ and $\mathbf{E}$ (the conductivity), it helps to simplify the formulas. I'll define ϖ as

$$\varpi \equiv \omega + i\frac{1}{\tau}. \tag{16.15}$$

ϖ equals ω at high frequencies and i/τ at low frequencies.* I'll also define $\sigma_D(\omega)$ as

$$\sigma_D(\omega) = \frac{ne^2\tau/m}{1 - i\omega\tau} = \frac{ine^2}{m\varpi}. \tag{16.16}$$

σ_D is just the zero-field frequency-dependent Drude conductivity, seen in in Eq. 8.21 or Eq. 4.8. (It was called just σ there.) With these definitions, I am able to write Eq. 16.14 in a way that resembles Ohm's law:

$$\mathbf{j} + i\frac{\omega_c}{\varpi}\mathbf{j}\times\hat{z} = \sigma_D\mathbf{E}. \tag{16.17}$$

---

* It might seem more direct to define a complex relaxation time as $\tilde{\tau} = \tau/(1 - i\omega\tau)$, making the Drude conductivity be $\sigma_D = ne^2\tilde{\tau}/m$. This definition connects well to the low-frequency conductivity. But shortly I'll be interested in frequencies near ω_c in the case where $\omega_c\tau \gg 1$ and the use of ϖ makes that discussion easier.

The components of Eq. 16.17 are

$$j_x + i\frac{\omega_c}{\varpi}j_y = \sigma_D E_x \tag{16.18x}$$

$$j_y - i\frac{\omega_c}{\varpi}j_x = \sigma_D E_y \tag{16.18y}$$

$$j_z = \sigma_D E_z, \tag{16.18z}$$

where the minus sign in Eq. 16.18y is a consequence of the right-hand rule in the cross product. Equation 16.18 tells me that if I apply an electric field in the $\hat{x}$ direction, I will get currents in both $\hat{y}$ and $\hat{x}$ directions. Which current is bigger depends on the value of $\omega_c\tau$.

I can immediately define a resistivity tensor $\overset{\leftrightarrow}{\rho}$ which relates electric field to current according to

$$\overset{\leftrightarrow}{\rho}\cdot\mathbf{j} = \mathbf{E}.$$

The resistivity tensor is

$$\overset{\leftrightarrow}{\rho} = \rho_D \begin{pmatrix} 1 & i\omega_c/\varpi & 0 \\ -i\omega_c/\varpi & 1 & 0 \\ 0 & 0 & 1 \end{pmatrix} \tag{16.19}$$

where $\rho_D = 1/\sigma_D$. I must remember that σ_D is a frequency-dependent complex quantity!

I also want to find the conductivity tensor $\overset{\leftrightarrow}{\sigma}$, for which

$$\mathbf{j} = \overset{\leftrightarrow}{\sigma}\cdot\mathbf{E}.$$

I could use some matrix inversion method if I want to be sophisticated, but I found it simpler just to go back to the component equations. I can solve Eq. 16.18y for j_y, substitute that into Eq. 16.18x, do a bit of algebra, and find

$$j_x = \frac{\sigma_D\varpi^2}{\varpi^2 - \omega_c^2}E_x - i\frac{\sigma_D\omega_c\varpi}{\varpi^2 - \omega_c^2}E_e y.$$

The coefficient of E_x is σ_{xx} and the coefficient of E_y is σ_{xy}. Evidently, and not unexpectedly, $\sigma_{xz} = 0$. Now I go back to Eq. 16.18x and solve for j_y and get something very similar. (There is a sign difference in the off-diagonal term.) The $\hat{z}$ current depends only on the $\hat{z}$ field, so I now know that the conductivity tensor is

$$\overset{\leftrightarrow}{\sigma} = \sigma_D \begin{pmatrix} \dfrac{\varpi^2}{\varpi^2 - \omega_c^2} & -i\dfrac{\omega_c\varpi}{\varpi^2 - \omega_c^2} & 0 \\ i\dfrac{\omega_c\varpi}{\varpi^2 - \omega_c^2} & \dfrac{\varpi^2}{\varpi^2 - \omega_c^2} & 0 \\ 0 & 0 & 1 \end{pmatrix}, \tag{16.20}$$

with $\sigma_D = ine^2/m\varpi = ne^2\tau/m(1 - i\omega\tau)$. (I remind myself that $\varpi = \omega + i/\tau$.)

Note that the conductivity tensor and the resistivity tensor are explicitly antisymmetric. Thus, there is no way to orient someting to make them diagonal.

16.2.1 Hall Effect

The Hall effect is a consequence of the off-diagonal terms in the magnetoconductivity tensor [19, 348]. Discovered by Edwin Hall [349] in 1879, the effect is the appearance of an electric field in a conductor at right angles to the electrical current in the conductor when there is a magnetic field perpendicular to the current. It is the basis of a large number of sensors; Hall devices are used to measure magnetic fields, make noncontact current measurements, and sense rotational speeds of engines.

The Hall effect as it usually is experienced is a low-frequency phenomenon, so I'll take $\omega\tau \ll 1$, making $\varpi = i/\tau$ and $\sigma_D = ne^2\tau/m$. The geometry of a Hall device is simple: it is a rectangular prism of thickness t, width w, and length L, often called a Hall bar. A current flows along the length of the device, in the $\hat{x}$ direction. (If the charge carriers are electrons, they flow in the $-\hat{x}$ direction.) A magnetic field is oriented along $\hat{z}$. The Lorentz force will accelerate the electrons in the $-\hat{y}$ direction. The electrons cannot escape from the device so they pile up along the $-y$ edge of the device, making that side negatively charged. The other side is depleted of electrons and becomes positively charged. The charge imbalance creates an electric field, the Hall field, from plus to minus; this field exerts a force on the electrons in the $+\hat{y}$ direction.* The charge buildup stops when the Hall electrical force balances the Lorentz magnetic force.

The Hall coefficient R_H is defined via

$$E_y = R_H B_0 j_x. \tag{16.21}$$

With current only in the $\hat{x}$ direction, the resistivity tensor gives the fields induced by this current. They are:

$$E_x = -\rho_D\, j_x$$
$$E_y = -\rho_D\omega_c\tau j_x.$$

Evidently

$$R_H B_0 = -\rho_D\omega_c\tau = -\frac{m}{ne^2\tau}\cdot\frac{eB_0\tau}{mc}.$$

There are some cancellations, after which I find

$$R_H = -\frac{1}{nec}.$$

R_H is negative because electrons are negative. The Hall effect allows measurement of the carrier density. Moreover, if the charge carriers are positive holes instead of negative electrons, R_H will be positive.

The electric field in the direction of current, E_x, is often called the longitudinal field. It is independent of magnetic field strength in this model.[†] The electric field in the transverse direction, the Hall field, grows without limit. Surfaces are not required to have the Hall

---

* Why, oh why, did Franklin not choose the other signs for vitreous and resinous electricity?
[†] A result contradicted by experiment.

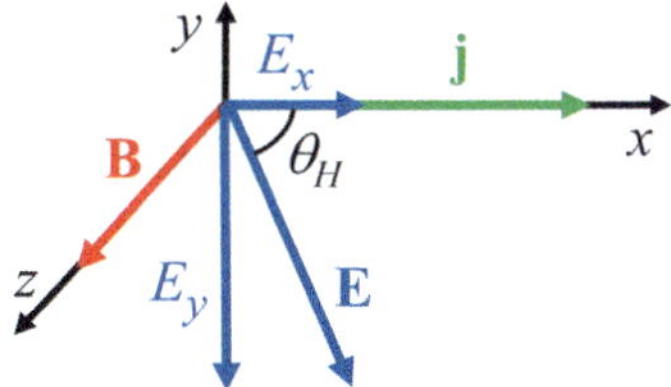

  The electric current and electric field in a metal (whose charge carriers are negative) with a strong magnetic field along the $\hat{z}$ axis.

field or a Hall current. In fact, I only put the surface in to make contact with the Hall bar geometry used to measure the Hall coefficient.

Figure 16.2 shows the relation between current and electric field inside the material. The external magnetic field is (as always) along $\hat{z}$. The current $\mathbf{j}$ is in the $\hat{x}$ direction and there is a component of the electric field E_x along the current. The Hall field in Fig. 16.2, along $-\hat{y}$ (because electrons are negatively charged), is larger than the longitudinal field; the diagram is drawn for $\omega_c \tau \approx 3$. The angle between $\mathbf{j}$ and $\mathbf{E}$ is the Hall angle, θ_H. The Hall angle satisfies $\tan \theta_H = \omega_c \tau$. As $\omega_c \tau$ increases from zero to a large value—by increasing B_0 or decreasing temperature*—the Hall angle increases from zero to $90°$.

This free electron model does a moderately good job of describing simple metals. For me to take it seriously, I'd like for experiment to observe a Hall coefficient that is independent of magnetic field and current, so that E_H is linear in B_0. I'd also like for R_H to be independent of temperature in metals and to agree with the thermal generation of free carriers for semiconductors. For the elements that are metals, one can list $-nec R_H$, which of course should be unity. To test whether it is, I need to know the carrier density n by other means. I can get it from transport (if I know τ). I can get it from chemistry. I can get it from electronic structure calculations. I can get it from the square of the plasma frequency. Here, I'll get it from Table 8.1. Then I find at high fields and low temperatures

$$
-nec R_H = \begin{cases}
1.0 \pm 0.1 & \text{Na, K, Cs, Rb} \\
1.4 \pm 0.2 & \text{Cu, Ag, Au} \\
-0.3 & \text{Al (at high fields).}
\end{cases}
$$

The alkali metals behave as I expect them to because I know that they lose their single valence electron easily and that their Fermi surfaces are close to spherical. The noble metals appear to have fewer carriers than their reputation as monovalent metals would suggest. To understand this deficit I have to look at their Fermi surfaces [19, 23] and notice the necks and the possibility of a number of "orbits" that the $\mathbf{k}$ vector can take in the field. Aluminum starts by having the wrong sign, suggesting positive carriers, and only one hole per atom. This result is a surprise to the free-electron picture but actually not uncommon. Many metals have positive Hall coefficients. It is a consequence of their electronic band structure. Another consequence (observed in aluminum among others) is that the Hall

---

* Increasing B_0 will *always* make $\omega_c \tau$ bigger; decreasing temperature may or may not do the trick. It depends on the cleanliness and the physics of the material.

coefficient has different values in low-field and high-field regimes, where the crossover occurs when $\omega_c \tau = 1$. Indeed, at high fields the Landau level spacing begins to dominate and Shubnikov-de Haas oscillations [19, 350] in resistivity and Hall effect can be seen.[*]

16.2.2 The Dielectric Tensor

Now that I know the conductivity tensor, I can think about how electromagnetic waves propagate in a metal subject to a static magnetic field. I'll keep the static field $\mathbf{B}_0$ along the $\hat{z}$ axis and orient the wave vector and electric field directions relative to that. The waves are governed by the complex frequency-dependent conductivity tensor, as written in Eq. 16.20, or, equivalently, the dielectric tensor. And, as ever and ever, by Maxwell's equations. I can use either the conductivity or the dielectric tensor[†] to develop the solutions for wave propagation. I'll make the same choice that I did in Chapter 2 and use the dielectric tensor. In terms of the tensor of Eq. 16.20, I write

$$\overset{\leftrightarrow}{\epsilon}(\omega) = \epsilon_c \overset{\leftrightarrow}{1} + \frac{4\pi i}{\omega} \overset{\leftrightarrow}{\sigma} \tag{16.22}$$

where ϵ_c is the (scalar) high-frequency contribution,[‡] $\overset{\leftrightarrow}{1}$ is the unit tensor, one on the diagonals, 0 on the off diagonals.

I'll now write the dielectric tensor explicitly. As before, I define a plasma frequency $\omega_p = \sqrt{4\pi n e^2/m}$ so that $\sigma_D = i\omega_p^2/4\pi\varpi$. After a bit of algebra I obtain

$$\overset{\leftrightarrow}{\epsilon} = \begin{pmatrix} \epsilon_c + \dfrac{\omega_p^2 \varpi}{\omega(\omega_c^2 - \varpi^2)} & -i\dfrac{\omega_p^2 \omega_c}{\omega(\omega_c^2 - \varpi^2)} & 0 \\[2ex] i\dfrac{\omega_p^2 \omega_c}{\omega(\omega_c^2 - \varpi^2)} & \epsilon_c + \dfrac{\omega_p^2 \varpi}{\omega(\omega_c^2 - \varpi^2)} & 0 \\[2ex] 0 & 0 & \epsilon_c - \dfrac{\omega_p^2}{\omega\varpi} \end{pmatrix}. \tag{16.23}$$

The first two diagonal components are equal, $\epsilon_{xx} = \epsilon_{yy}$. If $\omega_c \tau \gg 1$, these quantities have a sharp resonance at $\omega = \omega_c$; their real parts are positive below ω_c and negative above it. (They will cross zero again at $\omega_p/\sqrt{\epsilon_c}$.) The third diagonal component, ϵ_{zz}, has the dielectric function of a Drude metal, identical to Eq. 4.13.

The nonzero off-diagonal components are opposites: $\epsilon_{xy} = -\epsilon_{yx}$. They display the same resonance near ω_c as the diagonal terms do. At first glance, they may appear to be pure imaginary, but $\varpi = \omega + i/\tau$ is complex, so the off-diagonal terms are also complex.[§]

---

[*] I have neglected up to now to mention the very interesting and very well known quantum Hall effect. I'll just say that it is beyond the scope of this book, as are the Shubnikov-de Haas oscillations and related quantum oscillations.

[†] But not both!

[‡] In Section 16.5, I'll show that the dielectric function of an insulator (and ϵ_c is the simplest representation of an insulator I can think of) is also a tensor when a magnetic field is applied. For the time being, I'll pretend that this is not the case.

[§] Because $\epsilon_{xy} = -\epsilon_{yx}$, the imaginary part of one of the two off-diagonal terms will be negative. Generally, negative imaginary parts of dielectric functions – or negative real parts of conductivities – are forbidden by energy conservation. Here, because for example σ_{xy} relates the x component of current to the y component of electric field, the corresponding part of $\mathbf{j} \cdot \mathbf{E} = 0$.

16.3 Electromagnetic Wave Propagation

Now I will think about the propagation of electromagnetic waves in this metal. Making another visit to Maxwell's equations, I continue to write the electric and magnetic fields of the light as complex plane waves, as in Eqs. 2.5 and 2.6. I also have

$$\mathbf{D} = \overset{\leftrightarrow}{\epsilon} \cdot \mathbf{E},$$

so that $\mathbf{D}$ is not parallel to $\mathbf{E}$. I start with Maxwell's equations, Eqs. 2.4a–2.4d, in local, nonmagnetic, homogeneous, and linear materials. After substituting the plane-wave fields into Maxwell's equations, I have

$$\mathbf{q} \cdot \overset{\leftrightarrow}{\epsilon} \cdot \mathbf{E} = 0 \tag{16.24a}$$

$$\mathbf{q} \cdot \mathbf{H} = 0 \tag{16.24b}$$

$$\mathbf{q} \times \mathbf{E} = \frac{\omega}{c}\mathbf{H} \tag{16.24c}$$

$$\mathbf{q} \times \mathbf{H} = -\frac{\omega}{c}\overset{\leftrightarrow}{\epsilon} \cdot \mathbf{E}. \tag{16.24d}$$

I look at these equations for a bit and see that $\mathbf{H}$ is perpendicular to both $\mathbf{E}$ and $\mathbf{q}$ but that $\mathbf{E}$ and $\mathbf{q}$ are not perpendicular to each other.

The next step is to combine things in the usual way so as to eliminate $\mathbf{H}$. I take the cross product of $\mathbf{q}$ with Eq. 16.24c, use the $\mathbf{BAC}-\mathbf{CAB}$ rule on the triple cross product, and replace $\mathbf{q} \times \mathbf{H}$ using Eq. 16.24d:

$$\mathbf{q} \times (\mathbf{q} \times \mathbf{E}) = \frac{\omega}{c}\mathbf{q} \times \mathbf{H}$$

$$\mathbf{q}(\mathbf{q} \cdot \mathbf{E}) - q^2\mathbf{E} = -\frac{\omega^2}{c^2}\overset{\leftrightarrow}{\epsilon} \cdot \mathbf{E}. \tag{16.25}$$

In principle, there will be components of the propagation vector $\mathbf{q}$ and the electric field $\mathbf{E}$ along all three Cartesian axes. However, the only physics distinction is between $\hat{z}$ and perpendicular to $\hat{z}$. An electron running along $\hat{y}$ is deflected towards $\hat{x}$ and one running along $\hat{x}$ is deflected toward $-\hat{y}$. So, without loss of generality, I can rotate the x–y plane until $\mathbf{q}$ has only components along $\hat{x}$ and $\hat{z}$:

$$\mathbf{q} = q_x\hat{x} + q_z\hat{z} = q\sin\theta\,\hat{x} + q\cos\theta\,\hat{z}$$

where θ is the polar angle, as illustrated in Fig. 16.3. Note that I can *not* do what I did for anisotropic crystals (Section 15.6.3), which was to consider independently the cases of electric field parallel and perpendicular to the x–z plane. The dielectric tensor (or conductivity tensor) has intrinsically off-diagonal components, so there is always a component of $\mathbf{E}$ parallel to the current and a (Hall) component perpendicular to it. Thus I can only write

$$\mathbf{E} = E_x\hat{x} + E_y\hat{y} + E_z\hat{z}. \tag{16.26}$$

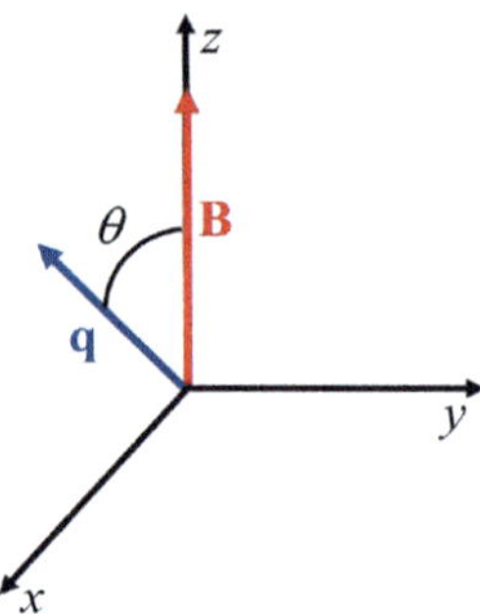

Fig. 16.3 The wave vector **q** is in the x–z plane, at an angle θ to the $\hat{z}$ axis.

Proceeding, I define the complex refractive index N via

$$\mathbf{q} = N\frac{\omega}{c}\sin\theta\,\hat{x} + N\frac{\omega}{c}\cos\theta\,\hat{z}. \tag{16.27}$$

I look at Eq. 16.25 and see that when Eqs. 16.26 and 16.27 are substituted, there will be a common factor of ω^2/c^2 everywhere. I do the substitutions, cancel this factor, collect the components, making use of $\epsilon_{yy} = \epsilon_{xx}$ and $\epsilon_{yx} = -\epsilon_{xy}$ in Eq. 16.23 to obtain

$$
\begin{aligned}
0 = &\,[(\epsilon_{xx} - N^2\cos^2\theta)E_x + \epsilon_{xy}E_y + N^2\cos\theta\sin\theta\,E_z]\hat{x} \\
&+ [-\epsilon_{xy}E_x + (\epsilon_{xx} - N^2)E_y]\hat{y} \\
&+ [N^2\cos\theta\sin\theta\,E_x + (\epsilon_{zz} - N^2\sin^2\theta)E_z]\hat{z}.
\end{aligned}
\tag{16.28}
$$

For a vector to be zero, all three components must be zero; thus, Eq. 16.28 is actually three equations. I use the $\hat{z}$ component to find a function representing E_z and the $\hat{y}$ component to do the same for E_y, substitute these into the $\hat{x}$ component, cancel E_x, and obtain an equation for N^2 in terms of θ, ϵ_{xx}, and ϵ_{xy}. At first it appears to be a cubic equation, but the N^6 terms add to zero, so it is quadratic:

$$
\begin{aligned}
0 = &\, N^4(\epsilon_{xx}\sin^2\theta + \epsilon_{zz}\cos^2\theta) \\
&- N^2[\epsilon_{xx}\epsilon_{zz}(1 + \cos^2\theta) + (\epsilon_{xx}^2 + \epsilon_{xy}^2)\sin^2\theta] \\
&+ (\epsilon_{xx}^2 + \epsilon_{xy}^2)\epsilon_{zz}.
\end{aligned}
\tag{16.29}
$$

Although I know how to solve quadratic equations, I find that I gain no insight in writing the general solution for N^2.* Instead, I will consider two special cases.

16.3.1 Faraday Geometry: Wave Vector Parallel to Field

First, I set $\hat{q} = \hat{z}$; $\theta = 0$. In this case, Eq. 16.29 becomes

$$0 = N^4 - 2N^2\epsilon_{xx} + \epsilon_{xx}^2 + \epsilon_{xy}^2$$

---

* I did check that if there is no magnetic field, so that $\overleftrightarrow{\epsilon} = \epsilon\overleftrightarrow{1}$, then $N^2 = \epsilon$.

the solutions to which are

$$N_\pm^2 = \epsilon_{xx} \pm i\epsilon_{xy}. \tag{16.30}$$

I use $N_\pm$ in Eq. 16.28 (the $\hat{\mathbf{y}}$ component) to find

$$E_y = \pm i E_x.$$

The $\hat{\mathbf{z}}$ component gives $E_z = 0$. Thus I have transverse waves ($\mathbf{q} \cdot \mathbf{E} = q\hat{\mathbf{z}} \cdot \mathbf{E} = 0$). I can go back to Eqs. 15.17 or 16.25, or even to Maxwell's equations, and find that the solutions are circular polarized

$$\mathbf{E}_\pm = E_0(\hat{\mathbf{x}} \pm i\hat{\mathbf{y}})e^{(qz-\omega t)}$$

with $q = (\omega/c)N_\pm$. See Eq. 15.4; the minus sign is for right circular polarization and the plus is for left circular polarization.*

16.3.2 Cyclotron Resonance and Helicons

The normal modes of the metal in a magnetic field and with propagation along the field direction are circularly polarized light. I'll define the dielectric functions of the circular polarized modes as

$$\epsilon_\pm \equiv N_\pm^2 = \epsilon_{xx} \pm i\epsilon_{xy},$$

where the components are given in Eq. 16.23. Writing these out,

$$\epsilon_\pm = \epsilon_c + \frac{\omega_p^2(\varpi \pm \omega_c)}{\omega(\omega_c^2 - \varpi^2)}.$$

This formula looks fine and has a nice resonant denominator, but I notice that $\omega_c^2 - \varpi^2 = (\omega_c + \varpi)(\omega_c - \varpi)$, allowing me to write the dielectric function for the left circular polarization as

$$\epsilon_+ = \epsilon_c + \frac{\omega_p^2}{\omega(\omega_c - \varpi)} = \epsilon_c + \frac{\omega_p^2}{\omega(\omega_c - \omega) - i\omega/\tau}, \tag{16.31}$$

while that for the right circular polarization is

$$\epsilon_- = \epsilon_c - \frac{\omega_p^2}{\omega(\omega_c + \varpi)} = \epsilon_c - \frac{\omega_p^2}{\omega(\omega_c + \omega) + i\omega/\tau}. \tag{16.32}$$

I have inserted in the rightmost formulas the definition $\varpi = \omega + i/\tau$. There are big differences in the behavior of ϵ_+ and ϵ_-, especially in the case where $\omega_c\tau \gg 1$. ϵ_+ is positive from $\omega = 0$ to $\omega = \omega_c$. Above ω_c it is negative and remains negative up to the screened plasma frequency. ϵ_- is negative at all frequencies up to the screened plasma frequency, just as in the case of zero magnetic field.

Figure 16.4 shows the real parts of the conductivity, σ_{1+}, and dielectric function, ϵ_{1+}, for a pure metal in zero field and in fields of 10, 30, 100, and 300 kG (1, 3, 10, and 30 T).

---

* See footnote in Section 15.2.3 about the conventions in use.

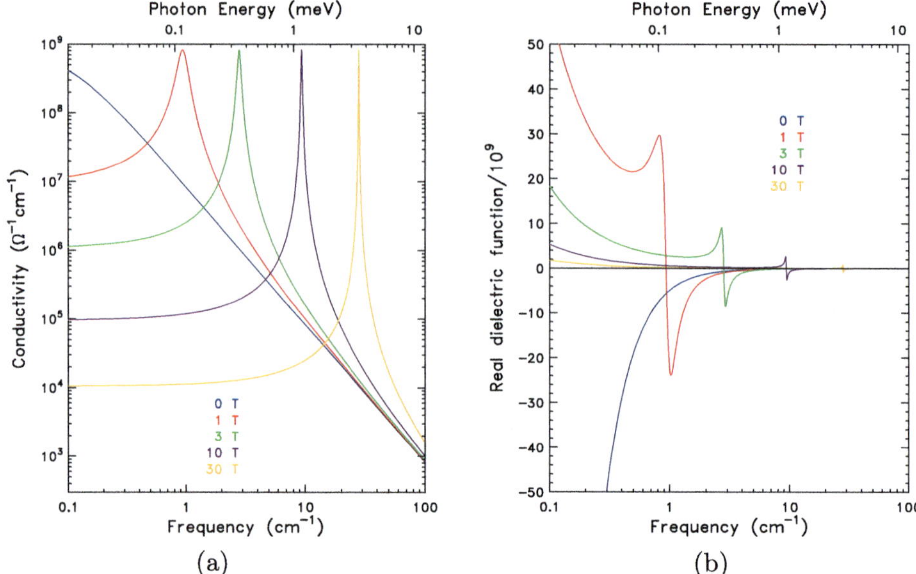

Fig. 16.4   (a) Faraday geometry conductivity for the $+$ polarization (left circular polarization) of a pure metal with $\hat{z}$-axis magnetic field. (b) Real part of the dielectric function in the same configuration. Note that the dielectric function values have been divided by 10^9 before plotting.

The metal is described by Eq. 16.31 with parameters chosen to represent silver or copper at low temperatures: $\omega_p = 70,000$ cm^{-1} (8.5 eV) and $1/\tau = 0.1$ cm^{-1}. The corresponding scattering time, $\tau = 50$ ps, gives a mean free path $\ell = 80$ μm and a resistivity ratio $RRR = 1,300$. The cyclotron frequencies of the conduction electrons are 0.93, 2.8, 9.3, and 28 cm^{-1} at the four fields so that the product $\omega_c\tau$ increases from 10 at 10 kG to 300 at 300 kG.

The zero-field conductivity and the zero-field $\epsilon_1(\omega)$ follow the Drude form (shown in Fig. 4.4). The conductivity rolls off as frequency increases, following $\sigma_1 \sim 1/\omega^2$ at high frequencies. The dielectric function is negative, also going as $\epsilon_1 \sim -1/\omega^2$. Things change dramatically when the field is applied. The low frequency conductivity is suppressed, more and more with increasing field. The suppression is consistent with the tensor conductivity of Eq. 16.20, although I need to remember that I am plotting only the real part of $\sigma_+ = \sigma_{xx} + i\sigma_{xy}$.

There is a strong resonance at $\omega = \omega_c$. This cyclotron resonance carries most of the spectral weight originally associated with the DC conductivity. The maximum conductivity is $\sigma_1(\omega_c) = \omega_p^2\tau/4\pi$, the same as the DC conductivity. The resonance becomes sharper as field increases, because $\omega_c\tau$ is linear in the field. The conductivity in field is close to that in zero field for $\omega \gg \omega_c$; this behavior continues to high frequencies. The real part of the dielectric function, ϵ_{1+}, has even more dramatic changes when the field is applied. It is positive for all $\omega < \omega_c$. A positive ϵ_{1+}, along with an ϵ_{2+} that is not too large, implies a region of transparency in the metal.

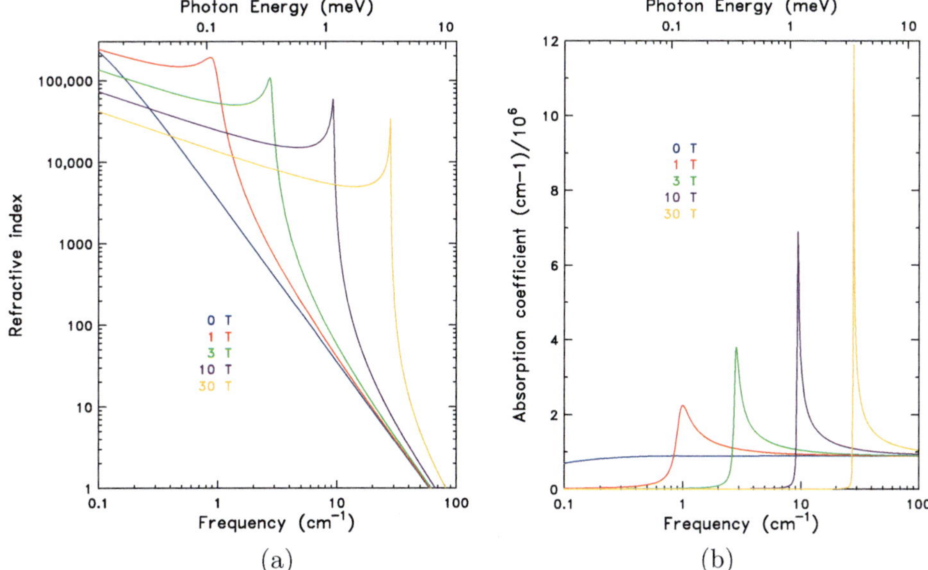

Fig. 16.5 (a) Refractive index for the + polarization (left circular polarization) of a pure metal in with $\hat{z}$-axis magnetic fields in the Faraday geometry. (b) Absorption coefficient in the same configuration. Note that the absorption coefficient values have been divided by 10^6 before plotting.

The optical constants that govern wave propagation are shown in Fig. 16.5. The left panel shows the real part of the refractive index and the right shows the absorption coefficient of a metal in a strong field. The zero-field curve is the same as in the Drude model discussion. (See Fig. 4.5. The zero-field absorption coefficient starts off as $\sqrt{\omega}$ and then flattens out when $\omega\tau \gg 1$.) In a strong field, the refractive index is large, of order $n \approx 10^4$–10^5. Consequently, electromagnetic wave propagation is slow, $v = c/n \approx 10^5$–10^6 cm/s.

The absorption coefficient is small for frequencies below the cyclotron frequency. There is a strong and narrow peak when $\omega = \omega_c$ (the cyclotron resonance absorption) and then a decrease to the zero field value.* The transmission of circularly polarized electromagnetic radiation at frequencies below the cyclotron resonance goes under several names: helicons, Alfvén waves, and whistlers [351]. Helicons have been seen in many clean metals, including potassium, sodium, indium, aluminum, and bismuth [352, 353].

Cyclotron resonance requires $\omega_c\tau > 1$ to give the appearance of a resonance in the spectrum. Figure 16.6 shows the effect of changing the scattering time τ from $\omega_c\tau = 30$ to $\omega_c\tau = 0.2$. If $\omega_c\tau \approx 1$, the resonance is not observed. If I wanted to take the maximum in α when $\omega_c\tau = 1$ as corresponding to the cyclotron frequency, I would obtain too large a value for the resonance frequency. The absorption curve sharpens for higher values of $\omega_c\tau$. (I observe the same thing in Fig. 16.5b, where I am varying ω_c and keeping τ constant.)

---

* Note that Fig. 16.5 shows a small absorption coefficient below the cyclotron frequency, but it is only small relative to zero field value. The parameters that I used give $\alpha(\omega = 1\ \mathrm{cm}^{-1}) = 300\ \mathrm{cm}^{-1}$. The transparency is relative; a slab of a 100 μm thickness might transmit 5% of the light that enters on the incident side to the far side. The reflection losses will be pretty impressive as well, given that $n \approx 10^5$!

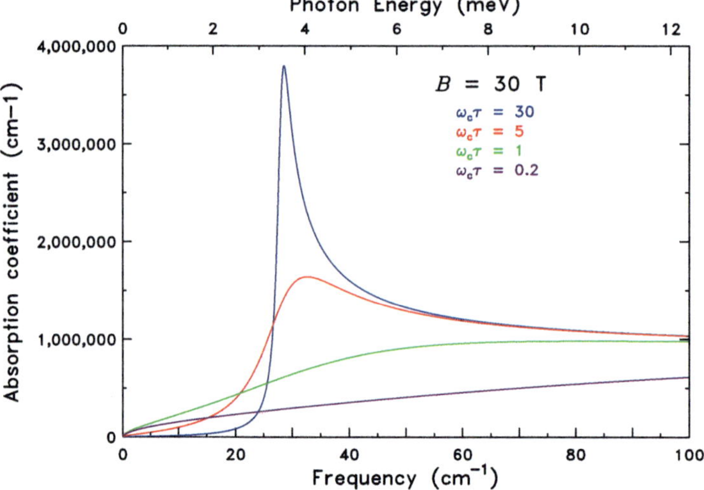

Fig. 16.6   Absorption coefficient of a metal for four values of the scattering time τ. The cyclotron frequency is $\omega_c = 28\,\text{cm}^{-1}$.

16.3.3 Cyclotron Resonance in Semiconductors

So far in this chapter, I have worked out and discussed the magneto-optics of metals. In actuality, the magneto-optics of semiconductors are equally if not more interesting and are more commonly measured [354, 355]. There are a number of reasons for this:

1. An important parameter of a semiconductor is the effective mass of the charge carriers, and measuring the cyclotron frequency, $\omega_c = eB/m^*c$, gives this unambiguously.
2. It is common for m^* to be smaller than the free electron mass, making the resonance frequency somewhat higher than for free electron metals and in a more accessible spectral region.*
3. The carrier density is three to four orders of magnitude smaller, making $\omega_p = \sqrt{4\pi ne^2/m^*}$ quite a bit smaller. The dielectric function and conductivity scale as ω_p^2 so the lower carrier density leads to smaller absorption, an effect somewhat counteracted by the smaller effective mass.†
4. Semiconductors may have carriers in several bands (electrons in the conduction band, light, heavy, and split-off holes in the valence band) and with anisotropic effective mass ellipsoids. Each has its own effective mass, adding richness to the cyclotron resonance spectra.
5. Although the doping process introduces impurities into the crystal lattice, semiconductors can have high mobilities, so that $\omega_c\tau >\!\gg 1$ can be achieved.

---

* For example, the conduction-band effective mass in GaAs is $m^*/m_e = 0.067$, making the 100 kG (10 T) cyclotron frequency $\omega_c = 139\ \text{cm}^{-1}$. Measurements at this frequency with commercial infrared spectrometers is easier than at $9.3\ \text{cm}^{-1}$, the resonant frequency for free-electron metals at this field.

† It is the ratio n/m^* that matters.

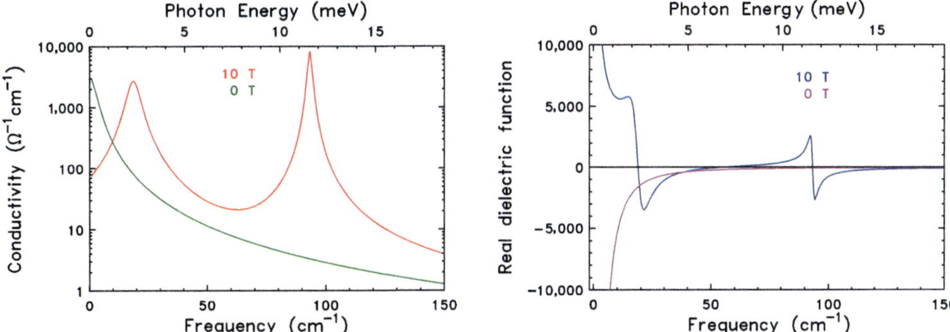

Fig. 16.7 (a) Faraday geometry conductivity for the − polarization (right circular polarization) of a hole-doped semiconductor with a 100 kG $\hat{z}$-axis magnetic field and with zero field. (b) Real part of the dielectric function in the same configuration. There are two peaks because the model includes holes with two different effective masses.

The DC conductivity of a semiconductor is often written in terms of the mobility. Suppose there are both electrons (number density n and mobility μ_e) and holes (number density p and mobility μ_h). Then the DC conductivity is

$$\sigma = ne\mu_e + pe\mu_h. \tag{16.33}$$

If I compare Eq. 16.33 with the Drude DC conductivity, Eq. 4.9, I see that $\mu = e\tau/m^*$, with subscript for electron or hole, as appropriate. Then, looking at Eq. 16.4, I notice that the product of cyclotron frequency and scattering time, $\omega_c\tau$, can be written

$$\omega_c\tau = \frac{eB_0\tau}{m^*c} = \frac{\mu B_0}{c}, \tag{16.34}$$

for each electron band and for each hole band in the semiconductor. The mass m^* written here and m in Eq. 16.4 both represent the effective mass of the charge carriers. Equation 16.34 then specifies $\mu B_0/c \gg 1$ as the requirement to see cyclotron resonance.*

To illustrate cyclotron resonance, I'll take the carrier density $n = 5 \times 10^{17}$ cm^{-3}, about 100,000 times lower than in a metal and about ten times lower than the concentration for the metal–insulator transition in doped silicon [356–358]. Effective masses in semiconductors range from 0.01 to nearly 1; I'll assume that the carriers live in two bands, one with $m^* = 0.1m_e$ and the other with $m^* = 0.5m_e$. These are representative values for the case where there are light and heavy holes bands at the top of the valence band. Finally, I'll do the calculation for a field of 100 kg (10 T) and for $\mu = 3 \times 10^4$ cm^2/volt-sec $= 1 \times 10^7$ cm^2/statvolt-sec. The results are shown in Fig. 16.7.

The optical conductivity has peaks at each of the two cyclotron resonance frequencies. The frequencies are set by the carrier effective mass. The lower peak is broader (because the cyclotron frequency is lower and I took identical scattering rates of $1/\tau = 3$ cm^{-1} for both carrier types) and has lower oscillator strength (because the carrier densities are identical but the effective masses are not). The real part of the dielectric function is positive below the

---

* See Appendix A for a discussion of units. The mobility here is expressed in cm^2/statvolt-sec, 300 times larger than mobility reported in cm^2/volt-sec in most handbooks. (And, the field is in Gauss!)

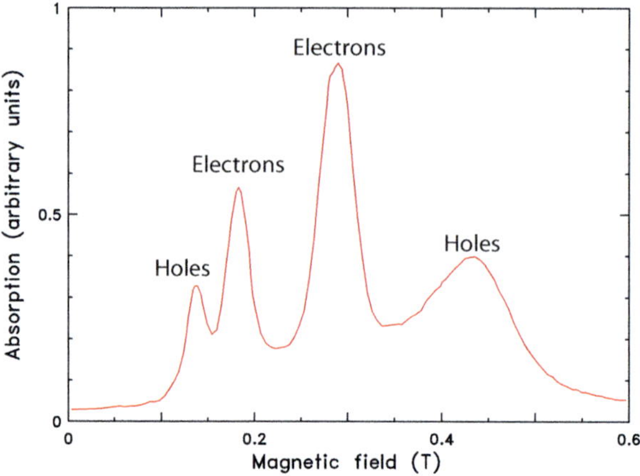

Fig. 16.8   Swept-frequency cyclotron resonance data for silicon. (After Ref. 354.)

lower resonance (and for some frequencies between the resonances). The semiconductor will be transparent (for thickness of a few millimeters) below the lower resonances as well as between the resonances and above the screened plasma frequency. Wave velocities in the semiconductor are smaller than in vacuum but not so small as in a metal. At $\omega = 1$ cm^{-1}, the parameters used give $n = 70$.

I've calculated spectra at (several) fixed field values. One way to study cyclotron resonance is to measure the absorption* spectrum at fixed field. In principal this method is best, because many quantities which affect the resonances could be field dependent, and the spectra can be measured at field up to the maximum available in the laboratory. Another method is to shine a monochromatic electromagnetic wave on the sample and sweep the magnetic field. There will be absorption as each and every type of charge carrier comes into resonance. An example [354] of such an experiment is shown in Fig. 16.8. If the incident radiation has left or right circular polarization, then only orbits due to electrons or holes will be seen. If it is linearly polarized,[†] then both will be seen, as I can decompose linear polarization into the sum of left and right circular polarizations. Figure 16.8 was measured with linear polarization.

Four absorption peaks are shown in Fig. 16.8. Two (the lowest field and the highest field) are associated with the light holes and heavy holes, respectively. The other two are for electrons. There are two frequencies for electrons because the conduction band has an anisotropic effective mass tensor. Cyclotron resonance played a big part in determining silicon's band structure.

16.3.4 Voigt Geometry

To use Voigt geometry I orient $\mathbf{q}$ perpendicular to the magnetic field. I look at Fig. 16.3 and Eq. 16.27, write $\mathbf{q} = \hat{\mathbf{x}} N\omega/c$, making $\theta = 90°$. Then, Eq. 16.28 becomes

$$0 = [\epsilon_{xx} E_x + \epsilon_{xy} E_y]\hat{\mathbf{x}} + [-\epsilon_{xy} E_x + (\epsilon_{xx} - N^2)E_y]\hat{\mathbf{y}} + [(\epsilon_{zz} - N^2)E_z]\hat{\mathbf{z}}. \qquad (16.35)$$

---

* Or transmission, or reflection, or conductivity.
† Or unpolarized, but see the discussion in Section 15.2.2.

Here, I *can* put the electric field either parallel or perpendicular to the $\hat{z}$ axis (the field direction). In the first case, $E_z \neq 0$ and $E_x = E_y = 0$, I get $N^2 = \epsilon_{zz}$. This is the same result I would get if there were no magnetic field. So, the wave with this polarization is insensitive to the magnetic field. Of course, I expect this result on physics grounds: the electric field E_z gives currents along $\hat{z}$ and the Lorentz force, proportional to $\mathbf{v} \times \mathbf{B}$, is zero.*

The case when the electric field is in the x–y plane is more interesting. I use the $\hat{x}$ component of Eq. 16.35 to find that the longitudinal field is related to the transverse field by $E_x = -(\epsilon_{xy}/\epsilon_{xx})E_y$. Next, I substitute E_x into the $\hat{y}$ component and obtain

$$N^2 \equiv \epsilon_\perp = \epsilon_{xx} + \frac{\epsilon_{xy}^2}{\epsilon_{xx}},$$

where $\epsilon_\perp$ is meant to indicate that this is the dielectric function for a wave traveling in the plane perpendicular to the static magnetic field. Now, the off-diagonal component is linear in ω_c, so the Voigt geometry gives an effect which is quadratic in the field. I insert the components from Eq. 16.23 and, after a bit of algebra, arrive at

$$\epsilon_\perp = \epsilon_c - \left(\frac{\omega_p^2}{\omega^2}\right)\frac{\tilde{\omega}_p^2 - \varpi^2}{\tilde{\omega}_p^2(\varpi/\omega) + \omega_c^2 - \varpi\omega},$$

where $\tilde{\omega}_p = \omega_p/\sqrt{\epsilon_c}$ is the screened plasma frequency.

There is a resonance near $\sqrt{\tilde{\omega}_p^2 + \omega_c^2}$ but a zero just below that, near $\tilde{\omega}_p$. To see the behavior a bit more clearly, I'll take the high frequency limit.[†] I set $\omega\tau \gg 1$, making $\varpi \approx \omega$. After further algebra, the dielectric function becomes:

$$\epsilon_\perp = \epsilon_c - \frac{\omega_p^2}{\omega^2} + \frac{\omega_c^2\omega_p^2}{(\tilde{\omega}_p^2 + \omega_c^2 - \omega^2)\omega^2}. \tag{16.36}$$

The first two terms in Eq. 16.36 are the dielectric function of free electrons (or the Drude model when $\omega\tau \gg 1$). The third term is a consequence of the magnetic field. It is proportional to B_0^2 through ω_c and has a resonance at $\omega = \sqrt{\tilde{\omega}_p^2 + \omega_c^2}$.

The wave in Voigt geometry has components parallel and perpendicular to the propagation direction. Taking $\hat{q} = \hat{x}$ I find

$$\frac{E_x}{E_y} = -\frac{\epsilon_{xy}}{\epsilon_{xx}} = i\,\frac{\tilde{\omega}_p^2(\omega_c/\omega)}{\tilde{\omega}_p^2(\varpi/\omega) + \omega_c^2 - \omega^2}.$$

This ratio has a maximum near $\sqrt{\omega_p^2 + \omega_c^2}$. At frequencies below the plasma frequency, I can simplify in one of two ways:

$$\frac{E_x}{E_y} = \begin{cases} i\dfrac{\omega_c}{\omega} & 1/\tau \ll \omega \ll \tilde{\omega}_p, \\[2ex] \omega_c\tau & \omega \ll 1/\tau. \end{cases}$$

and I'd want $\omega_c\tau \gg 1$ as usual.

---

* Remember: $\mathbf{j} = ne\mathbf{v}$.
† The plasma frequency is the highest frequency in the problem, followed by the cyclotron frequency, and then by the scattering rate.

16.4 Other Magneto-Optical Effects

Here I will mention briefly some of the variety of magneto-optical effects that one may observe. There are a number of these; in the main they are consequences of off-diagonal tensor components as seen in Eqs. 16.20 and 16.23. In turn, such components arise in magnetized material, either from an applied field or from intrinsic magnetization.

Faraday rotation. There is a rotation of the plane of polarization for linearly polarized light transmitted through transparent materials with $\hat{q}$ along the magnetic field direction.* The refractive indices for the two circular polarizations differ, and therefore so do the velocities. I will work out Faraday rotation in the next section.

Magnetic circular dichroism. The absorption coefficient for right and left circular polarization in the Faraday geometry differ. The dielectric function for cyclotron resonance and for helicons (Eqs. 16.31 and 16.32) is an extreme example of magnetic circular dichroism. See Fig. 16.5.

Linear magnetic birefringence. In the Voigt geometry, the material exhibits different refractive indices for polarizations parallel and perpendicular to the magnetic field.

Magnetic linear dichroism. The absorption differs for electric field parallel and perpendicular to the external field (in the Voigt geometry).

Magneto-optic Kerr effect. There is a change of polarization state when light is reflected from a material subject to a magnetic field. If I send in linearly polarized light, I may receive elliptically polarized light. But I may also observe linear polarization with the plane of polarization rotated. The experimental configurations may vary: one needs to distinguish the direction of the magnetization vector with respect to the reflecting surface, the direction of the magnetization vector with respect to the plane of incidence,[†] and the polarization direction of the incident light. The Kerr rotation is linear in the strength of the applied magnetic field.

Voigt effect. This effect is like the Kerr effect except that the rotation is quadratic in field strength. Because all these magneto-optical effects are small, the quadratic ones are very weak compared to the linear ones. The Voigt effect is sometimes called the Cotton-Mouton effect, but I think the Cotton-Mouton effect is really an effect of alignment of magnetic nanoparticles in a fluid.

Magnetic resonance. Electrons and nuclei have spin, and a magnetic dipole moment due to that spin degree of freedom. In a magnetic field, an electromagnetic field of the right frequency will induce a transition from a lower to a higher quantized energy level of the dipole. Among the possibilities are nuclear magnetic resonance [359], electron spin resonance (or electron paramagnetic resonance) [360, 361], ferromagnetic resonance [362, 363], and antiferromagnetic resonance [364]. In the latter two, the internal order of the spin system plays a big role.

---

* The geometry is the "Faraday geometry" that I used in Section 16.3.1.

† The plane of incidence is a plane containing the surface normal and the (non-normal) incident and reflected wave vectors. The electric field may be parallel or perpendicular to this plane.

Magnetoelectric effect. In the class of materials now called "multiferroics," there is coupling of ferroelectric and ferromagnetic degrees of freedom. An electric field will affect the magnetization and a magnetic field. Moreover, exotic spin excitations exist: electro-active magnons (or electromagnons), spin waves that are excited by electromagnetic fields and observed in the infrared spectrum [365–367].

These are all very interesting topics, but, except for Faraday rotation, I'll not discuss these (or other magneto-optical effects) further.

16.5 Faraday Rotation in an Insulator

Faraday rotation was discovered in 1845 when Michael Faraday observed the rotation of the polarization direction in plane-polarized light it traversed a glass rod in a magnetic field. The geometry had the light propagation direction parallel to the magnetic field direction. The amount of rotation increased with the field and with the length of the rod.

I can see three approaches to this problem. (1) I can say that the difference between a metal, which has at zero-field the Drude dielectric function given by Eq. 4.13, and an insulator, with the Lorentz dielectric function of Eq. 4.31, is the restoring force of the bound electrons. This restoring force leads to a resonant frequency at ω_e in the Lorentz oscillator. The metal has a pole at the origin and the insulator has one at the resonant frequency. So all I need to do is to shift the origin and put the resonant frequency into Eqs. 16.31 and 16.32. (2) A second approach would be essentially to duplicate the earlier sections of this chapter. Start with an equation of motion that includes the restoring force; find the dielectric tensor; use that tensor in Maxwell's equations for arbitrary $\hat{q}$ and $\hat{e}$ directions; and finally simplify to the geometry of the Faraday effect. (3) The third approach, which I will adopt, is to solve the equation of motion for the bound carriers to obtain the dielectric (or susceptibility) tensor, form the left and right circular combinations, and then consider the propagation of circularly-polarized light in the material.

16.5.1 The Susceptibility Tensor

I start with the equation of motion I used for bound electrons, Eq. 4.20, and include the Lorentz force from the static magnetic field $\mathbf{B}_0$ in the sum of forces on the right,

$$m\ddot{\mathbf{r}} = -m\omega_0^2\mathbf{r} - m\gamma\dot{\mathbf{r}} - e\mathbf{E} - \frac{e}{c}\dot{\mathbf{r}} \times \mathbf{B}_0. \tag{16.37}$$

As usual, I'll take an $e^{-i\omega t}$ time dependence for $\mathbf{r}$ and for $\mathbf{E}$, so that $\dot{\mathbf{r}} = -i\omega\mathbf{r}$ and $\ddot{\mathbf{r}} = -\omega^2\mathbf{r}$. Next, I will write the dipole moment/unit volume as $\mathbf{P} = -ne\mathbf{r}$, with n the number density of the bound electrons, as usual. Third, I will make the local field correction as done in Chapter 4 (Section 4.4). Here, this correction consists of replacing ω_0 with ω_e. With these three things done, I have this equation for the dipole moment/unit volume:

$$(\omega_e^2 - \omega^2 - i\omega\gamma)\mathbf{P} - \frac{i\omega e}{mc}\mathbf{P} \times \mathbf{B}_0 = \frac{ne^2}{m}\mathbf{E}. \tag{16.38}$$

Alertly, I recognize in Eq. 16.38 a cyclotron frequency $\omega_c = eB_0/mc$ and a plasma frequency (as in Eq. 4.31) $\omega_{pe} = \sqrt{4\pi n e^2/m}$. It helps to write the components of Eq. 16.38. I'll take, as I did earlier, $\mathbf{B}_0 = B_0\hat{z}$. Then the z component does not contain the magnetic field; in fact, it is identical with what I would get in the zero-field situation. The other two components do contain the field.*

$$(\omega_e^2 - \omega^2 - i\omega\gamma)P_x - i\omega\omega_c P_y = \frac{\omega_{pe}^2}{4\pi}E_x$$

$$(\omega_e^2 - \omega^2 - i\omega\gamma)P_y + i\omega\omega_c P_x = \frac{\omega_{pe}^2}{4\pi}E_y \qquad (16.39)$$

$$(\omega_e^2 - \omega^2 - i\omega\gamma)P_z = \frac{\omega_{pe}^2}{4\pi}E_z.$$

It is clear that I must write $\mathbf{P} = \overset{\leftrightarrow}{\chi} \cdot \mathbf{E}$, with $\overset{\leftrightarrow}{\chi}$ the electric susceptibility tensor. This tensor has the same format as Eqs. 16.20 and 16.23. There are four distinct elements; the diagonal elements satisfy $\chi_{xx} = \chi_{yy} \neq \chi_{zz}$ while the nonzero off-diagonal components obey $\chi_{yx} = -\chi_{xy}$.

16.5.2 The Refractive Index for Circular Polarization

Here, I want to consider the case where $\hat{q}$ is along $\hat{z}$, i.e., waves propagating along the magnetic field. This situation is the Faraday geometry, which I discussed for metals in Section 16.3.1. The analysis of field there applies here too: Maxwell's equations will give transverse plane waves, with $\mathbf{E}$ in the x–y plane.

However, instead of solving for P_x and P_y, I note that I see a sort of symmetry in Eq. 16.39, with $P_x - iF(\omega)P_y \propto E_x$ and $P_y + iF(\omega)P_x \propto E_y$, where F is the function $F(\omega) = i\omega\omega_c/(\omega_e^2 - \omega^2 - i\omega\gamma)$. Instead of plane polarized fields, let me take circular polarized fields. I will write the electric field and the polarization vector in the plane perpendicular to the magnetic field as $\mathbf{E}_\perp = \mathbf{E}_+ + \mathbf{E}_-$ and $\mathbf{P}_\perp = \mathbf{P}_+ + \mathbf{P}_-$ where

$$\mathbf{E}_+ = (\hat{x} + i\hat{y})E_+$$
$$\mathbf{E}_- = (\hat{x} - i\hat{y})E_-$$
$$\mathbf{P}_+ = (\hat{x} + i\hat{y})P_+$$
$$\mathbf{P}_- = (\hat{x} - i\hat{y})P_-,$$

where E_+, E_-, P_+, and P_- are the complex amplitudes of the four fields. By choice of these amplitudes and their phases, I can construct any specific xy-plane field. The components of these fields are

$$E_x = \hat{x} \cdot (\mathbf{E}_+ + \mathbf{E}_-) = E_+ + E_-$$
$$E_y = i(E_+ - E_-)$$
$$P_x = P_+ + P_-$$
$$P_y = i(P_+ - P_-).$$

---

* Actually they contain the cyclotron frequency.

Now, I substitute these components into the first two lines of Eq. 16.39, multiplying as I do the second by $-i$, to find

$$(\omega_e^2 - \omega^2 - i\omega\gamma)(P_+ + P_-) + \omega\omega_c(P_+ - P_-) = \frac{\omega_{pe}^2}{4\pi}(E_+ + E_-)$$

$$(\omega_e^2 - \omega^2 - i\omega\gamma)(P_+ - P_-) + \omega\omega_c(P_+ + P_-) = \frac{\omega_{pe}^2}{4\pi}(E_+ - E_-).$$

First I add these and then subtract them. A small amount of algebra leads me to

$$P_+ = \frac{\omega_{pe}^2/4\pi}{\omega_e^2 + \omega\omega_c - \omega^2 - i\omega\gamma}E_+ \equiv \chi_+ E_+$$

$$P_- = \frac{\omega_{pe}^2/4\pi}{\omega_e^2 - \omega\omega_c - \omega^2 - i\omega\gamma}E_- \equiv \chi_- E_-.$$

I will define as usual the dielectric function as $\epsilon = \epsilon_c + 4\pi\chi$ with ϵ_c the high frequency value of ϵ. Then,

$$\epsilon_+ = \epsilon_c + \frac{\omega_{pe}^2}{\omega_e^2 + \omega\omega_c - \omega^2 - i\omega\gamma}. \tag{16.40}$$

$$\epsilon_- = \epsilon_c + \frac{\omega_{pe}^2}{\omega_e^2 - \omega\omega_c - \omega^2 - i\omega\gamma}. \tag{16.41}$$

The effect of the field is to shift the resonance to higher frequencies for the E_+ circular polarization and to lower frequencies for the E_- circular polarization.

One almost always has $\omega_e \gg \omega_c$; the energy gap is probably in the near infrared or visible (0.5–5 eV) while the cyclotron frequency is in the far infrared (1 meV). One also wants ω_{pe} to be large enough to give a reasonable infrared refractive index.* Finally, $\omega_c \gg \gamma$ makes the absorption small at frequencies where the Faraday effect is important.[†] The optical frequency ω must clearly be below ω_e for small absorption. For simplicity, I can set $\gamma = 0$ and consider that $\epsilon_\pm$ is a real function representing the dispersion of the material for frequencies below ω_e, the electronic absorption edge.[‡]

With $\gamma = 0$, the dielectric function and refractive index are real. These functions differ for the two circular polarizations. The refractive indices are

$$n_+ = \sqrt{\epsilon_+} = \sqrt{\epsilon_c + \frac{\omega_{pe}^2}{\omega_e^2 + \omega\omega_c - \omega^2}} \tag{16.42}$$

for the left circular polarization (E_+) and

$$n_- = \sqrt{\epsilon_-} = \sqrt{\epsilon_c + \frac{\omega_{pe}^2}{\omega_e^2 - \omega\omega_c - \omega^2}} \tag{16.43}$$

for the right circular polarization (E_-). Clearly, $n_- > n_+$.

---

* As far as desirable values for the ratio of ω_{pe} to ω_e go, I'd look for anything from 1.1 to 2 or more.

[†] The restriction $\omega_c \gg \gamma$ is the same as $\omega_c\tau \gg 1$ in the metals case.

[‡] For a realistic model I should include one or more infrared oscillators representing optical phonons. These cause far-infrared absorption up to the highest longitudinal optical phonon frequency. There is transmission from there up to ω_e. They also have a small effect on numerical results but do not change the physics.

The refractive indices in Eqs. 16.42 and 16.43 are perfectly good for calculating wave propagation, but how they vary with the field strength B_0 (or ω_c) is not obvious. So I will use the fact that $\omega_c \ll \omega_e$ and expand the refractive index to first order in ω_c. I'll start by writing the expansion of $\epsilon_\pm$ and then notice that the leading term contains the zero-field frequency-dependent dielectric function, which I'll call ϵ_o for "ordinary:" $\epsilon_o(\omega) = \epsilon_c + \omega_{pe}^2/\omega_e^2 - \omega^2$.

$$\epsilon_\pm = \epsilon_c + \frac{\omega_{pe}^2}{\omega_e^2 - \omega^2}\left(1 \mp \frac{\omega\omega_c}{\omega_e^2 - \omega^2}\right) = \epsilon_o\left[1 \mp \left(\frac{\epsilon_o - \epsilon_c}{\epsilon_o}\right)\frac{\omega\omega_c}{\omega_e^2 - \omega^2}\right].$$

The refractive index is easy: take square root and expand:

$$n_\pm = n_o\left[1 \mp \left(\frac{\epsilon_o - \epsilon_c}{2\epsilon_o}\right)\frac{\omega\omega_c}{\omega_e^2 - \omega^2}\right], \tag{16.44}$$

where $n_o(\omega) = \sqrt{\epsilon_o(\omega)}$ is the frequency dependent refractive index in zero field.

16.5.3　Proof that the Plane of Polarization Is Rotated

I'll now work it out explicitly how the electric field direction changes as light travels through the material in the Faraday geometry. A sketch is in Fig. 16.9. I start with an $\hat{x}$-polarized electric field incident from the left on the crystal surface (located at $z = 0$) of the form

$$\mathbf{E}_l = \hat{x}E_0 e^{i(qz - \omega t)}.$$

Here, $q = n\omega/c$, with n the appropriate refractive index. As discussed in Section 15.2.3, plane-polarized light can be written as a sum of equal amplitudes of right and left circularly polarized light at some specific relative phase.* Thus the field (at $z = 0$ and $t = 0$) is

$$\mathbf{E}_l = \hat{x}E_0 = \frac{E_0}{2}(\hat{x} - i\,\hat{y}) + \frac{E_0}{2}(\hat{x} + i\,\hat{y}).$$

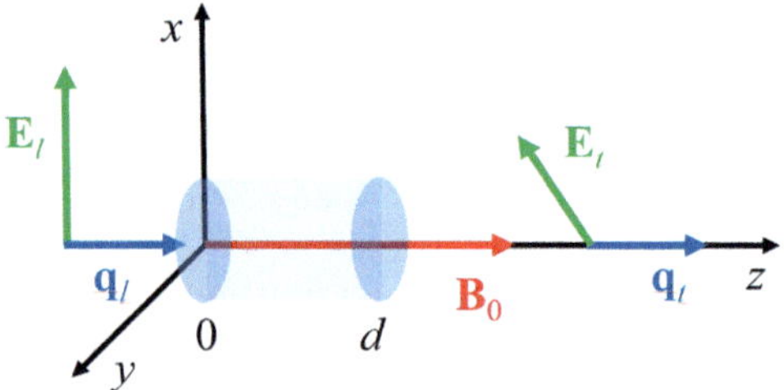

Fig. 16.9　The Faraday effect is the rotation of the plane of polarization of light as it passes through a crystal subjected to a static magnetic field $\mathbf{B}_0 = B_0\hat{z}$. The direction of propagation is along the magnetic field direction. The incident light, on the left, has its electric field parallel to $\hat{x}$ and the transmitted field has $\mathbf{E}$ rotated in the $x-y$ plane.

---

* If the relative phase is zero, I get polarization along the x axis.

The two circular polarizations have different velocities in the material because they have different refractive indices; there will be a phase difference between right and left circular polarizations that develops as the wave travels onward. At $z = +d$,

$$\mathbf{E}_t = \frac{E_0}{2}(\hat{\mathbf{x}} - i\,\hat{\mathbf{y}})e^{i(n_-)\frac{\omega}{c}d} + \frac{E_0}{2}(\hat{\mathbf{x}} + i\,\hat{\mathbf{y}})e^{i(n_+)\frac{\omega}{c}d}. \tag{16.45}$$

I factor the $E_0 e^{i(n_+)\frac{\omega}{c}d}$ term and collect together the $\hat{\mathbf{x}}$ and $\hat{\mathbf{y}}$ components:

$$\mathbf{E}_t = \frac{E_0}{2}e^{i(n_+)\frac{\omega}{c}d}\left[\hat{\mathbf{x}}\left(1 + e^{i(n_- - n_+)\frac{\omega}{c}d}\right) + i\,\hat{\mathbf{y}}\left(1 - e^{i(n_- - n_+)\frac{\omega}{c}d}\right)\right]. \tag{16.46}$$

Next, I define an overall phase $\delta \equiv (n_+)\omega d/c$ and a phase difference

$$\phi \equiv (n_- - n_+)\frac{\omega}{c}d. \tag{16.47}$$

Note that $n_- > n_+$ making ϕ a positive angle. The field can be simplified

$$\begin{aligned}
\mathbf{E}_t &= \frac{E_0}{2}e^{i\delta}\left[\hat{\mathbf{x}}\left(1 + e^{i\phi}\right) + i\,\hat{\mathbf{y}}\left(1 - e^{i\phi}\right)\right] \\
&= \frac{E_0}{2}e^{i(\delta+\phi/2)}\left[\hat{\mathbf{x}}\left(e^{-i\phi/2} + e^{i\phi/2}\right) + i\,\hat{\mathbf{y}}\left(e^{-i\phi/2} - e^{i\phi/2}\right)\right],
\end{aligned} \tag{16.48}$$

where I factored out another phase term. I could have done this back in Eq. 16.46 but it was not obvious that I needed to do so then. I recognize the complex exponential representation of cosine and sine in Eq. 16.48, so that I can write

$$\mathbf{E}_t = E_0 e^{i(\delta+\phi/2)}\left[\hat{\mathbf{x}}\cos(\phi/2) + \hat{\mathbf{y}}\sin(\phi/2)\right]. \tag{16.49}$$

The electric field started with only an $\hat{\mathbf{x}}$ component and after a distance d has both $\hat{\mathbf{x}}$ *and* $\hat{\mathbf{y}}$ components. But it is still linearly polarized light, at an angle θ_F to the x axis:

$$\tan\theta_F = \frac{\sin(\phi/2)}{\cos(\phi/2)} = \tan\left(\frac{\phi}{2}\right).$$

The rotation angle is half the phase difference for the two circular polarizations: $\theta_F = \phi/2$.

The Faraday rotator is said to be a "nonreciprocal" device: Sending the transmitted light back through in the opposite direction does not undo the change the beam suffered in the forward pass. To see how this works, let me imagine that the transmitted beam in Fig. 16.9 is reflected and returned to the rotator, incident from the right. This light is moving to the left, so the field is of the form

$$\mathbf{E}_r = \hat{\mathbf{x}}\,E_0 e^{-i(qz+\omega t)}.$$

It is convenient to shift the coordinate system, putting the origin of the z axis at right edge of the crystal and the left edge at $z = -d$. The light incident from the right can have both $\hat{\mathbf{x}}$ and $\hat{\mathbf{y}}$ components. I write the $\hat{\mathbf{x}}$ component at the new origin (and $t = 0$) as a sum of equal-amplitude right and left circular polarizations:

$$\mathbf{E}_{rx} = \hat{\mathbf{x}}\,E_{0x} = \frac{E_{0x}}{2}(\hat{\mathbf{x}} - i\,\hat{\mathbf{y}}) + \frac{E_{0x}}{2}(\hat{\mathbf{x}} + i\,\hat{\mathbf{y}}).$$

The two circular polarizations have different refractive indices and the light reaches the left side of the crystal (at $z = -d$) as

$$\mathbf{E}_{bx} = \frac{E_{0x}}{2}(\hat{x} - i\hat{y})e^{-i(n_-)\frac{\omega}{c}(-d)} + \frac{E_{0x}}{2}(\hat{x} + i\hat{y})e^{-i(n_+)\frac{\omega}{c}(-d)}. \tag{16.50}$$

Other than amplitude, Eq. 16.50 is *identical* to Eq. 16.45. Thus the rest of the algebra is also identical, and

$$\mathbf{E}_{bx} = E_{0x}e^{i(\delta+\phi/2)}\left[\hat{x}\cos(\phi/2) + \hat{y}\sin(\phi/2)\right]. \tag{16.51}$$

The x-component of the backward-going light leaving the Faraday rotator is plane polarized, pointing in the x–y plane at an angle $\theta_F = \phi/2$, with ϕ defined by Eq. 16.47.

If the backward-going incident field also has a y component, amplitude E_{0y}, I can follow it through as I did for the x component to find that this field is linearly polarized but with $\hat{x}$ as well as $\hat{y}$ components:

$$\mathbf{E}_{by} = E_{0y}e^{i(\delta+\phi/2)}\left[-\hat{x}\sin(\phi/2) + \hat{y}\cos(\phi/2)\right]. \tag{16.52}$$

The total backward-going light beam is the sum $\mathbf{E}_{bx} + \mathbf{E}_{by}$. The important case is when the incident backward-going light is reflected forward-going light. In this case,* Eq. 16.49 tells me that $E_{0x} = \cos(\phi/2)$ and $E_{0y} = \sin(\phi/2)$. With these as the components of $\mathbf{E}_r$, I get

$$\mathbf{E}_b = E_0e^{2i(\delta+\phi/2)}\left[\hat{x}\left(\cos^2(\phi/2) - \sin^2(\phi/2)\right) + 2\hat{y}\cos(\phi/2)\sin(\phi/2)\right].$$

Suppose $\phi/2$ is small. Then (ignoring phase) $\mathbf{E}_b = E_0[\hat{x}(1 - \phi^2/r) + \hat{y}\phi]$. The more interesting case is when $\phi/2 = \pi/4$ (45°). Then $\sin(\phi/2) = \cos(\phi/2) = 1\sqrt{2}$ and

$$\mathbf{E}_b = E_0e^{2i(\delta+\phi/2)}\hat{y}.$$

The light which has passed through the Faraday rotator twice leaves with polarization at right angles to the incident light.

16.5.4 The Verdet Constant

Using Eq. 16.44 for $n_\pm$, the phase difference between right and left circular polarizations becomes

$$\phi = \frac{\omega^2}{c}\left(\frac{\epsilon_o - \epsilon_c}{\epsilon_o}\right)\frac{\omega_c d}{\omega_e^2 - \omega^2} = \frac{e}{mc^2}\left(\frac{\epsilon_o - \epsilon_c}{\epsilon_o}\right)\frac{\omega^2}{\omega_e^2 - \omega^2}B_0 d, \tag{16.53}$$

where the cyclotron frequency $\omega_c = eB_0/mc$ has been used in the righthand quantity. The plane of polarization rotates through an angle $\theta_F = \phi/2$.

The ability of a material to rotate polarization depends on a number of its properties. It is characterized by the Verdet constant $\mathcal{V}$ of the substance, defined by

$$\theta_F \equiv \mathcal{V}B_0 d. \tag{16.54}$$

---

* I take the reflection coefficient to be 100%.

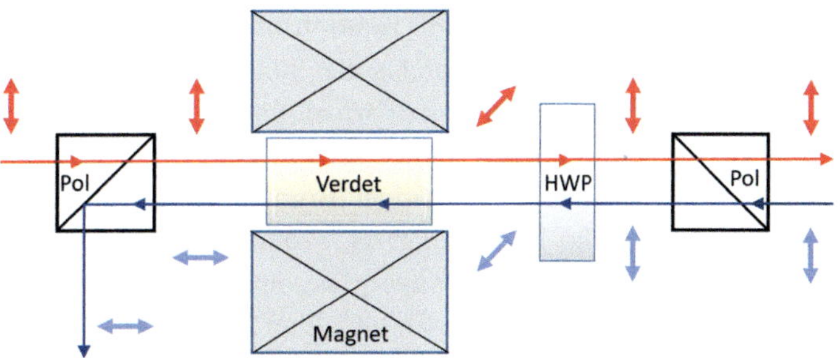

Fig. 16.10 Faraday isolator. The device consists of an input polarizer (Pol), a Faraday rotator, a half-wave plate (HWP), and an output polarizer. Light is incident from the left to right (red) and a return beam goes from right to left (blue). The arrows indicate the polarization direction in the plane perpendicular to the beam.

Comparison of Eq. 16.54 and Eq. 16.53 causes me to write

$$V = \frac{e}{2mc^2} \left(\frac{\epsilon_o - \epsilon_c}{\epsilon_o} \right) \frac{\omega^2}{\omega_e^2 - \omega^2}$$

Even in the most effective Faraday materials, the Verdet constant is quite small. The mechanism discussed above, which is from the $\mathbf{v} \times \mathbf{B}_0$ term in the equation of motion of the bound electrons, gives a strong frequency dependence to V. The Verdet constant increases as $V \sim \omega^2$. It has only weak dependence on temperature. Another mechanism exists in crystals and glasses with paramagnetic ions. These materials have a Verdet constant that is proportional to the paramagnetic susceptibility, X_p, which follows a Curie law, $X_p = C/T$. The Verdet constant therefore follows $1/T$. The frequency dependence is as above. As an example, the Verdet constant for terbium gallium garnet ($Tb_3Ga_5O_{12}$ or TGG), a paramagnetic material commonly used in near-infrared/visible Faraday isolators, at $\omega = 9{,}400$ cm^{-1} (the 1,064 nm wavelength of a Nd:YAG laser) is $V = -40$ μrad/G·cm (-40 rad/T·m). At $\omega = 15{,}800$ cm^{-1} (the 633 nm wavelength of a HeNe laser) it increases to $V = -134$ μrad/·cm. To achieve a rotation at Nd:YAG wavelength of $-45°$ (which is what I want for a Faraday isolator; see below) in a field of 10 kG (1 T) (which I can get from strong permanent magnets) the crystal needs to be 2 cm long. It could be about 30% of that length at the He–Ne wavelength. The TGG Verdet constant increases by a factor of 30 on cooling from 300 K to 10 K.

16.5.5 A Faraday Isolator

The Faraday effect allows the construction of a device where one observer can see a second observer while at the same time the second observer cannot see the first. The Faraday isolator, consisting of an input polarizer, the Faraday rotator, a half wave plate, and an output polarizer, is sketched in Fig. 16.10.

The Faraday rotator consists of a magnet, with magnetic field along the axis of the device, and a crystal (Verdet) with Verdet constant times length times magnetic field strength set to provide a 45° rotation of the plane of polarization. The half-wave plate is a reciprocal device that rotates the plane by −45° to align the polarization with the axes of the output polarizer. The half-wave plate is typically done for convenience, to make the output polarization in the same direction as the input polarization.

The beam (in red) is incident from the left side. If it is vertically polarized, it is transmitted by the input polarizer,* rotated to 45° by the Faraday rotator, rotated back to the vertical by the half-wave plate, and transmitted through the output polarizer. Any light incident from the right (in blue, due for example to reflections of the beam behind the isolator) is transmitted as vertically polarized light by the output polarizer, rotated to 45° by the half-wave plate, rotated further to 90° by the Faraday rotator, and diverted away from the path of the incident beam by the input polarizer, as shown. Betsy, on the right, can see Alan, on the left, but Alan cannot see Betsy.

---

* If it is not vertically polarized, some fraction is rejected by this polarizer; if it is unpolarized, that fraction is half of the incident power.

Many materials seen in everyday life are inhomogeneous: inks, paint, cement, ceramics, stained glass, etc. An example is shown in Fig. 17.1. These materials are mixtures or composites of two or more materials with different dielectric response functions. Some of these composite materials appear to be inhomogeneous to the human eye; the appearance (tint, brightness, saturation) varies from point to point. But others appear completely uniform. One of my goals here is to understand the ways in which the effective properties of an inhomogeneous medium depend upon the properties of its constituents and their geometric arrangement in the medium. I am mostly interested in materials in which the scale of the inhomogeneity is smaller than the wavelength, so that they appear to be uniform.* To understand the optical properties of such materials, I'll try to calculate the average or effective response of the mixture from knowledge (or assumptions) about the properties of the constituents, their dielectric functions, their size, their concentration, and perhaps some little knowledge of how they are arranged in the composite.

The field of inhomogeneous materials has been an active area of research for many years, both because of the intrinsic interest of the subject and due to possible technological applications of these materials. An area of basic interest is the insulator-to-conductor transition or "percolation transition" which occurs in these materials as a function of the concentrations of the constituents and which resembles in many ways a second-order thermodynamic phase transition. A possible important application is in the design of materials with useful infrared properties, such as efficient photothermal solar energy collectors.

I've identified the dielectric function ϵ (and permeability μ if magnetic) as specifying the optical properties of a material. In the homogeneous material, ϵ is the same at every point in the material. The situation is greatly complicated if the medium is not spatially homogeneous but instead has physical properties which vary throughout the material. The manner of the variation may be random, may be periodic, or may be ordered in some other way; in all cases the dielectric function and permeability become functions of position as well as of the frequency:

$$\epsilon = \epsilon(\mathbf{r}, \omega) \tag{17.1}$$

---

* The other extreme, materials which obviously differ as one looks here and there at them, is either trivial – just average the reflectance or transmittance of the components – or very difficult – the response is determined by scattering, diffraction, and interference of light striking the different parts and depends on the details of the organization of the composite.

Detail of a stained-glass window at the University of Virginia chapel. The red panels are "gold ruby" glass, a red glass made by adding gold salts or colloidal gold to molten glass. The process precipitates small (10 nm scale) gold particles in the glass. The blues come from cobalt and the other colors from the addition of a variety of metallic salts.

and

$$\mu = \mu(\mathbf{r}, \omega). \tag{17.2}$$

As an example of an inhomogeneous medium, I will consider a composite made from metal small particles and insulating grains. The dielectric response function of each constituent is assumed to be well understood. The metal could well be described by a Drude dielectric function Eq. 4.13 while the insulator has a dielectric function ϵ_{ins} that is mostly real and nearly constant.* Thus, the spatially varying dielectric function has two possible values:

$$\epsilon(\mathbf{r}, \omega) = \begin{cases} \epsilon_c - \dfrac{\omega_p^2}{\omega^2 + i\omega/\tau}, & \mathbf{r} \text{ in metal.} \\ \epsilon_{\text{ins}}, & \mathbf{r} \text{ in insulator.} \end{cases}$$

The local properties of this material vary dramatically depending upon whether $\mathbf{r}$ is in the metallic or insulating component of the composite.

Although it is true that the functions $\epsilon(\mathbf{r}, \omega)$ and $\mu(\mathbf{r}, \omega)$ completely describe the electromagnetic response of the medium, this description is not very useful because to employ it would require a knowledge of the exact geometric location of the constituents. Generally, this information is unavailable; in addition, the interesting quantity is the average response of the inhomogeneous medium to external fields, not the local response. Fortunately, so long as the scale on which measurements are made (the wavelength) is large compared to the scale of the fluctuations of the dielectric function (the particle size), the

---

* It could be vacuum, with $\epsilon = 1$.

inhomogeneous medium appears to be uniform in its response to external fields and can thus be described by an effective dielectric function $\epsilon_{eff}(\omega)$ and permeability $\mu_{eff}(\omega)$. The problem with which I am confronted, therefore, is in its most general terms the following: given an inhomogeneous medium that has a well-defined local dielectric function $\epsilon(\mathbf{r}, \omega)$ and permeability $\mu(\mathbf{r}, \omega)$, how can the effective response functions of the entire medium be determined?

17.1 The Effective Medium

Historically, the problem of calculating $\epsilon_{eff}(\omega)$ and $\mu_{eff}(\omega)$ has been addressed from two rather different points of view. The first is a molecular field model originally developed by Clausius, Mossotti, Lorentz, and Lorenz to calculate the local field in a crystal and later applied to the optical properties of an inhomogeneous medium by Garnett [368, 369]. Garnett wrote about the colors of stained glasses containing metallic inclusions. (See Fig. 17.1.) Because his full name was James Clerk Maxwell Garnett, this model has become known as the Maxwell-Garnett theory (MGT). The second point of view is a symmetrical effective medium approach, initially developed by Bruggeman [370] and rediscovered by Landauer [371]. The application of this model to optical properties was first suggested by Springett [372] and by Stroud [373]. This model is generally called the effective medium approximation (EMA), although both it and the MGT are in fact effective medium theories. Each provides an expression for the effective dielectric function of a homogeneous medium which has properties effectively identical to those of the inhomogeneous medium.

Each theory (and most subsequent developments) begins with a major simplification. The local dielectric function $\epsilon(\mathbf{r}, \omega)$ of the inhomogeneous medium is assumed to take on a limited number of discrete values (perhaps as small as two) rather than being allowed

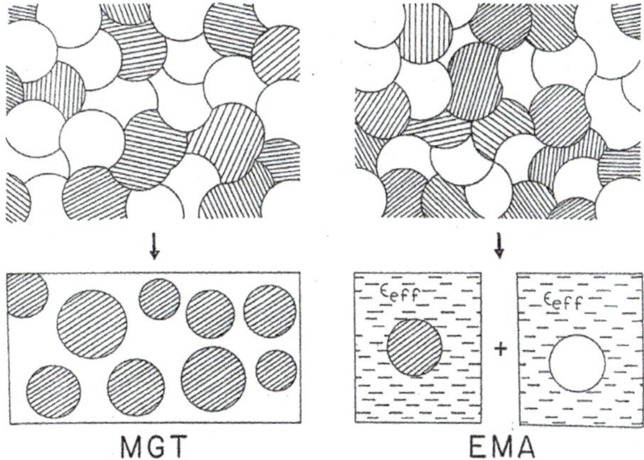

Fig. 17.2 The upper panels show a composite made of two types of materials, white grains and black grains. In the MGT, the composite is viewed as consisting of black grains embedded in an host of the white material. In the EMA, both white and black grains are viewed as surrounded by the effective medium and combined in a weighted average way.

to vary over a continuous range. This simplification corresponds to the example already discussed: the system is a disordered composite, a random mixture of many grains each of which by itself has spatially uniform properties. The composite is characterized by specifying the volume fraction, dielectric function, shape, and size of each grain type. Each grain is then taken to be a polarizable entity, with electric (and magnetic) dipole moment induced by electric (and magnetic) fields from external sources as well as from the fields of the polarized grains in the outside medium.

The principal difference between the MGT and the EMA is the way in which the medium surrounding the grain under consideration is treated. This difference is illustrated in Fig. 17.2. In the MGT it is assumed that the medium surrounding the grains is one of the constituents of the mixture (for example, the one with the largest volume fraction) while in the EMA it is assumed that the surrounding medium is characterized by the effective properties of the inhomogeneous medium.

17.2 Response of a Single Grain

17.2.1 Quasistatic Model: Electric Dipole Moment

The initial step in the solution for the effective properties of an inhomogeneous medium is to find the electromagnetic fields inside and outside of a single particle of radius a when that particle is subjected to a plane wave field. For simplicity I will assume that the particle is spherical and that I am in the long-wavelength limit, $\lambda \gg a$. Then, the applied fields appear to be spatially uniform, time varying fields. The particle in this quasistatic limit has only dipole moments; these moments are time varying so that electric current flows and energy can be absorbed.

I'll define the dipole moment in terms of the polarizability in a way similar to what I did for atoms in Eq. 4.1:

$$\mathbf{p} = \Omega \gamma_e \mathbf{E}_0 e^{-i\omega t} \tag{17.3}$$

where Ω is the volume and γ_e is the electric polarizability per unit volume of the particle, given by the solution to the static boundary value problem of a dielectric sphere in a uniform far field [2].

Defining the dipole moment does not tell me what it is. What I have to do (in this quasistatic limit) is to solve a boundary value problem: that of a sphere with dielectric function ϵ in a uniform far field.* Actually, I do not have to solve the problem; it is a standard problem in many textbooks [2]. I just need to write down the solution. The polarizability is

$$\gamma_e = \frac{3\epsilon_o}{4\pi} \frac{\epsilon_p - \epsilon_o}{\epsilon_p + 2\epsilon_o}, \tag{17.4}$$

---

* I can think of this problem as the task of finding the polarization in a small dielectric sphere located between the plates of a large capacitor. The capacitor plates are attached to a voltage source, producing far away from the sphere a field $\mathbf{E}$ between the plates.

where ϵ_p is the complex dielectric function of the particle and ϵ_o is the complex dielectric function of the medium outside the particle. The electric field outside the particle is that of a dipole with moment **p**. This electric dipole moment is the first term in the Mie [8, 374–376] solution for the multipole moments of the sphere.

The same solution finds that the electric field is

$$\mathbf{E}_p = \frac{3\epsilon_o}{\epsilon_p + 2\epsilon_o}\mathbf{E}_o e^{-i\omega t}. \tag{17.5}$$

Note that this field is uniform and parallel to the external field. The difference between $\mathbf{E}_o$ and $\mathbf{E}_p$ is called the depolarizing field.

17.3 Effective Medium Theories

To find the effective dielectric function of a composite material, I have to do some sort of averaging. It helps to use resistor networks as model systems in thinking about inhomogeneous systems. I suppose that the network contains a large number of resistors, with values varying over a wide range. The resistors are connected in a series/parallel configuration. The values of a particular resistor as well as how it is wired into the network are chosen at random. I inject current into the network and measure the voltage developed across it. If I probe around inside, I'll find that some resistors will carry a lot of current; some a small amount or even zero current. Similarly, there will be parts of the network with big voltage drops and other parts with small voltage drops. But the ratio of voltage across the network to the current flowing into or out of it is the effective resistance of the network.

A picture of the composite for which I will calculate the optical properties considers it to be composed of small grains of two or more materials. The grains are smaller than the wavelength of the light being used to examine them; consequently they are not resolved. All the grains are not the same (else the material would be homogeneous) but they are not all different either. I suppose that there are N_i grains of type i, each with average volume Ω_i. Then I can define the volume fraction f_i of type i grains as

$$f_i = \frac{N_i \Omega_i}{V} \tag{17.6}$$

where V is the sample volume. Clearly*

$$\sum_i f_i = 1.$$

17.3.1 Average Fields

I will define the effective dielectric function and permeability of the inhomogeneous medium in terms of volume-averaged fields, $\langle \mathbf{E} \rangle$, and responses, $\langle \mathbf{D} \rangle$. (It is also possible to work with the volume averaged dipole moment per unit volume, $\langle \mathbf{P} \rangle$, or current, $\langle \mathbf{j} \rangle$; the

---

* If there is empty (void) space in the composite, I will consider this to be one of the components of the inhomogeneous system.

choice is entirely a matter of individual preference.) If f_i is the volume fraction of the ith component of the medium, then the average electric field is

$$\langle \mathbf{E} \rangle = \sum_i f_i \mathbf{E}_i$$

with an analogous definition* for $\langle \mathbf{D} \rangle$.

For simplicity I will assume a two-component composite and label these constituents as type a and type b materials. If f is the volume fraction of type a, then $(1 - f)$ will be the volume fraction of type b, so that the average fields are:

$$\langle \mathbf{E} \rangle = f\mathbf{E}_a + (l - f)\mathbf{E}_b$$

$$\langle \mathbf{D} \rangle = f\epsilon_a \mathbf{E}_a + (l - f)\epsilon_b \mathbf{E}_b, \tag{17.7}$$

where I have used a local relationship $\mathbf{D}_i = \epsilon_i \mathbf{E}_i$. The effective dielectric function is defined in terms of the average fields:

$$\langle \mathbf{D} \rangle = \epsilon_{eff} \langle \mathbf{E} \rangle. \tag{17.8}$$

When I write Eq. 17.8, I take the point of view that the interesting property of the medium is the average response to external fields; the external field is just $\langle \mathbf{E} \rangle$ while $\langle \mathbf{D} \rangle$ is the response to those fields.

17.3.2　Maxwell-Garnett Dielectric Function

The Garnett theory (MGT) is implicitly a model for a dilute system. According to the MGT, a single grain of dielectric function ϵ_a is assumed to lie at the center of a cavity (the Lorentz cavity) carved out of the interior of the inhomogeneous medium. There may be other type a grains in this cavity; the remaining space is assumed to be filled with a medium with dielectric function e_b. Thus the local field at the grain is the superposition of the uniform applied field, the uniform field from the charge (polarization charge) located on the cavity surface, and the dipolar fields from the neighboring particles. As discussed by Landauer [21] in his review of inhomogeneous materials, I can exclude other grains from the cavity by making the cavity tightly to jacket the central grain. A more common approach is to argue that the dipolar fields from those neighboring grains which are inside a large cavity sum to zero in the vicinity of the central grain.† With this assumption, the electric field inside the grain is uniform and found by substitution of $\epsilon_p = \epsilon_a$ and $\epsilon_o = \epsilon_b$ into Eq. 17.5:

$$\mathbf{E}_a = \frac{3\epsilon_b}{\epsilon_a + 2\epsilon_b} \mathbf{E}_b e^{-i\omega t}. \tag{17.9}$$

---

* If I want to calculate the magnetic response, then I will also define $\langle \mathbf{H} \rangle$, and $\langle \mathbf{B} \rangle$. The derivations for effective permeability are perfectly analogous to those for effective dielectric function.

† If the arrangement is periodic, the contribution can be calculated; it often is zero. If the arrangement is random, one can argue that the field direction and magnitude at the cavity center will be random, and the random orientations of many contributions will tend to zero.

Combining Eqs. 17.7 and 17.9 gives, after some algebra, Garnett's expression for ϵ_{eff} which I label ϵ_{mgt}:

$$\epsilon_{\mathrm{mgt}} = \epsilon_b + \epsilon_b \frac{3f(\epsilon_a - \epsilon_b)}{(1 - f)(\epsilon_a - \epsilon_b) + 3\epsilon_b}. \tag{17.10}$$

Eq. 17.10 illustrates two properties which are characteristic of the MGT approach to inhomogeneous media. First, the equation is inherently asymmetric in its treatment of the two constituents: different values are obtained for ϵ_{mgt} depending on whether, for example, I regard the composite as consisting of metal grains embedded in insulator or the other way around. Second, the equation gives a smooth variation of ϵ_{mgt} with volume fraction, from ϵ_b when $f = 0$ to ϵ_a when $f = 1$.

That real physical systems do not always behave in this fashion may be seen by considering the low-frequency behavior of an inhomogeneous medium for which $\epsilon_a = 1$ and $\epsilon_b = 4\pi i \sigma_b / \omega$, i.e., material a is vacuum, with zero conductivity, and material b is a metal with DC conductivity σ_b. At $f = 0$ the MGT conductivity is just σ_b. Now I start carving or drilling holes in the material, increasing f. The presence of these holes cause the effective conductivity to decrease. According to the MGT, the conductivity will remain finite all the way to $f = 1$, when no conducting component remains. In reality, however, the medium is likely to stop conducting at a lower value of f, when the nonconducting regions occupy enough space to prevent current flow in the sample. This "percolation threshold" is not predicted by the MGT; it is, in fact, specifically excluded by the assumption that the embedded grains do not contact one another. In applications of the MGT the asymmetry and the absence of a percolation transition can be handled in an ad hoc fashion by assuming that as the volume fraction of one constituent increases from zero to unity, its role changes from that of inclusion to that of host.

Despite its shortcomings, the MGT is generally believed to work reasonably well in dilute mixtures, where the grains of the minority constituent are well separated. More generally, Hashin and Shtrikman [377] have shown that the MGT expression represents bounds to the actual conductivity of a two-component heterogeneous medium, an upper bound if $\sigma_a < \sigma_b$ and a lower bound if $\sigma_a > \sigma_b$.

The MGT approach to the properties of inhomogeneous media has been rediscovered and elaborated numerous times over the past 100 years [378–389]. As an example, let us consider the case of small metal particles embedded in a medium with dielectric constant $e_b = 1$. For the metal I assume a simple Drude dielectric function, Eq. 4.13, with $\epsilon_c = 1$:

$$\epsilon_a = 1 - \frac{\omega_p^2}{\omega^2 + i\omega/\tau}.$$

Under these assumptions, the MGT dielectric function becomes

$$\epsilon_{\mathrm{mgt}} = 1 + \frac{f\omega_p^2}{\omega_{\mathrm{mgt}}^2 - \omega^2 - i\omega/\tau}. \tag{17.11}$$

The effective dielectric function will exhibit a resonant absorption when the frequency of the light equals the Maxwell-Garnett resonant frequency, ω_{mgt}, equal to

$$\omega_{\mathrm{mgt}} = \omega_p \sqrt{\frac{1-f}{3}}.$$

This Maxwell-Garnett resonance arises physically from the fact that the Drude dielectric function of a metal, Eq. 4.13, is negative below the plasma frequency and can thus cancel in the electric polarizability, Eq. 17.4, the positive dielectric constant of the surrounding medium. In the example above, the cancellation occurs when $\epsilon_a = -2$. More sophisticated models, in which the outside medium has dielectric constant ϵ_b and the metal has a core contribution ϵ_c in its dielectric function, become algebraically more complex but retain the Lorentzian form of Eq. 17.11. The resonance frequency is shifted in this more general case to

$$\omega_{\mathrm{mgt}} = \omega_p \sqrt{\frac{1-f}{(1-f)\epsilon_c + (2+f)\epsilon_b}}. \tag{17.12}$$

Figure 17.3 shows the MGT frequency-dependent conductivity. I have taken the metallic portion of the composite to have a Drude dielectric function, Eq. 4.13, and the insulating portion to have a real, constant dielectric function. The conductivity is shown over the entire frequency range from DC to UV for metal volume fractions of 0.1, 0.3, 0.5, 0.7, and 0.9. A curve is also shown where the MGT is inverted, i.e., the metal was taken as host and

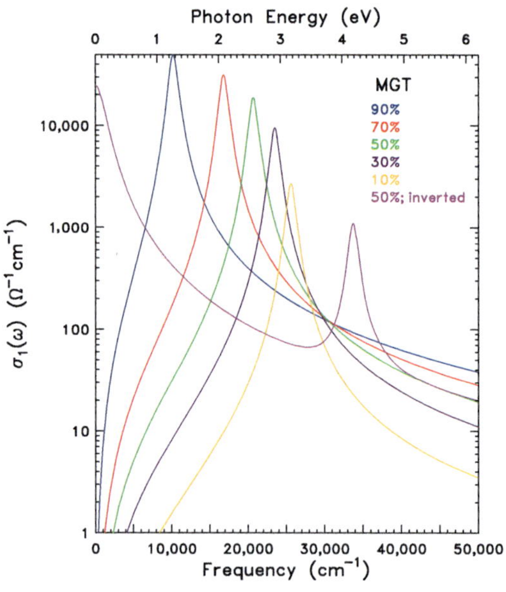

Fig. 17.3   Frequency-dependent conductivity calculated from the Garnett theory. The calculation used a Drude dielectric function for the inclusion ($\epsilon_a = \epsilon_{\mathrm{Drude}}$) with parameters meant to represent silver [$\epsilon_c = 4.0$, $\omega_p = 70{,}100\,\mathrm{cm}^{-1}$ (8.7 eV) and $1/\tau = 1{,}300\,\mathrm{cm}^{-1}$ (0.16 eV)]. The scattering rate is about $10\times$ that of pure silver, under an assumption that scattering from the surface of the particle determines the mean free path. For the insulator, $\epsilon_b = \epsilon_{ins} = 4$.

the insulator as inclusion, both with volume fractions of 0.5. Relatively short mean free paths (about 30 Å) were taken in calculating these plots; this choice leads to a rather broad MGT resonance.

There are several points to make about Fig. 17.3. There is no DC conductivity in the MGT when the metal is taken as inclusion so long as $f < 1$; there is always a finite DC conductivity when the metal is host. The low-frequency spectrum of the metal–host mixture is of Drude form. The MGT resonant frequency (given by Eq. 17.12), moves to lower frequencies as the concentration increases. The low-concentration value for the resonant frequency is $\omega_{\mathrm{mgt}} = \omega_p / \sqrt{\epsilon_c + 2\epsilon_b}$.

17.3.3 The Effective Medium Approximation

The second approach to the theory of inhomogeneous media, due to Bruggeman [370] and Landauer [371], is generally called the effective medium approximation (EMA). In this discussion I will again restrict myself to a two-component medium made up of spherical grains. The first component has dielectric function ϵ_a and is present with volume fraction f while the second has ϵ_b and the remaining volume fraction, $1 - f$.

In contrast to the MGT, the EMA has the attractive feature of treating all constituents of the medium in an equivalent way. It achieves this symmetry by regarding an individual grain (which may be either type of material) as being embedded in an otherwise homogeneous "effective" medium which is assumed to possess the average properties of the medium. When placed in an external field, the grain in question will be polarized; the field inside is given by Eq. 17.5, with ϵ_o equal to the effective dielectric function ϵ_{EMA}

$$\mathbf{E}_p = \frac{3\epsilon_{\mathrm{EMA}}}{\epsilon_p + 2\epsilon_{\mathrm{EMA}}} \mathbf{E}_{eff} e^{-i\omega t} \tag{17.13}$$

where $\mathbf{E}_{eff}$ is the effective field in the surrounding medium. This field is chosen in a self-consistent way so that the electric field in the medium, when summed over all the grains in the medium, equals $\mathbf{E}_{eff}$. This self-consistency condition may be incorporated in a number of ways; the simplest is to set $\mathbf{E}_{eff}$ equal to the definition of the average field, Eq. 17.7,

$$\mathbf{E}_{eff} = f\mathbf{E}_a + (l - f)\mathbf{E}_b$$

and then substitute Eq. 17.13 evaluated for the fields inside the two kinds of grains. After minimal algebra, this procedure yields a quadratic equation for ϵ_{EMA}, which is conventionally written

$$f\frac{\epsilon_a - \epsilon_{\mathrm{EMA}}}{\epsilon_a + 2\epsilon_{\mathrm{EMA}}} + (1 - f)\frac{\epsilon_b - \epsilon_{\mathrm{EMA}}}{\epsilon_b + 2\epsilon_{\mathrm{EMA}}} = 0. \tag{17.14}$$

This quadratic equation may also be obtained by using $\mathbf{D}_{eff} = \epsilon_{\mathrm{EMA}}\mathbf{E}_{eff}$, with $\mathbf{D}_{eff} = \langle \mathbf{D} \rangle$.

The solution to Eq. 17.14

$$\epsilon_{\mathrm{EMA}} = \frac{B}{4} \pm \frac{1}{4}\sqrt{B^2 + 8\epsilon_a\epsilon_b},$$

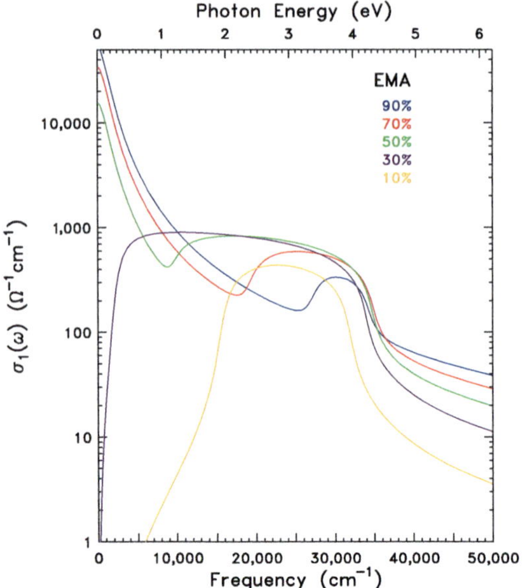

Fig. 17.4 Frequency-dependent conductivity calculated from the effective medium approximation (EMA).

where the sign of the square root is chosen such that Im $\epsilon_{\mathrm{EMA}} \geq 0$ and where

$$B = \epsilon_a(3f - 1) + \epsilon_b(2 - 3f).$$

The EMA differs from the MGT in two important ways. First, the equations treat each of the constituents of the medium on an equal basis. The EMA is thus a symmetric theory and is not restricted to a particular range of concentrations. Second, the EMA predicts a metal–insulator transition at a critical volume fraction (for spherical grains) of $f_c = 1/3$. If I reconsider the example of the metal into which I carve or drill holes (in Section 17.3.2 of this chapter), then the composite stops conducting when 67% of it is void and 33% is metal.*

To demonstrate this second point, I assume that the imaginary part of the complex dielectric function dominates the real part as $\omega \to 0$, so that $\epsilon_a = 4\pi i\sigma_{1a}/\omega \gg \epsilon_{1a}$. If, in addition, $\sigma_{1b} = 0$, Eq. 17.14 may be solved to find

$$\sigma_{1\mathrm{EMA}} = \begin{cases} 0 & f \leq 1/3 \\ \dfrac{3f - 1}{2}\sigma_{1a} & f > 1/3. \end{cases} \tag{17.15}$$

The EMA may readily be extended to systems more complicated than the one considered here. For metal–insulator composites with low metal volume fraction, the theory gives a broad absorption peak centered near the Maxwell-Garnett resonance frequency.

Figure 17.4 shows the EMA frequency-dependent conductivity over the entire frequency range from DC to UV for a metal–insulator composite. I have taken the metallic portion

---

* There are still metal grains touching other metal grains, but the chance of an infinite cluster of such grains spanning the entire sample is – within the EMA – zero.

of the composite to have a Drude dielectric function, Eq. 4.13, and the insulating portion to have a real, constant dielectric function with $\epsilon_b = 4$. The calculation was done for volume fractions f of 0.1, 0.3, 0.6, and 0.7. Relatively short mean free paths (about 30 Å) were taken in calculating this plot but the Drude width does not determine the width of the infrared band; the width of the EMA resonance is intrinsic to the theory.

There are several points to make about this figure. The symmetric EMA has a insulator–conductor transition when the metal volume fraction reaches 1/3, the critical concentration for percolation. Second, although the EMA does give a broad resonance centered near the MGT resonant frequency (a frequency given by Eq. 17.12), this resonance has much greater width than in the MGT. (See Fig. 17.3.) The low frequency edge of the EMA resonance moves to zero frequency at the percolation transition [390]; this behavior is seen in the curve for the $f = 0.3$ sample in Fig. 17.4. Third (although this point is hard to see from Fig. 17.4), the MGT and the EMA at small concentrations give nearly identical results for the low-frequency conductivity.

The concentration dependence of various characteristic frequencies in the EMA has been discussed by Stroud [390]. A plot is seen in Fig. 17.5. The frequencies ω_+ and ω_- are the impurity band limits as calculated in the EMA. $\tilde{\omega}_{\mathrm{imp}}$ and ω_{imp} are the impurity band centroids. There are two percolation concentrators marked. The first, $f = 1/3$, is where the metal component first forms a continuous path throughout the sample. The second, $f = 2/3$, marks where the insulating component first becomes discontinuous.

Figure 17.5 shows that the low-frequency and high-frequency limits move apart as f increases from zero to the metal percolation transition. At the transition, ω_- reaches zero frequency. At higher concentrations ω_- increases, leaving a gap between DC and the impurity band. The upper limit, ω_+ reaches ω_p at $f = 2/3$. As $f \to 1$ the width of the impurity band shrinks to zero. There is a percolating mode at zero frequency once $f > f_c$;

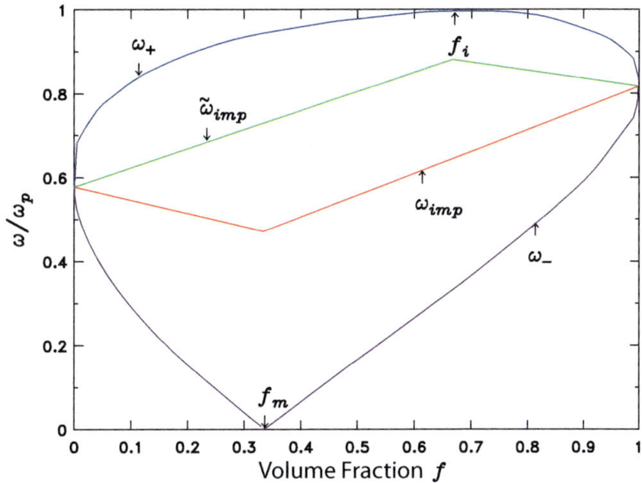

Fig. 17.5   Concentration (f) dependence of the characteristic frequencies of a metal–insulator composite. ω_+ and ω_- are the impurity band limits as calculated in the EMA. The concentrations where the metal and insulator components have their percolation transitions are shown by arrows. $\tilde{\omega}_{\mathrm{imp}}$ and ω_{imp} are the impurity band centroids. (From Ref. 390.)

it clearly can be seen in Fig. 17.4. The DC conductivity is zero for $f < 1/3$ and finite for $f > 1/3$. It grows with increasing metal concentration.

17.4 Other Approaches

17.4.1 Percolation Theory

Percolation theory addresses the properties of inhomogeneous materials by investigating an ordered lattice in which each site (or bond) is occupied with probability p or vacant with probability $1 - p$ and in which adjacent occupied sites (or bonds) are regarded as being electrically connected. The probability p in this model plays the role of volume fraction for a continuous medium. A computer is used to generate the lattices and to solve Kirchoff's laws for the conductivity of the lattice as a function of the occupation probability, the dimensionality, the nature of the interconnection, and the type of lattice employed.

Percolation theories always give a percolation transition, but the occupation probability p_c at which this transition occurs (the *critical probability*) varies from as low as 0.2 to as high as 0.7 depending on the coordination number and dimensionality. Several authors [391–393] have argued, however, that when the amount of space that can be filled by spheres (of equal size) in a particular lattice is taken into account, the critical volume fraction in three dimensions is always in the range $0.16 \leq f_c \leq 0.18$, a value which is in accord with experiment [394]. The electrical resistance of a percolating system has been widely addressed [395–405]. An important concept [406–412] relates the percolation transition to a second-order thermodynamic phase transition, leading to a scaling theory of the transition. The main predictions of this idea are that the critical probability p_c plays the role of the thermodynamic critical temperature T_c and that many properties of the inhomogeneous medium obey power-law dependences on the quantity $p - p_c$. For instance, the conductivity of the network, σ_e, goes to zero as $p \to p_c$ from above according to

$$\sigma_e \sim (p - p_c)^t \qquad p > p_c \tag{17.16}$$

while the static dielectric constant ϵ_e diverges below p_c as

$$\epsilon_e \sim (p_c - p)^{-s} \qquad p < p_c. \tag{17.17}$$

According to the scaling theory, quantities such as the percolation threshold and the proportionality constants in Eqs. 17.16 and 17.17 may vary from system to system depending upon the details of material and sample preparation, but the exponents s and t are universal; they depend only on the dimensionality. In three dimensions [413] $t \approx 1.7$ and $s \approx 0.6$. The latter estimate in particular is in remarkably good agreement with experiment [313, 414]. Note that both scaling exponents differ from those of the EMA, for which $s = t = 1$.

All I have done up to now is to write about the DC conductivity of a percolating system; I have not addressed the frequency dependence of this conductivity. Now, once I make a material inhomogeneous, it will contain both inductive and capacitive elements.

The inductive elements are there because currents will have to circulate around the nonconducting parts of the composite. Capacitive elements are there because charges will appear on the surfaces of these nonconducting parts. These elements will contribute to deviations from the Drude conductivity of the metal (at high metal concentrations) or the insulating dielectric response of the insulator (at low metal concentrations).

The low-frequency conductivity within the EMA varies as ω^2 for $f \ll f_c$ and as $\sqrt{(\omega)}$ for $f > fc$. Bergman and Imry [412] have suggested that the conductivity at $p = p_c$ obeys

$$\sigma_2 \sim \omega^\beta$$

with $\beta = 0.73$. A close result was obtained by Stroud and Bergman [415], who found that

$$\sigma_2 \sim \omega^{t/(t+s)}.$$

The exponent $t/(t + s) = 0.75$.

17.4.2 Eddy Currents: Magnetic Dipole Absorption

A conducting spherical particle in a time-varying magnetic field will have currents, commonly called eddy currents, induced in it. The magnetic field of an electromagnetic wave is such a time-varying field. If the diameter is small compared to the wavelength, these currents give the particle a time-varying magnetic dipole moment,

$$\mathbf{m} = \Omega \gamma_m \mathbf{H}_0 e^{-i\omega t},$$

where Ω is the volume of the sphere, $\mathbf{H}_0$ the amplitude of the field, and γ_m the magnetic polarizability. The solution for the latter may be found in Landau and Lifshitz [3] and Smythe [6]:

$$\gamma_m = \frac{3}{8\pi} \frac{j_2(ka)}{j_0(ka)} = \frac{3}{8\pi} \left[\frac{3}{(ka)^2} - 1 - \frac{3}{ka} \cot(ka) \right] \tag{17.18}$$

where j_0 and j_2 are spherical Bessel functions, a is the radius, and k is the wave vector for light propagating inside the material making up the sphere.* This wave vector is

$$k = \frac{\omega}{c} N = \frac{\omega}{c} \sqrt{\epsilon} = \frac{1}{\delta}(1 + i)$$

where the last equality is true for a metal in the low-frequency limit. δ is the skin depth.

The eddy currents are a consequence of the finite skin depth of the particle and are significant for the electromagnetic response of the inhomogeneous medium in the case where the radius is of the same order as the skin depth [416–419].

When light is traveling in a composite material composed of conducting metal spheres, there will be a bunch of magnetic dipoles in the composite; there is no magnetic dipole moment in the insulating regions. So I could carry through a Maxwell-Garnett or effective medium calculation for the effective magnetic permeability of the composite. I refer the reader to a few papers to see how this goes [417, 418, 420].

---

* I cannot say "propagating inside the sphere" because the sphere is small compared to the wavelength, making propagation not possible.

17.4.3 Mie Scattering

I have considered in this chapter only inhomogeneous media made up of particles whose size is smaller than the wavelength and skin depth. The case where the particles are comparable to the wavelength is the domain of Mie theory [374]. This theory, also developed by Debye [375], calculates the scattering by a spherical particle, essentially as a multipole expansion for the scattered electric and magnetic fields. The polarizabilities given by Eqs. 17.4 and 17.18 are related to the first two (dipolar) terms in this series. When the particle is nearly the same size as the wavelength, all terms are important. Stroud and Pan [421] applied the theory to the specific problem of electromagnetic wave propagation in composite media. A good discussion is in the book by van de Hulst [376].

First, let me start with common units used to measure frequency, photon energy,[*] and related quantities.

ω, angular frequency, in s^{-1} (radian/s)

f, frequency, in Hz $(f = \omega/2\pi)$

λ_0, vacuum wavelength[†] $(\lambda_0 f = c)$

ν, wavenumber (or "frequency"), in cm^{-1} $(\nu = f/c = 1/\lambda_0 = \omega/2\pi c)$ with λ_0 in cm.

E, photon energy, in eV $(E = hf = \hbar\omega = hc\nu)$

T, temperature, in K $(T = E/k_B)$

To convert:

$1\ cm^{-1}$ = 1.44 K = 1/8,066 eV = 0.124 meV = 30 GHz = $1.88 \times 10^{11}\ s^{-1}$

1 eV = 8,066 cm^{-1} = 11,600 K = 242 THz = $1.52 \times 10^{15}\ s^{-1}$ = 1/(1.24 μm)

red = 2 eV = 16,000 cm^{-1} = 1/(620 nm)

I am using cgs-Gaussian units for electromagnetic, quantum mechanical, and solid-state quantities.

t, time, in seconds (s)

ℓ, length, in cm

$c = 3 \times 10^{10}$ cm/s (more or less)

E field in statvolts/cm (but irrelevant, usually)

H field in gauss or oersted (but irrelevant, usually)

m, mass, in grams (g)

h and $\hbar$, Planck's constants, in erg-s[‡]

e, charge, in esu

The dielectric function, ϵ, permeability, μ, and refractive index, N, are dimensionless, a notable advantage.[§] The principal *measured* quantities that have units are the absorption coefficient, α, penetration depth, δ, and conductivity, σ.

---

[*] If, like the particle physicists, I set $\hbar = 1$, then (angular) frequency and energy are identical.

[†] The wavelength in a medium – even air – is of course different: $\lambda = \lambda_0/n$.

[‡] But it even more common to use electron volts, so that $\hbar = 6.58 \times 10^{-16}$ eV-s.

[§] In SI, it is the relative version, e.g., ϵ_r that has no dimensions; the dielectric constant has the units of the vacuum permittivity.

α, absorption coefficient, in cm^{-1} (even in SI, usually)

δ, penetration depth, in cm or μm or Å or even meters

σ, conductivity, is in s^{-1}, i.e., it is a *frequency*

In practical units, σ is measured in Ω^{-1}cm^{-1},* and

$\rho = 1/\sigma$ is in Ohm-cm or Ω-cm

To convert:

$$\sigma_{\mathrm{prac}} = \sigma_{\mathrm{esu}}/30c \approx \sigma_{\mathrm{esu}}/9 \times 10^{11}$$

For copper at 300 K, $\rho \approx 1.6$ $\mu\Omega$-cm, so $\sigma \approx 6 \times 10^5$ Ω^{-1}cm^{-1} $\approx 6 \times 10^{17}$ s^{-1}.

Note that the *equation* for conductivity is the same in both systems; in the Drude model at zero frequency it is given by Eq. 4.9:

$$\sigma_{dc} = ne^2\tau/m.$$

The factor of 9×10^{11} comes from using in cgs (SI) the following: esu (Coulombs) for the charge, grams (kg) for the mass, cm^{-3} (m^{-3}) for the carrier density, and then using the factor of 100 between between Ω^{-1}m^{-1} and Ω^{-1}cm^{-1}.

The equations for the plasma frequency are *not* the same in the two systems, although the units for it (s^{-1}) *are* the same. In cgs (see Eq. 4.12) it is

$$\omega_p = \sqrt{\frac{4\pi ne^2}{m}},$$

whereas in SI it is

$$\omega_p = \sqrt{\frac{ne^2}{\epsilon_0 m}},$$

where $\epsilon_0 = 8.854 \times 10^{-12}$ Farad/m is the permittivity of the vacuum.†

The impedance of free space, Z_0 is different both in equation and units in the two systems. It's defined by $Z_0 = (4\pi/c)(E/H) = 4\pi/c$ in cgs. Here E and H are respectively the amplitudes of the electric and magnetic fields, equal in vacuum. In SI, $Z_0 = E/H = \mu_0 c = 376.7$ Ω. See Eq. 14.12. It is often reported as 120π Ω, which it would be were $c = 3 \times 10^8$ m/s. The cgs value, $4\pi/c = 4.189 \times 10^{-10}$ sec/cm. The ratio of the two is $30c$. (Surprise!)

The absorption coefficient α and the optical conductivity σ_1 are similar quantities. It is often nice to go back and forth, especially in cases of weak absorption bands that do not affect the refractive index very much. Now,

$$\alpha = \frac{2\omega}{c}\kappa \quad \text{and} \quad \epsilon_2 = 2n\kappa$$

---

* To be fully SI-ized, one should report σ in Ohm^{-1}m^{-1}, and some do. There is a factor of 100 between Ω^{-1}cm^{-1} and Ω^{-1}m^{-1}. I'll let the reader figure out which way it goes.

† Persons who wonder about the reasons why the vacuum has a permittivity (and permeability) that differs from unity might want to read Casimir's discussion of the volumetric constant of empty space [422].

where n and κ are respectively the real and imaginary parts of the refractive index. The factor of two on the left comes because the absorption coefficient is defined in terms of the exponential decay of intensity as light passes through the material and the one on the right comes from writing the imaginary part of $\epsilon = N^2$. Then, because $4\pi\sigma_1/\omega = \epsilon_2$, I can write $\sigma_1 = nc\alpha/4\pi$. This equation gives esu values for the conductivity. To convert to SI, I (as usual) divide by $30c$. But then I notice the 120π in the denominator and recognize it as the impedance of free space. So, I write

$$\sigma_1 = \frac{n\alpha}{Z_0}$$

with n the refractive index, α the absorption coefficient, and $Z_0 = 376.7\ \Omega$ the impedance of free space. σ_1 appears in $\Omega^{-1}\mathrm{cm}^{-1}$ if α is in cm^{-1}.

The equations for the classical skin depth differ in the two systems. According to Eq. 4.38 and the discussion at the beginning of Chapter 14 (in Section 14.1), the classical skin depth is

$$\delta = \sqrt{\frac{c^2}{2\pi\omega\sigma_{\mathrm{dc}}}} = \sqrt{\frac{2c^2}{\omega_p^2\omega\tau}},$$

where σ_{dc} is the DC conductivity, measured (esu) in $\sec^{-1}$. The calculation will yield δ in cm. In SI, with the conductivity σ_{dc} in $\Omega^{-1}\mathrm{m}^{-1}$, the skin depth is

$$\delta = \sqrt{\frac{2\epsilon_0 c^2}{\omega\sigma_{\mathrm{dc}}}} = \sqrt{\frac{2c^2}{\omega_p^2\omega\tau}},$$

where you would use $c = 3 \times 10^8$ m/s so that δ emerges in m. I may also use $\delta = \sqrt{2/\omega\sigma_{\mathrm{dc}}\mu_0}$ with $\mu_0 = 4\pi \times 10^{-7}$ N/A, the permeability of free space, again with σ_{dc} in $\Omega^{-1}\mathrm{m}^{-1}$ and δ in m.

The mobility is also an optical/electrical materials property with units. The conductivity σ is written in terms of mobility as

$$\sigma = ne\mu$$

where n is the carrier concentration, e the electronic charge, and μ the mobility. If the semiconductor is in the regime where a Drude model applies, one has $\mu = e\tau/m^*$. Here τ is the mean time before the next collision and m^* is the effective mass. The units in which mobilities are usually reported are $\mathrm{cm}^2/\mathrm{V\cdot s}$. This set is neither-fish-nor-foul; in pure SI, the units should be $\mathrm{m}^2/\mathrm{V\cdot s}$. If the mobility is $\mu = 1{,}000\ \mathrm{cm}^2/\mathrm{V\cdot}$, it is $\mu = 0.1\ \mathrm{m}^2/\mathrm{V\cdot s}$. To convert to cgs-Gaussian, where the units are $\mathrm{cm}^2/\mathrm{statV\cdot s}$, multiply the mobility in $\mathrm{cm}^2/\mathrm{V\cdot s}$ by 300. (Or multiply the mobility in $\mathrm{m}^2/\mathrm{V\cdot s}$ by 3×10^6.)

Maxwell's equations look a bit different in SI. For example, Eq. 2.1 is simpler looking in SI:[*]

$$\nabla \cdot \mathbf{D} = \rho_{\text{ext}} \tag{B.1a}$$

$$\nabla \cdot \mathbf{B} = 0 \tag{B.1b}$$

$$\nabla \times \mathbf{E} = -\frac{\partial \mathbf{B}}{\partial t} \tag{B.1c}$$

$$\nabla \times \mathbf{H} = \mathbf{j}_{\text{free}} + \frac{\partial \mathbf{D}}{\partial t}. \tag{B.1d}$$

The Lorentz law (Eq. 2.2) is

$$\mathbf{F} = q(\mathbf{E} + \mathbf{v} \times \mathbf{B}), \tag{B.2}$$

where v is the particle's velocity.[†]

In vacuum, the SI version of Eq. 2.3 becomes $\mathbf{D} = \epsilon_0 \mathbf{E}$ and $\mathbf{B} = \mu_0 \mathbf{H}$, where ϵ_0 and μ_0 are the permittivity and permeability of vacuum, with values

$$\epsilon_0 = 8.854 \times 10^{-12} \quad \text{Farad/m}$$

$$\mu_0 = 4\pi \times 10^{-7} \quad \text{Henry/m}.$$

In vacuum, one finds plane-wave solutions, with velocity

$$\frac{1}{\sqrt{\epsilon_0 \mu_0}} \equiv c.$$

In macroscopic media, the auxiliary fields are

$$\mathbf{D} = \epsilon_0 \mathbf{E} + \mathbf{P}$$

$$\mathbf{H} = \frac{1}{\mu_0} \mathbf{B} - \mathbf{M} \tag{B.3}$$

which, in local, linear, isotropic, and homogeneous media, become

$$\mathbf{D} = \epsilon_r \epsilon_0 \mathbf{E}$$

$$\mathbf{H} = \frac{\mathbf{B}}{\mu_r \mu_0}$$

---

[*] Although I think that it is nice to see the speed of light appearing in the equations of electromagnetism.

[†] This one gives no indication that speeds need to approach lightspeed before magnetic forces and electric forces become comparable (for, say, fields with the same electric and magnetic energy densities).

with ϵ_r and μ_r dimensionless numbers, called respectively the relative permittivity and relative permeability, identical to the dielectric constant ϵ and permeability μ in cgs-Gaussian units.

The SI energy density is

$$u = 1/2(\mathbf{E} \cdot \mathbf{D} + \mathbf{B} \cdot \mathbf{H}).$$

(In vacuum, $u = 1/2(\epsilon_0 E^2 + B^2/\mu_0)$.) The Poynting vector is

$$\mathbf{S} = \mathbf{E} \times \mathbf{H}.$$

SI units for S are particularly easy (and, I must admit, easier than in cgs) because the electric field is in V/m and the magnetic field is in A/m, making the units of the Poynting vector (and of the intensity, equal to the magnitude of time-averaged S) be W/m^2.

Let's write a complex plane wave $\mathbf{A}$ as

$$\mathbf{A} = \mathbf{A}_0 e^{i(\mathbf{q}\cdot\mathbf{r}-\omega t)} = \left(A_{0x}\hat{x} + A_{0y}\hat{y} + A_{0z}\hat{z}\right) e^{iq_x x} e^{iq_y y} e^{iq_z z} e^{-i\omega t}$$

where $\mathbf{A}_0$ is a constant vector specifying the direction, magnitude, and phase of $\mathbf{A}$, and the wave vector $\mathbf{q} = q_x\hat{x} + q_y\hat{y} + q_z\hat{z}$ is normal to the planes of constant phase.

Now take the partial with respect to time of $\mathbf{A}$. The only term containing the time in the vector is $e^{-i\omega t}$ and

$$\frac{\partial}{\partial t} e^{-i\omega t} = -i\omega e^{-i\omega t}$$

so that

$$\frac{\partial}{\partial t}\mathbf{A} = -i\omega\mathbf{A}.$$

Now, I want to act on $\mathbf{A}$ with the vector operator ∇ where

$$\nabla = \hat{x}\frac{\partial}{\partial x} + \hat{y}\frac{\partial}{\partial y} + \hat{z}\frac{\partial}{\partial z}.$$

First, I'll compute the divergence $\nabla \cdot \mathbf{A}$. If I do this, and write out the dot product of ∇ with $\mathbf{A}_0$, I get

$$\nabla \cdot \mathbf{A} = \left(A_{0x}\frac{\partial}{\partial x} + A_{0y}\frac{\partial}{\partial y} + A_{0z}\frac{\partial}{\partial z}\right) e^{iq_x x} e^{iq_y y} e^{iq_z z} e^{-i\omega t}.$$

Each partial derivative only acts on one of the complex exponentials in $\mathbf{A}$, so that

$$\nabla \cdot \mathbf{A} = \left(iq_x A_{0x} + iq_y A_{0y} + iq_z A_{0z}\right) e^{iq_x x} e^{iq_y y} e^{iq_z z} e^{-i\omega t}$$

or

$$\nabla \cdot \mathbf{A} = i\mathbf{q} \cdot \mathbf{A}.$$

In a similar way, I can show that

$$\nabla \times \mathbf{A} = i\mathbf{q} \times \mathbf{A},$$
$$\nabla^2\mathbf{A} = -q^2 \cdot \mathbf{A},$$

and

$$\nabla A = i\mathbf{q}A.$$

In the last of these A is a scalar.

The great thing about the complex exponential solutions to Maxwell's equations is that they allow me to replace differential equations with algebraic ones.

It is not surprising that electromagnetic waves satisfy a wave equation. Many textbooks derive a wave equation from Maxwell's equations and use that to find the wave-like fields of propagating electromagnetic waves. Instead, in Chapter 2 I showed that the wavelike solutions satisfied directly Maxwell's equations.

The wave equation approach is certainly worth mentioning, so I do that here. I'll start with Eqs. 2.4a–b. I'll take the medium to be uncharged, linear, isotropic, and homogeneous, so that $\mathbf{D} = \epsilon_1 \mathbf{E}$, $\mathbf{j} = \sigma_1 \mathbf{E}$, and $\mathbf{B} = \mu \mathbf{H}$. With these definitions, I have

$$\nabla \cdot \mathbf{E} = 0 \tag{D.1a}$$

$$\nabla \cdot \mathbf{H} = 0 \tag{D.1b}$$

$$\nabla \times \mathbf{E} = -\frac{\mu}{c} \frac{\partial \mathbf{H}}{\partial t} \tag{D.1c}$$

$$\nabla \times \mathbf{H} = \frac{4\pi}{c} \sigma_1 \mathbf{E} + \frac{\epsilon_1}{c} \frac{\partial \mathbf{E}}{\partial t}. \tag{D.1d}$$

Now, I take the curl of Eq. D.1c, interchanging the order of derivatives on the right:

$$\nabla \times (\nabla \times \mathbf{E}) = -\frac{\mu}{c} \frac{\partial \nabla \times \mathbf{H}}{\partial t}.$$

I use a vector identity on the left and replace $\nabla \times \mathbf{H}$ on the right with the right side of Eq. D.1d

$$\nabla(\nabla \cdot \mathbf{E}) - \nabla^2 \mathbf{E} = -\frac{4\pi}{c^2} \sigma_1 \mu \frac{\partial \mathbf{E}}{\partial t} - \frac{\epsilon_1 \mu}{c^2} \frac{\partial^2 \mathbf{E}}{\partial t^2}.$$

Now $\nabla \cdot \mathbf{E} = 0$. Rearranging slightly the rest, I find

$$\nabla^2 \mathbf{E} - \frac{4\pi}{c^2} \sigma_1 \mu \frac{\partial \mathbf{E}}{\partial t} - \frac{\epsilon_1 \mu}{c^2} \frac{\partial^2 \mathbf{E}}{\partial t^2} = 0. \tag{D.2}$$

Eq. D.2 is the wave equation for the electric field in a conducting and magnetic solid.

If I take the curl of Eq. D.1d and follow exactly the same procedure, I obtain a wave equation for the magnetic field:

$$\nabla^2 \mathbf{H} - \frac{4\pi}{c^2} \sigma_1 \mu \frac{\partial \mathbf{H}}{\partial t} - \frac{\epsilon_1 \mu}{c^2} \frac{\partial^2 \mathbf{H}}{\partial t^2} = 0. \tag{D.3}$$

I can employ *either* Eq. 2.5 or Eq. 2.6 in these to show that

$$q^2 = \frac{\omega^2}{c^2} \epsilon_1 \mu + \frac{4\pi i}{c^2} \omega \sigma_1 \mu = \frac{\omega^2}{c^2} \epsilon \mu,$$

exactly as in Eq. 3.35.

I derived a formula for the reflectance at normal incidence in Chapter 3 (Section 3.6). (See Eq. 3.33.) Here, I'll consider the reflectance at nonnormal incidence. Two neat effects will appear: zero reflectance at Brewster's angle and zero transmittance in the regime of total internal reflection. I'll first discuss the relations amongst the angles of incident, reflected, and transmitted rays and then go on to obtain the transmission and reflection coefficients.

I'll work out the angle relations in two ways: first using Fermat's principle and second by the boundary conditions. Fermat's Principle of Least Time may be stated as "light travels between two given points along the path of shortest time." In a single medium, this statement is equivalent to saying that light travels in straight lines.* Fermat's principle may be regarded as an axiom, as an experimental observation, or as the results of solutions to Maxwell's equations.

Reflection and Refraction

In Fig. E.1, an incident light ray IO strikes (at point O) the interface between two media of complex refractive indices $N_a = n_a + i\kappa_a$ and $N_b = n_b + i\kappa_b$. Part of the ray is reflected as ray OR and part transmitted as ray OT. The angles that the incident, reflected, and transmitted rays make to the normals of the interface are θ_i, θ_r, and θ_t, respectively.

Light in medium a travels with speed $v_a = c/n_a$. The time it takes for the light to go from point I to point R via point O is the length of the path taken divided by the speed:

$$T = T_i + T_r = \frac{\sqrt{y^2 + h_i^2}}{v_a} + \frac{\sqrt{(\ell - y)^2 + h_r^2}}{v_a}$$

with the lengths defined in Fig. E.1. Now the light could in principal take any path that starts at I, ends at O, and includes the interface.† For instance, it could go horizontally to the right to the interface and then slant up to R, so that $y = 0$. There are an infinite number of such paths and Fermat tells me that the one taken has the minimum value of T. If so,

$$\frac{dT}{dy} = 0 = \frac{y}{v_a\sqrt{y^2 + h_i^2}} - \frac{\ell - y}{v_a\sqrt{(\ell - y)^2 + h_r^2}}.$$

---

* I ignore the lensing of light by strong gravitational fields. But see Ref. 423.

† Although I did not specify it, the interface must be smooth on the scale if the wavelength; this is the case of *specular* reflection. If the surface is rough, light can travel from point I to point R via many spots on the interface. This second case is called diffuse reflectivity (or transmissivity) and is what you want for illumination. Metal mirrors are specular; white paper is diffuse.

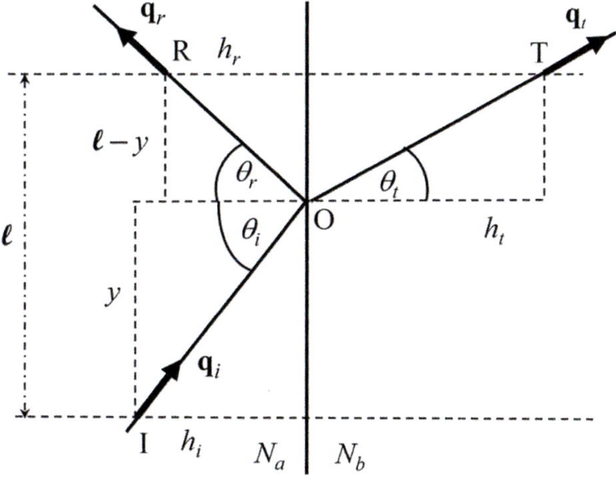

Fig. E.1 Non-normal incidence at the interface between two media. The wave vectors $\mathbf{q}_i$, $\mathbf{q}_r$, and $\mathbf{q}_t$ for the incident, reflected and transmitted waves are shown. The points I and R are separated along the interface (vertically here) by distance ℓ and are distances h_i and h_r above the interface. The point T has the same vertical distance from O as does R, but is h_t *below* the interface.

I look at the triangles in the figure and see that $\sin\theta_i = y/\sqrt{y^2 + h_i^2}$ and $\sin\theta_r = (\ell - y)/\sqrt{(\ell - y)^2 + h_r^2}$. Hence, after substitution and a modest amount of algebra, I find that the angle of the light leaving the surface is the same as the arrival angle. This result is the law of reflection,*

$$\theta_i = \theta_r. \tag{E.1}$$

I drew Fig. E.1 with points I, O, R in the plane containing also the normal to the surface. I should consider whether the light could take some skewed path and reach R by going out of the plane defined by I, O, R and the normal. But a little thought convinces me that such a path, which would go via a point on the interface O' that is above or below point O, is longer than the one shown and so does not satisfy Fermat's Principle of Least Time. The plane containing the wave vector of the incident light ($\mathbf{q}_i$) and the surface normal is called the *plane of incidence* and the law of reflection says that this plane also contains the reflected wave vector $\mathbf{q}_r$. By the same argument, the transmitted wave vector $\mathbf{q}_t$ is also in this plane.

When light is transmitted across the interface, its speed changes to $v_b = c/n_b$. The time taken for light to travel from I to T is now

$$T = T_i + T_t = \frac{\sqrt{y^2 + h_i^2}}{v_a} + \frac{\sqrt{(\ell - y)^2 + h_i^2}}{v_b}.$$

---

* I understand the diffuse reflection from a rough surface as satisfying the law of reflection locally, but the roughness makes the local normal, when averaged over the surface, point in many of the possible directions.

I set the derivative to zero and note that $\sin\theta_t = (\ell - y)/\sqrt{(\ell - y)^2 + h_t^2}$. So after a modicum of algebra, including substituting for the velocities, I get Snell's law:

$$n_a \sin\theta_i = n_b \sin\theta_t. \tag{E.2}$$

Let me now consider what the boundary conditions on the fields tell me. I'll orient the coordinates so that the x axis is along the interface normal, into medium b. I'll put the y axis in the interface and in the plane of incidence and the z axis in the interface perpendicular to the plane of incidence. For these coordinates, the real parts of the three wave vectors are $\mathrm{Re}\,\mathbf{q}_i = (\omega/c)n_a(\cos\theta_i\,\hat{\mathbf{x}} + \sin\theta_i\,\hat{\mathbf{y}})$, $\mathrm{Re}\,\mathbf{q}_r = (\omega/c)n_a(-\cos\theta_r\,\hat{\mathbf{x}} + \sin\theta_r\,\hat{\mathbf{y}})$, and $\mathrm{Re}\,\mathbf{q}_t = (\omega/c)n_b(\cos\theta_t\,\hat{\mathbf{x}} + \sin\theta_t\,\hat{\mathbf{y}})$. The fields in medium a are the superposition of incident field and reflected field; in medium b the transmitted field. The boundary conditions to consider now are the continuity of tangential electric field and tangential magnetic field, written in the forms of Eqs. 2.5 and 2.6. These generically may be written (at $x = 0$) as

$$\mathbf{F}_i e^{i\left(\frac{\omega_i}{c}n_a y\sin\theta_i - \omega_i t\right)} + \mathbf{F}_r e^{i\left(\frac{\omega_r}{c}n_a y\sin\theta_r - \omega_r t\right)} = \mathbf{F}_t e^{i\left(\frac{\omega_t}{c}n_b y\sin\theta_t - \omega_t t\right)}, \tag{E.3}$$

where $\mathbf{F}$ is either the electric or magnetic field of the light wave. I have put subscripts on the frequency to allow me to think whether the frequencies of all three waves need to be the same.

I can address the frequency issue quickly. So that the boundary conditions remain satisfied at all times, given that they are satisfied at one instant of time, the three harmonic terms must vary in time at the same rate. This conclusion follows from the linear nature of Maxwell's equations. Hence, all fields vary in time as $e^{-i\omega t}$.

Now, the boundary conditions must be satisfied at all points on the interface as well. Suppose that they are satisfied at one particular point, such as $y = z = 0$. Then, in order that they remain true for other points on the interface, each term in Eq. E.3 must vary in the same fashion I move along the y axis and in the plane of the interface. This requirement means that the coefficients of y must be equal. I get

$$\frac{\omega}{c}n_a \sin\theta_i = \frac{\omega}{c}n_a \sin\theta_r = \frac{\omega}{c}n_b \sin\theta_t.$$

The first pair reduce to the law of reflection $\theta_i = \theta_r$ and the first and third give Snell's law $n_a \sin\theta_i = n_b \sin\theta_t$.

Absorbing Media

The propagation in strongly absorbing materials of waves incident at oblique angles to interfaces is complicated. Let me pose the problem and discuss it qualitatively. For the moment, I'll take the material in which the incident and reflected waves are traveling as nonabsorbing.*

---

* A semi-infinite absorbing medium poses conceptual dissonance. For a finite amount of energy to arrive at the interface, it must have started with infinite energy.

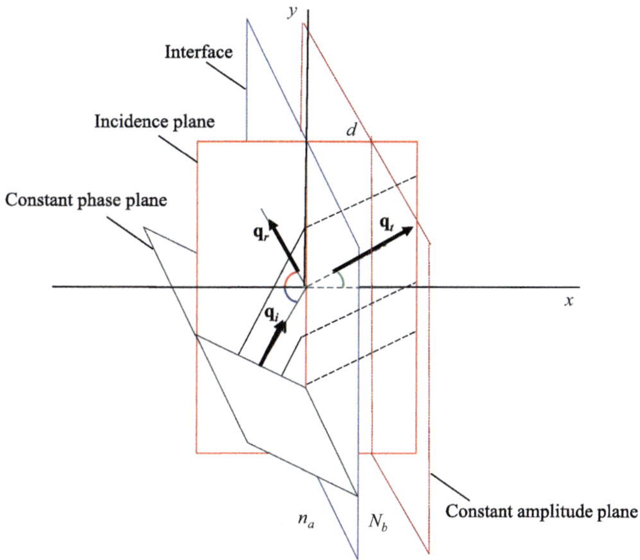

Fig. E.2 Nonnormal incidence at the interface between two media, one with refractive index n_a and the other with complex index N_b. The wave vectors for the incident, reflected and transmitted waves are shown. Also shown are the plane of incidence, the plane of constant phase for the incident wave, and the plane of constant amplitude for the transmitted wave.

The geometry is shown in Fig. E.2. The incident wave arrives at the interface at an angle to the normal. The planes of constant phase, to which the incident wave vector q_i are normal, are also inclined. Because medium a is nonabsorbing, the intensity at the interface (or just before the interface) is everywhere the same.

Light reaching the interface is refracted to continue as the transmitted wave. The planes of constant phase for the refracted wave, perpendicular to q_t, are also inclined to the interface.

What about the planes of constant amplitude? If the fields were to be written as $e^{i q_t \cdot \mathbf{r} - \omega t}$ (Eqs. 3.11 and 3.12), the planes of constant amplitude would also be orthogonal to q_t. But this claim is not correct. I see where the misconception comes from by looking at the four rays shown as dashed lines in Fig. E.2. When they arrive at a plane perpendicular to q_t, the rays will have gone different distances in the absorbing medium, with the one at the bottom of the diagram having gone the farthest and the one at the top the shortest distance. Now, after crossing the interface, each ray or pencil of light does follow Eqs. 3.11 and 3.12 and, after reaching a depth d in medium b, has amplitude $e^{-\kappa_b \omega d / c \cos \theta_t}$. The planes of constant amplitude are parallel to the interface, as indicated in the figure. A wave where the planes of constant phase and the planes of constant amplitude are not the same is an *inhomogeneous wave* [8, 424, 425]. The details are worked out in Born and Wolf [8]. To evade this complication, I will take the refractive indices as real. The interesting phenomena and physics are in transparent media anyway.

For further ease, I'll take the materials to have $\mu = 1$. The equations with different values of permeability may be seen in Jackson [2].

Amplitude Equations

It turns out that the oblique-angle reflection depends both on the angle of incidence of the light and on its polarization (**E**-vector direction). The equations are called the "Fresnel equations" and it is conventional (and correct) to work them out separately for the electric field parallel and perpendicular to the plane of incidence. Any other polarization direction can be worked out by superposition.

The terminology is diverse and often used without explanation. If the electric field lies in (is parallel to) the plane of incidence, I'll call it p polarization, for "parallel." The English word for perpendicular starts with same letter* so the polarization perpendicular to the plane of incidence is called s polarization (from "senkrecht," German for perpendicular). Other terms used for p polarization are transverse-magnetic (TM), pi-polarized, π-polarized, tangential-plane-polarized, and ∥-polarized; s polarization can be called transverse-electric (TE), sigma-polarized σ-polarized, sagittal-plane-polarized, or ⊥-polarized. Finally, the plane of incidence, to which the field is either perpendicular or parallel, is the plane defined by the normal to the flat interface and the incident wave vector, $\mathbf{q}_i$.[†] The boundary conditions are that the normal components of **D** and **B** and the tangential components of **E** and **H** are continuous across the interface. (With $\mu = 1$, $\mathbf{B} = \mathbf{H}$.) I'll use the coordinate system of Fig. E.2, with the x axis perpendicular to the interface. The y and z directions lie in the interface, with y also in the plane of incidence and z perpendicular to it.

Electric Field Perpendicular to the Plane of Incidence (s)

The coordinate system and field directions are shown in Fig. E.3. I write the fields (at the origin and at $t = 0$) as

$$\mathbf{E}_i = \hat{z}$$
$$\mathbf{E}_r = r_s\hat{z}$$
$$\mathbf{E}_t = t_s\hat{z} \tag{E.4}$$
$$\mathbf{H}_i = n_a(\hat{x}\sin\theta_i - \hat{y}\cos\theta_i)$$
$$\mathbf{H}_r = n_a r_s(\hat{x}\sin\theta_i + \hat{y}\cos\theta_i)$$
$$\mathbf{H}_t = n_b t_s(\hat{x}\sin\theta_t - \hat{y}\cos\theta_t)$$

---

* One source of confusion.

† It is *not* the plane of the interface. Also, normal incidence has no plane of incidence (or has all planes as the plane of incidence) and the subscripts p and s need not be used.

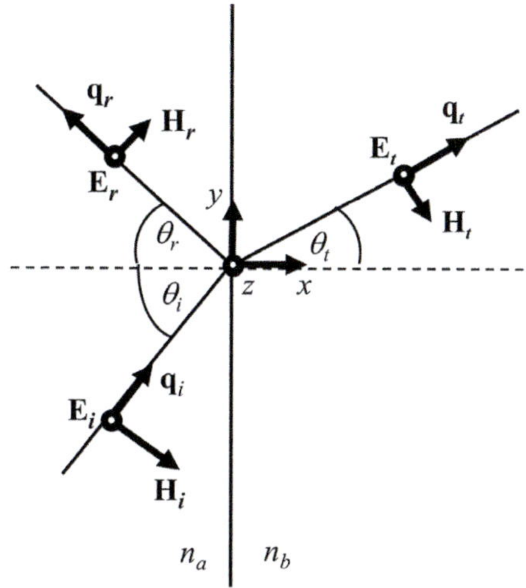

Fig. E.3 Geometry of transmission and reflection for s-polarized light. The interface between the two media (with refractive indices n_a and n_b) is the y–z plane. The plane of the paper is the plane of incidence.

where I took the amplitude of the incident field as unity, r_s is the amplitude of the reflected field, and t_s is the amplitude of the transmitted field. For the magnetic fields, I used $\hat{q} \times \hat{e}$ for the direction and Eq. 3.14 ($H = NE$) for the magnitudes.

Now, I apply the boundary conditions. Tangential $\mathbf{E}$ is continuous. I dot the $\mathbf{E}$ field with $\hat{z}$ and equate the fields on left and right side of the interface:

$$1 + r_s = t_s, \tag{E.5}$$

because the electric field is purely tangential. Tangential $\mathbf{H}$ is continuous. The magnetic field has both tangential and normal components. For the former, I dot with $\hat{y}$ yielding

$$-n_a \cos\theta_i + n_a r_s \cos\theta_i = -n_b t_s \cos\theta_t. \tag{E.6}$$

I substitute t_s from Eq. E.5 into Eq. E.6, and solve for the reflectivity

$$r_s = \frac{n_a \cos\theta_i - n_b \cos\theta_t}{n_a \cos\theta_i + n_b \cos\theta_t}. \tag{E.7}$$

Now, I put r_s from Eq. E.7 back into Eq. E.5. It cleans up nicely:

$$t_s = \frac{2n_a \cos\theta_i}{n_a \cos\theta_i + n_b \cos\theta_t}. \tag{E.8}$$

The angle $\theta_i = 0$ at normal incidence as is $\theta_t = 0$, by Snell's law. Equations E.7 and E.8 reduce to Eqs. 3.30 and 3.31 at normal incidence, as they should. The other limit, $\theta_i = 90°$, requires* me to specify that $n_a < n_b$, in which case $r_s = 1$ and $t_s = 0$.

I used two of the four boundary conditions. What about the other two? Well, $\mathbf{D} \cdot \hat{x} = 0$ for all three fields, so the normal $\mathbf{D}$ condition is trivially satisfied. As to normal $\mathbf{B}$, I use

---

* Otherwise Snell's law requires an angle for which the sine is greater than 1.

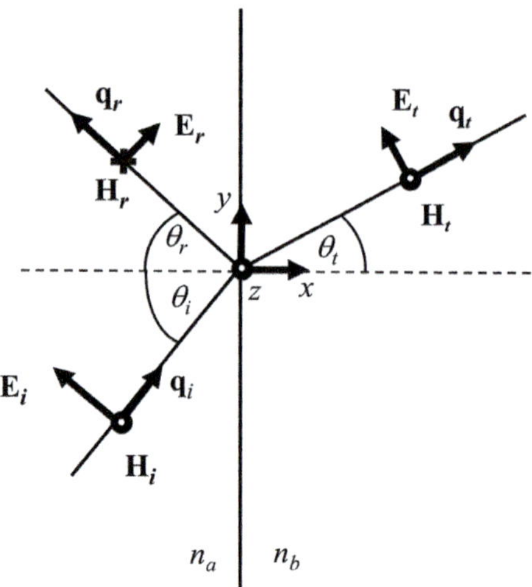

Fig. E.4   Geometry of transmission and reflection for p-polarized light. The interface between the two media (with refractive indices n_a and n_b) is the y–z plane.

$\mu = 1$ for both materials, so taking the dot product of the magnetic fields in Eq. E.4 with $\hat{x}$ gives

$$n_a \sin \theta_i + r_s n_a \sin \theta_i = t_s n_b \sin \theta_t.$$

But, Eq. E.5 gives $1 + r_s = t_s$. By use of this boundary condition, I rederive Snell's law.

Electric Field Parallel to the Plane of Incidence (p)

The coordinate system and field directions are shown in Fig. E.4. I've set the field directions so that the case where $\theta_i = \theta_r = 0$ in Fig. E.4 uses the same convention as in Fig. 3.2: the electric fields are parallel and the magnetic fields antiparallel. I should warn you that about half the time I see a different convention. Some books [3, 5] use this convention; others [2, 4, 8] use the opposite one.

I write the fields (at the origin and at $t = 0$) as

$$\begin{aligned}
\mathbf{E}_i &= -\hat{x} \sin \theta_i + \hat{y} \cos \theta_i \\
\mathbf{E}_r &= r_p (\hat{x} \sin \theta_i + \hat{y} \cos \theta_i) \\
\mathbf{E}_t &= t_p (-\hat{x} \sin \theta_t + \hat{y} \cos \theta_t) \\
\mathbf{H}_i &= n_a \hat{z} \\
\mathbf{H}_r &= -n_a r_p \hat{z} \\
\mathbf{H}_t &= n_b t_p \hat{z}
\end{aligned} \tag{E.9}$$

where I took the amplitude of the incident field as unity, r_p is the amplitude of the reflected field, and t_p is the amplitude of the transmitted field. I used $\hat{q} \times \hat{e}$ for the direction of the magnetic fields and Eq. 3.14 ($H = NE$) for their magnitudes.

Now, I apply the boundary conditions. Tangential $\mathbf{E}$ is continuous. I dot the $\mathbf{E}$ field with $\hat{y}$ and equate the fields on left and right side of the interface:

$$\cos\theta_i + r_p \cos\theta_i = t_p \cos\theta_t. \tag{E.10}$$

Tangential $\mathbf{H}$ is continuous. The magnetic field has only tangential components. I dot with $\hat{z}$ yielding

$$n_a - n_a r_p = n_b t_p. \tag{E.11}$$

I substitute t_p from Eq. E.11 into Eq. E.10, and solve for the reflectivity

$$r_p = \frac{n_a \cos\theta_t - n_b \cos\theta_i}{n_a \cos\theta_t + n_b \cos\theta_i}. \tag{E.12}$$

Now, I put r_p from Eq. E.12 back into Eq. E.11. It cleans up nicely:

$$t_p = \frac{2n_a \cos\theta_i}{n_a \cos\theta_t + n_b \cos\theta_i}. \tag{E.13}$$

Eqs. E.12 and E.13 reduce to Eq. 3.30 and 3.31 at normal incidence, as they should. The other limit, $\theta_i = 90°$, again requires me to specify that $n_a < n_b$, in which case $r_p = 1$ and $t_p = 0$.

As far as the other two boundary conditions go, the condition on $\mathbf{B}_n$ is trivially satisfied and the condition on $\mathbf{D}_n$ gives Snell's law.

Equations E.7, E.8, E.12, and E.13 all contain both $\cos\theta_i$ and $\cos\theta_t$. These quantities are not independent; Snell's law ($n_a \sin\theta_i = n_b \sin\theta_t$) relates them. When I need to, I can use

$$\cos\theta_t = \sqrt{1 - \frac{n_a^2}{n_b^2} \sin^2\theta_i} \tag{E.14}$$

to eliminate $\cos\theta_t$. Meantime, however, I'll calculate the reflected amplitudes in 4 cases: both polarizations with $n_b > n_a$ and both polarizations with $n_b < n_a$. The results are in Fig. E.5.

The left panel shows the results for the case where the light is incident from the "less dense" (smaller n) material onto the "more dense" (larger n) material. At normal incidence $\theta_i = 0$, the reflectivity is negative, $r = -1/3$ (for $n_a = 1$ and $n_b = 2$) in both the s and p cases* The reflectivity is basically flat for the first $5°$ or so but the two curves separate at larger angles. The s-polarized reflectivity becomes more negative (and hence larger in magnitude) eventually reaching $r_s = -1$ at $90°$ incident angle. The p-polarized reflectivity becomes smaller in magnitude and crosses zero at an incident angle (for the refractive indices used) just above $60°$. Then it grows in magnitude, reaching $r_p = +1$ at $90°$ incident angle. Thus both polarizations have identical reflectivities at normal and at grazing incidence but nowhere in between.

There are differences and similarities when the light is incident from the "more dense" (larger n) material onto the "less dense" (smaller n) material. The normal incidence reflectivity is positive rather than negative, with the same magnitude as the case I discussed above. The p-polarized reflectivity crosses zero at a smaller incident angle, just above

---

* As it should be. There is no plane of incidence when $\theta_i = 0$. (Or, there is an infinite number of them.)

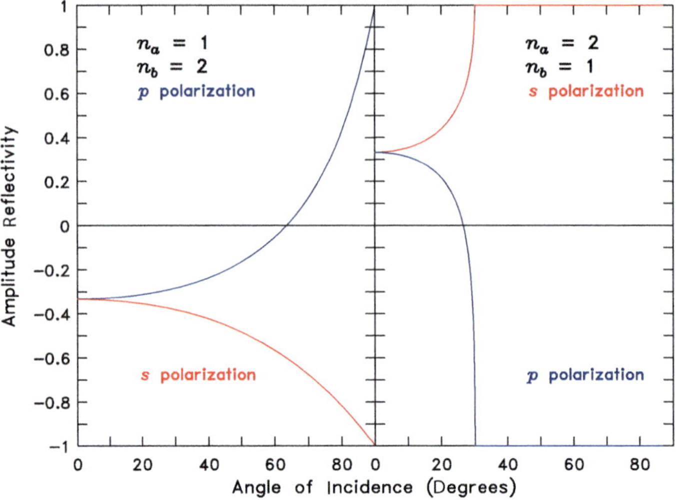

Fig. E.5   Amplitude reflectivity for p- and s-polarized light as a function of angle of incidence. Left panel: The media have $n_a = 1$ and $n_b = 2$. Right panel: The media have $n_a = 2$ and $n_b = 1$.

$25°$ for the parameters used. The reflectivity reaches ± 1, but at angles well below grazing incidence. Both s-polarized and p-polarized reflectivities reach $r_s = +1$ and $r_p = -1$ at $30°$ incident angle.

Special Angles

There is one angle for a given n_a and n_b when the value of r_p goes to zero.* All the p-polarized incident light is purely refracted into the second medium. To find the angle, I set $r_p = 0$ in Eq. E.12 to find

$$n_b \cos \theta_i = n_a \cos \theta_t = n_a \sqrt{1 - \frac{n_a^2}{n_b^2} \sin^2 \theta_i},$$

where I've used Eq. E.14 on the right to eliminate θ_t. I square both sides, use $1 = \sin^2 \theta_i + \cos^2 \theta_i$, collect terms, take a square root, use $\tan = \sin / \cos$, and find that the angle of zero reflectivity, θ_B obeys

$$\tan \theta_B = \frac{n_b}{n_a}. \tag{E.15}$$

---

* The angle at which r_p changes sign! This statement is true for nonabsorbing media. In contrast, if there is absorption, there is a complex refractive index and the amplitude reflectivity is complex in turn. The magnitude of the complex reflectivity is always positive. It can get small at some angle, but will not be zero.

This angle is known as Brewster's angle. It is $63.4°$ for the left panel of Fig. E.5 and $26.6°$ for the right panel. It is $56°$ for an interface between glass ($n = 1.5$) and vacuum or air ($n = 1$). Finally for a silicon ($n = 3.48$)/vacuum interface, $\theta_B = 74°$.

When I calculate the angle between the reflected wave and the transmitted wave (the angle $\pi - \theta_r - \theta_t$ in Fig. E.4 when $\theta_r = \theta_i = \theta_B$), I find that the angle is $90°$. I can see this in several ways: I can compute the angle using Eqs. E.2 and E.14 as well as some trigonometric identities and find that the angle is $\pi/2$. An alternate way is to compute the angle for which $\mathbf{q}_r \cdot \mathbf{q}_t = 0$ and find that it is indeed θ_B.

There is some physics in the fact that the reflectance is zero when the reflected and transmitted rays are at right angles to each other. I can ask "why is there light reflected from a surface and what are the sources of this light?" Electromagnetic radiation has as its sources accelerated charges. The transmitted field in medium b induces a time-varying electric dipole moment/unit volume $\mathbf{P}_b = \chi_b \mathbf{E}_t = (\epsilon_b - 1)\mathbf{E}_t/4\pi$ where χ_b is the complex susceptibility and ϵ_b is the complex dielectric function. The dipole moment/unit volume is composed of a very large number of oscillating dipoles, each of which emits dipole radiation. The reflected light is the superposition of the emitted fields from all these dipoles. The dipoles are oriented along $\mathbf{E}_t$, which is perpendicular to $\mathbf{q}_t$ and hence is, for p-polarized light, parallel to $\mathbf{q}_r$. But the radiated power is zero along the axis of the dipole; hence, the reflectivity is zero.

At Brewster's angle, reflected light is 100% s-polarized and transmitted light is preferentially p-polarized. A variety of devices take advantage of this selectivity. One of the simplest is a polarizer oriented for p polarization, and therefore suppressing reflected light from windows in photographs or optical instruments. A "pile of plates," at Brewster's angle, each reflecting a few percent of the s-polarized light, may be used to provide highly polarized transmitted light over a broad range of wavelengths.

The other interesting angle, known as the *critical angle*, appears in the right panel of Fig. E.5; it is the angle of incidence at and beyond which the reflectivity is ± 1. The phenomenon is known as "total internal reflection." Whereas Brewster's angle occurs only for p-polarized light and is found independent of which index is larger, the critical angle is found only if $n_b > n_a$ and occurs for both polarizations.

Snell's law, Eq. E.2, can be written $\sin \theta_t = (n_a/n_b) \sin \theta_i$. If $n_a > n_b$, then $\theta_t > \theta_i$. There is then some incident angle for which $\theta_t = 90°$. This, the critical angle, is then given by

$$\theta_c = \arcsin\left(\frac{n_b}{n_a}\right).$$

All light is reflected and $|r_s| = |r_p| = 1$. The critical angle is $41.8°$ for a glass ($n_a = 1.5$) to air or vacuum ($n_b = 1$) interface. For silicon ($n = 3.48$) to vacuum, $\theta_c = 16.7°$. Recall that the angles are measured with respect to the normal, so light does not need to be sent to the interface at near grazing incidence to have total internal reflection.

The effect of total internal reflection is to contain the light inside the medium. It is the basis of fiber optics and other dielectric wave guides or light pipes.

That there is no transmitted light does not mean that the fields are zero in medium b. In fact, there are exponentially decaying evanescent fields in the second medium [2, 8].

Reflectance and Transmittance

The Poynting vectors of incident, reflected, and transmitted waves tell me the intensities of the light in these beams. The reflectance $\mathscr{R}$ and transmittance $\mathscr{T}$ represent the fraction of the incident energy that is reflected or transmitted by the interface. Because intensity is energy/area, I'll have to consider the change of area of the beams.

For reflectance it is easy: the incident and reflected waves propagate in the same medium and make the same angle with the normal to the surface. Consequently the reflectance is

$$\mathscr{R} = |r|^2 .$$

The transmittance $\mathscr{T}$ is generally not equal to $|t|^2$ for two reasons. First, the Poynting vector includes a factor of the refractive index n, just at it did in Chapter 3 (Section 3.5). Second, because the light is refracted into a different direction, the cross-sectional areas of a pencil or rays is different in the two media. The transmittance is

$$\mathscr{T} = \frac{n_b \cos\theta_t}{n_a \cos\theta_i} \, |t|^2 . \tag{E.16}$$

The factor of n_b/n_a comes from the ratio of Poynting vectors. The factor of $\cos\theta_t / \cos\theta_i$ represents the change in area of the pencil of rays.

The reflectance for s-polarized light is

$$R_s = \left| \frac{n_a \cos\theta_i - n_b \cos\theta_t}{n_a \cos\theta_i + n_b \cos\theta_t} \right|^2 = \left| \frac{n_a \cos\theta_i - n_b \sqrt{1 - \frac{n_a^2}{n_b^2} \sin^2\theta_i}}{n_a \cos\theta_i + n_b \sqrt{1 - \frac{n_a^2}{n_b^2} \sin^2\theta_i}} \right|^2 ,$$

while the reflectance for p-polarized light is

$$R_p = \left| \frac{n_a \cos\theta_t - n_b \cos\theta_i}{n_a \cos\theta_t + n_b \cos\theta_i} \right|^2 = \left| \frac{n_a \sqrt{1 - \frac{n_a^2}{n_b^2} \sin^2\theta_i} - n_b \cos\theta_i}{n_a \sqrt{1 - \frac{n_a^2}{n_b^2} \sin^2\theta_i} + n_b \cos\theta_i} \right|^2 .$$

To obtain the second form of each equation I eliminated $\cos\theta_t$ using Eq. E.14.

I can calculate the transmittances for the two polarizations from Eqs. E.16, E.8, and E.13. As an alternative, I can invoke the conservation of energy and then the transmission coefficients are given by

$$\mathscr{T}_s = 1 - \mathscr{R}_s, \tag{E.17}$$

and

$$\mathscr{T}_p = 1 - \mathscr{R}_p. \tag{E.18}$$

What follows is a list and brief discussion of the optical functions used in studying the optical properties of solids.

Basic Functions

I'll start with the idea that I have measured the reflectance $\mathcal{R}$ and the phase shift on reflection ϕ (by Kramkers-Kronig, ellipsometry, analysis of thin film data, or other means. Then, the complex refractive index N comes from the inversion of Eq. 3.32. (See also Eq. 10.24.)

$$N = \frac{1 - \sqrt{\mathcal{R}}e^{i\phi}}{1 + \sqrt{\mathcal{R}}e^{i\phi}}. \tag{F.1}$$

1. *Refractive index.* The real part of N is

$$n = \mathrm{Re}(N). \tag{F.2}$$

2. *Extinction coefficient.* The imaginary part of N is

$$\kappa = \mathrm{Im}(N). \tag{F.3}$$

Continuing, the complex dielectric function according to Eq. 3.10 is $\epsilon = N^2$.

3. *Real dielectric function.*

$$\epsilon_1 = \mathrm{Re}(\epsilon). \tag{F.4}$$

4. *Imaginary dielectric function.*

$$\epsilon_2 = \mathrm{Im}(\epsilon). \tag{F.5}$$

Equation 3.6 relates the complex dielectric function and complex conductivity. In cgs the conductivity and frequency are both measured in sec^{-1} and the relation is

$$\epsilon = 1 + \frac{4\pi i\sigma}{\omega}, \tag{F.6}$$

whereas in SI the conductivity is in $\Omega^{-1}\mathrm{m}^{-1}$, and it is

$$\epsilon = 1 + \frac{i\sigma}{\omega\epsilon_0}, \tag{F.7}$$

with ϵ_0 the dielectric function of empty space. In lab units, frequencies are in cm^{-1} and conductivities are in $\Omega^{-1}\mathrm{cm}^{-1}$, and the relation becomes

$$\epsilon = 1 + \frac{60i\sigma}{\omega}. \tag{F.8}$$

5. *Conductivity (real part).* The real part of the conductivity (in $\Omega^{-1}\mathrm{cm}^{-1}$) is then

$$\sigma_1 = \frac{\omega\epsilon_2}{60}, \tag{F.9}$$

with ω in cm^{-1}. This equation* can be compared to the expression in cgs:

$$\sigma_1 = \frac{\omega\epsilon_2}{4\pi}, \tag{F.10}$$

with ω in $\sec^{-1}$ and σ_1 also in $\sec^{-1}$. The cgs conductivity is 9×10^{11} times larger than the SI conductivity.

6. *Conductivity (imaginary part).* The imaginary part of the conductivity (in $\Omega^{-1}\mathrm{cm}^{-1}$) is

$$\sigma_2 = -\frac{\omega(\epsilon_1 - 1)}{60}. \tag{F.11}$$

Sometimes it is helpful to employ a partial dielectric function to eliminate ϵ_c (including Maxwell's displacement current) and/or other contributions:

$$\sigma_2 = -\frac{\omega(\Delta\epsilon_1)}{60}, \tag{F.12}$$

where $\Delta\epsilon_1 = \epsilon_1 - \epsilon_c$.

7. *Absorption coefficient.* The absorption coefficient (in cm^{-1}) is

$$\alpha = 4\pi\omega\kappa, \tag{F.13}$$

correct for ω in cm^{-1}. In cgs, this is

$$\alpha = 2\omega\kappa/c, \tag{F.14}$$

where ω is in $\sec^{-1}$.

8. *Skin depth.* In the skin effect the electric field goes as $e^{-x/\delta}$ but of course the wave governed by a refractive index N has a field which decays as $e^{-\kappa\omega x/c}$ (ω in rad/s). Hence:

$$\delta = \frac{10^8}{2\pi\omega\kappa}, \tag{F.15}$$

with ω in cm^{-1}. The factor of 10^8 makes the units of δ be Å.

---

* The same factor of 60 appears if I calculate σ_{dc} from the plasma frequency and relaxation rate,

$$\sigma_{dc} = \frac{\omega_p^2}{60(1/\tau)},$$

with ω_p and $1/\tau$ in cm^{-1} and σ_{dc} in $\Omega^{-1}\mathrm{cm}^{-1}$. In reality, $60 = 377\,\Omega/2\pi$. If I want more accuracy, I have to use $Z_0 = 376.73\,\Omega = \sqrt{\mu_0/\epsilon_0}$ with μ_0 and ϵ_0 respectively the magnetic permeability and electric permittivity of the vacuum. The factor changes a small amount: $60 \rightarrow 59.958$.

9. *Loss function.* The energy loss function governs the transmission of fast electrons through the material. These depend on the longitudinal response of the dielectric. It is typically written as $-\,\mathrm{Im}(1/\epsilon)$; here I use L.

$$L = \frac{\epsilon_2}{(\epsilon_1^2 + \epsilon_2^2)} = -\,\mathrm{Im}\,\frac{1}{\epsilon}. \tag{F.16}$$

10. *Surface loss function.* The "surface" energy loss function is

$$L_s = \frac{\epsilon_2}{((\epsilon_1 - 1)^2 + \epsilon_2^2)} = -\,\mathrm{Im}\,\frac{1}{\epsilon - 1}. \tag{F.17}$$

Sum Rules

11. *Conductivity sum rule.* This is the most important sum rule function. The f-sum rule for solids can be written

$$\int_0^\infty \sigma_1(\omega')d\omega' = \frac{\omega_{p,tot}^2}{8}, \tag{F.18}$$

where $\omega_{p,tot} = \sqrt{4\pi n_{tot} e^2 / m}$ is the plasma frequency for all the electrons in the solid. I may write a partial sum rule (for the conduction band, say) as

$$\int_0^\omega \sigma_1(\omega')d\omega' = \frac{\pi n e^2}{2m^*}, \tag{F.19}$$

where m^* is the average effective mass of the band. If the upper limit is above the free-carrier contribution and below the onset of the interband transitions, the right hand side becomes $\omega_p^2/8$ with ω_p the conduction band plasma frequency.

Next, write $n = N_{eff}/V_c$ and let m be the free electron mass. The density n is in the number of effective electrons per unit cell volume (or formula volume). Then, calculate

$$N_{eff}\frac{m}{m^*} = \frac{2m V_c}{\pi e^2} \int_0^\omega \sigma_1(\omega')d\omega', \tag{F.20}$$

with m the mass of an electron and m^* the effective mass of the charge carriers.

12. *Energy loss function sum rule.* I may write a partial sum rule (for the conduction band, say) of the loss function as

$$\frac{1}{4\pi} \int_0^\omega \omega' L(\omega')d\omega' = \frac{\pi n e^2}{2m^*}, \tag{F.21}$$

where m^* is the average effective mass of the band. If the upper limit is above the free-carrier contribution and below the onset of the interband transitions, the right hand side becomes $\omega_p^2/8$ with ω_p the conduction band plasma frequency. (The factor of $1/4\pi$ is the same as appears in $\omega\epsilon_2/4\pi = \sigma_1$.) Thus,

$$N_{eff}\frac{m}{m^*} = \frac{m V_c}{2\pi^2 e^2} \int_0^\omega \omega' L(\omega')d\omega'. \tag{F.22}$$

13. *Surface loss function sum rule.* It is the same as Eqs. F.21 and F.22, but for L_s.

Frequency-Dependent Scattering, Self-Energy, and Effective Mass

The dielectric function of perfectly free (noninteracting) carriers with density n, charge $\pm e$, and mass m is given by

$$\epsilon(\omega) = \epsilon_c - \frac{\omega_p^2}{\omega^2}, \tag{F.23}$$

with ω_p, the plasma frequency, (in cgs)

$$\omega_p = \sqrt{\frac{4\pi n e^2}{m}}. \tag{F.24}$$

If I allow scattering of free carriers from impurities, defects, and surfaces with a mean free time τ or scattering rate $1/\tau$, independent of the frequency, the dielectric function has the Drude form:

$$\epsilon(\omega) = \epsilon_c - \frac{\omega_p^2}{\omega^2 + i\omega/\tau}. \tag{F.25}$$

As discussed in Section 13.1, interactions make the scattering rate $1/\tau$ a function of frequency (and probably also of temperature). However, to respect the Kramers–Kronig relations, one must consider the scattering rate to be a complex function or to have the effective mass m^* also vary with frequency.

Theorists have devised a number of such functions: a complex scattering (or memory) function Γ, a complex scattering rate $1/\tau^*$ and effective mass m^*, a complex scattering rate $1/\tau^*$ and mass enhancement factor λ, and a complex self-energy function Σ.* Each and all of these quantities are calculable from the complex dielectric function. It is necessary to know the unrenormalized plasma frequency and ϵ_c. (See below.)

14. *Damping via memory function.* The frequency dependent and complex damping is often written in an extended Drude model by replacing $1/\tau$ by a "memory function" Γ. I write

$$\epsilon = \epsilon_c - \frac{\omega_p^2}{\omega(\omega + i\Gamma)}, \tag{F.26}$$

with $\Gamma = \Gamma_1 + i\Gamma_2$, $\Delta\epsilon = \epsilon - \epsilon_c$ then

$$\Gamma_1 = -\frac{\omega_p^2}{\omega} \operatorname{Im} \frac{1}{\Delta\epsilon}. \tag{F.27}$$

15. *Imaginary part of the memory function.* The imaginary part of Γ is

$$\Gamma_2 = \frac{\omega_p^2}{\omega} \operatorname{Re} \frac{1}{\Delta\epsilon} + \omega. \tag{F.28}$$

---

* There are actually two of these.

16. Scattering rate, $1/\tau*$.

$$1/\tau* = -\omega \frac{\text{Im}(\Delta\epsilon)}{\text{Re}(\Delta\epsilon)}. \tag{F.29}$$

17. Effective mass.

$$\frac{m^*}{m} = -\frac{\omega_p^2}{\omega^2} \text{Re}\left(\frac{1}{\Delta\epsilon}\right). \tag{F.30}$$

18. Mass enhancement factor $\lambda(\omega)$. The effective mass may be written in terms of a mass enhancement factor $\lambda(\omega)$:

$$\frac{m^*}{m} = 1 + \lambda(\omega). \tag{F.31}$$

Hence,

$$\lambda(\omega) = -\frac{\omega_p^2}{\omega^2} \text{Re}\left(\frac{1}{\Delta\epsilon}\right) - 1. \tag{F.32}$$

19. Imaginary part of the (one-particle) self-energy. In some theories (such as the marginal Fermi liquid) the imaginary part of the self energy Σ plays the role of a scattering rate, with the dielectric function

$$\epsilon(\omega) = \epsilon_c - \frac{\omega_p^2}{\omega[\omega - \Sigma(\omega)]}. \tag{F.33}$$

Here, ω_p is the (unrenormalized)* plasma frequency and ϵ_c the limiting high frequency value of the dielectric function. The imaginary part of Σ is related to the quasiparticle lifetime through $1/\tau^*(\omega) = -\text{Im}\,\Sigma(\omega)(m_b/m^*(\omega))$ with m^* the effective mass and m_b the band mass. Note that $\Sigma_2 = -\Gamma_1$.

Compute $-\text{Im}\,\Sigma$ via

$$-\text{Im}\,\Sigma(\omega) = -\frac{\omega_p^2}{\omega} \text{Im}\,\frac{1}{\Delta\epsilon}, \tag{F.34}$$

with $\Delta\epsilon = \epsilon - \epsilon_c$. One must know the plasma frequency and ϵ_c. Or, one may subtract a model dielectric function that describes the noninteracting components to remove optical phonons and interband transitions.

20. Real (one-particle) self-energy. The real part of Σ is related to the effective mass m^* of the interacting carriers by

$$m^*(\omega)/m_b = 1 - \text{Re}\,\Sigma(\omega)/\omega. \tag{F.35}$$

Hence, compute

$$\text{Re}\,\Sigma(\omega) = \frac{\omega_p^2}{\omega} \text{Re}\,\frac{1}{\Delta\epsilon} + \omega. \tag{F.36}$$

Note that $\Sigma_1 = +\Gamma_2$.

---

* Unrenormalized means "what it would be if there were no interactions."

21. *Imaginary part of the (two-particle) self-energy.* A slightly different approach recognizes that optical excitations are created in pairs (electron and hole) and that the self-energy should be a two-particle process. Still, the imaginary part of the self energy Σ plays the role of a scattering rate, with the dielectric function written as

$$\epsilon(\omega) = \epsilon_c - \frac{\omega_p^2}{\omega[\omega - 2\Sigma(\omega/2)]} \tag{F.37}$$

with ω_p the (unrenormalized) plasma frequency and ϵ_c the limiting high frequency value. The factors of 2 occur because the excitations are created in pairs (electron-hole pairs or quasiparticle pairs). (These factors do not appear in all theoretical treatments. Indeed the "one-particle" versions are much more commonly seen.) The imaginary part of Σ is related to the quasiparticle lifetime through $1/\tau^*(\omega) = -2\,\mathrm{Im}\,\Sigma(\omega/2)(m_b/m^*(\omega))$ with m^* the effective mass and m_b the band mass.

Compute $-\,\mathrm{Im}\,\Sigma$ via

$$-\,\mathrm{Im}\,\Sigma = -\frac{\omega_p^2}{2\omega}\,\mathrm{Im}\,\frac{1}{\Delta\epsilon} \tag{F.38}$$

with $\Delta\epsilon = \epsilon - \epsilon_c$. The (unrenormalized) plasma frequency and ϵ_c must be known. Note that optical phonons and interband transitions may occur; one is allowed to subtract these from ϵ.

After computing the right hand side of the above equation, one should divide the frequencies by 2, so that the dielectric function value at 400 cm^{-1} (for example) generates the $-\,\mathrm{Im}\,\Sigma$ value at 200 cm^{-1}.

22. *Real part of the (two-particle) self-energy.* The real part of Σ is related to the effective mass m^* of the interacting carriers by

$$m^*(\omega)/m_b = 1 - 2\,\mathrm{Re}\,\Sigma(\omega/2)/\omega. \tag{F.39}$$

Hence, compute

$$\mathrm{Re}\,\Sigma(\omega/2) = \frac{\omega_p^2}{2\omega}\,\mathrm{Re}\,\frac{1}{\Delta\epsilon} + \omega/2. \tag{F.40}$$

After computing the right hand side of the above equation, divide the frequencies by 2, so that the dielectric function value at 400 cm^{-1} (for example) generates the $\mathrm{Re}\,\Sigma$ value at 200 cm^{-1}.

Specialized Functions, Mostly for Superfluids

23. *Superfluid plasma frequency.* A superfluid has a dielectric function that is

$$\epsilon = \epsilon_c - \frac{\omega_{ps}^2}{\omega^2}, \tag{F.41}$$

with ω_{ps} the superfluid plasma frequency. Calculate

$$\omega_{ps} = \sqrt{\omega^2 |\,\mathrm{Re}(\Delta\epsilon)|}, \tag{F.42}$$

where the absolute value is used instead of a minus sign to avoid numerical problems. I have to be sure that ϵ is negative when invokeing this function! If ω_{ps} is (nearly) constant in frequency, the behavior is that of a superfluid.

24. *Number of "free carriers."* Again, writing

$$\epsilon = \epsilon_c - \frac{\omega_{ps}^2}{\omega^2},\tag{F.43}$$

with ω_{ps} the superfluid plasma frequency. Then, write $\omega_{ps}^2 = 4\pi N_s(\omega)e^2/V_c m$ so that:

$$N_s(\omega) = \omega^2 V_c(m/4\pi e^2)|\operatorname{Re}(\Delta\epsilon)|,\tag{F.44}$$

$N_s(\omega)$ should be nearly constant as a function of frequency for the concept to be valid.

25. *London penetration depth.* The dielectric function in Eqs. F.41 or F.43 is that of a London superconductor, so one may also calculate λ_L as

$$\lambda_L = \frac{10^8}{2\pi\omega\sqrt{|\operatorname{Re}(\Delta\epsilon)|}},\tag{F.45}$$

with ω in cm^{-1}. The factor of 10^8 makes the units of $\lambda_L(\omega)$ in Å. $\lambda_L(\omega)$ should be nearly constant as a function of frequency for the concept to be valid.

Other Specialized Functions

28. *Optical resistivity (real part).* The optical resistivity is just the inverse of the optical conductivity, so the real part is

$$\rho_1 = \operatorname{Re}\left(\frac{1}{\sigma}\right).\tag{F.46}$$

Because σ is complex, $\rho_1 = \sigma_1/\left(\sigma_1^2 + \sigma_2^2\right)$.

29. *Optical resistivity (imaginary part).* The imaginary part of the optical resistivity is

$$\rho_2 = \operatorname{Im}\left(\frac{1}{\sigma}\right).\tag{F.47}$$

30. *Absolute value of the dielectric function.* There are examples where the magnitude of the complex dielectric function $|\epsilon|$ displays power law behavior. Hence, calculate

$$|\epsilon| = \sqrt{\epsilon_1^2 + \epsilon_2^2}.\tag{F.48}$$

Note that this should be the entire function, so ϵ_c is not generally subtracted.

31. *Absolute value of the conductivity.* This quantity is

$$|\sigma| = \sqrt{\sigma_1^2 + \sigma_2^2}.\tag{F.49}$$

32. *Tan(δ).* The loss tangent is commonly used in the analysis of RF and microwave experiments. It is

$$\tan\delta = \frac{\epsilon_2}{\epsilon_1}.\tag{F.50}$$

This calculation uses the entire function, so ϵ_c is not generally subtracted.

Data analysis programs for working with optical spectra are available at www.cambridge.org/tanner. These run in Windows (7 through 10 at least). They are not true Windows applications, though, lacking the message boxes, drop-down menus, radio buttons, tool tips, animated paper clips, etc. that we've come to know and love.

Installation is accomplished by extracting the zip file into a directory on your path. Then read "man.pdf," the manual. Run the programs from a command window. You can also add links on the desktop; most will accept drag/drop from Windows Explorer.

The zip file includes the following programs:

Data manipulation

`cmp.exe`	Compute with spectra: $+$, $-$, $*$, $/$ and more.
`interp.exe`	Interpolate at evenly spaced points.
`fts.exe`	Fourier-transform smoothing of data.
`gtf.exe`	Glover–Tinkham conductivity of a film from $\mathscr{R}/\mathscr{T}$.
`mav.exe`	Merge and average overlapping data sets.
`pik.exe`	Pick sections out of a data file.
`q.exe`	Quickly make plot on the screen.
`t2a.exe`	$\mathscr{T}$ to α. Allows for reflections at the surfaces.

Kramers–Kronig analysis

`kk.exe`	Kramers–Kronig analysis of $\mathscr{R}$ or $\mathscr{T}$. Requires `xro` for extrapolation.
`op.exe`	Optical properties from Kramers–Kronig results.
`xro.exe`	Compute extrapolated x-ray reflectance from atomic scattering data.

Fitting

`dff.exe`	Dielectric function fit. Least square to models.
`dfc.exe`	Dielectric function calculate. Companion to `dff`/`tff`.
`pfit.exe`	Power series fit, to up to x^8.
`tff.exe`	Thin film fit. Least-square to $\mathscr{R}$ or $\mathscr{T}$. Many models.

Utilities

`j2w.exe`	Converts JCAMP files to two column data.
`w2l.exe`	Convert from wavenumber to wavelength.
And others	Read the manual.

The Kramers–Kronig (reflectance) code was written by Claus Jacobsen. The least-squares minimization uses a routine by Bevington. Most of the other important routines were written by Charles Porter; the rest is by David Tanner. The programs are free to all to use and to redistribute, subject only to a request that users acknowledge the principal author, Charles Porter.

References

[1] J. Clerk Maxwell, "A dynamical theory of the electromagnetic field," *Phil. Trans. Royal Soc. (London)* **155**, 459–512 (1865).

[2] J.D. Jackson, *Classical Electrodynamics* (John Wiley & Sons, New York, 1975).

[3] L.D. Landau and E.M. Lifshitz, *Electrodynamics of Continuous Media* (Pergamon, New York, 1960).

[4] M.A. Heald and J.B. Marion, *Classical Electromagnetic Radiation* (Saunders College Publishing, Philadelphia, 1995).

[5] D.J. Griffiths, *Introduction to Electrodynamics* (Prentice Hall, Upper Saddle River, New Jersey, 1999).

[6] W.R. Smythe, *Static and Dynamic Electricity* (McGraw-Hill, New York, 1950).

[7] M.V. Klein and T.E. Furtak, *Optics* (Wiley, New York, 1986).

[8] M. Born and E. Wolf, *Principles of Optics* (Pergamon, Oxford, 1970).

[9] V.G. Veselago, "The electrodynamics of substances with simultaneously negative values of ϵ and μ," *Soviet Physics Uspekhi* **10**, 509–514 (1968).

[10] M.W. McCall, A. Lakhtakia, and W.S. Weiglhofer, "The negative index of refraction demystified," *Eur. J. Phys.* **23**, 353–359 (2002).

[11] J. Pendry, "Optics: Positively negative," *Nature* **423**, 22–23 (2003).

[12] S.A. Ramakrishna, "Physics of negative refractive index materials," *Rep. Prog. Phys.* **68**, 449–521 (2005).

[13] V.M. Shalaev, "Optical negative-index metamaterials," *Nat. Photon.* **1**, 41–48 (2007).

[14] R.A. Shelby, D.R. Smith, and S. Schultz, "Experimental verification of a negative index of refraction," *Science* **292** 77–79 (2001).

[15] L.I. Schiff, *Quantum Mechanics* (McGraw-Hill, New York, 1968).

[16] P. Drude, "Zur Elektronentheorie der Metalle" ("The electron theory of metals"), *Ann. Phys. (Leipzig)* **1**, 566 (1900).

[17] www.pdi-berlin.de/outreach/paul-drude. His important papers and books published in German in 1890, 1894, 1900, 1902, and 1904 are listed here. See also Paul Drude, *The Theory of Optics* (Dover, Mineola, 2005).

[18] D.C. Johnston, "The puzzle of high temperature superconductivity in layered iron pnictides and chalcogenides," *Adv. Phys.*, **59**, 803–1061 (2010).

[19] N.W. Ashcroft and N.D. Mermin, *Solid State Physics* (Holt, Rinehart and Winston, Philadelphia, 1976).

[20] E. Hagen and H. Rubens, "Über die Beziehung des Reflexions- und Emissionsvermögens der Metalle zu ihrem elektrischen Leitvermögen" ("On the relation

of reflectivity and emissivity of metals with their electrical conductivity"), *Ann. Phys. (Leipzig)* **4**, 873 (1903).

[21] R. Landauer, in *Electrical Transport and Optical Properties of Inhomogeneous Media*, edited by J.C. Garland and D.B. Tanner (The American Institute of Physics, New York, 1978), p. 1.

[22] B.L. Henke, E.M. Gullikson, and J.C. Davis, "X-ray interactions: Photoabsorption, scattering, transmission, and reflection at $E = 50$–30000 eV, $Z = 1$–92," *Atom. Data Nucl. Data Tables* **54**, 181–342 (1993).

[23] G. Burns, *Solid State Physics* (Academic Press, San Diego, 1985).

[24] C. Kittel, *Introduction to Solid State Physics* (Wiley, New York, November 2004). The 3rd and 5th editions are worth looking at also.

[25] N.A. Spaldin, "A beginner's guide to the modern theory of polarization," *J. Solid State Chem.* **195**, 2–10 (2012).

[26] D. Vanderbilt, "Soft self-consistent pseudopotentials in a generalized eigenvalue formalism," *Phys. Rev. B,* **41**, 7892 (1990).

[27] K. Laasonen, A. Pasquarello, R. Car, C. Lee, and D. Vanderbilt, "Car-Parrinello molecular dynamics with Vanderbilt ultrasoft pseudopotentials," *Phys. Rev. B,* **47**, 10142 (1993).

[28] M. Tinkham, *Group Theory and Quantum Mechanics* (Dover, Mineola, New York, 2003).

[29] G. Burns, *Introduction to Group Theory with Applications* (Academic Press, New York, 1977).

[30] M.S. Dresselhaus, "Applications of Group Theory to the Physics of Solids," MIT (2002). web.mit.edu/course/6/6.734j/www/group-full02.pdf.

[31] Wikipedia, the free encyclopedia, "Pole–zero plot," (2018). http://en.wikipedia.org/wiki/Pole-zero_plot.

[32] R. Lyddane, R. Sachs, and E. Teller, "On the polar vibrations of alkali halides," *Phys. Rev.* **59**, 673–676 (1941).

[33] W. Cochran and R.A. Cowley, "Dielectric constants and lattice vibrations," *J. Phys. Chem. Solids* **23**, 447–450 (1962).

[34] A.S. Barker, Jr., "Long-wavelength soft modes, central peaks, and the Lyddane-Sachs-Teller relation," *Phys. Rev. B* **12**, 4071–4084 (1975).

[35] A.J. Sievers and J.B. Page, "Generalized Lyddane-Sachs-Teller relation and disordered solids," *Phys. Rev. B* **41**, 3455–3459 1990.

[36] A.S. Barker and M. Tinkham, "Far-infrared ferroelectric vibration mode in SrTiO$_3$," *Phys. Rev.* **125**, 1527–1530 (1962).

[37] W.G. Spitzer, R.C. Miller, D.A. Kleinmann, and L.E. Howarth, "Far infrared dielectric dispersion in BaTiO$_3$, SrTiO$_3$, and TiO$_2$," *Phys. Rev.* **126**, 1710–1721 (1962). 20

[38] K. Kamarás, K.-L. Barth, F. Keilmann, R. Henn, M. Reedyk, C. Thomsen, M. Cardona, J. Kircher, P.L. Richards, and J.-L. Stehlé, "The low-temperature infrared optical functions of SrTiO$_3$ determined by reflectance spectroscopy and spectroscopic ellipsometry," *J. Appl. Phys.* **78**, 1235–1240 (1995).

[39] R. Henn, A. Wittlin, M. Cardona, and S. Uchida, "Dynamics of the c-polarized infrared-active modes in $La_{2-x}Sr_x CuO_4$," *Phys. Rev. B* **56**, 6295–6301 1997.

[40] R.B. Sanderson, "Far infrared optical properties of indium antimonide," *J. Phys. Chem. Solids* **26**, 803–810 (1965).

[41] A.G. Shenstone, "The arc spectrum of silver," *Phys. Rev.* **57**, 894 (1940).

[42] J.C. Pickering and V. Zilio, "New accurate data for the spectrum of neutral silver," *Eur. Phys. J. D* **13**, 181 (2001).

[43] E.D. Palik, *Handbook of Optical Constants of Solids I and II* (Academic Press, Orlando, Florida, 1985 and 1991).

[44] E.A. Taft and H.R. Philipp, "Optical constants of silver," *Phys. Rev.* 121, 1100–1103 1961.

[45] H. Ehrenreich and H.R. Philipp, "Optical properties of Ag and Cu," *Phys. Rev.* **128**, 1622 (1962).

[46] H. Bennett, M. Silver, and E. Ashley, "Infrared reflectance of aluminum evaporated in ultra-high vacuum," *J. Opt. Soc. Am.* **53**, 1089–1095 (1963).

[47] J.M. Bennett, E.J. Ashley, "Infrared reflectance and emittance of silver and gold evaporated in ultrahigh vacuum," *Appl. Optics* **4**, 221–224 (1965).

[48] B. Dold and R. Mecke, "Über Gangsmetallen und deren Legierungen im Infrarot" ("About transition metals and their alloys in the infrared"), *Optik* 22, 435 (1965).

[49] G. Irani, T. Huen, and F. Wooten, "Optical constants of silver and gold in the visible and vacuum ultraviolet," *J. Opt. Soc. Am.* **61**, 128–129 (1971).

[50] H.J. Hagemann, W. Gudat, and C. Kunz, "Optical constants from the far infrared to the x-ray region: Mg, Al, Cu, Ag, Au, Bi, C, and Al2O3," *J. Opt. Soc. Am.* **65**, 742 (1975). See also DESY Report SR-74/7, Hamburg, 1974.

[51] P. Winsemius, F.F. van Kampen, H.P. Lengkeek, and C.G. van Went, "Temperature dependence of the optical properties of Au, Ag, and Cu," *J. Phys. F*, **6**, 1583 (1976).

[52] G. Leveque, C.G. Olson, and D.W. Lynch, "Reflectance spectra and dielectric functions for Ag in the region of interband transitions," *Phys. Rev. B* **27**, 4654 (1983).

[53] H.U. Yang, J. D'Archangel, M.L. Sundheimer, E. Tucker, G.D. Boreman, and M.B. Raschke, "Optical dielectric function of silver," *Phys. Rev. B* **91**, 235137 (2015).

[54] F. Wooten, *Optical Properties of Solids* (Academic Press, New York, 1972).

[55] S. Roberts, "Optical properties of copper," *Phys. Rev.* **118**, 1509 (1960).

[56] A.D. Rakić, A.B. Djurišic, J.M. Elazar, and M.L. Majewski, "Optical properties of metallic films for vertical-cavity optoelectronic devices," *Appl. Opt.* **37**, 5271–5283 (1998).

[57] "Optical constants of Cu, Ag, and Au revisited," S. Babar and J. H. Weaver, *Appl. Opt.* **54**, 477–481 (2015).

[58] L.R. Canfield, G. Hass, and W.R. Hunter, "The optical properties of evaporated gold in the vacuum ultraviolet from 300 Å to 2000 Å ," *J. de Physique* **25**, 124 (1964).

[59] H. Ehrenreich, H.R. Philipp, and B. Segall, "Optical properties of aluminum," *Phys. Rev.* **132**, 1918–1928 (1963).

[60] E. Shiles, T. Sasaki, M. Inokuti, and D.Y. Smith, "Self-consistency and sum-rule tests in the Kramers-Kronig analysis of optical data: Applications to aluminum," *Phys. Rev. B* **22**, 1612 (1980).

[61] D.Y. Smith and B. Segall, "Intraband and interband processes in the infrared spectrum of metallic aluminum," *Phys. Rev. B* **34**, 5191–5198 (1986).

[62] D. Brust, "Electronic structure effects in the Drude and interband absorption of aluminum," *Phys. Rev. B* **2**, 818–825 (1970).

[63] H.H. Li, "Refractive index of silicon and germanium and its wavelength and temperature derivatives," *J. Phys. Chem. Ref. Data* **9**, 561–658 (1980).

[64] D.F. Edwards and E. Ochoa, "Infrared refractive index of silicon," *Appl. Optics* **19**, 4130–4131 (1980).

[65] W.R. Thurber, R.L. Mattis, Y.M. Liu, and J. J. Filliben, "Resistivity-dopant density relationship for phosphorus-doped silicon," *J. Electrochem. Soc.* **127**, 1807–1812 (1980).

[66] H.R. Philipp and E.A. Taft, "Optical constants of silicon in the region 1 to 10 eV," *Phys. Rev.* **120**, 37–38 (1960).

[67] H.R. Philipp and H. Ehrenreich. "Optical properties of semiconductors," *Phys. Rev.* **129**, 1550–1560 (1963).

[68] D.E. Aspnes and A.A. Studna, "Dielectric functions and optical parameters of Si, Ge, GaP, GaAs, GaSb, InP, InAs, and InSb from 1.5 to 6.0 eV," *Phys. Rev. B* **27**, 985–1009 (1983).

[69] G.E. Jellison Jr., "Optical functions of silicon determined by two-channel polarization modulation ellipsometry," *Opt. Mat.* **1**, 41–47 (1992).

[70] C.M. Herzinger, B. Johs, W.A. McGahan, J.A. Woollam, and W. Paulson, "Ellipsometric determination of optical constants for silicon and thermally grown silicon dioxide via a multi-sample, multi-wavelength, multi-angle investigation," *J. Appl. Phys.* **83**, 3323–3336 (1998).

[71] D. Chandler-Horowitz and P.M. Amirtharaj, "High-accuracy, midinfrared ($450 \, \text{cm}^{-1} \leq \omega \leq 4000 \, \text{cm}^{-1}$) refractive index values of silicon," *J. Appl. Phys.* **97**, 123526/1–8 (2005).

[72] H. Ehrenreich, H.R. Philipp, and J.C. Phillips, "Interband transitions in groups 4, 3-5, and 2-6 semiconductors," *Phys. Rev. Lett.* **8**, 59–61 (1962).

[73] J.C. Phillips, "Band structure of silicon, germanium, and related semiconductors," *Phys. Rev.* **125**, 1931–1936 (1962).

[74] D. Brust, "Electronic spectra of crystalline germanium and silicon," *Phys. Rev.* **134**, A1337–A1353 (1964).

[75] J.P. Walter and M.L. Cohen "Frequency- and wave-vector-dependent dielectric function for silicon," *Phys. Rev. B* **5**, 3101–3110 (1972).

[76] J.R. Chelikowsky and M.L. Cohen, "Electronic structure of silicon," *Phys. Rev. B* **10**, 5095–5107 (1974).

[77] O.S. Heavens, *Optical Properties of Thin Solid Films* (Dover, Mineola, NY, 1955).

[78] K.B. Blodgett, "Interference colors in oil films on water," *J. Opt. Soc. Am.* **24**, 313–315 (1934).

[79] E. Shevtsova, C. Hansson, D.H. Janzen, and J. Kjærandsen, "Stable structural color patterns displayed on transparent insect wings," *Proc. Nat. Acad. Sci.* **108**, 668–673 (2011).

[80] B.P. Abbott *et al.* (LIGO Scientific Collaboration), "LIGO: The Laser Interferometer Gravitational-wave Observatory," *Rep. Prog. Phys.* **72**, 076901/1–25 (2009).

[81] H. Kogelnik and T. Li, "Laser beams and resonators," *Appl. Optics*, **5**, 1550–1567 (1966).

[82] A.E. Siegman, *Lasers* (University Science Books, Mill Valley, Calif., 1986).

[83] F. Abelés, "La théorie générale des couches minces," ("The general theory of thin films,"), *J. Phys. Radium* **11**, 307–310 (1950).

[84] R.E. Glover and M. Tinkham, *Phys. Rev.*, **107**, 844 (1956); *ibid.*, *Phys. Rev. B* **108**, 1175 (1957).

[85] L.H. Palmer and M. Tinkham, *Phys. Rev.* **165**, 588 (1968).

[86] D.R. Karecki, G.L. Carr, S. Perkowitz, D.U. Gubser, and S.A. Wolf, *Phys. Rev. B* **27**, 5460 (1983).

[87] F. Gao, G.L. Carr, C.D. Porter, D.B. Tanner, G.P. Williams, C.J. Hirschmugl, B. Dutta, X.D. Wu, and S. Etemad, "Quasiparticle damping and the coherence peak in $YB_{a_2}Cu_3O_{7-\delta}$," *Phys. Rev. B* **54**, 700–710 (1996).

[88] H. Tashiro, J.M. Graybeal, D.B. Tanner, E.J. Nicol, J.P. Carbotte, and G.L. Carr, "Unusual thickness dependence of the superconducting transition of α-MoGe thin films," *Phys. Rev. B* **78**, 014509/1–7 (2008).

[89] A. Sommerfeld and H. Bethe, "Elektronentheorie der Metalle" ("Electron Theory of Metals"), *Handbuch der Physik* **24**, 333–622 (1933).

[90] N.F. Mott and H. Jones, *The Theory of the Properties of Metals and Alloys* (Oxford University Press, London and New York, 1936).

[91] A.H. Wilson, *The Theory of Metals* (Cambridge University Press, Cambridge, 1953).

[92] T. Holstein, *Phys. Rev.*, **96**, 535 (1954); *Ann. Phys. (N.Y.)* **29**, 410 (1964).

[93] Handbook of Electronic Tables and Formulas; www.powerstream.com/Wire_Size.htm.

[94] P. Debye, "Zur Theorie der spezifischen Wärme," ("The theory of specific heat"), *Ann. der Physik (Leipzig)* **39**, 789 (1912).

[95] D.B. Tanner and D.C. Larson, "Electrical resistivity of silver films," *Phys. Rev.*, **166**, 652–655 (1968).

[96] J. Cochran and M. Yaqub, "The mean free path of electrons in very pure gallium," *Phys. Lett.* **5**, 307–309 (1963).

[97] K. Huang, *Statistical Mechanics*, 2nd ed. (John Wiley & Sons, New York, 1987).

[98] M.L. Cohen and S.G. Louie, *Fundamentals of Condensed Matter Physics* (Cambridge University Press, Cambridge, 2016).

[99] M.P. Marder, *Condensed Matter Physics* (John Wiley & Sons, New York, 2010).

[100] J.M. Ziman, *Principles of the Theory of Solids*, 2nd ed. (Cambridge University Press, Cambridge, 1972).

[101] J.M. Ziman, *Electrons and Phonons: The Theory of Transport Phenomena in Solids* (Clarendon Press, Oxford, 1960).

[102] R. de L. Kronig, "On the theory of the dispersion of X-rays," *J. Opt. Soc. Am.* **12**, 547–557 (1926).

[103] H.A. Kramers, "La diffusion de la lumiere par les atomes" ("The diffusion of light by atoms"), *J. Atti Cong. Intern. Fisici, (Transactions of Volta Centenary Congress) Como* **2**, 545–557 (1927); "Die dispersion und absorption von Röntgenstrahlen" ("The dispersion and absorption of X-rays"), *Phys. Z.* **30**, 522–523 (1929).

[104] C. Kittel *Elementary Statistical Physics* (Wiley, New York, 1958).

[105] M. Dressel and G. Grüner, *Electrodynamics of Solids: Optical Properties of Electrons in Matter* (Cambridge University Press, Cambridge, 2002).

[106] J.S. Toll, "Causality and the dispersion relation: Logical foundations," *Phys. Rev.* **104**, 1760–1770 (1956).

[107] R. Kubo and M. Ichimura, "Kramers-Kronig relations and sum rules," *J. Math. Phys.* **13**, 1454 (1972).

[108] B.Y.-K. Hu, "Kramers-Kronig in two lines," *Am. J. Phys.* **57**, 821–821 (1989).

[109] K.-E. Peiponen and E.M. Vartiainen, "Kramers-Kronig relations in optical data inversion," *Phys. Rev. B* **44**, 8301 (1991).

[110] G.W. Milton, D.J. Eyre, and J.V. Mantese, "Finite frequency range Kramers-Kronig relations: Bounds on the dispersion," *Phys. Rev. Lett.* **79**, 3062–3065 (1997).

[111] F.W. King, "Efficient numerical approach to the evaluation of Kramers-Kronig transforms," *J. Opt. Soc. Am. B* **19**, 2427–2436 (2002).

[112] F.W. King, "Alternative approach to the derivation of dispersion relations for optical constants," *J. Phys. A: Math. Gen.* **39**, 10427 (2006).

[113] K.-E. Peiponen and J.J. Saarinen, "Generalized Kramers-Kronig relations in nonlinear optical- and THz-spectroscopy," *Rep. Prog. Phys.* **72**, 056401 (2009).

[114] C.F. Bohren, "What did Kramers and Kronig do and how did they do it?" *Eur. J. Phys.* **31**, 573–577 (2010).

[115] A.L. Cauchy, "Oeuvres complètes," Ser. 1, 4, Paris (1890).

[116] F.C. Jahod, "Fundamental absorption of barium oxide from its reflectivity spectrum," *Phys. Rev.* **107**, 1261–1265 (1957).

[117] H.R. Phillip and E.A. Taft, "Kramers-Kronig analysis of reflectance data for diamond," *Phys. Rev.* **136**, A1445–A1448 (1964).

[118] E.A. Taft and H.R. Philipp, "Optical properties of graphite," *Phys. Rev.* **138**, A197–A202 (1965).

[119] S. Tongay, J. Hwang, H.K. Pal, D.B. Tanner, D. Maslov, and A.F. Hebard, "Super-metallic conductivity in bromine-intercalated graphite," *Phys. Rev. B* **81**, 115428/1–6 (2010).

[120] C.S. Jacobsen, D.B. Tanner, A.F. Garito, and A.J. Heeger, "Single-crystal reflectance studies of tetrathiofulvalene tetracyanoquinodimethane," *Phys. Rev. Lett.* **33**, 1559–1562 (1974).

[121] C.S. Jacobsen, D.B. Tanner, and K. Bechgaard, "Dimensionality crossover in the organic superconductor tetramethyltetraselenafulvalene hexafluorophosphate [(TMTSF)$_2$PF$_6$]," *Phys. Rev. Lett.* **46**, 1142–1145 (1981).

[122] M. Dressel, A. Schwartz, G. Grüner, and L. Degiorgi, "Deviations from Drude response in low-dimensional metals: electrodynamics of the metallic state of (TMTSF)$_2$PF$_6$," *Phys. Rev. Lett.* **77**, 398–402 (1996).

[123] A.R. Beal and H.P. Hughes, "Kramers-Kronig analysis of the reflectivity spectra of 2H-MoS$_2$, 2H-MoSe$_2$ and 2H-MoTe$_2$," *J. Phys. C: Solid State Phys.* **12**, 881 (1979).

[124] L.H. Greene, D.B. Tanner, A.J. Epstein, and Joel S. Miller, "Optical properties of the cation-deficient platinum chain salt K$_{1.75}$Pt(CN)$_4$ · 1.5H$_2$O," *Phys. Rev. B* **25**, 1331–1338 (1982).

[125] W.A. Challener and P.L. Richards, "Far infrared optical properties of NbSe$_3$," *Solid State Comm.* **52**, 117–121 (1984).

[126] C.R. Fincher Jr, M. Ozaki, M. Tanaka, D. Peebles, L. Lauchlan, A.J. Heeger, and A.G. MacDiarmid. "Electronic structure of polyacetylene: Optical and infrared studies of undoped semiconducting (CH)$_x$ and heavily doped metallic (CH)$_x$," *Phys. Rev. B* **20**, 1589–1598 (1979).

[127] S. Stafström, J.L. Brédas, A.J. Epstein, H.S. Woo, D.B. Tanner, W.S. Huang, and A.G. MacDiarmid, "Polaron lattice in highly conducting polyaniline: Theoretical and optical studies," *Phys. Rev. Lett.* **59**, 1464–1467 (1987).

[128] K. Lee, S. Cho, S.H. Park, A.J. Heeger, C.-W. Lee, and S.-H. Lee, "Metallic transport in polyaniline," *Nature* **441**, 65–68 (2006).

[129] D.A. Bonn, J.E. Greedan, C.V. Stager, T. Timusk, M.G. Doss, S.L. Herr, K. Kamarás, and D.B. Tanner, "Far-infrared conductivity of the high-T_c superconductor YB$_{a_2}$Cu$_3$O$_7$," *Phys. Rev. Lett.* **58**, 2249–2250 (1987).

[130] M. Reedyk, D.A. Bonn, J.D. Garrett, J.E. Greedan, C.V. Stager, T. Timusk, K. Kamarás, and D.B. Tanner, "Far-infrared optical properties of Bi$_2$Sr$_2$CaCu$_2$O$_8$," *Phys. Rev. B* **38**, 11,981–11,984 (1988).

[131] K. Kamarás, S.L. Herr, C.D. Porter, N. Tache, D.B. Tanner, S. Etemad, T. Venkatesan, E. Chase, A. Inam, X.D. Wu, M.S. Hegde, and B. Dutta, "In a clean high-T_c superconductor you do not see the gap," *Phys. Rev. Lett.* **64**, 84–87 (1990).

[132] S.L. Cooper, G.A. Thomas, J. Orenstein, D.H. Rapkine A.J. Millis, S.-W. Cheong, A.S. Cooper, and Z. Fisk, "Growth of the optical conductivity in the Cu-O planes," *Phys. Rev. B* **41**, 11,605–11,608 (1990).

[133] F. Gao, D.B. Romero, D.B. Tanner, J. Talvacchio, and M.G. Forrester, "Infrared properties of epitaxial La$_{2-x}$Sr$_x$CuO$_4$ thin films in the normal and superconducting states," *Phys. Rev. B* **47**, 1036–1052 (1993).

[134] S.L. Cooper, D. Reznik, A. Kotz, M.A. Karlow, R. Liu, M.V. Klein, W.C. Lee, J. Giapintzakis, D.M. Ginsberg, B. Veal, and A.P. Paulikas, "Optical studies of the a-, b-, and c-axis charge dynamics in YBa$_2$ Cu$_3$O$_{6+x}$," *Phys. Rev. B* **47**, 8233–8248 (1993).

[135] C.C. Homes, T. Timusk, R. Liang, D.A. Bonn, and W.N. Hardy, "Optical conductivity of c axis oriented YBa$_2$Cu$_3$O$_{6.70}$: Evidence for a pseudogap," *Phys. Rev. Lett.* **71**, 1645 (1993).

[136] D.N. Basov, A.V. Puchkov, R.A. Hughes, T. Strach, J. Preston, T. Timusk, D.A. Bonn, R. Liang, and W.N. Hardy, "Disorder and superconducting-state conductivity of single crystals of $YBa_2Cu_3O_{6.95}$," *Phys. Rev. B* **49**, 12,165–12,169 (1994).

[137] D.N. Basov, R. Liang, D.A. Bonn, W.N. Hardy, B. Dabrowski, M. Quijada, D.B. Tanner, J.P. Rice, D.M. Ginsberg, and T. Timusk, "In-plane anisotropy of the penetration depth in $YBa_2Cu_3O_{7-x}$ and $YBa_2Cu_4O_8$ superconductors," *Phys. Rev. Lett.* **74**, 598–601 (1995).

[138] M.A. Quijada, D.B. Tanner, R.J. Kelley, M. Onellion, and H. Berger, "Anisotropy in the *ab*-plane optical properties of $Bi_2Sr_2CaCu_2O_8$ single-domain crystals," *Phys. Rev. B* **60**, 14,917–14,934 (1999).

[139] A.V. Puchkov, P. Fournier, T. Timusk, and N.N. Kolesnikov, "Optical conductivity of high Tc superconductors: From underdoped to overdoped," *Phys. Rev. Lett.* **77**, 1853 (1996).

[140] S.G. Kaplan, M.A. Quijada, H.D. Drew, D.B. Tanner, G.C. Xiong, R. Ramesh, C. Kwon, and T. Venkatesan, "Optical evidence for the dynamic Jahn-Teller effect in $Nd_{0.7}Sr_{0.3}MnO_3$," *Phys. Rev. Lett.* **77**, 2081–2084 (1996).

[141] Y. Murakami, H. Kawada, H. Kawata, M. Tanaka, T. Arima, Y. Moritomo, and Y. Tokura "Direct observation of charge and orbital ordering in $La_{0.5}Sr_{1.5}MnO_4$," *Phys. Rev. Lett.* **80**, 1932–1935 (1998).

[142] K.H. Kim, J.H. Jung, and T.W. Noh, "Polaron absorption in a perovskite manganite $La_{0.7}Ca_{0.3}MnO_3$," *Phys. Rev. Lett.* **81**, 1517–1520 (1998).

[143] G. Li, W.Z. Hu, J. Dong, Z. Li, P. Zheng, G.F. Chen, J.L. Luo, and N.L. Wang, "Probing the superconducting energy gap from infrared spectroscopy on a $Ba_{0.6}K_{0.4}Fe_2$ As_2 single crystal with $T_c = 37$ K," *Phys. Rev. Lett.* **101**, 107004 (2008).

[144] A. Lucarelli, A. Dusza, F. Pfuner, P. Lerch, J.G. Analytis, J.-H. Chu, I.R. Fisher, and L. Degiorgi, "Charge dynamics of Co-doped $BaFe_2As_2$," *New J. Phys.* **12**, 073036 (2010).

[145] N. Barišić, D. Wu, M. Dressel, L. J. Li, G. H. Cao, and Z. A. Xu, "Electrodynamics of electron-doped iron pnictide superconductors: Normal-state properties," *Phys. Rev. B* **82**, 054518 (2010).

[146] B. Cheng, B.F. Hu, R.Y. Chen, G. Xu, P. Zheng, J.L. Luo, and N.L. Wang, "Electronic properties of 3d transitional metal pnictides: A comparative study by optical spectroscopy," *Phys. Rev. B* **86**, 134503 (2012).

[147] Y.M. Dai, B. Xu, B. Shen, H.H. Wen, J.P. Hu, X.G. Qiu, and R.P.S.M. Lobo, "Pseudogap in underdoped $Ba_{1-x}K_xFe_2As_2$ as seen via optical conductivity," *Phys. Rev. B* **86**, 100501(R) (2012).

[148] S.J. Moon, A.A. Schafgans, M.A. Tanatar, R. Prozorov, A. Thaler, P.C. Canfield, A.S. Sefat, D. Mandrus, and D.N. Basov, "Interlayer coherence and superconducting condensate in the *c*-axis response of optimally doped $Ba(Fe_{1-x}Co_x)_2As_2$ high-T_c superconductor using infrared spectroscopy," *Phys. Rev. Lett.* **110**, 097003/1–4 (2013).

[149] Y.M. Dai, B. Xu, B. Shen, H.H. Wen, X.G. Qiu, and R.P.S.M. Lobo, "Optical conductivity of $Ba_{0.6}K_{0.4}Fe_2As_2$: The effect of in-plane and out-of-plane doping in the superconducting gap," *Europhys. Lett.* **104**, 47006 (2013).

[150] D.A. Bonn, J.D. Garrett, and T. Timusk "Far-infrared properties of URu_2Si_2, *Phys. Rev. Lett.* **61**, 1305–1308 (1988).

[151] A. Pimenov, T. Rudolf, F. Mayr, A. Loidl, A.A. Mukhin, and A.M. Balbashov, "Coupling of phonons and electromagnons in $GdMnO_3$," *Phys. Rev. B* **74**, 100403(R) (2006).

[152] X.S. Xu, M. Angst, T.V. Brinzari, R.P. Hermann, J.L. Musfeldt, A.D. Christianson, D. Mandrus, B.C. Sales, S. McGill, J.-W. Kim, and Z. Islam, "Charge order, dynamics, and magnetostructural transition in multiferroic $LuFe_2O_4$," *Phys. Rev. Lett.* **101**, 227602 (2008).

[153] K.H. Miller, P.W. Stephens, C. Martin, E. Constable, R.A. Lewis, H. Berger, G.L. Carr, and D.B. Tanner, "Infrared phonon anomaly and magnetic excitations in single-crystal $Cu_3Bi(SeO_3)_2O_2Cl$," *Phys. Rev. B* **86**, 174104/1–11 (2012).

[154] K.H. Miller, X.S. Xu, H. Berger, V. Craciun, Xiaoxiang Xi, C. Martin, G.L. Carr, and D.B. Tanner, "Infrared phonon modes in multiferroic single-crystal $FeTe_2O_5Br$," *Phys. Rev. B* **87**, 224108/1–8 (2013).

[155] A.D. LaForge, A. Frenzel, B.C. Pursley, T. Lin, X. Liu, J. Shi, and D.N. Basov, "Optical characterization of Bi_2Se_3 in a magnetic field: Infrared evidence for magnetoelectric coupling in a topological insulator material," *Phys. Rev. B* **81**, 125120 (2010).

[156] Ana Akrap, Michaël Tran, Alberto Ubaldini, Jérémie Teyssier, Enrico Giannini, Dirk van der Marel, Philippe Lerch, and Christopher C. Homes "Optical properties of Bi_2Te_2Se at ambient and high pressures," *Phys. Rev. B* **86**, 235207 (2012).

[157] C. Martin, E.D. Mun, H. Berger, V.S. Zapf, and D.B. Tanner, "Quantum oscillations and optical conductivity in Rashba spin-splitting BiTeI," *Phys. Rev. B* **87**, 041104(R)/1–5 (2013).

[158] D.M. Roessler, "Kramers-Kronig analysis of non-normal incidence reflection," *Br. J. Appl. Phys.* **16**, 1359 (1965).

[159] R.K. Ahrenkiel, "Modified Kramers-Kronig analysis of optical spectra," *J. Opt. Soc. Am.* **61**, 1651–1655 (1971).

[160] J.A. Bardwell and M.J. Dignam, "Extensions of the Kramers-Kronig transformation that cover a wide range of practical spectroscopic applications," *J. Chem. Phys.* **83**, 5468–5478 (1985).

[161] V. Lucarini, J.J. Saarinen, K.-E. Peiponen, and E.M. Vartiainen *Kramers-Kronig Relations in Optical Materials Research* (Springer-Verlag, Berlin Heidelberg, 2005).

[162] A.B. Kuzmenko, "Kramers-Kronig constrained variational analysis of optical spectra," *Rev. Sci. Instrum.* **76**, 083108 (2005).

[163] D. Crandles, F. Eftekhari, R. Faust, G. Rao, M. Reedyk, and F. Razavi, "Kramers-Kronig-constrained variational dielectric fitting and the reflectance of a thin film on a substrate," *Appl. Opt.* **47**, 4205–4211 (2008).

[164] J.S. Plaskett and P.N. Schatz, "On the Robinson and Price (Kramers-Kronig) method of Interpreting reflection data taken through a transparent window," *J. Chem.Phys.* **38**, 612 (1963).

[165] G. Andermann, A. Caron, and David A. Dows, "Kramers-Kronig dispersion analysis of infrared reflectance bands." *J. Opt. Soc. Am.* **55**, 1210–1212 (1965).

[166] D.B. Tanner, "Use of x-ray scattering functions in Kramers-Kronig analysis of reflectance," *Phys. Rev. B* **91**, 035123 (2015).

[167] "X-Ray Interactions With Matter," http://henke.lbl.gov/optical_constants/.

[168] P.A. Doyle and P.S. Turner, "Relativistic Hartree-Fock X-ray and electron scattering factors," *Acta Cryst.* **A24**, 390–397 (1968).

[169] D.T. Cromer and J.B. Mann, "X-ray scattering factors computed from numerical Hartree-Fock wave functions," *Acta Cryst.* **A24**, 321–324 (1968); Los Alamos Report LA–3816

[170] J.H. Hubbell, W.J. Veigele, E.A. Briggs, R.T. Brown, D.T. Cromer, and R.J. Howerton, "Atomic form factors, incoherent scattering functions, and photon scattering cross sections," *J. Phys. Chem. Ref. Data* **4**, 471–538 (1975).

[171] B.L. Henke, "Low energy x-ray interactions: Photoionization, scattering, specular and Bragg reflection," in *Low Energy X-ray Diagnostics* edited by D.T. Attwood and B.L. Henke (American Institute of Physics Conf. Proc. **75**, New York, 1981), pp. 146–155.

[172] P. Dreier, P. Rabe, W. Malzfeldt, and W. Niemann, "Anomalous X-ray scattering factors calculated from experimental absorption spectra," *J. Phys. C: Solid State Phys.* **17**, 3123–3136 (1984).

[173] B.L. Henke, J.C. Davis, E.M. Gullikson, and R.C.C. Perera, "A preliminary report on x-ray photoabsorption coefficients and atomic scattering factors for 92 elements in the 10–10,000 eV region," Lawrence Berkeley National Laboratory Report LBL-26259 (1988).

[174] R.E. Burge, "The interaction of x-rays," in *X-ray Science and Technology* edited by A.G. Michette and C.J. Buckley (Institute of Physics, Bristol, 1993), Chapter 5, pp. 160–206.

[175] C.T. Chantler, "Detailed Tabulation of Atomic Form Factors, Photoelectric Absorption and Scattering Cross Section, and Mass Attenuation Coefficients in the Vicinity of Absorption Edges in the Soft X-Ray ?$Z = 30$–36, $Z = 60$–89, $E = 0.1$ keV–10 keV?, Addressing Convergence Issues of Earlier Work," *J. Phys. Chem. Ref. Data* **29**, 597–1048 (2000).

[176] K. Kozima, W. Suetaka, and P.N. Schatz, "Optical constants of thin films by a Kramers-Kronig method," *J. Opt. Soc. Am.* **56**, 181–184 (1966).

[177] D.B. Romero, C.D. Porter, D.B. Tanner, L. Forro, D. Mandrus, L. Mihaly, G.L. Carr, and G.P. Williams, "Quasiparticle damping in $Bi_2Sr_2CaCu_2O_8$ and $Bi_2Sr_2CaCuO_6$," *Phys. Rev. Lett.* **68**, 1590–1593 (1992).

[178] Z. Wu, Z. Chen, X. Du, J.M. Logan, J. Sippel, M. Nikolou, K. Kamarás, J.R. Reynolds, D.B. Tanner, A.F. Hebard, and A.G. Rinzler, "Transparent, conductive nanotube Films," *Science* **305**, 1273–1276 (2004).

[179] F. Borondics, K. Kamarás, M. Nikolou, D.B. Tanner, Z.H. Chen, and A.G. Rinzler, "Study of charge dynamics in transparent single-walled carbon nanotube films," *Phys. Rev. B* **74**, 045431/1–6 (2006).

[180] P.C. Martin "Sum rules, Kramers-Kronig relations, and transport coefficients in charged systems," *Phys. Rev.* **161**, 143–155 (1967).

[181] W.M. Saslow, "Two classes of Kramers-Kronig sum rules," *Phys. Lett.* **33**, 157–158 (1970).

[182] M. Altarelli, D.L. Dexter, H.M. Nussenzveig, and D.Y. Smith, "Superconvergence and sum rules for the optical constants," *Phys. Rev. B* **6**, 4502–4509 (1972).

[183] M. Altarelli and D.Y. Smith "Superconvergence and sum rules for the optical constants: Physical meaning, comparison with experiment, and generalization," *Phys. Rev. B* **9**, 1290–1298 (1974).

[184] D.Y. Smith and E. Shiles, "Finite-energy f-sum rules for valence electrons," *Phys. Rev. B* **17**, 4689 (1978).

[185] F.W. King, "Some bounds for the absorption coefficient of an isotropic nonconducting medium," *Phys. Rev. B* **25**, 1381 (1982).

[186] G. Lévque, "Augmented partial sum rules for the analysis of optical data," *Phys. Rev. B* **34**, 5070 (1986).

[187] H.K. Onnes, "Further experiments with liquid Helium G. On the electrical resistance of pure metals, etc. VI. On the sudden change in the rate at which the resistance of mercury disappears," *Comm. Phys. Lab. Univ. Leiden* **122b**, 818 (1911).

[188] T.G. Berlincourt and R.R. Hake, "Superconductivity at high magnetic fields," *Phys. Rev.* **131**, 140 (1963).

[189] B.T. Matthias, T.H. Geballe, S. Geller, E. Corenzwit *Phys. Rev.* **95**, 1435 (1954).

[190] G. Bednorz and K.A. Mueller, "Possible high T_C superconductivity in the Ba-La-Cu-O system," *Z. Physik B.* **64**, 189–193 (1986).

[191] J. Nagamatsu, N. Nakagawa, T. Muranaka, Y. Zenitani, and J. Akimitsu, "Superconductivity at 39 K in magnesium diboride," *Nature* **410**, 63 (2001).

[192] D. Jerome, A. Mazaud, M. Ribault and K. Bechgaard, *J. Phys. Lett.* **41**, L95 (1980).

[193] V.V. Walatka, Jr. M. M. Labes, and J.H. Perlstein, "Polysulfur Nitride—A one-dimensional chain with a metallic ground state," *Phys. Rev. Lett.* **31**, 1139–1142 (1973).

[194] Julian Eisenstein, "Superconducting elements," *Rev. Mod. Phys.* **26**, 277–291 (1954).

[195] B.T. Matthias, "Superconductivity in the periodic system," *Prog. Low Temp. Phys.* **2**, 138–150 (1957).

[196] B. Matthias, T. Geballe, and V. Compton, "Superconductivity," *Rev. Mod. Phys.* **35**, 1–22 (1963).

[197] B.W. Roberts, "Survey of superconductive materials and critical evaluation of selected properties," *J. Phys. Chem. Ref. Data* **5**, 581 (1976).

[198] D. Jerome and H.J. Schultz, *Adv. Phys.* **51**, 293–479 (2002).

[199] John Singleton and Charles Mielke, "Quasi-two-dimensional organic superconductors: A review," *Contemp. Phys.* **43**, 63–96 (2002).

[200] T.H. Geballe, "The never ending search for high temperature superconductivity," *J. Supercond. Nov. Magn.* **19**, 261–276 (2006).

[201] A.G. Lebed (Ed.), "The Physics of Organic Superconductors and Conductors," (Springer Series in Materials Science , Vol. 110, 2008).

[202] A.A. Kordyuk, "Iron-based superconductors: Magnetism, superconductivity, and electronic structure," *Low Temp. Phys.*, **38**, 888–899 (2012).

[203] K. Kitazawa, "Superconductivity: 100th anniversary of its discovery and its future," *Jap. J. Appl. Phys.* **51**, 010001/1–14 (2012).

[204] Y. Li, J. Hao, H. Liu, Y. Li, and Y. Ma, "The metallization and superconductivity of dense hydrogen sulfide," *J. Chem. Phys.* **140**, 174712 (2014).

[205] D. Duan et al., "Pressure-induced metallization of dense $(H_2S)_2sb2$ with high-T_c superconductivity." *Sci. Rep.* **4**, 6968 (2014).

[206] A. Drozdov, M. Eremets, I. Troyan, V. Ksenofontov, and S. Shylin, "Conventional superconductivity at 203 Kelvin at high pressures in the sulfur hydride system," *Nature* **525**, 73–76 (2015).

[207] F. Capitani, B. Langerome, J.-B. Brubach, P. Roy, A. Drozdov, M. I. Eremets, E. J. Nicol, J. P. Carbotte, and T. Timusk, "Spectroscopic evidence of a new energy scale for superconductivity in H_3S," *Nat. Phys.* **14**, 000–000 (2017).

[208] W. Meissner, "Messungen mit Hilfe von fliissigem Helium. V. Supraleitfähigkeit von Kupfersulfid" ("Measurements with the aid of liquid helium. V. Superconductivity of copper sulfide"), *Z. fur Phys.*, **58**, 570 (1929).

[209] W. Meissner and R. Ochsenfeld, "Ein neuer Effekt bei Eintritt der Supraleitfähigkeit" ("A new effect at the onset of superconductivity"), *Naturwissenschaften* **21**, 787–788 (1933).

[210] J. Bardeen, L.N. Cooper, and J.R. Schrieffer, "Theory of superconductivity," *Phys. Rev.* **108**, 1175 (1957).

[211] J. Robert Schrieffer *Theory of Superconductivity* (W.A. Benjamin, New York, 1964; Perseus Books, 1999).

[212] L.D. Landau and V.L. Ginzburg, "On the theory of superconductivity," *Pis'ma Zh. Eksp. Teor. Fiz.* **20**, 1064 (1950).

[213] P.G. de Gennes, *Superconductivity of Metals and Alloys* (W.A. Benjamin, New York. 1966; Perseus Books 1999).

[214] L.P. Gorkov, "Microscopic derivation of the Ginzburg-Landau equations in the theory of superconductivity," *Sov. Phys. JETP*, **9**, 1364–1367 (1959).

[215] Michael Tinkham *Introduction to Superconductivity* (McGraw-Hill, New York, 1975; Robert E. Krieger, Malabar, 1980)

[216] P. Townsend and J. Sutton, "Investigation by electron tunneling of the superconducting energy gaps in Nb, Ta, Sn and Pb," *Phys. Rev.* **128**, 591–595 (1962).

[217] R.L. Frank, C. Hainzl, R. Seiringer, and J.P. Solovej, "Microscopic derivation of Ginzburg-Landau theory," *J. Amer. Math. Soc.*, **25**, 667 (2012).

[218] F. London and H. London, "The electromagnetic equations of the supraconductor," *Proc. Roy. Soc.* **A149**, 71–88 (1935).

[219] A.B. Pippard, "The surface energies of superconductors," *Math. Proc. Camb. Phil. Soc.*, **47**, 617–625 (1951); A.B. Pippard, *Proc. Roy. Soc.* **A216**, 547 (1953).

[220] T.A. Fulton and G.J. Dolan, "Observation of single-electron charging effects in small tunnel junctions," *Phys. Rev. Lett.* **59**, 109–112 (1987).

[221] X.X. Xi, J. Hwang, C. Martin, D.B. Tanner, and G.L. Carr, "Far-infrared conductivity measurements of pair breaking in superconducting $Nb_{0.5}Ti_{0.5}N$ thin-films induced by an external magnetic field," *Phys. Rev. Lett.* **105**, 257006/1–4 (2010).

[222] D.C. Mattis and J. Bardeen, "Theory of the anomalous skin effect in normal and superconducting metals," *Phys. Rev.* **111**, 412–417 (1958).

[223] S.B. Nam, "Theory of electromagnetic properties of superconducting and normal systems. I," *Phys. Rev.* **156**, 470–486 (1967).

[224] S.B. Nam, "Theory of electromagnetic properties of strong-coupling and impure superconductors. II," *Phys. Rev.* **156**, 487–93 (1967).

[225] P.B. Miller, "Surface impedance of superconductors," *Phys. Rev.* **118**, 928–934 (1960).

[226] S. Skalski, O. Betbeder-Matibet, and P.R. Weiss, "Properties of superconducting alloys containing paramagnetic impurities," *Phys. Rev.* **136**, A1500–A1518 (1964).

[227] W. Zimmermann, E.H. Brandt, M. Bauec, E. Seider, and L. Genzel, "Optical conductivity of BCS superconductors with arbitrary purity," *Phys. C: Supercond.* **183**, 99–104 (1991).

[228] D.R. Tilley and J. Tilley, *Superfluidity and Superconductivity* (Adam Hilger Ltd, Bristol, 1990).

[229] X.X. Xi, *Conventional and Time-Resolved Spectroscopy of Magnetic Properties of Superconducting Thin Films* (Thesis, University of Florida, 2011).

[230] D.J. Scalapino, "A common thread: The pairing interaction for unconventional superconductors," *Rev. Mod. Phys.* **84**, 1383–1417 (2012).

[231] M.R. Norman, "The challenge of unconventional superconductivity," *Science* **332**, 196–200 (2011).

[232] G.R. Stewart, "Unconventional superconductivity," *Adv. Phys.* **66**, 75–196 (2017).

[233] M.K Wu, J.R. Ashburn, C.J. Torng, P.H. Hor, R.L. Meng, L. Gao, Z.J. Huang, Y.Q. Wang, and C.W. Chu, "Superconductivity at 93 K in a new mixed-phase Y-Ba-Cu-O compound system at ambient pressure," *Phys. Rev. Lett.* **58**, 908 (1987).

[234] T. Timusk and D.B. Tanner, "Infrared properties of high-T_c superconductors," in *Physical Properties of High Temperature Superconductors I,* edited by D.M. Ginsberg (World Scientific, Singapore, 1989) p. 339.

[235] D.B. Tanner and T. Timusk, "Optical properties of high-temperature superconductors," in *Physical Properties of High Temperature Superconductors III,* edited by D.M. Ginsberg (World Scientific, Singapore, 1992) p. 363.

[236] T. Timusk, "Infrared properties of exotic superconductors," *Physica C* 317–318, 18–29, 1999.

[237] D.N. Basov and T. Timusk, "Electrodynamics of high-Tc superconductors," *Rev. Mod. Phys.* **77**, 721 (2005).

[238] D.B. Tanner, "Optical properties of one-dimensional systems," in *Extended Linear Chain Compounds, Vol. 2,* edited by Joel S. Miller (Plenum Press, New York, 1982) pp. 205–258.

[239] L. Degiorgi, "The electrodynamic response of heavy-electron compounds," *Rev. Mod. Phys.* 71: 687–734, 1999

[240] T. Ishiguro, K. Yamaji, and G. Saito. *Organic Superconductors* (Springer-Verlag, Heidelberg, 2002).

[241] M. Dressel and N. Drichko, "Optical properties of two-dimensional organic conductors: Signatures of charge ordering and correlation effects," *Chem. Rev.* **104**, 5689–5716 (2004).

[242] D.C. Johnston, "The puzzle of high temperature superconductivity in layered iron pnictides and chalcogenides," *Adv. Phys.* **59**, 803–1061 (2010).

[243] P.J. Hirschfeld, M.M. Korshunov, and I.I. Mazin, "Gap symmetry and structure of Fe-based superconductors," *Rep. Progress Phys.* **74**, 124508 (2011).

[244] M. Reedyk, D.A. Bonn, J.D. Garrett, J.E. Greedan, C.V. Stager, T. Timusk, K. Kamarás, and D.B. Tanner, "Far-infrared optical properties of $Bi_2Sr_2CaCu_2O_8$," *Phys. Rev. B*, **38**, 11,981 (1988).

[245] A. Puchkov, P. Fournier, D.N. Basov, T. Timusk, A. Kapitulnik, amd N.N. Kolesnikov, "Evolution of the pseudogap state of high-T_c superconductors with doping," *Phys. Rev. Lett.* **77**, 3212 (1996).

[246] A.V. Puchkov, D.N. Basov, and T. Timusk, "The pseudogap state in high-T_c superconductors: An infrared study," *J. Phys.: Condens. Matter* **8**, 10049 (1996).

[247] J. Hwang, J.P. Carbotte, and T. Timusk, "Evidence for a pseudogap in underdoped $Bi_2Sr_2CaCu_2O_{8+\delta}$ and $YBa_2Cu_3O_{6.50}$ from in-plane optical conductivity measurements," *Phys. Rev. Lett.* **100**, 177005 (2008).

[248] P.B. Allen, W.E. Pickett, and H. Krakauer, "Band-theory analysis of anisotropic transport in La_2CuO_4-based superconductors," *Phys. Rev. B* **36**, 3926 (1987).

[249] A. Umezawa, G.W. Crabtree, J.Z. Liu, T.J. Moran, S.K. Malik, L.H. Nunez, W.L. Kwok, and C.H. Sowers, "Anisotropy of the lower critical field, magnetic penetration depth, and equilibrium shielding current in single-crystal $YBa_2Cu_3O_{7-\delta}$," *Phys. Rev. B* **38**, 2843 (1988).

[250] K. Kamarás, S.L. Herr, C.D. Porter, N. Tache, D.B. Tanner, S. Etemad, T. Venkatesan, E. Chase, A. Inam, X.D. Wu, M.S. Hegde, and B. Dutta, "In a clean high-T_c superconductor you do not see the gap," *Phys. Rev. Lett.* **64**, 84–87 (1990).

[251] J.I. Pankove, *Optical Processes in Semiconductors* (Dover, New York, 1971).

[252] P.Y. Yu and M. Cardona, *Fundamentals of Semiconductors: Physics and Materials Properties* (Springer, New York, 1996).

[253] R.E. Peierls, *Quantum Theory of Solids* (Clarendon Press, Oxford, England, 1955).

[254] M. Grundmann *The Physics of Semiconductors, An Introduction including Nanophysics and Applications* (Springer, Heidelberg, 2016).

[255] Wikipedia, the free encyclopedia, "List of semiconductor materials," http://en.wikipedia.org/wiki/List_of_semiconductor_materials.

[256] M. Ohring, *Reliability and Failure of Electronic Materials and Devices* (Academic Press, San Diego, 1998), p. 310.

[257] Y. Tao, J.M. Boss, B.A. Moores, and C.L. Degen, "Single-crystal diamond nanomechanical resonators with quality factors exceeding one million," *Nat. Comm.* **5**, 3638 (2014).

[258] A.K. Abass and N.H. Ahmad, "Indirect band gap investigation of orthorhombic single crystals of sulfur,"*J. Phys. Chem. Solids* **47**, 143 (1986).

[259] D.A. Evans, A.G. McGlynn, B.M. Towlson, M. Gunn, D. Jones, T.E. Jenkins, R. Winter, N.R.J. Poolton, "Determination of the optical band-gap energy of cubic and hexagonal boron nitride using luminescence excitation spectroscopy," *J. Phys.: Cond. Matt.* **20**, 75233 (2008).

[260] C.F. Klingshirn *Semiconductor Optics* (Springer, Heidelberg, 1997), p. 127.

[261] S. Banerjee et al. "Physics and chemistry of photocatalytic titanium dioxide: Visualization of bactericidal activity using atomic force microscopy," *Curr. Sci.* **90**, 1378 (2006).

[262] O. Madelung, U. Rössler, M. Schulz (ed.), "Cuprous oxide (Cu_2O) band structure, band energies." in Group III Condensed Matter. Numerical Data and Functional Relationships in Science and Technology. 41C: Non-Tetrahedrally Bonded Elements and Binary Compounds I. doi:10.1007/10681727_62.

[263] B.G. Yacobi, *Semiconductor Materials: An Introduction to Basic Principles* (Springer, 2003) ISBN 0-306-47361-5; C. Hodes; Ebooks Corporation, *Chemical Solution Deposition of Semiconductor Films* (CRC Press, 2002), p. 319.

[264] Y.N. Rajakarunanayake, *Optical Properties of Si-Ge Superlattices and Wide Band Gap II-VI Superlattices*, Dissertation (Ph.D.), California Institute of Technology, 1991

[265] H. Ehrenreich, "Band structure and electron transport of GaAs," *Phys. Rev.* **120**, 1951–1963 (1960).

[266] M. Rohlfing, P. Krüger, and J. Pollmann, "Quasiparticle band-structure calculations for C, Si, Ge, GaAs, and SiC using Gaussian-orbital basis sets," *Phys. Rev. B* **48**, 17791–17805 (1993).

[267] S. Richard, F. Aniel, and G. Fishman, "Energy-band structure of Ge, Si, and GaAs: A thirty-band $\mathbf{k} \cdot \mathbf{p}$ method," *Phys. Rev. B* **70**, 235204 (2004).

[268] W.G. Spitzer and J.M. Whelan, "Infrared absorption and electron effective mass in n-type gallium arsenide," p. 114, 59–63 (1959).

[269] M.D. Sturge, "Optical absorption of gallium arsenide between 0.6 and 2.75 eV," *Phys. Rev.* **127**, 768–773 (1962).

[270] H.C. Casey, D.D. Sell, and K.W. Wecht, "Concentration dependence of the absorption coefficient for n- and p- type GaAs between 1.3 and 1.6 eV," *J. Appl. Phys.* **46**, 250 (1975).

[271] D.E. Aspnes, S.M. Kelso, R.A. Logan, and R. Bhat, "Optical properties of $Al_xGa_{1-x}As$," *J. Appl. Phys.* **60**, 754–767 (1986).

[272] D. Brust, J. C. Phillips, and F. Bassani, "Critical points and ultraviolet reflectivity of semiconductors," *Phys. Rev. Lett.* **9**, 94–97 (1962).

[273] E.O. Kane, "Energy band structure in p-type germanium and silicon," *J. Phys. Chem. Solids* **1**, 82–99 (1956).

[274] M. Cardona and F.H. Pollak, "Energy-band structure of germanium and silicon: The $k \cdot p$ method", *Phys. Rev.* **142**, 530–543 (1966).

[275] E.O. Kane, "Band structure of silicon from an adjusted Heine-Abarenkov calculation," *Phys. Rev.* **146**, 558–567 (1966).

[276] W. Saslow, T.K. Bergstresser, and Marvin L. Cohen, "Band structure and optical properties of diamond," *Phys. Rev. Lett.* **16**, 354–357 (1966).

[277] B.P. Zakharchenya and R.P. Selsyan "Diamagnetic excitons in semiconductors," *Sov. Phys. Uspekhi* **12**, 70 (1969).

[278] R.P. Seisyan, "Spectroscopy of Diamagnetic Excitons," Science, Moscow (1984).

[279] P.W. Baumeister, "Optical absorption of cuprous oxide," *Phys. Rev.* **121**, 359 (1961).

[280] R.A. Faulkner, "Higher donor excited states for prolate-spheroid conduction bands: A reevaluation of silicon and germanium," *Phys. Rev.* **184**, 713 (1969).

[281] A. Baldereschi and N.O. Lipari, "Spherical model of shallow acceptor states in semiconductors," *Phys. Rev. B* **8**, 2697 (1973).

[282] A. Baldereschi and N.O. Lipari "Cubic contributions to the spherical model of shallow acceptor states," *Phys. Rev. B* **9**, 1525 (1974).

[283] J.W. Allen and J.C. Mikkelsen, "Optical properties of CrSb, MnSb, NiSb, and NiAs," *Phys. Rev. B* **15**, 2952 (1977).

[284] W. Götze and P. Wölfle, "Homogeneous dynamical conductivity of simple metals," *Phys. Rev. B* **6**, 1226 (1972).

[285] P.B. Littlewood and C.M. Varma, "Phenomenology of the normal and superconducting states of a marginal Fermi liquid," *J. Appl. Phys.* **69**, 4979 (1991).

[286] J.P. Carbotte, E. Schachinger, and J. Hwang, "Boson structures in the relation between optical conductivity and quasiparticle dynamics," *Phys. Rev. B* **71**, 054506 (2005).

[287] J. Hwang, T. Timusk, and G.D. Gu, "Doping dependent optical properties of $Bi_2Sr_2CaCu_2O_{8+\delta}$," *J. Phys.: Cond. Matt.* **19**, 125208 (2007).

[288] P.B. Allen, "Electron self-energy and generalized Drude formula for infrared conductivity of metals," *Phys. Rev. B* **92**, 054305 (2015).

[289] R.N. Gurzhi, "Mutual electron correlations in metal optics," *Sov. Phys. JETP* **8**, 673–675 (1959).

[290] R.N. Gurzhi, "Hydrodynamic effects in solids at low temperature," *Sov. Phys. Uspekhi* **11**, 255–270 (1968).

[291] J.B. Smith and H. Ehrenreich, "Frequency dependence of the optical relaxation time in metals," *Phys. Rev. B* **25**, 923–930 (1982).

[292] A. Rosch, "Optical conductivity of clean metals," *Ann. Phys.* 15, 526–534 (2006).

[293] D.L. Maslov and A.V. Chubukov, "First-Matsubara-frequency rule in a Fermi liquid. II. Optical conductivity and comparison to experiment," *Phys. Rev. B* **86**, 155137 (2012).

[294] C. Berthod, J. Mravlje, X. Deng, D. van der Marel, and A. Georges, "Non-Drude universal scaling laws for the optical response of local Fermi liquids," *Phys. Rev. B* **87**, 115109 (2013).

[295] D.L. Maslov and A.V. Chubukov, "Optical response of correlated electron systems," *Rep. Prog. Phys.* **80**, 026503 (2016).

[296] C.M. Varma, P.B. Littlewood, S. Schmitt-Rink, E. Abrahams, and A.E. Ruckenstein, "Phenomenology of the normal state of Cu-O high-temperature superconductors," *Phys. Rev. Lett.* **63**, 1996 (1989).

[297] F. Slakey, S.L. Cooper, M.V. Klein, J.P. Rice, E.D. Bukowski, and D.M. Ginsberg, "Gap anisotropy and phonon self-energy effects in single-crystal $YBa_2Cu_3O_{7-\delta}$," *Phys. Rev. B* **38**, 11,934 1989.

[298] C.G. Olson, R. Liu, D.W. Lynch, R.S. List, A.J. Arko, B.W. Veal, Y.C. Chang, P.Z. Jiang and A.P. Paulikas, "High-resolution angle-resolved photoemission study of the Fermi surface and the normal-state electronic structure of $Bi_2Sr_2CaCu_2O_8$," *Phys. Rev. B* **42**, 381 (1990).

[299] Y. Hwu, L. Lozzi, M. Marsi, S. La Rosa, M. Winokur, P. Davis, M. Onellion, H. Berger, F. Gozzo, F. Lévy, and G. Margaritondo, "Electronic spectrum of the high-temperature superconducting state," *Phys. Rev. Lett.* **67**, 2573 (1991).

[300] M. Gurvitch and A.T. Fiory, "Resistivity of $La_{1.825}Sr_{0.175}CuO_4$ and $YBa_2Cu_3O_7$ to 1100 K: Absence of saturation and its implications," *Phys. Rev. Lett.* **59**, 1337–1340 (1987).

[301] S. Martin, A.T. Fiory, R.M. Fleming, L.F. Schneemeyer, and J.V. Waszczak, "Temperature dependence of the resistivity tensor in superconducting $Bi_2Sr_{2.2}Ca_{0.8}Cu_2O_8$ crystals," *Phys. Rev. B* **41**, 846 (1990).

[302] J. Hubbard, "Electron correlation in narrow energy bands," *Proc. Roy. Soc. (London)* **A276**, 238–257 (1963).

[303] N.F. Mott, "Metal-insulator transition," *Rev. Mod. Phys.* **40**, 677–683 (1968).

[304] E.H. Lieb and F.Y.Wu, "Absence of Mott transition in an exact solution of the short-range one-band model in one dimension," *Phys. Rev. Lett.* **20**, 1445–1448 (1968).

[305] J. Hubbard, "Generalized Wigner lattices in one dimension and some applications to tetracyanoquinodimethane (TCNQ) salts," *Phys. Rev. B* **17**, 494 (1978).

[306] D.B. Tanner, C.S. Jacobsen, A.A. Bright, and A.J. Heeger, "Infrared studies of the energy gap and electron-phonon interaction in potassium-tetracyanoquinodimethane (K-TCNQ)," *Phys. Rev. B* **16**, 3283–3290 (1977).

[307] M.J. Rice. "Towards the experimental determination of the fundamental microscopic parameters of organic ion-radical compounds," *Solid State Commun.* **31**, 93–98 (1979).

[308] A. Brooks Harris and Robert V. Lange. "Single-particle excitations in narrow energy bands," *Phys. Rev.* **157**, 295–314 (1967).

[309] A. Painelli and A. Girlando, "Electron-molecular vibration (e-mv) coupling in charge-transfer compounds and its consequences on the optical spectra: A theoretical framework," *J. Chem. Phys.* **84**, 5655–5671 (1986).

[310] J.J. Hopfield, "Infrared properties of transition metals (Electron phonon coupling and IR optical constants relationship to superconductivity in transition metals)," in *Superconductivity in d- and f- band metals*, edited by H. Suhl and M.B. Maple (Plenum, New York, 1972).

[311] J.J. Hopfield, "Infrared properties of transition metals," *AIP Conference Proceedings* **4**, 358–366 (1972).

[312] P.B. Allen, "Electron-phonon effects in the infrared properties of metals," *Phys. Rev. B* **3**, 305–320 (1971).

[313] R.R. Joyce and P.L. Richards, "Phonon contribution to the far-infrared absorptivity of superconducting and normal lead," *Phys. Rev. Lett.* **24**, 1007–1011 (1970).

[314] G. Brändli and A.J. Sievers, "Absolute measurement of the far-infrared surface resistance of Pb," *Phys. Rev. B* **5**, 3550 (1972).

[315] B. Farnworth and T. Timusk, "Far-infrared measurements of the phonon density of states of superconducting lead," *Phys. Rev. B* **10**, 2799–2802 (1974).

[316] B. Farnworth and T. Timusk, "Phonon density of states of superconducting lead," *Phys. Rev. B* **14**, 5119 (1976).

[317] W. Shaw and J.C. Swihart, "Theory of phonon contribution to infrared absorptivity of superconducting and normal metals," *Bull. Am. Phys. Soc.* **16**, 352 (1971).

[318] H. Scher, "Far-infrared absorptivity of normal lead," *Phys. Rev. Lett.* **25**, 759–762 (1970).

[319] M.J. Rice, "Organic linear conductors as systems for the study of electron-phonon interactions in the organic solid state," *Phys. Rev. Lett.* **37**, 36–39 (1976).

[320] M.J. Rice, L. Pietronero, and P. Brüesch, "Phase phonons and intramolecular electron-phonon coupling in the organic linear chain semiconductor TEA(TCNQ)$_2$," *Solid State Commun.* **21**, 757–760 (1977).

[321] M.J. Rice, N.O. Lipari, and S. Strässler, "Dimerized organic linear-chain conductors and the unambiguous experimental determination of electron-molecular-vibration coupling constants," *Phys. Rev. Lett.* **39**, 1359–1362 (1977).

[322] M.J. Rice, V.M. Yartsev, and C.S. Jacobsen, "Investigation of the nature of the unpaired electron states in the organic semiconductor N-methyl-N-ethylmorpholinium-tetracyanoquinodimethane," *Phys. Rev. B* **21**, 3437–3446 (1980).

[323] A. Brau, P. Brüesch, J.P. Farges, W. Hinz, and D. Kuse, "Polarized optical reflection spectra of the highly anisotropic organic semiconductor TEA(TCNQ)$_2$," *Phys. Stat. Sol. (b)* **62**, 615–623 (1974).

[324] D.B. Tanner, J.E. Deis, A.J. Epstein, and J.S. Miller, "Optical properties of the semiconducting 'metal-like' complex (NMe$_3$H)(I)(TCNQ)," *Solid State Commun.* **31**, 671–675 (1979).

[325] R.P. McCall, D.B. Tanner, J.S. Miller, and A.J. Epstein, "Optical properties of the 1:2 compound of dimethylferrocenium with tetracyanoquinodimethanide: [(Me$_2$Fc)(TCNQ)$_2$]," *Phys. Rev. B* **35**, 9209–9217 (1987).

[326] J.L. Musfeldt, C.C. Homes, M. Almeida, and D.B. Tanner, "Temperature dependence of the infrared and optical properties of N-dimethyl thiomorpholinium-(tetracyanoquinodimethane)$_2$," *Phys. Rev. B* **46**, 8777–8789 (1992).

[327] Eugene J. Mele and Michael J. Rice. "Vibrational excitations of charged solitons in polyacetylene," *Phys. Rev. Lett.* **45**, 926–929 (1980).

[328] S. Etemad, A. Pron, A. J. Heeger, A. G. MacDiarmid, E. J. Mele, and M. J. Rice, "Infrared-active vibrational modes of charged solitons in $(CH)_x$ and $(CD)_x$," *Phys. Rev. B* **23**, 5137–5141 (1981).

[329] E. Ehrenfreund, Z. Vardeny, O. Brafman, and B. Horovitz, "Amplitude and phase modes in trans-polyacetylene: Resonant Raman scattering and induced infrared activity," *Phys. Rev. B* **36**, 1535–1553 (1987).

[330] Y.H. Kim, and A.J. Heeger, "Infrared-active vibrational modes of heavily doped 'metallic' polyacetylene," *Phys. Rev. B* **40**, 8393–8398 (1989).

[331] Ke-Jian Fu, William L. Karney, Orville L. Chapman, Shiou-Mei Huang, Richard B. Kaner, François Diederich, Károly Holczer, and Robert L. Whetten "Giant vibrational resonances in A_6C_{60} compounds," *Phys. Rev. B* **46**, 1937 (1992).

[332] H.-Y. Choi and M.J. Rice, "Infrared absorption by charged phonons in doped C60," in *Electronic Properties of High-T_c Superconductors* (Springer, Berlin Heidelberg, 1993) pp. 501–506.

[333] T. Timusk and D.B. Tanner, "Evidence for strong bound-electron phonon interaction at 52 meV in $YB_{a2}Cu_3O_7$," *Physica C* **169**, 425–428 (1990).

[334] T. Timusk, C.D. Porter, and D.B. Tanner, "Strong electron-phonon interaction in the high-T_c superconductors: Evidence from the infrared," *Phys. Rev. Lett.* **66**, 663–666 (1991).

[335] M.J. Rice, V.M. Yartsev and C.S. Jacobsen, "Investigation of the nature of the unpaired electron states in the organic semiconductor N-methyl-N-ethylmorpholinium-tetracyanoquinodimethane," *Phys. Rev. B* **21**, 3437 (1980).

[336] Y. Iida, "Electronic spectra and charge-transfer absorption of solid anion radical salts of 2,3-dicyano-p-benzoquinone and 2,3-dichloro-5,6-dicyano-p-benzoquinone," *Bull. Chem. Soc. Jap.* **51**, 631–632 (1978).

[337] J. Tanaka, M. Tanaka, T. Kawai, T. Takabe and O. Maki, "Electronic spectra and electronic structure of TCNQ complexes," *Bull. Chem. Soc. Jap.* **49**, 2358 (1976).

[338] A.B. Pippard, "The surface impedance of superconductors and normal metals at high frequencies. II. The anomalous skin effect in normal metals," *Proc. R. Soc. Lond. A* **191**, 385–399 (1947).

[339] R.G. Chambers, "Anomalous skin effect in metals," *Nature* **165**, 239–240 (1950).

[340] H.B.G. Casimir and J. Ubbink, "The skin effect: II. The skin effect at high frequencies," *Philips Tech. Rev.* **28**, 300–315 (1967).

[341] E.H. Land, "Some aspects on the development of sheet polarizers," *J. Opt. Soc. Am.* **41**, 957–963 (1951).

[342] B.P. Abbott *et al.* (LIGO Scientific Collaboration), "LIGO: The Laser Interferometer Gravitational-wave Observatory," *Rep. Prog. Phys.* 72, 076901/1–25 (2009).

[343] W.S. Bickel and W.M. Bailey, "Stokes vectors, Mueller matrices, and polarized scattered light," *Am. J. Phys.* **53**, 468 (1985).

[344] S.N. Savenkov, "Jones and Mueller matrices: Structure, symmetry relations and information content," *Light Scatter. Rev.* **4**, 71 (2009).

[345] "The Physics Hypertextbook, 'Refraction,' " http://physics.info/refraction/.

[346] J.L. Duarte, J.A. Sanjurjo, and R.S. Katiyar, "Off-normal infrared reflectivity in uniaxial crystals: α-LiIO$_3$ and α-quartz," *Phys. Rev. B* **36**, 3368–3372 (1987).

[347] L. Landau, "Diamagnetismus der Metalle" ("Diamagnetism of metals"), *Z. Phys.* **64**, 629 (1930).

[348] Colin M. Hurd, *The Hall Effect in Metals and Alloys* (Plenum Press, New York, 1972).

[349] E.H. Hall, "On a new action of the magnet on electric currents," *American Journal of Mathematics* **2**, 287–292 (1879).

[350] D. Schoenberg, *Magnetic Oscillations in Metals* (Cambridge University Press, London, 1984).

[351] B.W. Maxfield, "Helicon waves in solids," *Am. J. Phys.* **37**, 241 (1969).

[352] J.M. Goodman, "Helicon waves, surface-mode loss, and the accurate determination of the hall coefficients of aluminum, indium, sodium, and potassium," *Phys. Rev.* **171**, 641 (1968).

[353] R. Blewitt and A.J. Sievers, "Harmonic series of cyclotronlike resonances in Bismuth," *Phys. Rev. Lett.* **30**, 1041 (1973).

[354] G. Dresselhaus, A.F. Kip and C. Kittel, "Cyclotron resonance of electrons and holes in silicon and germanium crystals," *Phys. Rev.* **98**, 368 (1955).

[355] J. Kono, "Cyclotron resonance," in *Methods in Materials Research*, edited by E.N. Kaufmann *et al.* (Wiley, 1999).

[356] T.F. Rosenbaum, K. Andres, G.A. Thomas, and R.N. Bhatt, "Sharp metal-insulator transition in a random solid," *Phys. Rev. Lett.* **45**, 1723 (1980).

[357] W.N. Shafarman, D.W. Koon, and T.G. Castner, "DC conductivity of arsenic-doped silicon near the metal-insulator transition," *Phys. Rev. B* **40**, 1216 (1989).

[358] N. Mott, "On metal-insulator transitions," *J. Solid State Chem.* **88** 5–7 (1990).

[359] I.I. Rabi, J.R. Zacharias, S. Millman, and P. Kusch, "A new method of measuring nuclear magnetic moment," *Phys. Rev.*, 53, 318 (1938).

[360] G. Feher and A.F. Kip, "Electron spin resonance absorption in metals. I. Experimental," *Phys. Rev.*, **98**, 337 (1955).

[361] F.J. Dyson "Electron spin resonance absorption in metals. II. Theory of electron diffusion and the skin effect," *Phys. Rev.* **98**, 349 (1955).

[362] J.H.E. Griffiths, "Anomalous high-frequency resistance of ferromagnetic metals," *Nature* **158**, 670 (1946).

[363] C. Kittel, "On the theory of ferromagnetic resonance absorption," *Phys. Rev.* **73**, 155 (1948).

[364] C. Kittel, "Theory of antiferromagnetic resonance," *Phys. Rev.* **82**, 565 (1951); F. Keffer and C. Kittel "Theory of antiferromagnetic resonance," *Phys. Rev.* **85**, 329 (1952).

[365] A. Pimenov, A. A. Mukhin, V. Yu. Ivanov, V. D. Travkin, A.M. Balbashov, and A. Loidl, "Possible evidence for electromagnons in multiferroic manganites," *Nature Physics* **2**, 97–100 (2006).

[366] T. Yildirim, L.I. Vergara, Jorge Íñiguez, J.L. Musfeldt, A.B. Harris, N. Rogado, R.J. Cava, F. Yen, R.P. Chaudhury, and B. Lorenz, "Phonons and magnetoelectric interactions in $Ni_3V_2O_8$," *J. Phys.: Cond. Mat.* **20**, 434214 (2008).

[367] A. Shuvaev, V. Dziom, Anna Pimenov, M. Schiebl, A. A. Mukhin, A. C. Komarek, T. Finger, M. Braden, and A. Pimenov, "Electric field control of terahertz polarization in a multiferroic manganite with electromagnons," *Phys. Rev. Lett.* **111**, 227201 (2013).

[368] J.C.M. Garnett, "Colours in metal glasses and in metallic films," *Phil. Trans. Roy. Soc.* **A203**, 385 (1904).

[369] J.C.M. Garnett, "Colours in metal glasses, in metallic films, and in metallic solutions–II," *Phil. Trans. Roy. Soc.* **A205**, 237 (1906).

[370] D.A.G. Bruggeman, "Berechnung verschiedener physikalischer Konstanten von heterogenen Substanzen. I. Dielektrizitätskonstanten und Leitfähigkeiten der Mischkörper aus isotropen Substanzen" ("Calculation of different physical constants of heterogeneous substances. I. Dielectric constants and conductivities of the mixed bodies of isotropic substances"), *Ann. Physik (Leipz.)* **24**, 636 (1935).

[371] R. Landauer, "The electrical resistance of binary metallic mixtures," *J. Appl. Phys.* **23**, 779 (1952).

[372] B.E. Springett, "Effective-medium theory for the AC behavior of a random system," *Phys. Rev. Lett.* **31**, 1463 (1973).

[373] D. Stroud, "Generalized effective-medium approach to the conductivity of an inhomogeneous material," *Phys. Rev. B* **12**, 3368 (1975).

[374] G. Mie, "Beiträge zur Optik trüber Medien, speziell kolloidaler Metallösungen (Contributions to optics via media, especially colloidal metal compounds)," *Ann. Physik* **25**, 377 (1908).

[375] P. Debye, "Der Lichtdruck auf Kugeln von beliebigem Material" ("The light pressure on spheres of any material"), *Ann. Physik* **30**, 57 (1909).

[376] H.C. Van de Hulst, *Light Scattering by Small Particles* (Dover, New York, 1981).

[377] Z. Hashin and S. Shtrikman, *J. Appl. Phys.* **33**, 3125 (1962).

[378] W.T. Doyle, "Optical absorption by *F* centers in alkali halides," *Phys. Rev.* **111**, 1072 (1958).

[379] F.L. Galeener, "Submicroscopic-void resonance: The effect of internal roughness on optical absorption," *Phys. Rev. Lett.* **27**, 421 (1971).

[380] F.L. Galeener, "Optical evidence for a network of cracklike voids in amorphous germanium," *Phys. Rev. Lett.* **27**, 1716 (1971).

[381] L. Genzel and T.P. Martin, "Infrared absorption by surface phonons and surface plasmons in small crystals," *Surf. Sci.* **34**, 33 (1973).

[382] A.S. Barker, Jr. "Infrared absorption of localized longitudinal-optical phonons," *Phys. Rev. B* **7**, 2507 (1973).

[383] J.H. Weaver, R.W. Alexander, L. Teng, R.A. Mann, and R.J. Bell, "Infrared absorption of small silicon particles with oxide overlayers," *Phys. Stat. Sol. (a)* **20**, 321 (1973).

[384] R.W. Cohen, G.D. Cody, M.D. Coutts, and B. Abeles, "Optical properties of granular silver and gold films," *Phys. Rev. B* **8**, 3689 (1973).

[385] B. Abeles, H.L. Pinch, and J.I. Gittleman, "Percolation conductivity in W-Al$_2$O$_3$ granular metal films," *Phys. Rev. Lett.* **35**, 247 (1975).

[386] C.G. Granqvist and O. Hunderi, "Retardation effects on the optical properties of ultrafine particles," *Phys. Rev. B* **16**, 1353 (1977).

[387] C.G. Granqvist and O. Hunderi, "Optical properties of ultrafine gold particles," *Phys. Rev. B* **16**, 3513 (1977).

[388] I. Webman, J. Jortner and M.H. Cohen, "Theory of optical and microwave properties of microscopically inhomogeneous materials," *Phys. Rev. B* **15**, 5712 (1977).

[389] W. Lamb, D.M. Wood, and N.W. Ashcroft, "Long-wavelength electromagnetic propagation in heterogeneous media," *Phys. Rev. B* **21**, 2248 (1980).

[390] D. Stroud, "Percolation effects and sum rules in the optical properties of composities," *Phys. Rev. B* **19**, 1783 (1979).

[391] H. Scher and R. Zallen, "Critical density in percolation processes," *J. Chem. Phys.* **53**, 3759 (1970).

[392] Richard Zallen, "The percolation model," in *The Physics of Amorphous Solids* (Wiley, 1983), pp. 135–204.

[393] M.J. Powell, "Site percolation in randomly packed spheres," *Phys. Rev. B* **20**, 4194 (1979).

[394] D.M. Grannan, J.C. Garland, and D.B. Tanner, "Critical behavior of the dielectric constant of a random composite near the percolation threshold," *Phys. Rev. Lett.* **46**, 375 (1981).

[395] S. Kirkpatrick, "Classical transport in disordered media: Scaling and effective-medium theories," *Phys. Rev. Lett.* **27**, 1722 (1971).

[396] S. Kirkpatrick, "Percolation and conduction," *Rev. Mod. Phys.* 45, 574–588 (1973).

[397] S. Kirkpatrick, "Percolation phenomena in higher dimensions: Approach to the mean-field limit," *Phys. Rev. Lett.* **36**, 69 (1976).

[398] S. Kirkpatrick, "Percolation thresholds in Ising magnets and conducting mixtures," *Phys. Rev. B* **15**, 1533 (1977).

[399] B.J. Last and D.J. Thouless, "Percolation theory and electrical conductivity," *Phys. Rev. Lett.* **27**, 1719 (1971).

[400] V. Ambegaokar, S. Cochran, and J. Kurkijarvi, "Conduction in random systems," *Phys. Rev. B* **8**, 3682 (1973).

[401] B.P. Watson, and P.L. Leath, "Conductivity in the two-dimensional-site percolation problem," *Phys. Rev. B* **9**, 4893 (1974).

[402] I. Webman, and J. Jortner, "Numerical simulation of electrical conductivity in microscopically inhomogeneous materials," *Phys. Rev. B* **15**, 2885 (1975).

[403] M.E. Levinshtein, B.I. Shklovskii, M.S. Shur, and A.L. Efros, "The relation between the critical exponents of percolation theory," *Sov. Phys JETP*, **42**, 197 (1975).

[404] G. Deutscher, R. Zallen, and J. Adler eds., *Percolation Structures and Processes* (Adam Hilger, Bristol, 1983).

[405] D.S. Stauffer and A. Ahorony, *Introduction to Percolation Theory* (Taylor and Francis, 1994).

[406] J.P. Straley, "Critical phenomena in resistor networks," *J. Phys. C: Solid State Phys.* **9**, 783 (1976).

[407] J.P. Straley, "Critical exponents for the conductivity of random resistor lattices," *Phys. Rev. B* **15**, 5733 (1977).

[408] J.P. Straley, "Random resistor tree in an applied field," *J. Phys. C: Solid State Phys.* **10**, 3009 (1977).

[409] R.B. Stinchcombe and B.P. Watson, "Renormalization group approach for percolation conductivity," *J. Phys. C: Solid State Phys.* **9**, 3221 (1976).

[410] A.L. Efros, and B.I. Shklovskii, "Critical behaviour of conductivity and dielectric constant near the metal-non-metal transition threshold," *Phys. Stat. Sol. (B)* **76**, 475 (1976).

[411] A.B. Harris and R. Fisch, "Critical behavior of random resistor networks," *Phys. Rev. Lett.* **38**, 796 (1977).

[412] D.J. Bergman and Y. Imry, "Critical behavior of the complex dielectric constant near the percolation threshold of a heterogeneous material," *Phys. Rev. Lett.* **39**, 1222 (1977).

[413] J.P. Straley, in *Electrical Transport and Optical Properties of Inhomogeneous Media,* edited by J.C. Garland and D.B. Tanner (The American Institute of Physics, New York, 1978), p. 118.

[414] J. Wu and D.S. McLachlan, "Percolation exponents and thresholds obtained from the nearly ideal continuum percolation system graphite-boron nitride," *Phys. Rev. B* **56**, 1236 (1997).

[415] D. Stroud and D.J. Bergman, "Frequency dependence of the polarization catastrophe at a metal-insulator transition and related problems," *Phys. Rev. B* **25**, 2061 (1982).

[416] D.B. Tanner, A.J. Sievers, and R.A. Buhrman, "Far-infrared absorption in small metallic particles," *Phys. Rev. B* **11**, 1330–1341 (1975).

[417] N.E. Russell, J.C. Garland, and D.B. Tanner, "Absorption of far-infrared radiation by random metal particle composites," *Phys. Rev. B* **23**, 632 (1981).

[418] G.L. Carr, R.L. Henry, N.E. Russell, J.C. Garland, and D.B. Tanner, "Anomalous far-infrared absorption in random small-particle composites," Phys. Rev. B., **24**, 777 (1981).

[419] S.-I. Lee, T.W. Noh, J.R. Gaines, Y.-H. Ko, and E.R. Kreidler, "Optical studies of porous glass media containing silver particles," *Phys. Rev. B* **37**, 2918 (1988).

[420] G.L. Carr, S. Perkowitz, and D.B. Tanner, "Far-infrared properties of inhomogeneous materials," in *Infrared and Millimeter Waves, Vol. 13,* edited by Kenneth J. Button (Academic Press, Orlando, 1985), pp. 171–263.

[421] D. Stroud and F.P. Pan, "Self-consistent approach to electromagnetic wave propagation in composite media: Application to model granular metals," *Phys. Rev. B* **17**, 1602 (1978).

[422] H.B.G. Casimir, "On electromagnetic units," *Helvetica Phys. Acta*, **41**, 741–742 (1968) This article is reprinted in Robert L. Weber and Eric Mendoza, *A Random Walk in Science* (Institute of Physics Publishing, Bristol, 1973).

[423] R. Nityananda and J. Samuel,"Fermat's principle in general relativity," *Phys. Rev. D* **45**, 3862 (1992).

[424] M.A. Dupertuis, M. Proctor, and B. Acklin, "Generalization of complex Snell-Descartes and Fresnel laws," *J. Opt. Soc. Am. A* **11**, 1159–1166 (1994).

[425] P.C.Y. Chang, J.G. Walker, and K.I. Hopcraft "Ray tracing in absorbing media," *J. Quant. Spect. Rad. Trans.* **96**, 327–341 (2005).

Index